Long-Term Field Studies of Primates

Peter M. Kappeler • David P. Watts
Editors

Long-Term Field Studies of Primates

Editors

Prof. Dr. Peter M. Kappeler
Department of Behavioral Ecology
and Sociobiology
German Primate Center
Kellnerweg 4
37077 Göttingen
Germany
pkappel@gwdg.de

Dr. David P. Watts
Yale University
Department of Anthropology
Sachem Street 10
06520-8277 New Haven
Connecticut
USA
david.watts@yale.edu

ISBN 978-3-642-22513-0   e-ISBN 978-3-642-22514-7
DOI 10.1007/978-3-642-22514-7
Springer Heidelberg Dordrecht London New York

Library of Congress Control Number: 2011944014

© Springer-Verlag Berlin Heidelberg 2012
This work is subject to copyright. All rights are reserved, whether the whole or part of the material is concerned, specifically the rights of translation, reprinting, reuse of illustrations, recitation, broadcasting, reproduction on microfilm or in any other way, and storage in data banks. Duplication of this publication or parts thereof is permitted only under the provisions of the German Copyright Law of September 9, 1965, in its current version, and permission for use must always be obtained from Springer. Violations are liable to prosecution under the German Copyright Law.
The use of general descriptive names, registered names, trademarks, etc. in this publication does not imply, even in the absence of a specific statement, that such names are exempt from the relevant protective laws and regulations and therefore free for general use.

Printed on acid-free paper

Springer is part of Springer Science+Business Media (www.springer.com)

In memoriam Toshisada Nishida (1941–2011)

Preface

This edited volume features a collection of essays discussing the virtues and challenges of conducting long-term field projects on wild primate populations in Madagascar, the Americas, Asia, and Africa. All of these projects have been ongoing for a significant portion of the respective species' life cycle; some of them already cover multiple generations. Taken together, the contributions to this volume represent all major primate radiations and therefore provide a representative sample of taxon-specific opportunities and challenges of this type of research. Still, not all important long-term studies could be included in this volume, with the absence of specific chapters on nocturnal primates perhaps the most important omission. In contrast, some taxa, such as sifakas, capuchin monkeys, and chimpanzees, are covered in more than one chapter. In each case, the comparison revealed striking differences between study sites and populations, emphasizing the perhaps not so obvious fact that single long-term studies are apparently not sufficient to document the full range of species-typical life history adaptations and variation in social systems.

Continuous observations of habituated and individually recognizable primates originated with Imanishi's project on Japanese macaques in the late 1940s. Some of the projects described in this volume originated in the 1960s. Gombe, Amboseli, Berenty, Caratinga, Koshima, and Ketambe are names familiar to all primatologists today because these projects have contributed unique insights into the natural histories and life histories of wild primates. Such long-term studies are especially necessary in primatology, because primates have slower life histories than most other terrestrial mammals. In many cases, several decades are necessary to document the timing of important life history milestones, and in most cases this is not even sufficient to garner large enough samples to analyze the adaptive basis of life history variation. Variability in key social parameters and documentation of behavioral development provide additional justification for a long-term approach. Moreover, rare but important events, such as predation, infanticide, or dispersal, also necessitate continuous observation for many years to obtain samples large enough for satisfactory empirical analyses. Finally, most long-term primate field studies today have positive effects on local conservation efforts; this alone is enough to justify their continuation in the view of many primatologists.

The authors of the chapters on specific projects covered in this volume were asked to summarize results and insights that were only possible because of the long-term nature of their studies and thereby to provide concrete examples for the scientific necessity and benefits of this kind of research. Because long-term projects of this kind also face numerous practical problems and challenges, especially with respect to data management and continuous funding, we also asked all authors to furnish summaries of the history and logistics of their projects, which we hope will be valuable for historians of our discipline and for colleagues initiating new projects, respectively. Because this volume presents the first collective summary of unique datasets from several influential long-term primate field studies along with new research results, we hope that it will interest not just primatologists, but also anthropologists interested in the value of comparative research on nonhuman primates for understanding human evolution and the behavior of modern humans and behavioral ecologists involved in long-term projects on other vertebrates. If examples included in this volume can provide arguments and examples that convince academic administrations and funding agencies of both the scientific value and the conservation importance of such field projects, it would have achieved another important purpose.

Göttingen and New Haven
Peter M. Kappeler
David P. Watts

Acknowledgements

This volume is largely based on contributions to a conference in Göttingen (Germany) in December 2009, the *VII. Göttinger Freilandtage*. We thank the Deutsche Forschungsgemeinschaft (DFG), the Deutsches Primatenzentrum (DPZ), the City of Göttingen and the University of Göttingen for their support of this conference. We subsequently solicited several additional contributions for this volume to cover studies not presented during the conference. We are particularly grateful to those colleagues for contributing chapters to this volume at much shorter notice, and we appreciate their professional collegiality. Every chapter of this volume was peer-reviewed, and we thank all authors and Mitch Irwin for their helpful comments on at least one chapter. Special thanks to Ulrike Walbaum for formatting chapters, figures and tables, and for carefully double-checking all references, as well as Anette Lindqvist for her editorial support.

Contents

Part I Introduction

1 The Values and Challenges of Long-Term Field Studies   3
Peter M. Kappeler, Carel P. van Schaik, and David P. Watts

Part II Madagascar

2 Berenty Reserve, Madagascar: A Long Time in a Small Space . . .   21
Alison Jolly

**3 Beza Mahafaly Special Reserve: Long-Term Research on Lemurs
in Southwestern Madagascar**   45
Robert W. Sussman, Alison F. Richard, Joelisoa Ratsirarson, Michelle
L. Sauther, Diane K. Brockman, Lisa Gould, Richard Lawler, and Frank
P. Cuozzo

**4 Long-Term Lemur Research at Centre Valbio, Ranomafana
National Park, Madagascar**   67
Patricia C. Wright, Elizabeth M. Erhart, Stacey Tecot, Andrea L. Baden,
Summer J. Arrigo-Nelson, James Herrera, Toni Lyn Morelli, Marina
B. Blanco, Anja Deppe, Sylvia Atsalis, Steig Johnson, Felix Ratelolahy
Chia Tan, and Sarah Zohdy

**5 A 15-Year Perspective on the Social Organization and Life
History of Sifaka in Kirindy Forest**   101
Peter M. Kappeler and Claudia Fichtel

xi

Part III America

6 **The Northern Muriqui (*Brachyteles hypoxanthus*): Lessons on Behavioral Plasticity and Population Dynamics from a Critically Endangered Species** 125
Karen B. Strier and Sérgio L. Mendes

7 **The Lomas Barbudal Monkey Project: Two Decades of Research on *Cebus capucinus*** 141
Susan Perry, Irene Godoy, and Wiebke Lammers

8 **Tracking Neotropical Monkeys in Santa Rosa: Lessons from a Regenerating Costa Rican Dry Forest** 165
Linda M. Fedigan and Katharine M. Jack

9 **The Group Life Cycle and Demography of Brown Capuchin Monkeys (*Cebus [apella] nigritus*) in Iguazú National Park, Argentina** 185
Charles Janson, Maria Celia Baldovino, and Mario Di Bitetti

Part IV Asia

10 **Social Organization and Male Residence Pattern in Phayre's Leaf Monkeys** 215
Andreas Koenig and Carola Borries

11 **White-Handed Gibbons of Khao Yai: Social Flexibility, Complex Reproductive Strategies, and a Slow Life History** 237
Ulrich H. Reichard, Manoch Ganpanakngan, and Claudia Barelli

Part V Africa

12 **The Amboseli Baboon Research Project: 40 Years of Continuity and Change** 261
Susan C. Alberts and Jeanne Altmann

13 **The 30-Year Blues: What We Know and Don't Know About Life History, Group Size, and Group Fission of Blue Monkeys in the Kakamega Forest, Kenya** 289
Marina Cords

14 **Long-Term Research on Chimpanzee Behavioral Ecology in Kibale National Park, Uganda** 313
David P. Watts

15 **Long-Term Field Studies of Chimpanzees at Mahale Mountains National Park, Tanzania** 339
Michio Nakamura and Toshisada Nishida

Contents   xiii

16 Long-Term Studies of the Chimpanzees of Gombe National Park, Tanzania 357
Michael L. Wilson

17 Long-Term Research on Grauer's Gorillas in Kahuzi-Biega National Park, DRC: Life History, Foraging Strategies, and Ecological Differentiation from Sympatric Chimpanzees 385
Juichi Yamagiwa, Augustin Kanyunyi Basabose, John Kahekwa, Dominique Bikaba, Chieko Ando, Miki Matsubara, Nobusuke Iwasaki, and David S. Sprague

18 Long-Term Studies on Wild Bonobos at Wamba, Luo Scientific Reserve, D. R. Congo: Towards the Understanding of Female Life History in a Male-Philopatric Species 413
Takeshi Furuichi, Gen'ichi Idani, Hiroshi Ihobe, Chie Hashimoto, Yasuko Tashiro, Tetsuya Sakamaki, Mbangi N. Mulavwa, Kumugo Yangozene, and Suehisa Kuroda

Part VI   Summary

19 Long-Term, Individual-Based Field Studies 437
Tim Clutton-Brock

Erratum to Chapter 4: Long-Term Lemur Research at Centre Valbio, Ranomafana National Park, Madagascar E1

Erratum to Chapter 13: The 30-Year Blues: What We Know and Don't Know About Life History, GroupSize, and Group Fission of Blue Monkeys in the Kakamega Forest, Kenya E3

Index 451

Contributors

Susan C. Alberts Department of Biology, Duke University, Durham, NC, USA, alberts@duke.edu

Jeanne Altmann Department of Ecology and Evolutionary Biology, Princeton University, Princeton, NJ, USA, altj@princeton.edu

Chieko Ando Department of Zoology, Kyoto University, Kyoto, Japan

Summer J. Arrigo-Nelson Department of Biological and Environmental Sciences, University of California in Pennsylvania, California, PA, USA, arrigonelson@calu.edu

Sylvia Atsalis San Diego Zoological Society, The San Diego Zoo's Institute for Conservation Research, San Diego, CA, USA, satsalis55@gmail.com

Andrea L. Baden IDPAS, Stony Brook University, Stony Brook, NY, USA, andrea.baden@gmail.com

Maria Celia Baldovino CONICET, Instituto de Biología Subtropical, Universidad Nacional de Misiones, Puerto Iguazú, Argentina, macelia@arnet.com.ar

Claudia Barelli Department of Reproductive Biology, German Primate Center, Göttingen, Germany; Tropical Biodiversity Section, Trento Science Museum, Trento, Italy

Augustin Kanyuni Basabosa International Gorilla Conservation Program, Mont Goma, Democratic Republic of Congo

Dominique Bikaba Pole Pole Foundation, Miti, Democratic Republic of Congo

Marina B. Blanco Department of Anthropology, University of Massachusetts, Amherst, MA, USA, mbblanco@anthro.umass.edu

Carola Borries Department of Anthropology, Stony Brook University, Stony Brook, NY, USA, cborries@notes.sunysb.edu

Diane K. Brockman Department of Anthropology, University of North Carolina, Charlotte, NC, USA, dkbrockm@uncc.edu

Tim Clutton-Brock Department of Zoology, University of Cambridge, Cambridge, UK, thcb@cam.ac.uk

Marina Cords Department of Ecology, Evolution and Environmental Biology, Columbia University, New York, NY, USA, mc51@columbia.edu

Frank P. Cuozzo Department of Anthropology, University of North Dakota, Grand Forks, ND, USA, frank.cuozzo@und.nodak.edu

Anja Deppe IDPAS, Stony Brook University, Stony Brook, NY, USA, deppeam@hotmail.com

Mario Di Bitetti CONICET, Instituto de Biología Subtropical, Universidad Nacional de Misiones, Puerto Iguazú, Argentina, dibitetti@yahoo.com.ar

Elizabeth M. Erhart Department of Anthropology, Texas State University, San Marcos, TX, USA, berhart@txstate.edu

Linda M. Fedigan Department of Anthropology, University of Calgary, Calgary, Canada, fedigan@ucalgary.ca

Claudia Fichtel Department of Behavioral Ecology & Sociobiology, German Primate Center, Göttingen, Germany, claudia.fichtel@gwdg.de

Takeshi Furuichi Primate Research Institute, Kyoto University, Inuyama, Aichi, Japan, furuichi@pri.kyoto-u.ac.jp

Manoch Ganpanakngan Khao Yai National Park, Bangkok, Thailand

Irene Godoy Department of Anthropology, Centre for Behavior, Evolution and Culture, University of California, Los Angeles, CA, USA, godoy@ucla.edu

Lisa Gould Department of Anthropology, University of Victoria, Victoria, BC, Canada, lgould@uvic.ca

Chie Hashimoto Primate Research Institute, Kyoto University, Inuyama, Aichi, Japan

James Herrera IDPAS, Stony Brook University, Stony Brook, NY, USA, jherrera84@gmail.com

Gen'ichi Idani Wildlife Research Center, Kyoto University, Kyoto, Japan

Hiroshi Ihobe School of Human Sciences, Sugiyama Jogakuen University, Nisshin, Aichi, Japan

Nobusuke Iwasaki National Institute for Agro-Environmental Sciences, Tsukuba, Ibaragi, Japan

Katherine M. Jack Department of Anthropology, Tulane University, New Orleans, LA, USA, kjack@tulane.edu

Charles Janson Division of Biological Sciences, University of Montana, Missoula, MT, USA, charles.janson@mso.umt.edu

Steig Johnson Department of Anthropology, University of Calgary, Calgary, Canada, steig.johnson@ucalgary.ca

Alison Jolly Department of Biology, University of Sussex, Brighton, UK, ajolly@sussex.ac.uk

John Kahekwa Pole Pole Foundation, Miti, Democratic Republic of Congo

Peter M. Kappeler Department of Sociobiology/Anthropology and CRC Evolution of Social Behavior, University of Göttingen, Göttingen, Germany; Behavioral Ecology & Sociobiology Unit, German Primate Center, Göttingen, Germany, pkappel@gwdg.de

Andreas Koenig Department of Anthropology, Stony Brook University, Stony Brook, NY, USA, akoenig@notes.cc.sunysb.edu

Suehisa Kuroda School of Human Cultures, The University of Shiga Prefecture, Hikone, Shiga, Japan

Wiebke Lammers Proyecto de Monos, Bagaces, Costa Rica, wiebkelammers@gmail.com

Richard Lawler Department of Anthropology, James Madison University, Harrisonburg, VA, USA, lawlerr@jmu.edu

Miki Matsubara Primate Research Institute, Kyoto University, Inuyama, Aichi, Japan

Sérgio L. Mendes Departamento de Ciências Biológicas, Universidade Federal do Espírito Santo, Vítoria, Brazil, slmendes1@gmail.com

Tony Lyn Morelli Museum of Vertebrate Zoology, Department of Environmental Sciences, Policy and Management, U.C. Berkeley, Berkeley, CA, USA, morellitlm@gmail.com

Mbangi N. Mulavwa Research Center of Ecology and Forestry, Ministry of Scientific Research, Bikolo, Equateur, Democratic Republic of Congo

Michio Nakamura Wildlife Research Center, Kyoto University, Kyoto, Japan, nakamura@wrc.kyoto-u.ac.jp

Toshisada Nishida Japan Monkey Centre, Kanrin, Inuyama, Japan

Susan Perry Department of Anthropology, Centre for Behavior, Evolution and Culture, University of California, Los Angeles, CA, USA, sperry@anthro.ucla.edu

Felix Ratelolahy Wildlife Conservation Society, Antananarivo, Madagascar, Jeaf_ratel@yahoo.fr

Joelisoa Ratsirarson School of Agronomy, Department of Water and Forestry (ESSA/Forêts), University of Antananarivo, Antananarivo, Madagascar, ratsirarson@gmail.com

Ulrich Reichard Department of Anthropology, Southern Illinois University, Carbondale, IL, USA, ureich@siu.edu

Alison F. Richard Professor Emerita, Yale University, New Haven, CT, USA, alisonfrichard@gmail.com

Tetsuya Sakamaki Primate Research Institute, Kyoto University, Inuyama, Aichi, Japan

Michelle L. Sauther Department of Anthropology, University of Colorado, Boulder, CO, USA, michelle.sauther@colorado.edu

Karen B. Strier Department of Anthropology, University of Wisconsin-Madison, Madison, WI, USA, kbstrier@wisc.edu

Robert W. Sussman Department of Anthropology and Environmental Science, Washington University, St. Louis, MO, USA, rwsussma@artsci.wustl.edu

Chia Tan The San Diego Zoo's Institute for Conservation Research, San Diego, CA, USA, ctan@sandiegozoo.org

Yasuko Tashiro Hayashibara Great Ape Research Institute, Tamano, Okayama, Japan

Stacey Tecot School of Anthropology, University of Arizona, Tucson, AZ, USA, stecot@email.arizona.edu

David Sprague National Institute for Agro-Environmental Sciences, Tsukuba, Ibaragi, Japan

Carel P. van Schaik Anthropological Institute and Museum, University of Zürich, Zürich, Switzerland, vschaik@aim.unizh.ch

David P. Watts Department of Anthropology, Yale University, New Haven, CT, USA, david.watts@yale.edu

Michael L. Wilson Departments of Anthropology and Ecology, Evolution and Behavior, University of Minnesota, Minneapolis, MN, USA, wilso198@umn.edu

Patricia C. Wright Institute for the Conservation of Tropical Environments, Stony Brook University, Stony Brook, NY, USA, patchapplewright@gmail.com

Juichi Yamagiwa Department of Zoology, Kyoto University, Kyoto, Japan, yamagiwa@jinrui.zool.kyoto-u.ac.jp

Kumugo Yangozene Research Center of Ecology and Forestry, Ministry of Scientific Research, Bikolo, Equateur, Democratic Republic of Congo

Sarah Zohdy University of Helsinki, Helsinki, Finland, sarahzohdy@gmail.com

Part I
Introduction

Chapter 1
The Values and Challenges of Long-Term Field Studies

Peter M. Kappeler, Carel P. van Schaik, and David P. Watts

Abstract In this chapter, we review some of the benefits and challenges of long-term primate field studies. We define long-term studies as those that cover a significant part of the study species' life cycle; in reality, many studies have already extended over multiple generations. We first provide a brief overview of the historical beginnings of modern primate field studies, most of which lay in the 1950s and early 1960s. Next, we identify a number of biological constraints and scientific questions that necessitate and justify a long-term approach to studying wild primate populations. Most research questions in this context are related to fitness determinants and outcomes and can be broadly classified as addressing either aspects of behavior, life history and demography, or the possible interactions among them. Positive side effects of long-term field projects on the conservation of the study site or the study species have recently become additional important reasons for the continuation of these projects. Studying individually known primates over years and decades also poses some unique challenges, however, especially with respect to data management and funding. We close this chapter by summarizing some of the unique insights about primate social systems and life history only made possible by the long-term nature of the studies, focusing on the chapters making up the remainder of this volume.

P.M. Kappeler (✉)
Department of Sociobiology/Anthropology and CRC Evolution of Social Behavior, University of Göttingen, Göttingen, Germany

Behavioral Ecology & Sociobiology Unit, German Primate Center, Göttingen, Germany
e-mail: pkappel@gwdg.de

C.P. van Schaik
Anthropological Institute and Museum, University of Zürich, Zürich, Switzerland
e-mail: vschaik@aim.unizh.ch

D.P. Watts
Department of Anthropology, Yale University, New Haven, CT, USA
e-mail: david.watts@yale.edu

P.M. Kappeler and D.P. Watts (eds.), *Long-Term Field Studies of Primates*,
DOI 10.1007/978-3-642-22514-7_1, © Springer-Verlag Berlin Heidelberg 2012

1.1 Introduction

No matter whether you travel to the swamp forests of Indonesia or the bizarre spiny forests of Madagascar, whether you visit an East African savanna or a remote rainforest on a small tributary of the Amazon, you may run into a primatologist studying one of the local primate species in their remaining natural habitats. In all likelihood, you will find this fictive colleague of ours to be part of a team composed of local and foreign researchers, field assistants, and students working out of some more or less primitive research camp or field station. Moreover, the project is likely to have a history of several years or even decades. You may ask yourself why anybody would want to travel to the end of the world, leaving all the amenities of modern civilization behind for months or even years, only to follow a group of primates from dusk to dawn in habitats full of parasites, predators, or even rebels. In this chapter, we summarize some of the scientific reasons that motivate an increasing number of primatologists to initiate or contribute to field studies that extend well over the duration of an average PhD project, i.e., the values of long-term field studies.

Primatologists were not the first to study individually known vertebrates over significant parts of their life cycles or even over several generations (see Clutton-Brock 2012), but we focus here on those projects that have aimed at unraveling the natural and life histories of our closest biological relatives. As always, it is useful to begin with a short historical review to put current activities in perspective. We will then discuss theoretical reasons requiring or justifying a long-term approach to studying primate behavior and ecology, and we touch upon some problems and challenges unique to this type of research. We close by highlighting some insights that could only be obtained from long-term research.

1.2 A Brief History of Early Primate Field Studies

The study of the behavior of animals in their natural habitat had to wait a long time to acquire some scientific respectability. During most of the nineteenth century, fieldwork consisted of collecting for taxonomic study or comparative anatomy, and no serious biologist came up with the idea to study the behavior of wild primates. The main reason was that science was something one did in the lab, not in the field: observing spontaneously behaving animals outside strictly controlled laboratory conditions was considered utterly unscientific (e.g., Skinner 1938). The success of Darwin's natural-history-based theory of evolution did little to change this. Even psychology, the study of behavioral mechanisms, largely developed as a laboratory-based science, mainly because of the rigorous control this allowed.

Only in the early twentieth century did a few eccentric Central European and British naturalists take to the field and begin to study bird behavior, thus paving the way for the ethologists, led by Konrad Lorenz and Niko Tinbergen, who built the scientific study of behavior from the ground up during the 1930s, using a mix of observations (often from blinds) and field experimentation. More or less simultaneously, and independently, in the USA, Robert Yerkes (1876–1956) at Yale University motivated some of his students to conduct the first field studies of wild primate populations. Among them, Henry Nissen conducted the first systematic study of feeding and social behavior of wild chimpanzees (*Pan troglodytes*) over 9 weeks in French Guinea (Nissen 1931). Clarence Ray Carpenter clearly stands out in retrospect for his pioneering field studies of mantled howler monkeys (*Alouatta palliata*; Carpenter 1934), white-handed gibbons (*Hylobates lar;* Carpenter 1940), and rhesus monkeys (*Macaca mulatta;* Carpenter 1942).

The history and tradition of modern field studies was established on December 3, 1948, when Kinji Imanishi, a most remarkable person and influential scientist (Matsuzawa and McGrew 2008), and two of his students (Jun'ichiro Itani and Shunzo Kawamura) went to Koshima Island to study wild Japanese macaques (*Macaca fuscata*). They established a new style of studying wild primates based on habituation (facilitated by provisioning), individual identification, and long-term observations that still provides the methodological standard for most primatological fieldwork. In the following 7 years, Imanishi and his students exported this new and successful method to 19 other populations of Japanese macaques, most of which are still being studied today. In 1958, Imanishi and Itani also embarked on an expedition to Central Africa, paving the ground for Japanese-led field studies of chimpanzees at Mahale in 1965 (Nishida 1968; Nakamura and Nishida 2012) and bonobos (*Pan paniscus*) in the Congo basin (Furuichi et al. 2012).

During the 1950s, several more primate field studies were initiated, some of which continue until today. North American primatologists whose backgrounds were in anthropology (notably Sherwood Washburn) were particularly active in this respect, but zoological and ethological influences were also important there and in Europe. Irven DeVore, one of Washburn's students, began studying savanna baboons (*Papio cynocephalus*) in Amboseli and other East African sites in the late 1950s (Washburn and DeVore 1961). At about the same time, Phyllis Jay, another Washburn student, initiated a study of Hanuman langurs (*Semnopithecus entellus*) in India (Jay 1962), and Charles Southwick started research on wild rhesus monkeys (*Macaca mulatta*) at another Indian site (Southwick and Siddiqi 1965). George Schaller (1963) conducted the first detailed field study of mountain gorillas in 1959–1960; this research provided the basis for his PhD in Zoology at the University of Wisconsin. European primatologists went to the field in the late 1950s to study hamadryas baboons in Ethiopia (*Papio hamadryas*: Kummer and Kurt 1963), chimpanzees in the Congo (Kortlandt 1962), and lemurs in Madagascar (Petter 1962) in their natural habitats. By 1960, technological advances had made tropical field sites both more accessible and medically safer for researchers from temperate zone countries. Jane Goodall began her field study of chimpanzees at what was then Gombe Stream Game Reserve in 1960 (Goodall 1963). In 1962, Vernon and Frances Reynolds spent 9 months studying chimpanzees in Budongo Forest, Uganda (Reynolds and Reynolds 1965); Vernon Reynolds resumed this project in 1990 (Reynolds 2005), and it continues until today.

The field projects initiated in the 1950s employed more systematic methodologies than earlier studies. Also, they usually studied relatively small social groups to facilitate identification of individuals, something considered necessary for proper understanding of social dynamics. Stuart Altmann (1962) pioneered systematic study of larger groups in his research on rhesus macaques on Cayo Santiago in the early 1960s; individual recognition was possible because each monkey was distinctively marked. By now, researchers standing on the shoulders of these early pioneers have studied at least one representative of virtually every primate genus in the wild.

In sum, the current tradition of long-term field studies of known and well-habituated individuals, with its focus on observation more than on experiment, arose from the combination of the Japanese zoologists' emphasis on individual recognition and the anthropologist's penchant for patient long-term observation. Next, we turn to the reasons that motivated these early studies and contrast them with current ones.

1.3 Why Long-Term Field Studies of Primates Are Needed

The early field studies of primates were primarily motivated by the insight that comparative studies of primate behavior can inform the study of human evolution (Washburn 1951; Hooton 1954; Crook and Gartlan 1966). Later, developing interest in the social lives and ecological adaptations of wild primates *per se* contributed to the characterization of broad patterns of adaptive variation (Eisenberg et al. 1972; Clutton-Brock and Harvey 1977). For both of these purposes, it was generally assumed that a field study lasting a few months would be sufficient to identify all species-typical behaviors (Strum and Fedigan 2000). Accordingly, none of these early studies were designed as long-term projects; in fact, most were just long enough to allow data collection for a PhD project. The fact that today, we have dozens of field projects that have been ongoing for a decade or more is motivated by several specific scientific reasons as well as by conservation concerns that have only explicitly emerged since the early 1980s.

The scientific need and justification for long-term field studies can be separated into aspects related to the study of social behavior and life history, respectively. The fact that primates live longer than other mammals of similar body size (Allman et al. 1993) provides a practical justification for research in both of these domains. Whereas mouse lemurs (*Microcebus* spp.) and other small-bodied strepsirrhines may typically live only a year or two, their average longevity exceeds that of other similarly sized small mammals (Austad and Fischer 1992), and some individuals survive until age 10 (Kappeler et al. unpublished data). Other larger primates that are still small in absolute terms, such as Verreaux's sifakas (*Propithecus verreauxi*), can live up to 30 years in the wild (Richard et al. 2002). Orangutans (*Pongo* spp.) and other great apes are so long-lived that their average life expectancy exceeds the duration of most academic careers (Wich et al. 2004). Additionally, long-term data on life history variation that is independent of body size can provide important insights into adaptive strategies, as Cords (2012) nicely illustrates with her comparative analysis of guenon life histories.

We must emphasize that the adaptive significance of social behavior and life history cannot be assessed with captive populations. The only way to identify the selective agents maintaining particular traits is to study primates in their natural habitats. Moreover, captivity also can induce changes in many aspects of life histories. Primates mature more rapidly and grow larger in captivity because of the abundance of food, absence of predators, and protection from immunological and other stressors (Austad and Fischer 1992; Smith and Jungers 1997; see also Künzel et al. 2003), and may also get older in captivity for these reasons (e.g., Picq 1992). Thus, field data are necessary to obtain more realistic measures of the behavioral tactics and the timing of life history events as they occur in the ecological context under which they evolved.

1.3.1 Long-Term Studies and Behavior

Long lives generate many interesting questions about primate social behavior, several of which we briefly consider below. Most of these have general importance because they are directly linked to fitness determinants (Clutton-Brock and Sheldon 2010a), and they relate both to phenomena that play out over long periods and to events that are rare but important.

Individual behavioral development provides obvious examples in the first category. Infancy and juvenility, where much social and individual learning takes place (Rapaport and Brown 2008; see also Thornton and Clutton-Brock 2011), extends over several years in most primate species (Jaeggi et al. 2010). Perhaps unsurprisingly, only a few studies of behavioral ontogeny have been conducted on wild primates (e.g., Pereira 1988; Altmann 1998; Watts and Pusey 1993; van Noordwijk and van Schaik 2005). Study of development is all the more important because events early in life can have lasting consequences (Clutton-Brock and Sheldon 2010a) that can only be understood if we have the full behavioral histories of individuals. Similarly, many behavioral patterns and individual tactics, such as coalition formation or rank challenges, can only be interpreted in the context of known social group histories. More interestingly, some of these consequences may only become apparent in older individuals (Suomi 1997) and can only be tracked in long-term studies.

Second, short-term studies often yield significant correlations between variables relevant to fitness but leave open the question of whether these actually consistently lead to differences in lifetime fitness. A classic example concerns dominance in those species in which females and/or males typically form dominance hierarchies. Rank is generally positively correlated with instantaneous measures of reproductive success in many taxa, including many primates. However, we need to know how well instantaneous measures predict lifetime reproductive success. Also, some influences on reproductive success are characterized by time lags. In Assamese macaques (*M. assamensis*), for example, male coalitions affect reproductive success only years later (Schülke et al. 2010). Indeed, identifying the determinants of lifetime reproductive success requires that we record the changes in social position and social strategies (the career trajectory; van Noordwijk and van Schaik 2001) of both males and females over their lifetimes (Alberts 2012; Pusey 2012). This is a prerequisite for illuminating the adaptive function of different social strategies. Long-term data on dispersal, mating behavior, paternity success, and morphology of a remarkably large sample of male sifakas at Beza Mahafaly (Sussman et al. 2012) are noteworthy in this regard. For example, they show that while annual fertility rates have the strongest influence on male lifetime reproductive success, lifespan also has an important effect, something that is probably true for many primates. Thus, the questions about possible trade-offs between instantaneous and lifetime reproductive success and the resulting optimum career profile can only be addressed by long-term studies.

Third, many primate species show remarkable behavioral variation, all the way down to the group level, where local traditions and conventions form and new behavior patterns spread within and between groups (Whiten and van Schaik 2007; Nakamura and Nishida 2012). Most of the work on such variation has consisted of cross-sectional mappings of its occurrence (e.g. Whiten et al. 1999), but better insight into its origins and possible adaptive value requires detailed research on transmission and possible fitness payoffs. Japanese macaques have provided several well-studied examples of the origin and spread of novel behaviors over the decades (Kawai 1963; Leca et al. 2010). Long-term studies of white-faced capuchins (*Cebus capucinus*) have also provided remarkable examples of the temporal dynamics of the origin, spread, and extinction of novel behaviors within a population (Perry et al. 2012). These comparative data can inform the study of similar processes during human evolution.

Fourth, long-term studies provide unique opportunities to study systematic interactions between primates and their habitats. Changes in demography and habitat may provide elegant, semi-experimental tests of socio-ecological models. For instance, resource access and defendability play central roles in theoretical models of socio-ecology. Territoriality is one expected outcome when resources are potentially defendable against neighboring social units, but not all primates defend territories. Variation in whether they do along with variation in territory or home-range size and location in response to long-term demographic and ecological variation (e.g., Jolly 2012; Yamagiwa et al. 2012) can therefore provide valuable data for tests of socio-ecological predictions. In addition, only long-term data provide the information needed to understand the interactions among neighboring groups or communities (Wilson 2012; Watts 2012). Feeding ecology and socio-ecology should also respond to changes in population density, but little is known of such effects. For instance, the density of mountain gorillas at Karisoke has more than doubled over the past 30 years, and investigating the consequences is bound to improve our understanding of gorilla socio-ecology (Grüter and Robbins in preparation).

1 The Values and Challenges of Long-Term Field Studies

Fifth, primates are renowned for their social complexity. Kinship, permanent association of males and females from overlapping generations, and sex-specific reproductive strategies generate complex social networks in different behavioral domains (Sueur et al. 2011). One main goal of socio-ecology is to identify the patterns and dynamics of these networks and the proximate behavioral mechanisms mediating them. Achieving this goal requires long-term monitoring of dyadic relationships, ideally in multiple groups. A distinct major goal of socio-ecology is the identification of the fitness consequences of different social tactics. Because the fitness of long-lived vertebrates is best estimated by estimating lifetime reproductive success, and because social relationships can affect longevity (Silk et al. 2010), achieving this goal requires documentation of individual life histories and complete reproductive careers.

Finally, the fitness of individual primates can be severely affected by rare but important events. Formal assessment of the relative importance of such events requires long-term study of multiple individuals or groups to yield large enough sample sizes. Predation and some forms of conspecific aggression, notably infanticide, compromise individual survival, which is one determinant of fitness. We have compelling reasons to think that behavioral adaptations to mitigate the potential consequences of these attacks exist, but data with which to test functional hypotheses necessarily accumulate slowly (Janson et al. 2012). Moreover, population differences in predation rates may elicit differences in life-history traits, social organization, and mating tactics, as indicated by a comparison between different sifaka populations (Kappeler and Fichtel 2012). Another compelling example comes from research on chimpanzees in the Taï Forest, Côte d'Ivoire, where lethal intergroup aggression was first seen only after 23 years of study (Boesch et al. 2008). Lethal aggression is more common in other chimpanzee populations (Mitani et al. 2010), and its documentation at Taï helps to generate a more nuanced discussion of its functions and especially of how variation in ecology and associated variation in social strategies should lead to variation in its frequency (Boesch et al. 2008).

Long-lived species may also face rare ecological crises, for example prolonged drought or cyclones, that provide opportunities to study the effects of these events on individual behavior and the limits on behavioral flexibility under natural condition (e.g., Sussman et al. 2012). Habitats may change over time too, both due to climate change and to successional processes. Long-term primate field studies have often kept detailed and systematic long-term records of habitat composition and productivity and on climate that are invaluable for addressing questions about behavioral and demographic variation (e.g., Wright et al. 2012). For example, Bronikowski and Altmann (1996) used 10 years of data on behavior, ecology, rainfall, and temperature to document plasticity in the foraging behavior of Amboseli baboons and to show that responses to environmental variation differed among social groups in ways that led to differences in predation risk. Such differences among groups in single populations are probably widespread among primates (*ibid.*), but confirming their existence depends on comparably extensive data sets.

1.3.2 Long-Term Studies and Life Histories

Time is the crucial variable in studying primate life histories. Understanding species-typical life histories is important because they express the evolutionary integration of adaptations involving the resolution of trade-offs for particular combinations of social, ecological, and demographic variables (Stearns 2000). Life histories therefore provide particularly deep insights into evolutionary processes and mechanisms (Clutton-Brock and Sheldon 2010a). Below, we discuss some concrete examples.

First, the description of key life-history parameters provides the basis for the comparative analysis of life-history strategies and vital population statistics (Alberts and Altmann 2003). The collection of reliable data on these variables is time-consuming. Following individuals of known ages until the age of first reproduction, for example, may require up to 15 years (Wich et al. 2004). Subsequent interbirth intervals may also play out over several years. Moreover, for animals with such long lifetimes, even 30 or 40 years might be insufficient to document maximal lifespan with a sample large enough to allow the calculation of reliable descriptive statistics. Enough long-term demographic data now exist for a few species to allow investigation of senescence in the wild (Bronikowski et al. 2002, 2011), but serious research on this topic is just beginning.

Second, life-history schedules can be adapted to extreme climatic fluctuations or catastrophes, such as cyclones, droughts, El Nino effects, or, more recently, global warming (e.g., Pavelka et al. 2007; Wright 2007; Dunham et al. 2010; Sussman et al. 2012; Wright et al. 2012). If solid baseline data exist, such natural experiments can contribute to better understanding of causal relationships in life-history evolution, the identification of trade-offs, and the limits of phenotypic plasticity. Because most of the independent variables in the natural experiments are rare and unpredictable, long-term monitoring is essential.

Third, because of the effects of individual life histories on demography, local population densities fluctuate over the years. Population density can influence the behavior of individuals, such as their dispersal strategies, but also group-level phenomena, such as territoriality, and a long-term approach allows studies of behavioral plasticity in response to changes in population density (Strier and Mendes 2012). Monitoring population size over years also constitutes a crucial prerequisite for conservation planning and action. The remarkable stability of many primate populations, relative to those of the other taxa, may reflect the slowness of primate life histories because of its buffering effects on fluctuations in survival rates (Morris et al. 2011).

Fourth, the size and composition of individual social units can fluctuate, especially in group-living species. In some species, groups have short life cycles tied directly to changes in male membership (Steenbeek et al. 2000; Koenig et al. 2012). Similarly, pair-living species may exhibit long-term fluctuations in the number of adult males and females (Reichard et al. 2012). Identifying how new groups form, whether maturing males are tolerated in their natal groups, and whether males immigrate or instead take over groups is vital for full understanding of socio-ecology. These factors depend on individual dispersal decisions, which are often sex-, context-, and condition-dependent (e.g., Janson et al. 2012). Infrequent events such as dispersals are easy to miss, especially if multiple social units are being studied, and demographic data may become biased if dispersal cannot be distinguished from mortality, leading to spurious conclusions. Again, years of observation of known individuals are required to identify rules and regularities in the sum of independent dispersal events and maturation schedules. Fedigan and Jack (2012) provide an excellent example. In their chapter, they note that they needed many years of data to determine mean age at natal dispersal for male white-faced capuchins and to confirm that males disperse throughout their lives and that females are not exclusively philopatric. Thus, long-term studies of demography and population dynamics can provide crucial context to interpret the social structure and organization of a species.

Fifth, slow life histories also generate constraints that necessitate a long-term approach. Genetic analyses, in particular, depend on access to suitable sample sizes. Given enough patience and luck, such analyses have enabled studies of fundamental topics in behavioral ecology, such as kin-based altruism, reproductive skew, and inbreeding avoidance (see e.g., Alberts and Altmann 2012).

1.4 Possible Drawbacks of Long-Term Studies

A long-term field study in practice always requires that the subjects are fully habituated. While this has made it possible to collect the important information specified above, this standard technique of primate fieldwork has possible drawbacks.

One is that we could impair the health of our subjects. The habituation process is likely to be stressful to the animals because wild animals almost universally see humans as predators. Indeed, some recent results indicate that unhabituated orangutans show a stress response during the habituation process, but this ends once they are habituated (Williamson and Feistner 2011; see also Marty et al. unpublished data). More insidiously, nonhuman primates and humans are vulnerable to many of the same diseases, and researchers and local field assistants can unwittingly infect their study subjects. This likelihood is especially high for species phylogenetically close to humans, notably the great apes, and evidence exists for transmission of serious diseases to great apes via close contact with humans (Köndgen et al. 2008). Fortunately, careful precautions, such as inoculations and wearing facemasks, can prevent transmission, and the protection that long-term research presence affords against other threats can outweigh the risk of disease transmission.

Another unintended consequence is more subtle. The presence of humans tends to discourage the activity of some classes of predators, in particular felids. Likewise, habituated groups meeting unhabituated ones tend to gain access to potentially contested resources in overlap areas. Both effects of habituation may lead to an increase of the size of the habituated group or community. Habituating neighboring groups solves the second problem, but there is no easy solution to the first.

1.5 Unique Problems and Future Challenges to Long-Term Studies

Studying wild primates continuously for several years or decades poses some other problems that apply less forcibly, if at all, to shorter studies or surveys. First, long-term research requires reliable individual identification. In some species, this is facilitated by variation in visual traits on which human observers can focus; sexual dimorphism, sexual dichromatism, variable sexual swellings, scars of healed wounds, and age-related changes in phenotypic traits can also be useful. In some species, capture and individual marking is possible (e.g., Wright et al. 2012; Kappeler and Fichtel 2012).

Second, long-term studies face several unique logistical challenges. If animals are not individually marked, individual recognition needs to be assured across generations of students and field assistants, and interobserver reliability requires particular attention to achieve uniform methodology across seasons and years (see e.g., Perry et al. 2012). One way to guarantee long-term accuracy of recognition is to reanalyze genetic samples of individuals, especially if individuals undergo dramatic changes in appearance. Among orangutans, for instance, it may be difficult to recognize individually known unflanged males once they have become flanged. Genetic analysis confirmed that one particular male had remained unflanged for two decades (Utami et al. 2002); this observation would have been doubted in the absence of genetic analysis. Most contributors to this volume discuss the practical aspects of providing suitable working, living, and laboratory facilities required for successful field research over months or years at a time. The chapters by Koenig et al. (2012) and by Perry et al. (2012) also provide particularly useful information about logistical and methodological prerequisites and standards for setting up and operating long-term primate field projects, as does a comprehensive guide edited by Setchell and Curtis (2011).

Third, behavioral and demographic data accumulate quickly, and specific strategies and policies for data management and storage are required. The data problem should not be underestimated because resources to create these databases and to enter and edit the accumulating data are rarely available, but the productivity of many projects, measured as papers produced per year, is a positive function of project duration (Clutton-Brock and Sheldon 2010a). This economy of scale can justify the needed investment in such large, long-term endeavors.

Finally, and perhaps most seriously, long-term studies extend past the typical grant duration of most funding agencies, which creates several problems. First, all the good scientific reasons for studying wild primates listed above cannot generate tangible results with a typical funding period of 3 or so years, and grant proposals are necessarily handicapped because they cannot promise results on the underlying big questions in behavioral ecology. Second, if the chain of grants on which the infrastructure of the project depends is interrupted only once, the project is fundamentally jeopardized (see Koenig et al. 2012 for a sad example). Thus, an increasing awareness of the benefits and risks of long-term field projects is required to eventually also change the available funding mechanisms (Clutton-Brock and Sheldon 2010b). Third, it is difficult to find grant support for data management and analysis, an essential but less popular part of the fieldwork production cycle. Only long-term institutional support and commitment may provide a viable solution to all these problems in the long run (see e.g., Kappeler and Fichtel 2012).

1.6 Unintended Benefits from Long-Term Studies

Apart from the numerous scientific virtues discussed above, the main unexpected benefit of long-term study sites concerns their positive impact on conservation. In most regions where wild primates live, protected areas are not necessarily safe from exploitation or conversion. Recent work has quantified what many fieldworkers have maintained for decades, namely that areas with ongoing long-term field projects suffer less habitat disturbance than other protected areas and have higher wildlife densities (Wrangham and Ross 2008; Campbell et al. 2011). This is because resident researchers report transgressions to the authorities and they employ local people who would lose their livelihoods if the projects were to fold due to loss of habitat or animals; it is also because risk of discovery can discourage transgressors from entering areas where the probability of direct encounters with researchers or local assistants is high.

Other positive impacts are economic. Long-term projects tend to become major employers in the surrounding areas because by their very nature, they are usually located in regions with virtually no opportunities for employment or cash incomes outside of limited cash cropping or dangerous (and often illegal) extractive activities, such as logging or collection of non-timber forest products. In several cases, the name recognition created by media coverage of the research has led to increased ecotourism at the same sites or in nearby areas that, despite its attendant problems, has increased local employment and lifted those lucky enough to be steadily employed out of poverty (see e.g., Jolly 2012). In addition, several field assistants have gone on to become local politicians or government officials based on the insights gained from contact with outsiders and the educational opportunities brought by employment in long-term projects.

Finally, another impact of long-term field studies is that many students, especially but not exclusively from the host countries, get involved. In some countries, such opportunities would hardly be available without the foreign-funded projects. The impact of such training opportunities is difficult to measure but almost certainly highly positive. Unlike many collecting or recording studies, primate fieldwork involves long-term stays at a single site. Long-term exposure to a particular area tends to produce intensive familiarity and therefore love and respect for the area and its biota; it also allows participants to witness the changes in and around the area and thus to appreciate conservation problems much better. Long-term residence at a field site also inevitably leads to intense contact with local people, far more than any traveler could experience. This engenders realistic appreciation of local attitudes toward conservation and of the problems (usually related to economic security and access to information) that can impede conservation even when attitudes are positive in principle. Not surprisingly, many of the most determined and effective tropical conservationists, whether from host countries or western countries, received their initial training by participating in long-term primate field studies.

References

Alberts SC (2012) Magnitude and sources of variation in male reproductive performance. In: Mitani J, Call J, Kappeler PM, Palombit RA, Silk JB (eds) The evolution of primate societies. University of Chicago Press, Chicago

Alberts SC, Altmann J (2003) Matrix models for primate life history analysis. In: Kappeler PM, Pereira ME (eds) Primate life history and socioecology. University of Chicago Press, Chicago, pp 66–102

Alberts SC, Altmann J (2012) The Amboseli Baboon Research Project: Forty years of continuity and change. In: Kappeler PM, Watts DP (eds) Long-term field studies of primates. Springer, Heidelberg

Allman JM, McLaughlin T, Hakeem A (1993) Brain structures and life span in primate species. Proc Natl Acad Sci U S A 90:3559–3563

Altmann SA (1962) A field study of the sociobiology of rhesus monkeys, *Macaca mulatta*. Ann N Y Acad Sci 102:338–435

Altmann SA (1998) Foraging for survival: yearling baboons in Africa. University of Chicago Press, Chicago

Austad SN, Fischer KE (1992) Primate longevity: its place in the mammalian scheme. Am J Primatol 28:251–261

Boesch C, Crockford C, Herbinger I, Wittig RM, Moebius Y, Normand E (2008) Intergroup conflicts among chimpanzees in Taï National Park: lethal violence and the female perspective. Am J Primatol 70:519–532

Bronikowski AM, Altmann J (1996) Foraging in a variable environment: weather patterns and the behavioral ecology of baboons. Behav Ecol Sociobiol 39:11–25

Bronikowski AM, Alberts SC, Altmann J, Packer C, Carey KD, Tatar M (2002) The aging baboon: comparative demography in a non-human primate. Proc Natl Acad Sci USA 99:9591–9595

Bronikowski AM, Altmann J, Brockman DK, Cords M, Fedigan LM, Pusey AE, Stoinski T, Morris WF, Strier KB, Alberts SC (2011) Aging in the natural world: comparative data reveal similar mortality patterns across primates. Science 331:1325–1328

Campbell G, Kuehl H, Diarrassouba A, N'Goran PK, Boesch C (2011) Long-term research sites as refugia for threatened and over-harvested species. Biol Lett 7; doi:10.1098/rsbl.2011.0155

Carpenter CR (1934) A field study of the behavior and social relations of howling monkeys (*Alouatta palliata*). Comp Psychol Monogr 10:1–168

Carpenter CR (1940) A field study in Siam of the behavior and social relations of the gibbon (*Hylobates lar*). Comp Psychol Monogr 16:1–212

1 The Values and Challenges of Long-Term Field Studies

Carpenter CR (1942) Sexual behavior of free-ranging rhesus monkeys (*Macaca mulatta*). I. Specimens, procedures and behavioral characteristics of estrous. J Comp Psychol 33:113–142

Clutton-Brock TH (2012) Long-term, individual-based field studies. In: Kappeler PM, Watts DP (eds) Long-term field studies of primates. Springer, Heidelberg

Clutton-Brock TH, Harvey PH (1977) Primate ecology and social organization. J Zool Lond 183:1–39

Clutton-Brock TH, Sheldon BC (2010a) Individuals and populations: the role of long-term, individual-based studies of animals in ecology and evolutionary biology. Trends Ecol Evol 25:562–573

Clutton-Brock TH, Sheldon BC (2010b) The seven ages of *Pan*. Science 327:1207–1208

Cords M (2012) The thirty year blues: What we know and don't know about life-history, group size, and group fission in blue monkeys in the Kakamega Forest, Kenya. Kappeler PM, Watts DP (eds) Long-term field studies of primates. Springer, Heidelberg

Crook JH, Gartlan JS (1966) Evolution of primate societies. Nature 210:1200–1203

Dunham AE, Erhart EM, Wright PC (2010) Global climate cycles and cyclones: consequences for rainfall patterns and lemur reproduction in southeastern Madagascar. Global Change Biol 17:219–227

Eisenberg JF, Muckenhirn NA, Rudran R (1972) The relation between ecology and social structure in primates. Science 176:863–874

Fedigan LM, Jack KM (2012) Tracking Neotropical monkeys in Santa Rosa: Lessons from a regenerating Costa Rican dry forest. In: Kappeler PM, Watts DP (eds) Long-term field studies of primates. Springer, Heidelberg

Furuichi T, Idani G, Ihobe H, Hashimoto C, Tashiro Y, Sakamaki T, Mulavwa MN, Yangozene K, Kuroda S (2012) Long-term studies on wild bonobos at Wamba, Luo Scientific Reserve D.R. Congo: Towards the understanding of female life Hhstory in a male-philopatric species. In: Kappeler PM, Watts DP (eds) Long-term field studies of primates. Springer, Heidelberg

Goodall J (1963) Feeding behaviour of wild chimpanzees: a preliminary report. Symp Zool Soc Lond 10:39–47

Hooton E (1954) The importance of primate studies in anthropology. Hum Biol 26:179–188

Jaeggi AV, Dunkel LP, van Noordwijk MA, Wich SA, Sura AAL, van Schaik CP (2010) Social learning of diet and foraging skills by wild immature Bornean orangutans: implications for culture. Am J Primatol 72:62–71

Janson C, Baldovino MC, Bitetti MD (2012) The group life cycle and demography of brown capuchin monkeys (*Cebus [apella] nigritus*) in Iguazú National Park, Argentina. In: Kappeler PM, Watts DP (eds) Long-term field studies of primates. Springer, Heidelberg

Jay PC (1962) Aspects of maternal behavior among langurs. Ann N Y Acad Sci 102:468–476

Jolly A (2012) Berenty Reserve, Madagascar: a long time in a small space. In: Kappeler PM, Watts DP (eds) Long-term field studies of primates. Springer, Heidelberg

Kappeler PM, Fichtel C (2012) A 15-Year perspective on the social organization and life history of Sifaka in Kirindy forest. In: Kappeler PM, Watts DP (eds) Long-term field studies of primates. Springer, Heidelberg

Kawai M (1963) On the newly-acquired behaviors of the natural troop of Japanese monkeys on Koshima Island. Primates 4:113–115

Koenig A, Borries C, Koenig A, Borries C (2012) Social organization and male residence pattern in Phayre's leaf monkeys. In: Kappeler PM, Watts DP (eds) Long-term field studies of primates. Springer, Heidelberg

Köndgen S, Kühl H, N'Goran PK, Walsh PD, Schenk S, Ernst N, Biek R, Formenty P, Mätz-Rensing K, Schweiger B, Junglen S, Ellerbrok H, Nitsche A, Briese T, Lipkin WI, Pauli G, Boesch C, Leendertz FH (2008) Pandemic human viruses cause decline of endangered great apes. Curr Biol 18:260–264

Kortlandt A (1962) Chimpanzees in the wild. Sci Am 206:128–138

Kummer H, Kurt F (1963) Social units of a free-living population of Hamadryas baboons. Folia Primatol 1:4–19

Künzel C, Kaiser S, Meier E, Sachser N (2003) Is a wild mammal kept and reared in captivity still a wild animal? Horm Behav 43:187–196

Leca J-B, Gunst N, Huffman MA (2010) Indirect social influence in the maintenance of the stone-handling tradition in Japanese macaques, *Macaca fuscata*. Anim Behav 79:117–126

Matsuzawa T, McGrew WC (2008) Kinji Imanishi and 60 years of Japanese primatology. Curr Biol 18:R587–R591

Mitani JC, Watts DP, Amsler SJ (2010) Lethal intergroup aggression leads to territorial expansion in wild chimpanzees. Curr Biol 20:R507–R508

Morris WF, Altmann J, Brockman DK, Cords M, Fedigan LM, Pusey AE, Stoinski TS, Bronikowski AM, Alberts SC, Strier KB (2011) Low demographic variability in wild primate populations: fitness impacts of variation, covariation, and serial correlation in vital rates. Am Nat 177:e14–e28. doi:10.1086/657443

Nakamura M, Nishida T (2012) Long-term field studies of chimpanzees at Mahale Mountains National Park, Tanzania. In: Kappeler PM, Watts DP (eds) Long-term field studies of primates. Springer, Heidelberg

Nishida T (1968) The social group of wild chimpanzees in the Mahali Mountains. Primates 9:167–224

Nissen H (1931) A field study of the chimpanzee: observations of chimpanzee behavior and environment in western French Guinea. Comp Psychol Monogr 8:1–122

Pavelka MSM, McGoogan KC, Steffens TS (2007) Population size and characteristics of *Alouatta pigra* before and after a major hurricane. Int J Primatol 28:919–929

Pereira ME (1988) Agonistic interactions of juvenile savannah baboons I. Fundamental features. Ethology 79:195–217

Perry S, Godoy I, Lammers W (2012) The Lomas Barbudal Monkey project: two decades of research on *Cebus capucinus*. In: Kappeler PM, Watts DP (eds) Long-term field studies of primates. Springer, Heidelberg

Petter J-J (1962) Recherches sur l'écologie et l'éthologie des lémuriens Malgaches. Mém Mus Nat Hist Nat A 27:1–146

Picq J-L (1992) Aging and social behaviour in captivity in *Microcebus murinus*. Folia Primatol 59:217–220

Pusey AE (2012) Magnitude and sources of variation in female reproductive performance. In: Mitani J, Call J, Kappeler PM, Palombit RA, Silk JB (eds) The evolution of primate societies. University of Chicago Press, Chicago

Rapaport LG, Brown GR (2008) Social influences on foraging behavior in young nonhuman primates: learning what, where, and how to eat. Evol Anthropol 17:189–201

Reichard UH, Ganpanakngan M, Barelli C (2012) White-handed gibbons of Khao Yai: Social flexibility, complex reproductive strategies, and a slow life history. In: Kappeler PM, Watts DP (eds) Long-term field studies of primates. Springer, Heidelberg

Reynolds V (2005) The chimpanzees of the Budongo Forest: ecology, behaviour, and conservation. Oxford University Press, Oxford

Reynolds V, Reynolds F (1965) Chimpanzees of the Budongo Forest. In: DeVore I (ed) Primate behavior: field studies of monkeys and apes. Holt Rinehardt and Winston, New York, pp 368–424

Richard AF, Dewar RE, Schwartz M, Ratsirarson J (2002) Life in the slow lane? Demography and life histories of male and female sifaka (*Propithecus verreauxi verreauxi*). J Zool Lond 256:421–436

Schaller GB (1963) The mountain gorilla: ecology and behavior. University of Chicago Press, Chicago

Schülke O, Bhagavatula J, Vigilant L, Ostner J (2010) Social bonds enhance reproductive success in male macaques. Curr Biol 20:2207–2210

Setchell JM, Curtis DJ (2011) Field and laboratory methods in primatology: a practical guide, 2nd edn. Cambridge University Press, Cambridge

Silk JB, Beehner JC, Bergman TJ, Crockford C, Engh AL, Moscovice LR, Wittig RM, Seyfarth RM, Cheney DL (2010) Strong and consistent social bonds enhance the longevity of female baboons. Curr Biol 20:1359–1361

Skinner BF (1938) The behaviour of organisms: an experimental analysis. Appleton Century, Oxford

Smith RJ, Jungers WL (1997) Body mass in comparative primatology. J Hum Evol 32:523–559

Southwick CH, Siddiqi MR (1965) Population changes of rhesus monkeys (*Macaca mulatta*) in India, 1959 to 1965. Primates 7:303–314

Stearns SC (2000) Life history evolution: successes, limitations, and prospects. Naturwissenschaften 87:476–486

Steenbeek R, Sterck EHM, de Vries H, van Hooff JARAM (2000) Costs and benefits of the one-male, age-graded, and all-male phases in wild Thomas's langur groups. In: Kappeler PM (ed) Primate males: causes and consequences of variation in group composition. Cambridge University Press, Cambridge, pp 130–145

Strier KB, Mendes SL (2012) The Northern Muriqui (*Brachyteles hypoxanthus*): lessons on behavioral plasticity and population dynamics from a critically endangered species. In: Kappeler PM, Watts DP (eds) Long-term field studies of primates. Springer, Heidelberg

Strum SC, Fedigan LM (2000) Changing views of primate society: a situated North American view. In: Strum SC, Fedigan LM (eds) Primate encounters: models of science, gender, and society. University of Chicago Press, Chicago, pp 3–49

Sueur C, Jacobs A, Amblard F, Petit O, King AJ (2011) How can social network analysis improve the study of primate behavior? Am J Primatol 73. doi:10.1002/ajp. 20915

Suomi SJ (1997) Early determinants of behaviour: evidence from primate studies. Br Med Bull 73:170–184

Sussman RW, Richard AF, Ratsirarson J, Sauther ML, Brockman DK, Gould L, Lawler R, Cuozzo FP (2012) Beza Mahafaly Special Reserve: Long-term research on lemurs in Southwestern Madagascar. In: Kappeler PM, Watts DP (eds) Long-term field studies of primates. Springer, Heidelberg

Thornton A, Clutton-Brock TH (2011) Social learning and the development of individual and group behaviour in mammal societies. Philos Trans R Soc B 366:978–987

Utami SS, Goossens B, Bruford MW, de Ruiter JR, van Hooff JARAM (2002) Male bimaturism and reproductive success in Sumatran orang-utans. Behav Ecol 13:643–652

van Noordwijk MA, van Schaik CP (2001) Career moves: transfer and rank challenge decisions by male long-tailed macaques. Behaviour 138:359–395

van Noordwijk MA, van Schaik CP (2005) Development of ecological competence in Sumatran orangutans. Am J Phys Anthropol 127:79–94

Washburn SL (1951) The new physical anthropology. Trans NY Acad Sci 13:298–304

Washburn SL, DeVore I (1961) The social life of baboons. Sci Am 204:61–72

Watts DP (2012) Long-term research on chimpanzee behavioral ecology in Kibale National Park, Uganda. In: Kappeler PM, Watts DP (eds) Long-term field studies of primates. Springer, Heidelberg

Watts DP, Pusey AE (1993) Behavior of juvenile and adolescent great apes. In: Pereira ME, Faibanks LA (eds) Juvenile primates: life history, development and behavior. University of Chicago Press, Chicago, pp 148–167

Whiten A, van Schaik CP (2007) The evolution of animal "cultures" and social intelligence. Phil Trans R Soc B 362:603–620

Whiten A, Goodall J, McGrew WC, Nishida T, Reynolds V, Sugiyama Y, Tutin CEG, Wrangham RW, Boesch C (1999) Cultures in chimpanzees. Nature 399:682–685

Wich SA, Utami-Atmoko SS, Mitra Setia T, Rijksen HD, Schürmann C, van Hooff JARAM, van Schaik CP (2004) Life history of wild Sumatran orangutans (*Pongo abelii*). J Hum Evol 47:385–398

Williamson EA, Feistner ATC (2011) Habituating primates: processes, techniques, variables and ethics. In: Setchell JM, Curtis DJ (eds) Field and laboratory methods in primatology: a practical guide. Cambridge University Press, Cambridge, pp 25–39

Wilson ML (2012) Long-term studies of the chimpanzees of Gombe National Park, Tanzania. In: Kappeler PM, Watts DP (eds) Long-term field studies of primates. Springer, Heidelberg

Wrangham RW, Ross E (2008) Science and conservation in African forests: the benefits of long-term research. Cambridge University Press, Cambridge

Wright PC (2007) Considering climate change effects in lemur ecology and conservation. In: Gould L, Sauther ML (eds) Lemurs: ecology and adaptation. Springer, New York, pp 385–401

Wright PC, Erhart EM, Tecot S, Baden AL, Arrigo-Nelson SJ, Herrera J, Morelli TL, Blanco M, Deppe A, Atsalis S, Johnson S, Ratelolahy F, Tan C, Zohdy S (2012) Long-term lemur research at Centre Valbio, Ranomafana National Park, Madagascar. In: Kappeler PM, Watts DP (eds) Long-term field studies of primates. Springer, Heidelberg

Yamagiwa J, Basabose AK, Kahekwa J, Bikaba D, Ando C, Matsubara M, Iwasaki N, Sprague DS (2012) Long-term research on Grauer's gorillas in Kahuzi-Biega National Park, DRC: life history, foraging strategies, and ecological differentiation from sympatric chimpanzees. In: Kappeler PM, Watts DP (eds) Long-term field studies of primates. Springer, Heidelberg

Part II
Madagascar

Chapter 2
Berenty Reserve, Madagascar: A Long Time in a Small Space

Alison Jolly

Abstract Berenty Reserve is a privately owned reserve established in 1936. At 200 ha, it holds the largest remaining gallery forest on the Mandrare River. Scientists of many nationalities have studied there: research follows their diverse interests rather than being coordinated overall. One finding which emerged from long-term monitoring concerns the importance of both within- and between-troop competition for female *Lemur catta* and their inheritance of territory in the female line, at least in this islanded population. This may play a role in the maintenance of female dominance over males. Another aspect of long-term study is the changes brought by introduced species, including *Leucaena leucocephala*, a favored, but toxic, forage tree. The growth of leucaena stands paralleled population growth of the *L. catta* troops with access to these stands, although highly affected females lost fur and had low infant survival; eradication produced a local population crash. Introduced *Eulemur rufus x collaris* have grown from about 16 individuals in 1975–1980 to almost 600 in 2009. They are taking over the central gallery forest. *L. catta* troops maintain their original sleeping areas in the gallery forest but increasingly forage on the periphery, recreating the niche separation described for *Lemur* and *Eulemur* in natural sympatry. Overall, Berenty Reserve is drying out, with closed-canopy gallery forest giving way to more open scrub. Research on Berenty's lemurs is thus the study of evolved adaptations confronted with a changing environment.

A. Jolly (✉)
Department of Biology, University of Sussex, Brighton, UK
e-mail: ajolly@sussex.ac.uk

P.M. Kappeler and D.P. Watts (eds.), *Long-Term Field Studies of Primates*,
DOI 10.1007/978-3-642-22514-7_2, © Springer-Verlag Berlin Heidelberg 2012

2.1 Introduction

Berenty Estate was founded in 1936 by the de Heaulme family in consultation with local Tandroy clans. It lies beside the Mandrare River in southern Madagascar at about S 25°00'E 46°18'. The estate comprises 6,000 ha: sisal fields, pasture, a tourist complex, and 1,000 ha of forest established as nature reserves long before conservation became fashionable, just because the forest was too beautiful to cut down. The largest reserve parcel, often just called Berenty Reserve, is 200 ha. It is continuous with about 400 further hectares of gallery and spiny forest. Berenty Reserve and the 100 ha Bealoka Reserve, which is also on Berenty Estate, are the two largest of only four remaining blocks of gallery forest on the Mandrare river below the steep tributaries of the headwaters. This is one of the most threatened forest types of Madagascar.

Lemur research at Berenty began with my arrival in 1963. Since then the de Heaulmes have welcomed scientists of all nationalities. There have been long gaps between studies, and many different projects and teams. Research follows individual interests and possibilities rather than being coordinated overall. This article focuses on two aspects of long-term interest: *Lemur catta* social behavior and the influence of introduced species on *L. catta* behavior and ecology. It also provides a bibliography to follow up other aspects of research at Berenty.

Much of what is ecologically interesting about Berenty relates to the fact that it is so small. This forest fragment serves as a scale model of large wilderness reserves. Of the various threats to a small reserve, much of the early theoretical literature focused on inbreeding depression, or on variation in reproductive success either stochastically or due to fluctuating weather, or else on edge effects (Soulé and Wilcox 1980; Frankel and Soulé 1981). At Berenty it has been the accidental or deliberate introduction of exotic species that has most profoundly changed the ecology of this limited space. Especially important have been brown lemurs, *Eulemur rufus x collaris*, and the nutritious but toxic tree *Leucaena leucocephala*. Lemur research at Berenty thus involves the study both of long-evolved species norms of behavior, and of how these adapt in the face of environmental challenges new to these species' history.

The wider importance of Berenty will be scarcely considered here. Berenty Estate is a kaleidoscope of human cultures. Local Tandroy people, traditional warrior-pastoralists, are now agriculturalists who submit to rules on forest use in return for employment that buffers them against the region's recurrent famines. Tourists (*Homo sapiens garbagedispersiensis*) provide income which ensures the reserve's support and survival. Most of the tourist personnel are multilingual Tanosy people from neighboring Fort Dauphin. The scientists themselves function like a clan or tribe: distinct from but dependent on the others. Berenty is the most televised spot in Madagascar. Nearly every foreign film features its parading ringtails and dancing Verreaux's sifaka. This may in fact be its most important contribution, as a show window for Madagascar. In 1992–1994, Helen Crowley became the first manager of the forest reserves, funded by the Wildlife Trust and by the de Heaulmes. Subsequent managers were Hajarimanitra Rambeloarivony and Sahoby Marin Raharison. The forest is now managed by Claire de Heaulme Foulon and her husband Didier Foulon (Jolly 2004, 2010).

2.2 Habitat Zones and Fauna

For a general description of Berenty Reserve see Jolly et al. (2006a Fig. 2.1). There is also a fundamental study by Sheila O'Connor and Mark Pigeon comparing Berenty and the 100 ha Bealoka Reserve. Bealoka was still grazed by goats and zebu until 1985, which gave it a much more open understory. Its one ringtail troop used the whole of the 100 ha forest, moving round from season to season with no identifiable core area – most unlike the defended, stable territories of Berenty. O'Connor went on to a career with WWF rather than publishing research, but her thesis remains a baseline for many Berenty ecological studies (O'Connor 1987).

The natural habitat zones of Berenty grade from rich alluvial Gallery forest through transitional Scrub to Spiny forest (Figs. 2.2 and 2.3). The Front zone has many introduced trees which supplement natural lemur diet (Fig. 2.4). These four zones form the 100-ha study area are called Malaza. The Ankoba zone is 70-year-old secondary forest, combining natural and introduced trees for the highest lemur population densities. There is a fivefold difference in ringtailed lemur density from about 500/km^2 in Front and Ankoba to about 100/km^2 in the Spiny zone (Jolly 1966a; Budnitz and Dainis 1975; Budnitz 1978; Mertl-Millhollen et al. 1979; Blumenfeld-Jones et al. 2006; Jolly et al. 2006a; Razafindramanana et al. 2008).

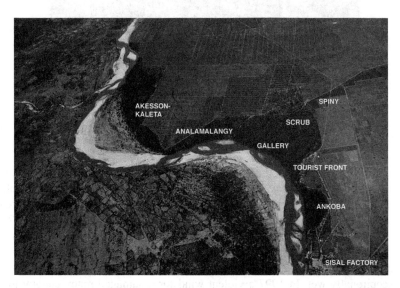

Fig. 2.1 Air photo of Berenty Reserve from the north, with habitat zones. Courtesy Barry Ferguson and the Libanona Ecoly Centre. From Jolly et al. (2006a)

Fig. 2.2 Gallery forest. Giant tamarind trees traditionally provide half the ringtails' feeding time, as well as highways and sleeping sites, though they are now largely occupied by brown lemurs. Photo © Cyril Ruoso

Fig. 2.3 Spiny forest. Verreaux's sifaka and ringtails can live without free water, gaining their food from leaves and stems. Photo © Cyril Ruoso

Southern Madagascar's climate alternates hot wet summers, with temperatures at or above 40°C at mid-day, and cold dry winters, when temperatures fall below 10°C at night. Rainfall varies erratically from 300 to 900 cm if calculated in years beginning Oct 1, which group all of the summer wet season together (Fig. 2.5). Even this masks some of the variation, as in 1991–1992, when two-thirds of the season's rain fell during a 3-day storm in January. El Niño years usually mean drought for the south of Madagascar as for southern Africa, but some El Niño years are exceptionally wet. In 1997 a violent windstorm damaged many canopy trees. Lemurs and other species are adapted to survive recurrent catastrophic years

Fig. 2.4 Front zone. Berenty's non-human primates are used to ignoring tourists. Photo © A. Jolly

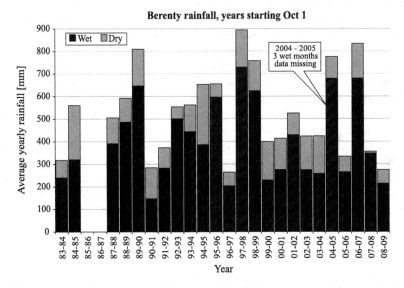

Fig. 2.5 Yearly rainfall at Berenty, which can vary threefold between years. Years begin Oct. 1, which keeps the rainy season as one block (data thanks to K. Blumenfeld-Jones, S. O'Connor, C. and A. Rakotomalala)

(Gould et al. 1999; Wright 1999; Rasamimanana et al. 2000; Richard et al. 2002; Jolly et al. 2006b).

Ringtailed lemurs (*L. catta*) numbered about 450 animals in the Malaza-Ankoba zones in 2009. The ecology of the different zones varies dramatically, with almost no overlap between plant species eaten by any lemurs in the Gallery and Spiny zones. Ringtailed lemurs rely more on leaves in the natural habitats than in the Front and Ankoba zones, since a higher proportion of fruit is available in the modified zones. Niche separation between ringtails and Verreaux's sifaka (*Propithecus verreauxi*) is clear, with the sifaka eating more leaves and flowers, and better able to detoxify secondary compounds, though the ringtails also have high tannin tolerance. Ringtails eat little prey: seasonal locusts, acacia army caterpillars, and the occasional chameleon. (Niche separation between ringtails and brown lemurs will be considered below, with the influence of introduced species; Rasamimanana and Rafidinarivo 1993; Pitts 1995; Oda 1996b; Rasamimanana 1999; Simmen et al. 1999, 2003a, b, 2005, 2006a, b, 2010; Rasamimanana et al. 2000, 2006; Rasamimanana and Anjaranantenaina 2008).

Berenty's five other lemur species each have populations of several hundred animals. Most research on Verreaux's sifaka has been done by Alison Richard and her colleagues at Beza Mahafaly (Sussman et al. 2011) further enlarged by the Kappeler team at Kirindy (Kappeler and Fichtel 2011). Sifaka research at Berenty has recently been revived by Norscia, Palagi, Simmen, Rasamimanana, and Fichtel with their colleagues and students. A strong male bias appeared in the 2006 Norscia–Palagi census: 62% of 206 animals counted were male. As Richard originally suggested, the varying sifaka sex ratios seen at Berenty may reflect greater stress on sifaka females during hard years. However, different methods so far give very different total numbers, from 206 in Malaza and Ankoba to Rasamimanana's census of 230 in Malaza alone (Richard 1978; Oda 1998; Charrier et al. 2007; Fichtel 2008; Norscia and Palagi 2008; Palagi et al. 2008; Norscia et al. 2009; Fichtel and Kappeler 2011; Rasamimanana personal communication).

Lepilemurs (*Lepilemur leucopus*) and mouse lemurs (*Microcebus murinus, Microcebus griseorufus*) have not been fully censused. It remains a puzzle how small-bodied lepilemurs survive the cold nights of winter on a diet of mature leaves (Charles-Dominique and Hladik 1971; Russell 1977). Grey mouse lemurs eat fruit and insects. Grey-and-red mouse lemurs (only recently identified as a separate species: Rasoloarison et al. 2000) are largely gummivorous during the dry season food shortages, with a complex pattern of sleeping associations (Génin 2001, 2007, 2008, 2010, in press; Génin et al. 2010). Hybrid brown lemurs (*Eulemur rufus x E. collaris*) were introduced in 1975, and will be discussed under the changing influence of exotic species (Sect. 2.4).

Berenty holds southern Madagascar's largest colony of the Madagascar giant fruit bat (*Pteropus rufus*). This colony seems to be diminishing from several thousand in 2000 to 927 in a census of 2006 to only 100 in a 2009 census. However, there are wide seasonal variations in numbers (Long 2002; Razafindramanana personal communication). Other mammals include Commerson's leaf-nosed bat *Hipposideros commersoni*, the rufus trident bat *Triaenops rufus, Miniopterus majori, Miniopterus manavi*, and the Malagasy mouse-eared bat *Myotis goudoti* (Fish 2010), as well as the spiny tenrec *Setifer setosus*, the large tenrec *Tenrec ecaudatus*, and the shrew-like tenrec *Microgale* spp., and the Madagascar tree-rat *Eliurus myoxinus*. There is an all-too flourishing population of *Rattus rattus*, the scourge of Malagasy small mammals (Crowley 1995; Goodman 1995). The forest is too small to hold *Cryptoprocta ferox*, the fossa, and there are no reports of the feral species of wildcat, but domestic dogs and cats, and possibly the Indian civet (*Viverricula indica*) take their place as potential lemur predators.

Fifty-two species of resident birds have been recorded out of a total of 99 species seen, of which 41% are endemic to Madagascar (Goodman et al. 1997). The genetics of the two male color morphs of the Madagascar paradise flycatcher (*Terpsiphone mutata*) have been studied in the Bealoka parcel of gallery forest (Mulder et al. 2002). Crested coua (*Coua cristata*) calls, especially alarm calls, are remarkably similar to ringtail vocalizations (McGeorge 1978a, b). Lemur predators include the harrier hawk (*Polyboroides radiatus*), the Madagascar buzzard (*Buteo madagascariensis*), and the black kite (*Milvus migrans*) (Crowley 1995; Karpanty and Goodman 1999).

Parasitism and disease are a neglected part of the ringtail story, with few published veterinary studies. The most obvious parasites are red ticks which cluster round the eyes of debilitated animals (Takahata et al. 1998; Koyama et al. 2008).

Two constants at Berenty shape much of ringtailed lemur ranging behavior: the prevalence of tamarind trees, and the year-round availability of water. *Tamarindus indica* are the dominant tree of the gallery forest. About 50% of gallery forest ringtailed lemur feeding time is on tamarind fruit pulp or leaves through all seasons. Population density largely mirrors the availability of tamarind trees, and in seasons where there is a dearth in the usual range, troops may make long excursions to fruiting trees (Rasamimanana and Rafidinarivo 1993; Mertl-Millhollen et al. 2003, 2004, 2006; Blumenfeld-Jones et al. 2006; Koyama et al. 2006; Soma 2006). This means that there are fixed points of great importance to ranging patterns, and these major food sources persist over many generations of lemurs. This is less true in the spiny forest, where not only is there less food overall, but the important food trees are smaller and more numerous in a given home range.

The Mandrare River dries up only in September–October of the worst drought years (historically about once every 10 years), and even then a few stagnant puddles remain near the bank. Gallery forest ringtails prefer to descend to the river in places where there are overhanging trees and shrubs that provide cover from aerial predators: again fixed points of great value. In the Scrub and Spiny zones water is obtained from the leaves of succulent plants. These may be in limited stands of the vine *Xerocysios* or introduced *Opuntia*, also points of value which may be worth defending, which influence daily ranging and inter-troop competition (Budnitz 1978).

2.3 Ringtailed Lemur Social Behavior

My early work focused on behavior of ringtailed lemurs, especially their near-absolute female dominance, their highly compressed mating season, and the fact that their social complexity seemed to far outweigh their interest in manipulating objects (Jolly 1966a, b, 1967). The long-term studies which followed have hugely enriched our knowledge, but not actually solved my initial questions.

2.3.1 Communication

Ringtailed lemur troops are multi-male and multi-female, ranging in size from 3 to 34 non-infants. They are highly social, with a wide range of vocal, visual, tactile, and scent communication. Vocal signals include the adult male howl or song, contact calls ranging from a soft mew to a loud meow (often given in chorus) locomotor and alarm calls (Oda 1996a). Like vervet monkeys, they use different calls toward different predators: clicks and mobbing toward snakes, yap chorus toward ground predators, and screams toward hawks, often followed by movement upward or downward as appropriate. Scent marking still has subtleties that we do not appreciate, but is done along with stereotyped posturing. A handstand during genital marking by either sex puts the mark at the height of other lemurs' noses. Male spur marking is done with side-to-side jerking of the upper body; male stink-fighting with anointing and waving the tail (Jolly 1966a; Mertl-Millhollen 2007). Urine-marking with raised tail and a "bottom-drag" along a substrate, differs from normal urination (Palagi et al. 2005a; Palagi and Norscia 2009). The scents and vocalizations are individually identifiable (Mertl 1977; Palagi and Dapporto 2006) but also are responded to between troops and even, in the case of alarm calls, between ringtails and sifaka (Oda 1996a, 1998, 1999, 2001; Oda and Masataka 1996). Mertl-Millhollen and Palagi independently conclude that females are more attentive to female marks (especially as inter-troop territorial communication), and males to both males and females (especially for mating opportunities). Mertl-Millhollen first identified the scent-marked ring of territorial boundary well within home range boundaries for both ringtails and Verreaux's sifaka in the gallery forest. The ring of intensive marking is a "battle zone" where troops confront others of their own species. (Mertl 1977; Mertl-Millhollen 1979, 1986, 1988, 2000a, b, 2004, 2006; Mertl-Millhollen et al. 1979). Visual attention, like olfactory attention, is largely directed within members of each sex, except during the brief mating season (Lane and Bard 2007, 2008).

2.3.2 Female Affiliation and Aggression

The core of each troop is a female matriline. Subgroups (cliques) reflect kinship, mainly mothers with adult daughters. Time spent in affiliative cuddling and grooming greatly outweighs time spent in aggressive behavior but aggression plays a large part in troop life (Nakamichi and Koyama 1997, 2000; Nakamichi et al. 1997; Sussman et al. 2003; Sussman and Garber 2004). Female dominance rank is highly contested and reversals may be violent (Koyama et al. 2001, 2005). The alpha female has a special role in vigilance and defense toward other troops. If she does not defend personally, her henchwomen, usually a daughter, takes a lead role (Gould 1996). An alpha female may have a relaxed style of dominance with little friction within the troop, but some are actively "spiteful," chasing subordinates and males from feeding sites even if the dominant does not feed there herself (Dubovick 1998). One may suppose that this behavior eventually pays off for the dominant in reproductive success, by imposing greater costs on her within-troop rivals. In the short term, it clearly costs the alpha energy in the chase and in lost feeding time for herself. If the observer knows enough of the troop history, the differences in alphas' style may sometimes be attributed to outside causes: for instance, if the alpha herself is an immigrant who has fought her way in from another troop, or if there is challenge from a rising sister or cousin, or if the pressure from successful adjacent troops is compressing troop range and thus exacerbating within-troop rivalries.

Larger troops, on average, have lower reproductive success, though optimal group size is smaller in the sparser areas of the reserve, and there are complex tradeoffs between group size and seasonality (Jolly et al. 2002; Pride 2003, 2005b; Pride et al. 2006). In the highest-density area of the reserve, the highest number of surviving infants was found among mid-ranked females in middle-sized and large groups with 4–9 adult females. Smaller groups with only 2–3 adult females had fewer surviving infants (Koyama et al. 2001, 2002, 2005). The somewhat lower success among the highest ranked females in large groups seems like an anomaly, but the variation in aggressivity and cortisol levels between different females, and the pressures of high-density territoriality, may need analysis on an individual by individual basis (Cavigelli et al. 2003; Pride 2003, 2005a, b).

Long-term, when a troop grows too large or when a grandmother dies or is deposed, the troop splits. A group fission involves active targeting by the dominant clique of a subordinate clique, their most distant cousins. Subordinates, once forced out, can fight to establish new territory by subdividing the original home range, or taking neighbor troops' ranges. The evicted group may remain nomadic through others' home ranges for up to 2 years before claiming discrete home range of its own with active defense of a part of this range as a territory. In 23 observed cases, the subordinate daughter troop remained in ranges adjacent to the dominants which had driven them out (Koechlin 1972; Koyama 1991; Hood and Jolly 1995; Koyama et al. 2002; Takahata et al. 2005; Ichino 2006; Ichino and Koyama 2006; Jolly et al. 2006b).

The alpha female often gives a soft call which alerts the troop that it is time to move, and nearly always leads the troop in progressions of over 30 m. A few anecdotal reports of alpha females leading troops on excursions well outside their normal range raise the suggestion whether the memory of old female lemurs is also a resource for their troops (Dolins and Jolly 2007; Miles and Rambeloarivony 2008).

Elsewhere, I argued that within-troop competition between female ringtails evolved within the constraints of a territorial system where it may be easier to evict your cousins than to expand a home range into areas used by neighbors. Frequency of encounters between troops varies with population density: Front troops may meet five or more times per day, in the Gallery forest only once per day, and in the Scrub in the 1990s only twice a week – though these figures have changed with recent population shifts. However, when troops do meet, they are equally likely to be aggressive whether at high or low densities – and much depends on the individual troop histories which neighbors are treated as the worst threats (Jolly et al. 1993, 2006b; Takahata et al. 2006). Pride showed that even in a case where owners were gaining less food per individual than raiders, the owners continued to actively defend their territory, suggesting that long-term ownership might be worth short-term loss (Pride et al. 2006). Scent marking not only indicates the territory boundaries, but serves as a long-term tradition to maintain those boundaries (Mertl-Millhollen 2000b, 2006).

One of the earliest Berenty studies was conducted by Peter Klopfer. He observed that my 1963 study troop was still in the same area in 1969, with many of the same territorial boundaries (Klopfer and Jolly 1970). Mertl-Millhollen brought order to all studies up to 1975, again noting that troops remained in the same ranges (Mertl-Millhollen et al. 1979). We now know that the ranges of subdivided troops at Berenty remain adjacent to each other, such that female descendents of the first known troops still occupy the same parts of the forest after 40 years (Jolly and Pride 1999; Jolly et al. 2006b).

2.3.3 Males, Infants, and Juveniles

Male transfer between ringtail troops during the birth season was first recorded at Berenty (Jones 1983). It has been much more extensively studied in the tagged population at Beza Mahafaly, as has male behavior (Sussman et al. 2011). Males thus choose new troops 6 months prior to actual mating. Females are only receptive for a few hours of 1 day, though the few that do not conceive may cycle again a month later. Within a troop, all females reach oestrus in a 2-week period, though there is "asynchrony within synchrony," such that no two are receptive on the same day. The troop's dominant male sometime, but far from always, has mating priority. Males, including extra-troop individuals, may vie in violent "jump-fights," which often lead to canine slashes to head or body. Oddly, the female retains absolute mate choice: if she does not like the current winner of a jump-fight she remains in the open, inviting other challengers – and she may mate with several males in succession. In short, behaviorally it seems that males gain little advantage from their year-round dominance contests, their jump-fights, and from their attempts to mate-guard (including copulatory plugs which can be removed by the next male). Actual male reproductive success at Berenty is now under active investigation (Ichino personal communication).

Infants are born in September–October, with most births within about 3 weeks. Births are generally at night or during siesta hours (Okamoto 1998; Takahata et al. 2001). Infants are precocial for lemurs, transferring to the mother's back at about 2 weeks, and are weaned in about February–March at the peak of the wet season (Gould 1990). Infants grow rapidly during in the wet season, and cease to grow in the following dry season, even under constant conditions in captivity (Pereira 1993b). Wild females normally give birth for the first time at age 3 or 4, though in captivity they give birth at 2 years. Twins are rare in the wild, common in the better-nourished conditions of captivity. Males have little to do with raising infants, but a mother's close kin may carry her infant or even let it suckle (Gould 1992; Koyama 1992; Koyama et al. 2006). Infanticide happens, though very rarely. In the few cases seen, some perpetrators were extra-troop or immigrant males, but others were females suppressing the reproduction of a subordinate (Hood 1994; Jolly et al. 2000; Ichino 2005). A quarter to a half of infants die in the first year, many during very early lactation when the mother is losing body condition at the end of the dry season (Jolly et al. 2002; Koyama et al. 2002). The mother stays with her fallen infant, or starts to follow the troop and returns as long as it can cry, but she eventually leaves it to rejoin the troop. Lemurs, unlike monkeys, do not have hands that can support a dead infant, though the mother may return to the corpse or site some hours later. The high mortality rate does not reflect maternal indifference (Nakamichi et al. 1996). What it does reflect is the extreme variability of yearly climate, such that infants die but the vast majority of females survive (Jolly et al. 2002, 2006b).

One major difference between ringtailed lemurs and anthropoid primates is in rough-and-tumble play. In most primates including humans, males have more physical contact play than females. In ringtails the sexes spend equal time in rough contact play. Juveniles of both sexes also have occasional serious wrestling bouts which lead to dominance decisions (Gould 1990; Pereira 1993a).

2.3.4 Female Dominance over Males

This is not the place for a full review of theories and studies relating to the evolution and maintenance of lemur female dominance over males. That will largely concern the differences between lemur species' intensity or expression of female dominance, first pointed out by Pereira and Kappeler (Pereira et al. 1990; Pereira and Kappeler 1997). Three lines of study at Berenty contribute to this still unsolved question.

First, it is clear that ringtailed lemur females can be highly aggressive, both within and between troops, as well as toward males. Pereira argues (and I agree) that this involves motivation to gain power per se (Pereira 2006). The adaptive advantages of such motivation are a different question. Lewis argues for "leverage" or "power" as an evolutionary correlate of female dominance (Lewis 2002). In species which are ecologically constrained to have slight or no sexual dimorphism, and where the male–female ratio is equal (or even irrelevant), females may be able to exert their leverage of mate choice over males, achieving behavioral dominance. Lewis' evolutionary reasoning may or may not apply to all the lemur species which show female dominance. The motivational aspect similarly does not apply to all species – in particular, female Verreaux's sifaka, which Lewis studied, have full female dominance but show very rare aggression within a troop, unlike the power-hungry ringtails.

A different line of argument suggested that lemur females as a whole have exceptional need for food, either physiologically or because of the highly erratic Malagasy climate (Jolly 1984; Dewar and Wallis 1999; Wright 1999; Richard et al. 2000, 2002; Dewar and Richard 2007). However, time and locomotion assessment of male and female Berenty ringtails showed that the two sexes actually consume similar amounts of protein and calories, though males' diet is more fibrous, and that the two sexes also expend similar amounts of energy (Rasamimanana and Rafidinarivo 1993; Rasamimanana 1999; Rasamimanana et al. 2006). A double-labeled water study of oxygen consumption during March (after lactation and before the mating season) also showed similar energy budgets between males and females of both ringtailed and brown lemurs (Simmen et al. 2010). It may be that seasonal changes and erratic year-to-year climate still impose extra costs on the females, since they fatten up very markedly during the rainy season in preparation for gestation during the dry season (Simmen et al. 2010), but this is also true of the males in preparation for energy expenditure during mating. Seasonality of breeding in relation to seasonality of food supply will be a part of the story, as first suggested by Hrdy, but female need is not a simple explanation for female dominance (Hrdy 1981; Pride 2005a).

The third argument is so far in need of much more evidence. Jolly et al. (2006b) suggested that inheritance of material property in the female line can increase the variance in female reproductive fitness over that of males. Males start over again in each generation with variance reflecting only their genes, their luck, and their bodily prowess. Female ringtails at Berenty inherit territory which must be defended in each generation, but which is hard to reclaim once lost. This territorial stability may be an artifact of Berenty's high density, though the aggression shown between troops at all densities, and the rough-and-tumble play of young females suggests that female competitive behavior is an evolutionary norm for ringtails. However, proving the importance of maternal inheritance of property needs mathematical rigor and actual data on males' reproductive variance as well as females', as well as further data on troop range, territoriality, and territorial inheritance in other species and at sites outside Berenty. Interestingly, E.O. Wilson now argues that the evolutionary origins of social insect society should be traced to defense of a nest or nest site, a rare and heritable resource. In Hymenoptera, with maternal inheritance, this arguably involves an extraordinary form of female dominance, while in Isoptera both the resident reproductives and the worker cast are male and female (Hölldobler and Wilson 2009; Nowak et al. 2010)

2.3.5 Complexity of Social Relations

The other interest of the very early studies was the comparison between lemur social relations and those of anthropoids. I believed that the simple composition of multi-male, multi-female troops would favor social intelligence (Jolly 1966b). I did point out that tripartite reactions in which one animal would threaten another while ensuring support from a third had not been seen in the lemurs (Kummer 1967). Since then, it seems that ringtailed lemur interactions are much more black and white than in many anthropoids – either affiliative or aggressive between any two animals, with minimal ambiguity, and no reconciliation after quarrels (Kappeler 1993). Reconciliation has recently been asserted for sifaka, ringtails and brown lemurs, using different measurements (Palagi et al. 2001, 2005b, 2008). More obvious in the field is the careful geometry of a troop, where distance between animals is a very good measure of affiliation or aggression, and arguably involves awareness of multiple individuals, not just dyads (Nakamichi et al. 1997). Complexity of social relations would be worth revisiting by someone very familiar with behavior of both monkeys and prosimians.

2.4 The Changing Ecology of Introduced Species

2.4.1 Nurse Trees and Food Trees

The concentration of fruit trees with different phenologies from the highly seasonal native forest is one of the main reasons for high lemur density in the Front zone (Rasamimanana and Rafidinarivo 1993). Introduced tree species from which lemurs obtain food and those that serve as forest nurse trees represent benign interventions, though they are far from natural. *Pithecellobium dulcii*, the "monkeypod" or "ape's earring" tree is a nurse tree, has promoted the regeneration of the Ankoba zone with its high density of all lemur species. Prickly pear, *Opuntia*, is found at the periphery, and is now spreading along the river bank. Too much would be disastrous, but it serves as an important water source for lemurs in both the Front and Scrub zones.

2.4.2 Cissus quadrangularis

Cissus, the "veldt grape" or "Devil's backbone," is a euphorb imported to Madagascar from India or Africa either as an ornamental, or for its curative qualities on broken bones and other ailments. In Berenty it is a smothering vine that blankets whole trees or sections of forest. Management campaigns have cleared sections of forest by hand, but it re-grows from tiny dropped fragments, so this is an endless process. There is nothing good to be said about *Cissus* in a forest reserve.

2.4.3 Leucaena leucocephala

Stands of leucaena were planted in or just before 1990, in the bottleneck between Ankoba and Malaza and also at the northern tip of Ankoba. They were an experiment, to see if they could provide extra fodder for cattle and ostriches. The rise in ringtail population in the Front zone tracked the growth of this fast-growing tree. Demography was studied by the Koyama team during this period of rapid population expansion. Unfortunately, leucaena contains mimosine, a non-protein amino acid that blocks cell division. Some troops became so dependent on leucaena that they ate it for 40–50% of their feeding time during dry season months. This produced "Bald Lemur Syndrome," or rather, naked lemur syndrome in which the loss of fur from body and tail mimicked chemotherapy (Fig. 2.6). Adults mostly recovered full pelage when the diet changed with the onset of rains. Embryos seemed to be buffered against the leucaena effect. The birth rate of highly affected females was not lower, but survival of infants with no fur to cling to, and of juveniles weaned onto the toxic tree fell markedly by the end of the leucaena period in 2006–2007. Nearly all of these trees have now been removed. However, the loss of a major foodstuff has also increased adult mortality in the Japanese study troops, which are concentrated at the bottleneck (Jolly 1980, 2009a, b; Crawford et al. 2006, 2008; Soma 2006, in press; Soma et al. 2008; Berg et al. 2009; Ichino et al. in press).

2.4.4 Brown Lemurs

By far the largest influence on ringtails has been the introduction of brown lemurs. About eight orphaned pet *E. rufus* from the Menabe in western Madagascar, escaped to the forest during a 1975 cyclone. About eight more *E. collaris* from the Fort Dauphin region were deliberately released in the years up to 1985. These two species have different chromosome numbers, and should not be able to breed, but no one told the lemurs. They are direct and dominant competitors with the

Fig. 2.6 Bald lemur. Photo © Wiebke Berg

ringtails for habitat and food, although their social structure is quite different, as is their cathemeral activity. The hybrid population of Malaza and Ankoba numbered 596 in 2009, compared to 462 ringtails (Figs. 2.6 and 2.7). The brown population grew exponentially from their introduction until a peak of 653 in 2007 (Fig. 2.7). At that point, scientists persuaded the management to stop providing water in artificial basins in the forest. This might have finally limited the brown lemur population in space, if not in numbers, since they do not spend much time in the sunny scrub zone. Ringtailed lemur troops have largely abandoned the rich gallery forest to the brown lemurs. Although they maintain a foothold, and sleep in their old ranges, much of the daytime foraging is now concentrated on the reserve's periphery. However, their numbers have not fallen (Pinkus et al. 2006; Tanaka 2007; Razafindramanana et al. 2008; Donati et al. 2009; Norscia and Palagi 2011; Palagi and Norscia 2011; Rasamimanana et al. in press).

It is still not clear whether brown and ringtailed lemurs will continue to co-exist stably. They have re-created the classic ecological niche separation described by Sussman (1974), with browns in the central shade, feeding on the gallery forest trees once favored by ringtails, and ringtails in the peripheral sun with a diet increasingly based on plants of scrub and spiny forest. There is no clear sign that the two species are over-browsing their habitat. However, one needs perhaps another 5 years to be sure of this conclusion (Fig. 2.8).

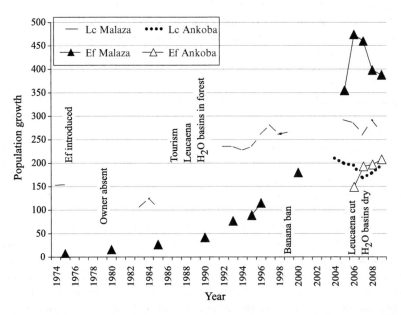

Fig. 2.7 Population growth of ringtailed and brown lemurs (Razafindramanana, Pinkus, O'Connor, Dainis, Jolly)

Fig. 2.8 Hybrid brown lemurs at Berenty. The male has the crest and eyebrows of *Eulemur rufus*, the full beard of *E. collaris*. The female has the grey head of *E. collaris* and the frontal stripe of *E. rufus*. Photo © Cyril Ruoso

2.5 The Future

Berenty is drying out. The loss of tamarind trees and closed canopy forest is clear over the decades (Blumenfeld-Jones et al. 2006; Ichino 2007). It is not clear why. One reason may be natural succession within the gallery forest. If the reserve were not so bounded there might be re-growth in other places. Another likely cause is the lowering of the water table. The Mandrare River suffers from deforestation in the headwaters, and loss of grass cover lower down; flood crests which used to pass in days now pass in hours.

Add to this the effects of climate change. Berenty has suffered repeated droughts in this decade. El Niño years commonly bring drought, and they seem to be growing commoner. All global climate models predict a greater intensity of cyclones. These can sometimes bring needed rain to the south, but a mis-timed cyclone ruins the years' crops. Madagascar lies on the intertropical convergence, in the latitudes most vulnerable to climate change. All types of native forest are much better buffered against extremes than are the annual crops, but progressive drying out could eventually destroy the forest.

Berenty Estate is also vulnerable to social change and unrest. Its reserve system has survived for 70 years. This is due to the care of the de Heaulme family, and to the fact that local people have had salaries that allied their interests with the estate. Any reserve's survival depends on the good will and enthusiasm of particular people, and on an enabling economy. When the reserve is as small as Berenty, the people are few, and outside income comes only from sisal and tourism.

Subsistence farming is at the mercy of the weather. The weather itself is at the mercy of southern deforestation and the northern greed for carbon (Koyama 2009).

Berenty has provided much, though certainly not all of our understanding of the behavior of ringtailed and other southern lemurs – especially the interplay of within- and between-troop aggression in a female-dominant species. It has served as a base for ecological research, but an ecologist at Berenty cannot simply produce a picture of the "environment of ecological adaptation" – it takes all the running a scientist can do to keep up with the changes brought by introduced species and climate fluctuations in such a small space.

The very things that make Berenty interesting, though – lemur social interactions in a region of maximum population density, and the constantly changing ecological background – also make it vulnerable. Its survival is by no means guaranteed, in the face of physical or social changes to come.

References

Berg W, Jolly A, Rambeloarivony H, Andrianome VN, Rasamimanana HR (2009) A scoring system for coat and tail condition in ringtailed lemurs, *Lemur catta*. Am J Primatol 71:183–190

Blumenfeld-Jones KC, Randriamboavonjy TM, Williams GW, Mertl-Millhollen AS, Pinkus S, Rasamimanana HR (2006) Tamarind recruitment and long-term stability in the gallery forest at Berenty, Madagascar. In: Jolly A, Sussman RW, Koyama N, Rasamimanana HR (eds) Ringtailed lemur biology: *Lemur catta* in Madagascar. Springer, New York, pp 69–85

Budnitz N (1978) Feeding behavior of *Lemur catta* in different habitats. In: Bateson PPG, Klopfer PH (eds) Perspectives in ethology, vol 3. Plenum, New York, pp 85–108

Budnitz N, Dainis K (1975) *Lemur catta*: ecology and behavior. In: Tattersall I, Sussman RW (eds) Lemur biology. Plenum, New York, pp 219–235

Cavigelli SA, Dubovick TH, Levash WA, Jolly A, Pitts A (2003) Female dominance status and fecal corticoids in a cooperative breeder with low reproductive skew: ring-tailed lemurs (*Lemur catta*). Horm Behav 43:166–179

Charles-Dominique P, Hladik CM (1971) Le lépilemur du sud de Madagascar: écologie, alimentation et vie sociale. Rev Écol 25:3–66

Charrier A, Hladik A, Simmen B (2007) Stratégie alimentaire et dominance des femelles propithèque de Verreaux (*Propithecus v. verreauxi*) dans la forêt à Didiereacea du sud de Madagascar. Rev Écol 62:257–263

Crawford GC, Andriafaneva L-E, Blumenfeld-Jones KC, Calaba G, Clarke L, Gray L, Ichino S, Jolly A, Koyama N, Mertl-Millhollen AS, Ostpak S, Pride RE, Rasamimanana HR, Simmen B, Soma T, Tarnaud L, Tew A, Williams GW (2006) Bald lemur syndrome and the miracle tree: alopecia associated with *Leucaena leucocephala* at Berenty Reserve, Madagascar. In: Jolly A, Sussman RW, Koyama N, Rasamimanana HR (eds) Ringtailed lemur biology: *Lemur catta* in Madagascar. Springer, New York, pp 332–342

Crawford GC, Ostapak SE, Davidson A, Baker T, Puschner B, Affolter L, Stalis I, Rasamimanana HR, Jolly A (2008) Bald lemur syndrome: the systematic effects of *Leucaena leucocephala* on *Lemur catta* at Berenty Reserve, Madagascar. Primate Eye 96 (special issue):187

Crowley HM (1995) Berenty reserve management plan. Wildlife Preservation Trust International, Philadelphia

Dewar RE, Richard AF (2007) Evolution in the hypervariable environment of Madagascar. Proc Natl Acad Sci USA 104:13723–13727

Dewar RE, Wallis JR (1999) Geographical patterning of interannual rainfall variability in the tropics and near tropics: an L-moments approach. J Climate 12:3457–3466

Dolins F, Jolly A (2007) Ranging behavior and social decision-making in two species of lemurs (*Lemur catta & Propithecus verreauxi verrreauxi*) at Berenty Reserve, Madagascar. EG-Meeting: Social organization and cognitive tools: general patterns in vertebrates? Konrad Lorenz Forchungsstelle, Grünau

Donati G, Baldi N, Morelli V, Ganzhorn JU, Borgognini-Tarli SM (2009) Proximate and ultimate determinants of cathemeral activity in brown lemurs. Anim Behav 77:317–325

Dubovick TH (1998) A historical, social and ecological analysis of three tourist ranging troops of Lemur catta, Berenty Reserve, Madagascar. BA thesis, University of Princeton/NJ

Fichtel C (2008) Ontogeny of conspecific and heterospecific alarm call recognition in wild Verreaux's sifakas (*Propithecus verreauxi verreauxi*). Am J Primatol 70:127–135

Fichtel C, Kappeler PM (2011) Variation in the meaning of alarm calls in Coquerel's and Verreaux's sifakas (*Propithecus coquereli, P. verreauxi*). Int J Primatol 32:346–361

Fish KD (2010) Niche separation between mouse lemurs (*Microcebus murinus*) and clutter-foraging bats at Berenty Private Reserve, Madagascar. PhD thesis, University of Colorado, Boulder/CO

Frankel OH, Soulé ME (1981) Conservation and evolution. Cambridge University Press, Cambridge

Génin F (2001) Gumnivory in mouse lemurs during the dry season in Madagascar. Folia Primatol 72:119–120

Génin F (2007) Energy-dependent plasticity of grey mouse lemur social systems: lessons from field and captive studies. Rev Écol 62:245–256

Génin F (2008) Life in unpredictable environments: first investigation of the natural history of *Microcebus griseorufus*. Int J Primatol 29:303–321

Génin F (2010) Who sleeps with whom? Sleeping associations and socio-territoriality in *Microcebus griseorufus*. J Mammal 91:942–951

Génin F (in press) Living in unpredictable environments: energy management and socio-spatial organization of *Microcebus griseorufus* in the southern spiny forest of Madagascar. In: Masters JC, Gamba M, Génin F (eds) Leaping ahead: advances in prosimian biology. Springer, New York

Génin F, Masters JC, Ganzhorn JU (2010) Gummivory in cheirogaleids: primitive retention or adaptation to hypervariable environments? In: Burrows AM, Nash LT (eds) The evolution of exudativory in primates. Springer, New York

Goodman SM (1995) *Rattus* on Madagascar and the dilemma of protecting the endemic rodent fauna. Conserv Biol 9:450–453

Goodman SM, Pidgeon M, Hawkins AFA, Schulenberg TS (1997) The birds of southern Madagascar. Fieldiana Zool 187:1–132

Gould L (1990) The social development of free-ranging infant *Lemur catta* at Berenty Reserve, Madagascar. Int J Primatol 11:297–318

Gould L (1992) Alloparental care in free-ranging *Lemur catta* at Berenty Reserve, Madagascar. Folia Primatol 58:72–83

Gould L (1996) Vigilance behavior during the birth and lactation season in naturally occurring ring-tailed lemurs (*Lemur catta*) at the Beza-Mahafaly Reserve, Madagascar. Int J Primatol 17:331–347

Gould L, Sussman RW, Sauther ML (1999) Natural disasters and primate populations: the effects of a two-year drought on a naturally occurring population of ring-tailed lemurs (*Lemur catta*) in southwestern Madagascar. Int J Primatol 20:69–84

Hölldobler B, Wilson EO (2009) The superorganism: the beauty, elegance, and strangeness of insect societies. Norton and Company, New York

Hood LC (1994) Infanticide among ringtailed lemurs (*Lemur catta*) at Berenty Reserve, Madagascar. Am J Primatol 33:65–69

2 Berenty Reserve, Madagascar: A Long Time in a Small Space

Hood LC, Jolly A (1995) Troop fission in female *Lemur catta* at Berenty Reserve, Madagascar. Int J Primatol 16:997–1015

Hrdy SB (1981) The woman that never evolved. Harvard University Press, Cambridge, MA

Ichino S (2005) Attacks on a wild infant ring-tailed lemur (*Lemur catta*) by immigrant males at Berenty, Madagascar: interpreting infanticide by males. Am J Primatol 67:267–272

Ichino S (2006) Troop fission in wild ring-tailed lemurs (*Lemur catta*) at Berenty, Madagascar. Am J Primatol 68:97–102

Ichino S (2007) The status and problems of lemur conservation in the Berenty Reserve, southern Madagascar [Japanese with English abstract]. Asian Afr Area Stud 6:197–214

Ichino S, Koyama N (2006) Social changes in a wild population of ringtailed lemurs (*Lemur catta*) at Berenty, Madagascar. In: Jolly A, Sussman RW, Koyama N, Rasamimanana HR (eds) Ringtailed lemur biology: *Lemur catta* in Madagascar. Springer, New York, pp 233–244

Ichino S, Soma T, Koyama N (in press) The alopecia syndrome of ring-tailed lemurs (*Lemur catta*) at Berenty Reserve, Madagascar: a preliminary report on impact on female reproductive parameters. In: Masters JC, Gamba M, Génin F (eds) Leaping ahead: advances in prosimian biology. Springer, New York

Jolly A (1966a) Lemur behavior: a Madagascar field study. University of Chicago Press, Chicago

Jolly A (1966b) Lemur social behavior and primate intelligence. Science 153:501–506

Jolly A (1967) Breeding synchrony in wild *Lemur catta*. In: Altmann SA (ed) Social communication among primates. University of Chicago Press, Chicago, pp 3–14

Jolly A (1980) A world like our own: man and nature in Madagascar. Yale University Press, New Haven

Jolly A (1984) The puzzle of female feeding priority. In: Small MF (ed) Female primates: studies by women primatologists. Liss, New York, pp 197–215

Jolly A (2004) Lords and lemurs: Mad scientists, kings with spears, and the survival of diversity in Madagascar. Houghton Mifflin, Boston

Jolly A (2009a) Coat condition of ringtailed lemurs, *Lemur catta*, at Berenty Reserve, Madagascar: I. Differences by age, sex, density and tourism 1996–2006. Am J Primatol 71:191–198

Jolly A (2009b) Coat condition of ringtailed lemurs, *Lemur catta*, at Berenty Reserve, Madagascar: II. Coat and tail alopecia associated with *Leucaena leucocephala*, 2001–2006. Am J Primatol 71:199–205

Jolly A (2010) The narrator's stance: story-telling and science at Berenty Reserve. In: MacClancy J, Fuentes A (eds) Centralizing fieldwork: critical perspectives from primatology, biological and social anthropology. Berghan Books, New York, pp 223–241

Jolly A, Pride RE (1999) Troop histories and range inertia of *Lemur catta* at Berenty, Madagascar: a 33-year perspective. Int J Primatol 20:359–373

Jolly A, Rasamimanana HR, Kinnaird MF, O'Brien TG, Crowley HM, Harcourt CS, Gardner S, Davidson JM (1993) Territoriality in *Lemur catta* groups during the birth season at Berenty, Madagascar. In: Kappeler PM, Ganzhorn JU (eds) Lemur social systems and their ecological basis. Plenum, New York, pp 85–109

Jolly A, Caless S, Cavigelli S, Gould L, Pereira ME, Pitts A, Pride RE, Rabenandrasana HD, Walker JD, Zafison T (2000) Infant killing, wounding and predation in *Eulemur* and *Lemur*. Int J Primatol 21:21–40

Jolly A, Dobson A, Rasamimanana HR, Walker J, O'Connor S, Solberg M, Perel V (2002) Demography of *Lemur catta* at Berenty Reserve, Madagascar: effects of troop size, habitat and rainfall. Int J Primatol 23:327–353

Jolly A, Koyama N, Rasamimanana HR, Crowley H, Williams GW (2006a) Berenty Reserve: a research site in southern Madagascar. In: Jolly A, Sussman RW, Koyama N, Rasamimanana HR (eds) Ringtailed lemur biology: *Lemur catta* in Madagascar. Springer, New York, pp 32–42

Jolly A, Rasamimanana HR, Braun MA, Dubovick TH, Mills CN, Williams GW (2006b) Territory as bet-hedging: *Lemur catta* in a rich forest and an erratic climate. In: Jolly A, Sussman RW, Koyama N, Rasamimanana HR (eds) Ringtailed lemur biology: *Lemur catta* in Madagascar. Springer, New York, pp 187–207

Jones KC (1983) Inter-troop transfer of *Lemur catta* males at Berenty, Madagascar. Folia Primatol 40:145–160

Kappeler PM (1993) Reconciliation and post-conflict behaviour in ringtailed lemurs, *Lemur catta* and redfronted lemurs, *Eulemur fulvus rufus*. Anim Behav 45:901–915

Kappeler PM, Fichtel C (2011) A 15-year perspective on the social organization and life history of Sifaka in Kirindy Forest. In: Kappeler PM (ed) Long-term field studies of primates. Springer, Heidelberg

Karpanty SM, Goodman SM (1999) Diet of the Madagascar harrier-hawk, *Polyboroides radiatus*, in southeastern Madagascar. J Raptor Res 4:313–316

Klopfer PH, Jolly A (1970) The stability of territorial boundaries in a lemur troop. Folia Primatol 12:199–208

Koechlin J (1972) Flora and vegetation of Madagascar. In: Battistini R, Richard-Vindard G (eds) Biogeography and ecology of Madagascar. Dr. W. Junk B.V Publishers, The Hague, pp 145–190

Koyama N (1991) Troop division and inter-troop relationships of ring-tailed lemurs (*Lemur catta*) at Berenty, Madagascar. In: Ehara A, Kimura T, Takenaka O, Iwamoto M (eds) Primatology today. Elsevier, Amsterdam, pp 173–176

Koyama N (1992) Multiple births and care-taking behavior of ring-tailed lemurs (*Lemur catta*) at Berenty, Madagascar. In: Yamagishi S (ed) Social structure of Madagascar higher vertebrates in relation to their adaptive radiation. Osaka City University, Osaka, pp 5–9

Koyama N (2009) Madagascar tou-Nishiindoyou chiikikennkyuu nyuumon (The Island of Madagascar: an introduction to area studies in the Western Indian Ocean Region) [in Japanese]. Tokaidaigaku Press, Kanagawa

Koyama N, Nakamichi M, Oda R, Miyamoto N, Ichino S, Takahata Y (2001) A ten-year summary of reproductive parameters for ring-tailed lemurs at Berenty, Madagascar. Primates 42:1–14

Koyama N, Nakamichi M, Ichino S, Takahata Y (2002) Population and social dynamics changes in ring-tailed lemur troops at Berenty, Madagascar between 1989–1999. Primates 43:291–314

Koyama N, Ichino S, Nakamichi M, Takahata Y (2005) Long-term changes in dominance ranks among ring-tailed lemurs at Berenty Reserve, Madagascar. Primates 46:225–234

Koyama N, Soma T, Ichino S, Takahata Y (2006) Home ranges of ringtailed lemur troops and the density of large trees at Berenty Reserve, Madagascar. In: Jolly A, Sussman RW, Koyama N, Rasamimanana HR (eds) Ringtailed lemur biology: *Lemur catta* in Madagascar. Springer, New York, pp 86–101

Koyama N, Aimi M, Kawamoto Y, Hirai H, Go Y, Ichino S, Takahata Y (2008) Body mass of wild ring-tailed lemurs in Berenty Reserve, Madagascar, with reference to tick infestation: a preliminary analysis. Primates 49:9–15

Kummer H (1967) Tripartite relations in hamadryas baboons. In: Altmann SA (ed) Social communication among primates. University of Chicago Press, Chicago, pp 63–71

Lane L, Bard KA (2007) Seasonal influences on visual social monitoring in free ranging *Lemur catta* at Berenty Reserve, Madagascar. Am J Primatol 69(suppl 1):32

Lane L, Bard KA (2008) Assessing dyadic social attention in *Lemur catta*. Primate Eye 96:107

Lewis RJ (2002) Beyond dominance: the importance of leverage. Quart Rev Biol 77:149–164

Long E (2002) The feeding ecology of *Pteropus rufus* in a remnant gallery forest surrounded by sisal plantations in south-east Madagascar. PhD thesis, University of Aberdeen

McGeorge LW (1978a) Circumvention of noise in the communication channel by the structure and timing of the calls of forest animals. PhD thesis, Duke University, Durham, NC

McGeorge LW (1978b) Influences on the structure of vocalizations of three Malagasy lemurs. In: Chivers DJ, Joysey KA (eds) Recent advances in primatology. Academic, London, pp 103–109

Mertl AS (1977) Habituation to territorial scent marks in the field by *Lemur catta*. Behav Biol 21:500–507

Mertl-Millhollen AS (1979) Olfactory demarcation of territorial boundaries by a primate – *Propithecus verreauxi*. Folia Primatol 32:35–42

2 Berenty Reserve, Madagascar: A Long Time in a Small Space

Mertl-Millhollen AS (1986) Territorial scent-marking by two sympatric lemur species. In: Duvall D, Muller-Schwarze D, Silverstein RM (eds) Chemical signals in vertebrates, vol 4, Ecology, evolution and comparative biology. Plenum, New York, pp 647–652

Mertl-Millhollen AS (1988) Olfactory demarcation of territorial but not home range boundaries by *Lemur catta*. Folia Primatol 50:175–187

Mertl-Millhollen AS (2000a) Components of scent marking behavior by male ringtailed lemurs (*Lemur catta*) at Berenty Reserve, Madagascar. Am J Primatol 51(suppl):74

Mertl-Millhollen AS (2000b) Ringtailed lemur (*Lemur catta*) over-marking as an example of resource defense. Am J Primatol 51(suppl):73–74

Mertl-Millhollen AS (2004) Primate scent marking as resource defence. Folia Primatol 75(suppl 1):49–50

Mertl-Millhollen AS (2006) Scent marking as resource defense by female *Lemur catta*. Am J Primatol 68:605–621

Mertl-Millhollen AS (2007) Lateral bias to the leading limb in an olfactory social signal by male ring-tailed lemurs. Am J Primatol 69:635–640

Mertl-Millhollen AS, Gustafson HL, Budnitz N, Dainis K, Jolly A (1979) Population and territory stability of the *Lemur catta* at Berenty, Madagascar. Folia Primatol 31:106–122

Mertl-Millhollen AS, Moret ES, Felantsoa D, Rasamimanana HR, Blumenfeld-Jones KC, Jolly A (2003) Ring-tailed lemur home ranges correlate with food abundance and nutritional content at a time of environmental stress. Int J Primatol 24:969–985

Mertl-Millhollen AS, Rambeloarivony H, Miles W, Rasamimanana HR (2004) Tamarind leaf quality and *Lemur catta* population density and behavior. Folia Primatol 75(suppl 1):157–158

Mertl-Millhollen AS, Rambeloarivony H, Miles W, Kaiser VA, Gray L, Dorn LT, Williams GW, Rasamimanana HR (2006) The influence of tamarind tree quality and quantity on *Lemur catta* behavior. In: Jolly A, Sussman RW, Koyama N, Rasamimanana HR (eds) Ringtailed lemur biology: *Lemur catta* in Madagascar. Springer, New York, pp 102–118

Miles W, Rambeloarivony H (2008) Initiation and leading of travel in *Lemur catta*. Lemur News 13:22–24

Mulder RA, Ramiarison R, Emahalala RE (2002) Ontogeny of male plumage dichromatism in Madagascar paradise flycatchers. J Avian Biol 33:342–348

Nakamichi M, Koyama N (1997) Social relationships among ring-tailed lemurs (*Lemur catta*) in two free-ranging troops at Berenty Reserve, Madagascar. Int J Primatol 18:73–93

Nakamichi M, Koyama N (2000) Intra-troop affiliative relationships of females with newborn infants in wild ring-tailed lemurs (*Lemur catta*). Am J Primatol 50:187–203

Nakamichi M, Koyama N, Jolly A (1996) Maternal responses to dead and dying infants in wild troops of ring-tailed lemurs at the Berenty Reserve, Madagascar. Int J Primatol 17:505–523

Nakamichi M, Rakototiana MLO, Koyama N (1997) Effects of spatial proximity and alliances on dominance relations among female ring-tailed lemurs (*Lemur catta*) at Berenty Reserve, Madagascar. Primates 38:331–340

Norscia I, Palagi E (2008) Berenty 2006: census of *Propithecus verreauxi* and possible evidence of population stress. Int J Primatol 29:1099–1115

Norscia I, Palagi E (2011) Do wild brown lemurs reconcile? Not always. J Ethol 29:181–185

Norscia I, Antonacci D, Palagi E (2009) Mating first, mating more: biological market fluctuation in a wild prosimian. PLoS One 4:e4679. doi:10.1371/journal.pone.0004679

Nowak MA, Tarnita CE, Wilson EO (2010) The evolution of eusociality. Nature 466:1057–1062

O'Connor SM (1987) The effect of human impact on vegetation and the consequences to primates in two riverine forests, southern Madagascar. PhD thesis, Cambridge University

Oda R (1996a) Effects of contextual and social variables on contact call production in free-ranging ringtailed lemurs (*Lemur catta*). Int J Primatol 17:191–205

Oda R (1996b) Predation on a chameleon by a ring-tailed lemur (*Lemur catta*) in the Berenty Reserve, Madagascar. Folia Primatol 67:40–43

Oda R (1998) The responses of Verreaux's sifakas to anti-predator alarm calls given by sympatric ring-tailed lemurs. Folia Primatol 69:357–360

Oda R (1999) Scent marking and contact call production in ring-tailed lemurs (*Lemur catta*). Folia Primatol 70:121–124

Oda R (2001) Lemur vocal communication and the origin of human language. In: Matsuzawa T (ed) Primate origins of human cognition and behavior. Springer, Tokyo, pp 115–134

Oda R, Masataka N (1996) Interspecific responses of ringtailed lemurs to playback of antipredator alarm calls given by Verreaux's sifakas. Ethology 102:441–453

Okamoto M (1998) The birth of wild ring-tailed lemurs at Berenty Reserve, Madagascar [Japanese with English Abstract]. Primate Res 14:25–34

Palagi E, Dapporto L (2006) Beyond odor discrimination: demonstrating individual recognition by scent in *Lemur catta*. Chem Senses 31:437–443

Palagi E, Norscia I (2009) Multimodal signaling in wild *Lemur catta*: economic design and territorial function of urine marking. Am J Phys Anthropol 139:182–192

Palagi E, Norscia I (2011) Scratching around stress: hierarchy and reconciliation make the difference in wild brown lemurs (*Eulemur fulvus*). Stress 14:93–97

Palagi E, Bastianelli E, Borgognini Tarli SM (2001) Social relationships, target aggressions and post-conflict behaviour in ringtailed lemurs (*Lemur catta*). Folia Primatol 72:129

Palagi E, Dapporto L, Borgognini Tarli SM (2005a) The neglected scent: on the marking function of urine in *Lemur catta*. Behav Ecol Sociobiol 58:437–445

Palagi E, Paoli T, Borgognini Tarli SM (2005b) Aggression and reconciliation in two captive groups of *Lemur catta*. Int J Primatol 26:279–294

Palagi E, Antonacci D, Norscia I (2008) Peacemaking on treetops: first evidence of reconciliation from a wild prosimian (*Propithecus verreauxi*). Anim Behav 76:737–747

Pereira ME (1993a) Agonistic interaction, dominance relations, and ontogenetic trajectories in ringtailed lemurs. In: Pereira ME, Fairbanks LA (eds) Juvenile primates: life history, development, and behavior. Oxford University Press, New York, pp 285–305

Pereira ME (1993b) Seasonal adjustment of growth rate and adult body weight in ringtailed lemurs. In: Kappeler PM, Ganzhorn JU (eds) Lemur social systems and their ecological basis. Plenum, New York, pp 205–221

Pereira ME (2006) Obsession with agonistic power. In: Jolly A, Sussman RW, Koyama N, Rasamimanana HR (eds) Ringtailed lemur biology: *Lemur catta* in Madagascar. Springer, New York, pp 245–270

Pereira ME, Kappeler PM (1997) Divergent systems of agonistic behaviour in lemurid primates. Behaviour 134:225–274

Pereira ME, Kaufman R, Kappeler PM, Overdorff DJ (1990) Female dominance does not characterize all of the Lemuridae. Folia Primatol 55:96–103

Pinkus S, Smith JNM, Jolly A (2006) Feeding competition between introduced *Eulemur fulvus* and native *Lemur catta* during the birth season at Berenty Reserve, southern Madagascar. In: Jolly A, Sussman RW, Koyama N, Rasamimanana HR (eds) Ringtailed lemur biology: *Lemur catta* in Madagascar. Springer, New York, pp 119–140

Pitts A (1995) Predation by *Eulemur fulvus rufus* on an infant *Lemur catta* at Berenty, Madagascar. Folia Primatol 65:169–171

Pride RE (2003) The socio-endocrinology of group size in *Lemur catta*. PhD thesis, Princeton University, NJ

Pride RE (2005a) High faecal glucocorticoid levels predict mortality in ring-tailed lemurs (*Lemur catta*). Biol Lett 1:60–63

Pride RE (2005b) Optimal group size and seasonal stress in ring-tailed lemurs (*Lemur catta*). Behav Ecol 16:550–560

Pride RE, Felantsoa D, Randriamboavonjy TM, Randriambelona R (2006) Resource defence in *Lemur catta*: the importance of group size. In: Jolly A, Sussman RW, Koyama N, Rasamimanana HR (eds) Ringtailed lemur biology: *Lemur catta* in Madagascar. Springer, New York, pp 208–232

Rasamimanana HR (1999) Influence of social organization patterns on food intake of *Lemur catta* in the Berenty Reserve. In: Rakotosamimananana B, Rasamimanana HR, Ganzhorn JU,

Goodman SM (eds) New directions in lemur studies. Kluwer Academic Press, Plenum Publishers, New York, pp 173–188

Rasamimanana HR, Anjaranantenaina S (2008) Influence of hierarchy on the social behaviour of female ring-tailed lemurs at Berenty Reserve, south Madagascar. Primate Eye 96 (special issue):91

Rasamimanana HR, Rafidinarivo E (1993) Feeding behavior of *Lemur catta* females in relation to their physiological state. In: Kappeler PM, Ganzhorn JU (eds) Lemur social systems and their ecological basis. Plenum, New York, pp 123–133

Rasamimanana HR, Ratovonirina JA, Pride E (2000) Storm damage at Berenty Reserve. Lemur News 5:7–8

Rasamimanana HR, Andrianome VN, Rambeloarivony H, Pasquet P (2006) Male and female ringtailed lemurs' energetic strategy does not explain female dominance. In: Jolly A, Sussman RW, Koyama N, Rasamimanana HR (eds) Ringtailed lemur biology: *Lemur catta* in Madagascar. Springer, New York, pp 271–295

Rasamimanana HR, Razafindramanana J, Mertl-Millhollen A, Blumenfeld-Jones KC, Raharison SM, Tsaramanana RD, Razolliharisoa V, Tarnaud L (in press) Berenty Reserve: interactions between the diurnal lemur species and their gallery forest. In: Masters JC, Gamba M, Génin F (eds) Leaping ahead: advances in prosimian biology. Springer, New York

Rasoloarison RM, Goodman SM, Ganzhorn JU (2000) Taxonomic revision of mouse lemurs (*Microcebus*) in the western portions of Madagascar. Int J Primatol 21:963–1019

Razafindramanana J, Jolly A, Rasamimanana HR (2008) Population dynamics and distribution of brown and ringtailed lemurs at Berenty Reserve, southeastern Madagascar. Primate Eye 96 (special issue):166

Richard AF (1978) Behavioral variation: case study of a Malagasy lemur. Bucknell University Press, Lewisburg, PA

Richard AF, Dewar RE, Schwartz M, Ratsirarson J (2000) Mass change, environmental variability and female fertility in wild *Propithecus verreauxi*. J Hum Evol 39:381–391

Richard AF, Dewar RE, Schwartz M, Ratsirarson J (2002) Life in the slow lane? Demography and life histories of male and female sifaka (*Propithecus verreauxi verreauxi*). J Zool Lond 256:421–436

Russell RJ (1977) The behavior, ecology, and environmental physiology of a nocturnal primate, *Lepilemur mustelinus* (Strepsirhini, Lemuriformes, Lepilemuridae). PhD thesis, Duke University, Durham, NC

Simmen B, Hladik A, Ramasiarisoa PL, Iaconellli S, Hladik CM (1999) Taste discrimination in lemurs and other primates, and the relationships to distribution of plant allelochemicals in different habitats of Madagascar. In: Rakotosamimananana B, Rasamimanana HR, Ganzhorn JU, Goodman SM (eds) New directions in lemur studies. Kluwer Academic Press, Plenum Publishers, New York, pp 201–219

Simmen B, Hladik A, Hladik CM, Ramasiarisoa PL (2003a) Occurrence of alkaloids and phenolics in Malagasy forests and responses by primates. In: Goodman SM, Benstead JM (eds) The natural history of Madagascar. University of Chicago Press, Chicago, pp 268–271

Simmen B, Hladik A, Ramasiarisoa PL (2003b) Food intake and dietary overlap in native *Lemur catta* and *Propithecus verreauxi* and introduced *Eulemur fulvus* at Berenty, southern Madagascar. Int J Primatol 24:949–968

Simmen B, Tarnaud L, Bayart F, Hladik A, Thiberge A-L, Jaspart S, Jeanson M, Marez A (2005) Richesse en métabolites secondaires des forêts de Mayotte et de Madagascar et incidence sur la consommation de feuillage chez deux espèces de lémurs (*Eulemur* spp.). Rev Écol 60:297–324

Simmen B, Peronny S, Jeanson M, Hladik A, Marez A (2006a) Diet quality and taste perception of plant secondary metabolites by *Lemur catta*. In: Jolly A, Sussman RW, Koyama N, Rasamimanana HR (eds) Ringtailed lemur biology: *Lemur catta* in Madagascar. Springer, New York, pp 160–183

Simmen B, Sauther ML, Soma T, Rasamimanana HR, Sussman RW, Jolly A, Tarnaud L, Hladik A (2006b) Plant species fed on by *Lemur catta* in gallery forests of the southern domain of

Madagascar. In: Jolly A, Sussman RW, Koyama N, Rasamimanana HR (eds) Ringtailed lemur biology: *Lemur catta* in Madagascar. Springer, New York, pp 55–68

Simmen B, Bayart F, Rasamimanana HR, Zahariev A, Blanc S, Pasquet P (2010) Total energy expenditure and body composition in two free-living sympatric lemurs. PLoS One 5:e9860. doi:10.1371/journal.pone.0009860

Soma T (2006) Tradition and novelty: *Lemur catta* feeding strategy on introduced tree species at Berenty Reserve. In: Jolly A, Sussman RW, Koyama N, Rasamimanana HR (eds) Ringtailed lemur biology: *Lemur catta* in Madagascar. Springer, New York, pp 141–159

Soma T (in press) The cause of alopecia? The difference of feeding ecology of two ring-tailed lemurs in relation to introduced plant species. In: Masters JC, Gamba M, Génin F (eds) Leaping ahead: advances in prosimian biology. Springer, New York

Soma T, Ichino S, Koyama N, Jolly A (2008) The influence of toxic *Leucaena leucocephala* on the demography of ring-tailed lemurs, *Lemur catta*, at Berenty Reserve, Madagascar. Primate Eye 96 (special issue):195–196

Soulé ME, Wilcox BA (1980) Conservation biology: an evolutionary-ecological perspective. Sinauer, Sunderland, MA

Sussman RW (1974) Ecological distinctions in sympatric species of *Lemur*. In: Martin RD, Doyle GA, Walker AC (eds) Prosimian biology. Duckworth, London, pp 75–108

Sussman RW, Garber PA (2004) Rethinking sociality: cooperation and aggression among primates. In: Sussman RW, Chapman AR (eds) The origins and nature of sociality. Aldine de Gruyter, New York, pp 161–190

Sussman RW, Andrianasolondraibe O, Soma T, Ichino S (2003) Social behavior and aggression among ringtailed lemurs. Folia Primatol 74:168–172

Sussman RW, Richard AF, Ratsirarson J, Sauther ML, Brockman DK, Gould L, Lawler R, Cuozzo FP, Mahafaly B (2011) Special reserve: long-term research on lemurs in Southwestern Madagascar. In: Kappeler PM (ed) Long-term field studies of primates. Springer, Heidelberg

Takahata Y, Kawamoto Y, Hirai H, Miyamoto N, Koyama N, Kitaoka S, Suzuki H (1998) Ticks found among the wild ringtailed lemurs at Berenty Reserve, Madagascar. Afr Stud Monogr 19:217–222

Takahata Y, Koyama N, Miyamoto N, Okamoto M (2001) Daytime deliveries observed for the ring-tailed lemurs of the Berenty Reserve, Madagascar. Primates 42:267–271

Takahata Y, Koyama N, Ichino S, Miyamoto N (2005) Inter- and within-troop competition of female ring-tailed lemurs: a preliminary report. Afr Stud Monogr 26:1–14

Takahata Y, Koyama N, Ichino S, Miyamoto N, Nakamichi M (2006) Influence of group size on reproductive success of female ring-tailed lemurs: distinguishing between IGFC and PFC hypotheses. Primates 47:383–387

Tanaka M (2007) Habitat use and social structure of a brown lemur hybrid population in the Berenty Reserve, Madagascar. Am J Primatol 69:1189–1194

Wright PC (1999) Lemur traits and Madagascar ecology: coping with an island environment. Yearbk Phys Anthropol 42:31–72

Chapter 3
Beza Mahafaly Special Reserve: Long-Term Research on Lemurs in Southwestern Madagascar

Robert W. Sussman, Alison F. Richard, Joelisoa Ratsirarson, Michelle L. Sauther, Diane K. Brockman, Lisa Gould, Richard Lawler, and Frank P. Cuozzo

Abstract The Beza Mahafaly Project in southwestern Madagascar was founded in 1975. It was established as a collaborative effort among the University of Madagascar (now University of Antananarivo), Washington University, Yale University, and the local communities for long-term training and research, biodiversity

---

R.W. Sussman (✉)
Department of Anthropology and Environmental Science, Washington University, St. Louis, MO, USA
e-mail: rwsussma@artsci.wustl.edu

A.F. Richard
Professor Emerita, Yale University, New Haven, CT, USA
e-mail: alisonfrichard@gmail.com

J. Ratsirarson
School of Agronomy, Department of Water and Forestry (ESSA/Forêts), University of Antananarivo, Antananarivo, Madagascar
e-mail: ratsirarson@gmail.com

M.L. Sauther
Department of Anthropology, University of Colorado, Boulder, CO, USA
e-mail: michelle.sauther@colorado.edu

D.K. Brockman
Department of Anthropology, University of North Carolina, Charlotte, NC, USA
e-mail: dkbrockm@uncc.edu

L. Gould
Department of Anthropology, University Of Victoria, Victoria, BC, Canada
e-mail: lgould@uvic.ca

R. Lawler
Department of Anthropology, James Madison University, Harrisonburg, VA, USA
e-mail: lawlerr@jmu.edu

F.P. Cuozzo
Department of Anthropology, University of North Dakota, Grand Forks, ND, USA
e-mail: frank.cuozzo@und.nodak.edu

P.M. Kappeler and D.P. Watts (eds.), *Long-Term Field Studies of Primates*,
DOI 10.1007/978-3-642-22514-7_3, © Springer-Verlag Berlin Heidelberg 2012

conservation, and socioeconomic development. Beza Mahafaly consists of two noncontiguous forest parcels separated by 10 km that became a protected area (Réserve Spéciale) in 1986: an 80-ha gallery forest and a 520-ha xerophytic spiny forest. The region has a diversity of habitats and a very diverse and highly endemic flora and fauna, including four species of lemurs found in or near the reserve. The ringtailed lemur (*Lemur catta*) and Verreaux's sifaka (*Propithecus verreauxi*) have been the subject of our long-term research. In this chapter we highlight some of the results of this research. Our multidisciplinary studies illustrate the feasibility of collecting long-term data on careers of individual animals and of obtaining large samples on numerous animals, across numerous social groups, in relatively isolated breeding populations. Thus, we can provide insights into many of the demographic, socioecological, anthropogenic, and epidemiological factors that shape the local ringtailed lemur and sifaka population. Here we summarize how ringtailed lemur demographic structure is affected by climatic perturbations (drought); how aspects of general health (parasite loads and dental health) are directly related to habitat, dietary, and anthropogenic factors; how tight birth seasonality in sifaka can elicit stress responses in males associated with increased male aggression, group takeovers, and infanticide risk; how life history schedules are related to evolutionary responses to extreme climatic fluctuations; and how directional selection among sifaka males leads to longer, stronger legs, but not to increase in male body mass relative to females.

3.1 The Beza Mahafaly Reserve

3.1.1 History of the Reserve

In the mid-1970s, R.W. Sussman received a phone call from Edward Steele of the Defenders of Wildlife (DOW) asking: "What can we do to save Madagascar's wildlife?" Steele (1975) had recently returned from Madagascar and had fallen in love with the country. Sussman, along with Alison Richard, then at Yale University, and Guy Ramanantsoa, then the Head of the Forestry Department of the School of Agronomy at the University of Madagascar (Ecole Supérieure des Sciences Agronomiques, ESSA/Forêts), had been discussing establishing a unique type of reserve in Madagascar. They envisaged a protected area for long-term research on biodiversity and conservation, but also to be used as a training and research center and springboard for socioeconomic development for local villagers.

Sussman explained this vision to Steele, who set up a meeting of DOW board members with Richard and Sussman. The board was impressed with the proposal. However, DOW worked mainly on litigation within the United States. Therefore, Richard Pough, a board member also on the Board of Directors of World Wildlife Fund, volunteered to present our ideas to WWF, which agreed to fund the project. In 1975, Ramanantsoa, with his collaborator the late Pothin Rakotomanga, set out to survey southwestern Madagascar, looking for a relatively accessible, undisturbed area with a representative faunal and floristic diversity. Another criterion was that local villagers agree to the project. He met the President of the Firaisana (Commune) of Beavoha, who proposed Anala Sakamena (forest bordering the Sakamena river) as this site. The site was named "Beza Mahafaly" in recognition of the commune President, who came from the village Beza Mahafaly, 8 km from the Reserve.

Local villagers were conscious of the vital importance of preserving this habitat and were enthusiastic about the project. In July 1978, the advisory committee of the local community agreed to grant two noncontiguous parcels of forest to the School of Agronomy. Thus, ESSA/Forêts began collaborative work with local communities, Yale and Washington Universities, as well as other national and international partners, to establish the Beza Mahafaly Reserve. Between 1978 and 1985, with funding mainly from WWF, reserve boundaries were demarcated, basic facilities were constructed, local guards were recruited and trained, and a field ecology school was developed for students from ESSA/Forêts.

On June 4, 1986, Beza Mahafaly was inaugurated as a Réserve Spéciale (Beza Mahafaly Special Reserve; BMSR) by government decree No. 86–168. Biodiversity research, especially on lemurs, had already begun and has continued to this day. Integration of conservation and development in villages surrounding the reserve, an integral component of the plan since the project's conception, continued, with increasing support from WWF and the U.S. Agency for International Development (USAID). In a coordinated effort with the local people, including their input and requests for assistance, community projects were begun to improve the production and marketing of crops (irrigation projects and road construction), education (construction of a local school), and health (water well construction). Students and faculty from ESSA/Forêts with local residents played an important role in the development of these activities, in partnership with Yale and Washington University.

In 1989, WWF took over management of the reserve, and in November 1995, ESSA/Forêts became the principal administrator. Research, training, and education programs as well as local development projects have continued to flourish. The site has hosted a multidisciplinary field course for fifth-year ESSA/Forêts students since 1986 (Ratsirarson 2003). Since 1994, additional support has been received from a large number of sources (see Acknowledgments).

In 2005, the management of the Reserve was transferred to Madagascar National Parks (MNP), like all protected areas in Madagascar. The School of Agronomy became the principal partner in research and training. ESSA's activities address the Reserve's management and development plan needs, and research specifically follows the annual work plan agreed by both parties (ESSA and MNP). The main objectives of the reserve continue to be long-term biodiversity monitoring and research on community ecology and conservation biology, long-term research on the lemur populations, education at the local, national, and international level and integrated research in the social and natural sciences with the goal of improving the lives of local communities while conserving the biodiversity in surrounding habitats. The main specific goals of our long-term lemur research are to provide a better understanding of behavioral ecology, demography, population genetics, reproductive biology, health status, and mating and life history strategies. Success at achieving these goals depends on, and results from, the connection between research and local development.

3.1.2 Physical Description of the Reserve

The reserve is located in southern Madagascar, 35 km to the northeast of the town of Betioky-Sud at $23°41'60''$S and $44°32'20''$E. Southern Madagascar is characterized by a long dry season and short wet season. During a non-drought, non-cyclone year, annual rainfall is about 700 mm, of which 600 mm falls during the austral summer between November and March. The Sakamena River is dry during the long dry season. Annual daily maximum temperatures average 25°C. Averages for wet season are high ambient temperatures (32°C), and daily maxima reach 46°C. Temperatures during the coolest months (July–August) usually range between 20°C and 30°C during the day, but can fall to 2°C at night (Sussman and Rakotozafy 1994; Ratsirarson 2003).

The reserve consists of two noncontiguous parcels separated by 10 km (Fig. 3.1). Parcel 1 is characterized by a gallery forest dominated by *Tamarindus indica*. It covers 80 ha of fenced and protected forest located on the banks of the Sakamena River, and is relatively flat with a slightly elevated plateau starting at the banks of the river (Fig. 3.2). The gallery forest is divided by marked transects whose paths intersect to form squares of 100×100 m. This parcel has been enclosed by barbed wire fence since 1979. Before this it was exposed to cattle and goats and used by the local people for various resources, as is the surrounding forest currently. Parcel 1 is surrounded by similar but unprotected and somewhat degraded gallery forest on the north and south. To the east is the Sakamena River and to the west is contiguous dry forest. The parcel is bounded on the south by the dirt road that runs from Betioky to the reserve and on to the small village of Analafaly 2 km east. The campsite and reception center is just south of the road. The infrastructure includes two wooden houses, a museum, an office building, a large open gazebo for courses and meetings, camping space, as well as solar energy and a water well.

Parcel 2 is a 520-ha area of xerophytic forest dominated by species adapted for the long dry season (Ratsirarson 2003, 2008). This parcel is often referred to as spiny forest. Between the two noncontiguous parcels, the vegetation transitional between the gallery and the xerophytic habitats is more degraded than the reserve, because it is used for grazing and for the collection of various forest products for food, medicines, building, etc.

3.1.3 Flora of Beza Mahafaly

The forest represented in Parcel 1 is western Malagasy dry deciduous forest (White 1983). It has an average of 369 trees of ≥ 2.5 cm DBH/1,000 m^2 (Sussman and

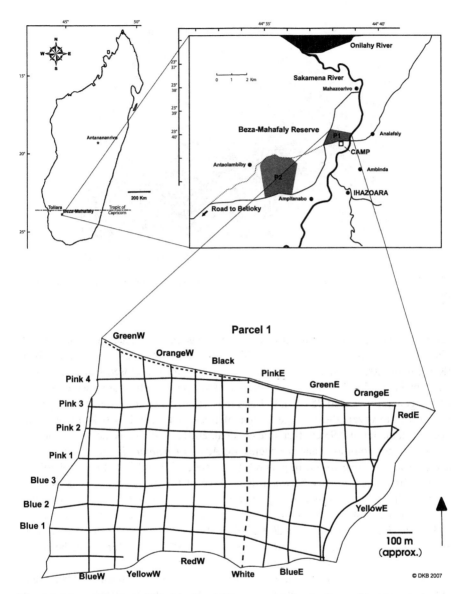

Fig. 3.1 Map of the Beza Mahafaly Special Reserve showing both parcels and marked path system in Parcel # 1 (provided by D. Brockman)

Rakotozafy 1994), which is typical of dry forests in Africa and the Neotropics (Gentry 1993). A soil moisture gradient exists, with soils becoming dryer farther from the river. Overall tree density does not decline with distance from the river, but the density of larger trees does; a uniformly closed upper canopy occurs on wet soil, but this grades into denser bush on dryer soil, where distinctions between the

Fig. 3.2 Aerial photograph of Parcel # 1 to the right of the dry Sakamena River. Photo taken from Northeast of the reserve

canopy strata are obscured and forest gradually passes into thicket. Tree height and diameter decrease progressively with increasing distance from the river (Ratsirarson 2003). On wet soils, the upper strata form a closed canopy, mostly uniform in height (15–20 m). Members of the upper stratum are species whose trunks generally exceed 25 cm DBH and may attain 50 cm or more, especially on wet soils. However, the average height of trees in Parcel 1 is 6.33 m, with an average diameter of 7.93 cm (Ratsirarson et al. 2001). *T. indica* is the dominant tree species. Other common canopy species are *Acacia rovumae*, *Euphorbia tirucalli*, and *Salvadora augustifolia*. In general, those species found in both microhabitats are not distributed equally between them. Five of the most common species are found mainly on wet soils and eight mainly on dry. Parcel 2 contains desert-like vegetation dominated by *Alluaudia procera*, *Cedrelopsis grevei*, *Commiphora* spp., and *Euphorbia* spp. (Ratsirarson 2003, 2008). Average height of trees is 4.5 m, with an average diameter of 6.5 cm (Ratsirarson et al. 2001). This forest has been the subject of fewer studies than the gallery forest.

Twelve permanent transects, each 1,000 m × 10 m, have been set inside and outside the two parcels to continuously monitor the density and distribution of plant species. All trees of ≥3 cm diameters were identified, tagged and their diameter as well as height measured. Overall, the flora of BMSR contains approximately 120 species and 49 families (Ratsirarson 2003). However, half the families are represented by a sole species. Euphorbiaceae and Mimosaceae are the most species-rich. A local reference herbarium and the Beza Mahafaly Osteological Collection are currently housed at the onsite Museum.

3.1.4 Fauna of Beza Mahafaly

There are four species of lemur at BMSR, two of them diurnal (Verreaux's sifaka, *Propithecus verreauxi*, and ringtailed lemurs, *Lemur catta*), and two nocturnal species (the white-footed sportive lemur, *Lepilemur leucopus*, and the grey-and-red mouse lemur, *Microcebus griseorufus*). We initially thought there were more species of *Microcebus*. However, recent genetic analysis revealed only one species with three color morphs (Heckman et al. 2006).

Three introduced carnivores (the domestic dog, *Canis lupus familiaris*, the small Indian civet, *Viverricula indica*, and the invasive wildcat, *Felis silvestris*) occur in the reserve (Brockman et al. 2008). The wildcat is semi-arboreal and distinguished from domestic cats by its large, pronounced ear lobes (Ratsirarson et al. 2001), brown and gray tabby pattern, larger size, and substantial sexual dimorphism (Brockman et al. 2008). Ratsirarson et al. (2001) saw an adult *Felis* carrying an infant ringtailed lemur in its mouth. The one endemic carnivore, the fossa, *Cryptoprocta ferox*, had not been directly observed in the reserve since 1993 (Brockman et al. 2008), but a camera trap image of a fossa moving through the reserve in 2008 and other recent observations confirm that it still exists in the area (Sauther et al. unpublished data). For a list of other mammalian species found in the Reserve, see Ratsirarson et al. (2001), Ratsirarson (2003), and Sussman and Ratsirarson (2006).

One hundred and two species of birds representing 43 families have been observed at Beza Mahafaly. Over half of the families are represented by only one species (Ratsirarson et al. 2001). The Reserve is home to at least 15 species of snakes, 17 species of lizards, one species of tortoise and fresh water turtle, and one species of crocodile; there are three species of amphibians. There also is a notable diversity of insects (Ratsirarson 2003).

3.1.5 Collection and Management of Long-Term Data

The long-term research focuses especially on the socioecology, population demography, life history, reproductive biology, mating behavior, socioendocrinology, feeding and nutritional ecology, population genetics, and health status of the diurnal lemurs (Fig. 3.3a, b). It also includes work on the conservation biology of other elements of the fauna and of the flora, monitoring of plant phenology, and research on the socioeconomics of the human population in relation to economic stability, development, and conservation. We maintain the education and training component of the reserve locally, regionally, and internationally. Each component of the project requires different strategies for data collection and management.

Fig. 3.3 (**a**) Collared ringtailed lemur (*Lemur catta*; photo © M. Sauther); (**b**) collared Verreauxi's sifaka (*Propithecus verreauxi*; photo © D. Brockman)

3.1.6 Future Plans for the Reserve

Beza Mahafaly Reserve was included among the sites of implementation of President Ravalomanana's "Durban vision" declared in 2003, where he set the goal of extending Madagascar's protected areas from 1.7 to 6 million ha, which is 10% of the national territory. BMSR's surface area was to increase from 500 to 4,600 ha. The two Parcels would be united into one protected zone, which would ensure better conservation of the region. Zones for controlled human utilization have been established around the protected zone in response to the needs of neighboring communities. Use of resources for commercial purposes is prohibited. However, strictly controlled harvesting of ligneous and non-ligneous products (honey, medicinal plants, wood for building, pastureland) is allowed. The right to utilize paths and trails located outside the protected zone is granted to inhabitants and livestock.

3.2 Socioecology of the Diurnal Lemurs

Two vital aspects of long-term data collection and management of socioecological information are: (1) Permanent identification of individual animals and regular censusing of these populations to document life history parameters and facilitate recensusing. Most adult ringtailed lemurs, all adult sifaka, and ~90% of all sifaka at least 1 year old are tranquilized and given color-coded collars or other marks that allow individual recognition and facilitate repeated censuses. Collection of these baseline data allows accumulation of data on known individuals and groups on topics including diet and nutrition, general and dental health status, and stress responses to demographic perturbations both inside and outside the reserve (Sussman and Ratsirarson 2006). (2) Obtaining baseline data on the flora within and surrounding the reserve. Using vegetation analysis and satellite images from BMSR as a baseline, we have estimated characteristics of vegetation cover and ringtailed lemur population density throughout this species' entire geographical range (Sussman et al. 2003, 2006). The relationship of population structure to habitat thus can be measured, as can changes over time under various conditions.

In the following, we provide some specific strategies for data collection and management in relationship to ringtailed lemur and sifaka socioecology. For more detailed information on phenological monitoring and vegetation analysis, see Sussman and Rakotozafy (1994) and Ratsirarson (2003, 2008). We also have collected data on the socioeconomy of the human population surrounding the Reserve (see Ratsirarson et al. 2001), but will not discuss these.

3.2.1 Collection and Management of Long-Term Data for Individual Identification and Population Censuses of Ringtailed Lemurs

Between June 1987 and February 1988, nine groups of *L. catta* were censused, 85 of the 88 adults in the groups were given collars for individual identification, and monitoring of the groups was initiated (Sussman 1991). This included all groups that ranged mainly in the 80-ha fenced portion of Parcel 1 and one group adjacent to it. In November, the groups were recensused and 155 individuals were identified, including infants and juveniles. Collaring of three of the groups was renewed between 2001 and 2004 (Gould et al. 2003, 2005). As of 2008, 218 individuals in nine groups had been collared; this represented virtually all of the approximately 225 individuals in the study area, which in 2003 was expanded to include 9 km^2 of fragmented forest south and west of the reserve (Cuozzo and Sauther 2006; Sauther et al. 2006). Beginning in May 2003, captured animals also received a subcutaneous microchip (PIT tag). The size and composition of these groups is monitored monthly, with new individuals captured and collared yearly (Sauther et al. 2006). Census data and data on individual identification have been computerized since 2003. Mean group size is 11.5 individuals, with a range of 3–21. Home ranges of 8–9 groups have a mean area of 25 ha and overlap extensively.

While animals are tranquilized, researchers collect systematic data on body mass, reproductive state, general physical condition, internal body temperature, ectoparasites, and dental condition. Body measurements, hair samples, and dental casts also are collected. Beginning in 2003, blood samples have been collected for white blood cell counts and measurement of packed cell volumes. Hair and blood samples also allow disease screening, isotope studies, and extraction of DNA for analyses of kinship and genealogical reconstruction. Data are recorded into the project's database on Excel. Initially, age grades were established through a combination of body weights, canine eruption patterns, general tooth wear, nipple length for females, and presence or absence of testes in males (Sauther et al. 2002). Since 2003, actual ages are known for most animals.

3.2.2 Collection and Management of Sifaka Long-Term Data

The ecology, life history, social behavior, reproductive biology, and population genetics of the sifaka population at BMSR have been studied continuously by Richard and colleagues since 1984, focusing on populations residing in, and adjacent to, Parcel 1. From 1984 to 2009, 718 individuals residing in 50–55 social groups have been captured, measured, and marked. The sifaka study population currently comprises ~280 marked and habituated individuals residing in 38 core social groups. Mean group size is 5–6 individuals, with a range of 2–16. The 38 social groups have 4–6 ha, overlapping home ranges within the boundaries of Parcel 1 (Richard et al. 1991, 2002; Lawler et al. 2009).

Individuals are immobilized using Telazol delivered from a Telinject blowgun dart (Richard et al. 2002). During the subsequent ~90-min processing period, individuals are marked and data similar to those on ringtailed lemurs are collected (see above). Data are recorded on individual capture sheets. At the end of the annual capture season, data sheets are copied and sent to Marion Schwartz, BMSR Sifaka Database Manager, who enters the capture, census, and morphological data into the sifaka database. During processing, each individual is given an identification collar and individuals are also ear-notched using a binary system that duplicates the tag number. Captures of unmarked juveniles born the previous year and of new immigrants typically occur during the austral July–August birth season, but birth season captures do not include likely pregnant females and those with dependent young. Initially, individuals were assigned to 1 of 5 age classes based on tooth wear, with the age classes calibrated from recaptures of individuals of known age (Richard et al. 1991). Today, the exact ages of most animals are known because they were born into the population.

Censuses were carried out annually between 1984 and 1991 and have been done at monthly intervals since, yielding an unparalleled computerized dataset. Also, life history data (age at first reproduction, fertility, longevity, etc.) derived from the Sifaka Database have been incorporated into the Primate Life History Database starting in 2007 (PLHD, Strier et al. 2010; http://demo.plhdb.org) to facilitate comparative analyses of species-specific mortality and fertility schedules in seven primate taxa and to test specific hypotheses about life history evolution (e.g. Morris et al. 2011).

3.3 Research Highlights

3.3.1 Ringtailed Lemurs

3.3.1.1 Demography and Life Histories

Long-term census data allowed us to determine who survived during a 2-year drought in 1991–1992, which deaths could be directly attributed to the drought, and how the population recovered (Gould et al. 1999, 2003). During this period, approximately half of the adult females died and 80% of the infants died in the second year of the drought. The population declined considerably up to 1994, 2 years after the drought ended. However, by 1997 it was recovering well. Annual reproduction, early sexual maturity, high birth rates, and dietary adaptability likely contributed to the population recovery.

Tracking which males disperse into which groups every year from the beginning of the project has contributed greatly to the analysis of patterns of male affiliative behavior (Sussman 1992; Gould 1997a,b) and has allowed us to assess migration status and to understand the strategies that males of different ages use to disperse successfully between groups (Gould 2006). Furthermore, the hormonal correlates of mating and post-mating behavior in adult males were investigated (Gould et al. 2005; Gould and Ziegler 2007). By knowing relative ages and which groups the older males had resided in over the previous years, Gould has determined how male tenure and the number of males in a group affect physiological stress levels, and how rank, tenure, and age affect testosterone levels and rates of aggression.

For example, results obtained from mating and post-mating seasons were compared to test Wingfield et al.'s (1990) "challenge hypothesis,", which predicts a strong positive relationship between male testosterone levels and investment in male–male competition for access to receptive females during the breeding season. Fecal testosterone (fT) levels and rates of intermale aggression were significantly higher during mating season compared to the post-mating period. Mean fT levels and aggression rates were higher in the first half of the mating season compared with the second half. The number of males in a group affected rates of intermale agonism, but not mean fT levels. The highest-ranking males exhibited higher mean fT levels than did lower-ranking males, and young males exhibited lower fT levels than prime-aged and old males. In the post-mating period, mean male fT levels did not differ between groups, nor were there rank or age effects. Thus, although male testosterone levels rose in relation to mating and heightened male–male aggression, fT levels fell to baseline breeding levels shortly after the early mating period, and to baseline non-breeding levels immediately after mating season had ended, offsetting the high cost of maintaining both high testosterone and high levels of male–male aggression in the early breeding period.

3.3.1.2 Effects of Habitat Fragmentation on Ringtailed Lemur Biology and Ecology

In 2003, Sauther and Cuozzo established the Beza Mahafaly Lemur Biology Project. This synergistic research program involves collaborative ties with zoos, toxicologists, virologists, and veterinarians and uses a broad perspective and a wide range of interdisciplinary approaches to study the effects of environmental change (both natural and human-induced) on lemur biology.

Health and disease ecology. Long-term monitoring of health parameters among ringtailed lemurs has allowed us to establish baseline health data and then to perform yearly assessments of how the lemurs are affected by ongoing environmental change and human disturbance. For example, Rainwater (2009) examined lemur blood for signs of environmental toxins and documented exposure of ringtailed lemurs at BMSR to multiple organochlorine pesticides and metals. The large number of captures has also allowed us to document a range of morphological variants within a natural population, including female virilization, microtia, and dental variants such as supernumerary teeth, maxillary incisor agenesis, and severe malocclusion. Such data are important for understanding natural variation in wild populations and provide a critical first step for assessing whether habitat change is creating abnormal patterns (Sauther and Cuozzo 2008).

Data on ringtailed lemurs show that habitat disturbance can have important effects on health in wild primates. For example, more degraded habitats are linked to lower body weights, reduced body fat, a higher incidence of tooth damage, and smaller body size (Sauther et al. 2006). Female dominance and residence patterns add complexity because males and females experience different pressures that result in differences in health and trauma (Sauther et al. 2006). For example, as males migrate from their natal groups, they often enter groups in which they lack established social relationships. The result can be reduced allogrooming, which can lead to serious health issues, including a higher incidence of parasite infestation, skin lesions, and hair loss (Sauther et al. 2006).

Long-term parasitological monitoring of the diurnal lemur species (Loudon et al. 2006; Loudon 2009) is revealing how socioecological variation interacts with anthropogenic change to affect the types and prevalence of parasites in each, and indicates that local domestic animals such as dogs and cattle may be transmitting parasites. Sifaka and ringtailed lemurs live in both intact and altered habitats, but ringtailed lemurs exhibit more nematodes and protistan parasites. Differences in each primate's parasite profile appear to be linked to host behavior and the ecological distribution of parasites. Ringtailed lemurs spend much more time on the ground than the sifaka and terrestrial substrate use provides greater opportunities for soil-transmitted parasites to acquire hosts. Ringtailed lemurs using the anthropogenically disturbed forests surrounding the reserve also harbor novel parasites that they may be acquiring via coprophagy or via physical contact.

Dental ecology. One of the most important health findings resulting from long-term research on ringtailed lemurs at BMSR is that severe tooth wear and loss is common. Sauther et al. (2002) initiated a systematic study of this phenomenon, which had been noted upon the initial examination of ringtailed lemurs in 1987/1988. Since then, they have documented exceptionally high frequencies of severe wear and antemortem tooth loss (greater than 20% of the population), largely due to processing the mechanically challenging fruit of the tamarind tree (*Tamarindus indica*), the dominant food and the essential fallback resource of ringtailed lemurs (Sauther and Cuozzo 2009). Severe tooth wear and tooth loss is more common in areas of degraded habitat, and, in some cases, is linked to exploitation of introduced foods (Sauther and Cuozzo 2009).

The ringtailed lemurs respond to decreases in food processing ability due to tooth loss and dental wear by changing their activity patterns and feeding behavior. For example, individuals who have lost teeth spend more time feeding and foraging throughout the day than individuals who are not dentally impaired, and in particular they forage more during early afternoon when other lemurs are resting. They also spend more time licking tamarind fruit to soften it before ingestion (Cuozzo and Sauther 2004, 2006; Millette et al. 2009).

Isotope ecology. Assessing stable isotope values has proven fruitful for understanding the ecology and habitat of living and fossil primates (e.g, Sponheimer et al. 2009). More specifically, stable isotopes have often been used to reconstruct the ecology of Madagascar's now-extinct "giant" lemurs (e.g., Crowley et al. 2011). Until recently, isotopic analyses of extinct lemur ecology have not been based on data from living lemur populations. Our comparative data from the BMSR ringtailed lemur population as well as from other sites are now providing points of comparison for interpreting the ecology of Madagascar's extinct lemurs. For example, Loudon et al. (2007) found that C_{13} and N_{13} signatures of ringtailed lemurs with poor dental health differ from other members of the population, possibly reflecting dietary changes resulting from severe wear and tooth loss, and that the C_{13} and N_{13} signatures of immigrant males differ from those in their original troops and resemble more closely those in their new groups, which exploit different resources. Also, comparisons between the ringtailed lemurs at two sites (Beza Mahafaly and Tsimanampesotse) reveal habitat-related differences in isotopic signatures. These data also have implications for the conservation biology of extant lemur species and as a tool for understanding changing environments in Madagascar.

3.3.2 Verreaux's Sifaka

3.3.2.1 Sifaka Behavioral Endocrinology

Brockman and colleagues have used a combination of endocrine data (derived from analysis of fecal steroids) and behavioral data to examine the impact of physical and social environments on reproduction and mating, social strategies, male life history, and demography in Verreaux's sifaka (Brockman 1994, 1999; Brockman and Whitten 1996; Brockman et al. 1998, 2001, 2009).

Brockman found that females have flexible mating strategies and that both sexes exercise mate choice, with the opportunity to do so enhanced by estrous synchrony within groups. She also documented intense androgen-mediated mating competition in both sexes; coercive mating tactics by males; and situation-dependent receptivity in which anovulatory females mate with immigrant males regardless of season (Brockman 1994, 1999; Brockman and Whitten 1996; Brockman et al. 1998).

Beginning in 1998, Brockman et al. (2001) used longitudinal data from individually marked sifaka to document age-specific patterns of male dispersal. They showed that males exhibit marked hormonal responses to socially disruptive events during the birth season, including substantially elevated fT concentrations in alpha males residing in unstable groups, in males making aggressive attempts to immigrate into neighboring groups, and in resident males evicting subordinates.

Previous studies examining fecal glucocorticoid (*fGC*) interactions with behavior in male sifaka at BMSR showed that high fGC levels are not a predictable cost of high rank during the birth season, and that elevated fGC concentrations coincide with specific behavioral traits and social contexts, including social instability (Brockman et al. 2001) and the aggressive eviction of subordinates by resident alpha males (Brockman et al. 1998). Using data on 124 males in 55 groups collected over several seasons, Brockman et al. (2009) found that fGC levels in males were unrelated to age, residence, group stability, or rank, but were substantially higher in males residing in groups containing infants than in those without infants. Also, annual variation in male fGC levels paralleled annual changes in infant birth rates (Brockman et al. 2009). These findings support the proposition that anticipation of relatively predictable future events, such as the birth of infants, can elicit GC responses. The entire birth season is probably a stressor, especially when births coincide with uncontrollable events such as increased intergroup male transfers and infanticides (ibid.). Of the five groups targeted for aggressive male transfers during this study, four suffered takeovers and three of those groups contained one or more infants which disappeared the following day or were mortally wounded by immigrant males. This research is the first to show that in seasonal plural breeding species, elevated fGC in males reflects specific events related to reproduction rather than states or social context during the birth season. These data provide new insights into the role of endocrine mechanisms in mediating male strategies to cope with natural sources of stress in wild lemur populations.

3.3.2.2 Sifaka Life History, Demography, and Population Genetics

Life history and demography. The combination of life history, demographic, phenotypic, and genetic datasets on the sifaka population provides powerful means to test major hypotheses in life history theory, to determine patterns of selection and adaptation, and to measure fitness. For example, Richard et al. (2000) documented how body mass influences female fertility by repeatedly capturing individual females and measuring changes in body mass over time, then combining this information with rainfall and other ecological data and data on reproduction. Male and female sifaka show seasonal fluctuations in body mass, which also is lower during drought years. When the primary productivity of the forest is low, body mass is particularly low. Females who were heavier during the previous mating season were significantly more likely than lighter females to give birth in the following birth season. Richard et al. (2000) showed specifically that (1) females lose more mass than males on a seasonal basis, (2) fertility is linked to body mass, and (3) gestation and lactation are uncorrelated with periods of high body mass. These findings suggest that sifaka females follow a strategy akin to "capital breeding," in which animals store energy for reproduction rather than immediately converting it into reproduction – that is, by decoupling energy acquisition and reproduction by storing energy to be used at a later, more adaptively advantageous time period (but see Brockman and van Schaik (2005) for a reassessment of this view).

Richard et al. (2002) also documented age-specific patterns of fertility, mortality, and dispersal. They showed that female sifaka reproduce later and live longer, in relation to body size, than females of any other primate species. The life history strategy associated with delayed reproduction and extended longevity is known as bet-hedging, where animals are selected to slow down the pace of reproduction and growth to mitigate the negative effects of stochastic fluctuations in the environment that influence animal livelihoods. Because climate determines food availability, Richard et al. (2002) argued that particular rainfall patterns in Madagascar were the main drivers of the life history schedules of Malagasy fauna, which (especially those of mammals) are characterized by extreme "fastness" or "slowness" as evolutionary responses to climatic fluctuations.

Subsequently, Dewar and Richard (2007) provided additional evidence that patterns of rainfall on Madagascar show uniquely high intra- and inter-annual variability compared to locales outside of Madagascar. The influence of rainfall on sifaka demography was modeled explicitly by Lawler et al. (2009). They showed that demographic parameters such as survival, reproductive value, and expected life span were depressed when annual rainfall was below 300 mm. Furthermore, a decrease in mean annual rainfall or an increase in the variance in annual rainfall resulted in negative population growth rates. These studies concur in finding that climate plays a key role in shaping both life history and demographic traits in the sifaka population.

Population genetics. Lawler et al. (2003) genotyped 444 sifaka to assess population structure and patterns of reproduction. The genetic structure of the population mimicked the dispersal and sex-ratio pattern first observed by Richard et al. (1991, 1993). Specifically, within social groups, both average and pair-wise genetic relatedness was higher for females than for males. This makes sense given that female sifaka are mostly philopatric, whereas males disperse from their natal groups on reaching sexual maturity (Richard et al. 1993; see also Kappeler and Fichtel 2011). This results in genetically distinct matrilines across social groups and creates a population more strongly structured through female lineages than male lineages. Similarly, offspring are more genetically distinct than adults across the population, because those in any given group can share the same father but have different mothers. Thus, offspring cohorts are united through paternal alleles within social groups, but across the population each cohort of offspring is genetically different.

However, as Richard et al. (1991, 2002) have documented, most of the infants that survive to 1 year are male (60–70% in most years studied). When males reach sexual maturity they disperse, somewhat randomly, into adjacent groups. At the genetic level, this randomizes the genetic structure of the offspring cohort and results in adults within social groups retaining less genetic structure than offspring. These results also show that the sex ratio, in addition to genetic relatedness, is an important factor with respect to the evolution of social behavior. For example, Richard et al. (2002) argued that the behavioral effects of a male-biased sex ratio include intense competition between females, heightened importance of female mate choice, and female social dominance.

Genetic data coupled with phenotypic and demographic data can elucidate reproductive strategies. Lawler et al. (2005) analyzed paternity and combined phenotypic data with information on reproductive success to measure the strength of intrasexual selection on male traits relevant to male–male competition during the mating season. They found that directional selection targets leg length and size, while stabilizing selection targets body mass. That is, males with longer, stronger legs, and an intermediate body size were more successful at reproduction. These results suggest that traits related to chasing are more important to male fitness than traits related directly to fighting involving physical contact (e.g., overall body size, canine size) and provide one possible explanation for the sexual monomorphism in body mass among male and female sifaka (Fig. 3.4). During male mating competition, there is no intrasexual directional selection acting to increase male body mass relative to that of females (see Lawler 2009).

Lawler (2007) also examined male reproductive success from a population-wide perspective. Specifically, he decomposed variation in male fitness (i.e., total reproductive output) into three components: reproductive lifespan, fertility (i.e. the per-year reproductive output), and infant survival. These components were estimated for males reproducing in their resident groups and those who reproduced outside of their resident groups. The results reveal that fertility makes the greatest contribution to variance in male fitness, followed by reproductive lifespan, and offspring survival. Factors that enhance opportunities for extra-group mating include female choice, a high density of social groups with overlapping home ranges, and a restricted mating season (Lawler 2007). Thus, long-term genetic studies provide a powerful complement to behavioral-ecological studies, since they can reveal the fitness consequences of particular behaviors that occur during an animal's lifetime.

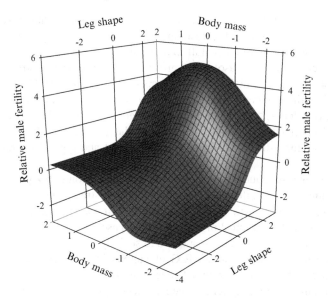

Fig. 3.4 Fitness surface showing relationship between male fertility and body mass and leg shape: Lawler et al. (2005) found that directional selection acts on leg length and thigh circumference, whereas stabilizing selection acts on body mass

3.4 Conclusions

Compared to other mammals, primates are characterized by delayed maturation, slow rates of growth, complex social behavior, and long lifespans. Because of these characteristics, many field studies of wild primates have traditionally focused on documenting behavioral interactions among one or a few individuals throughout their lives. This approach emphasizes the collection of longitudinal data on the careers of individuals rather than the collection of data on numerous individuals that comprise a population. The multidisciplinary studies of ringtailed lemurs and sifaka discussed above show that it is possible to obtain long-term data on careers of individual animals along with extensive data on many animals in multiple social groups and even on entire populations. These data have been, and continue to be, analyzed from various perspectives, thus providing important new insights into the demographic, socioecological, anthropogenic, and epidemiological factors that shape an evolving population.

The following are a few specific examples of research highlighting the benefits of long-term research at Beza Mahafaly:

1. Knowledge of the demographic structure of the ringtailed lemur population in the reserve over the long-term allowed us to determine the effects of a severe drought on the population and the factors that enabled the population to recover.

2. Research on hormone levels in ringtailed lemur males has revealed that male–male aggression and testosterone levels rise at the beginning of mating season. However, these levels fall to non-breeding baseline levels immediately after the early mating season, thus offsetting the high cost of stress related to high testosterone and high levels of aggression. Furthermore, fT levels are not affected by the number of males in a group, nor are there rank or age effects.
3. The parasite loads, types of parasites, and dental health of the ringtailed lemur and sifaka populations, both inside and out of the reserve, are affected directly by habitat structure, dietary differences, and amount and type of interaction with humans.
4. Female sifaka reproduce later and live longer, in relation to body size, than any other primate species, thus displaying a life history strategy in which the pace of reproduction and growth is slowed down to mitigate the negative effects of extreme fluctuations typical of southern Madagascar.
5. Male sifaka exhibit hormonal responses to socially disruptive events during the birth season, including marked fCG elevations in the presence of newborns associated with increased group takeovers and infanticide risk.
6. The combination of long-term genetic, phenotypic, and demographic studies allowed measurement of the strength of sexual selection in a wild primate. Specifically, sifaka males with longer, stronger legs were more successful at reproduction than were males with larger body mass, thus suggesting that traits related to chasing and locomotion were more important to male fitness than those related directly to fighting (e.g., body mass).

Acknowledgments We are grateful to the Government of Madagascar, Madagascar National Parks (MNP formerly ANGAP), the University of Antananarivo (The School of Agronomy (ESSA) especially the Forestry Department of this School (ESSA-Forêts) for permission to work at Beza Mahafaly Special Reserve. We especially appreciate the past and the present help, advice, and logistical support of our colleagues in Madagascar, notably Jo Ajimy, Krista Fish, James Loudon, Teague O'Mara, Jennifer Ness, Rafidisoa Tsiory, the late Rakotomanga Pothin, Ranaivoson Andrianasolo, Razakanirina Daniel, Ramanoelina Panja, Rasoarahona Jean, Rajoelison Gabrielle, Bruno Ramamonjisoa, Ranaivonasy Jeannin, Jessica Scott, and Youssouf Jacky. Our profound thanks go to the members of the BMSR Monitoring Team past and present, and in particular Enafa, Elahavelo, Emady Rigobert, Ellis Edidy, Efitiria, Eboroke Sylvain, Ranarivelo Ny Andry, Randrianarisoa Jeannicq, Ravelonjatovo Sylvia, Razanajaonarivaly Elyse, Ratsirarson Helian, and as well as to the Madagascar National Park team on the ground now led by Andry Randrianandrasana. Enafa's skill with the blow gun is extraordinary, and together the BMSR Monitoring Team not only made it possible to capture so many animals safely but also, thereafter, to census and monitor them regularly.

We are grateful to our U.S. colleagues who continue to make important contributions to research on sifaka at BMSR, most notably the members of the Beza Mahafaly Sifaka Research Consortium (BMSRC), Marion Schwartz (Sifaka Database Manager), Patricia L. Whitten, Laurie R. Godfrey, and Robert E. Dewar. For veterinary assistance we thank David Miller, Martha Weber, Scott Larsen, Anneke Moresco, Mandala Hunter, Heather Culbertson, Kerry Sondgeroth, Katie Eckert, Rachel Mills, Angie Simai, Jessica Kurek, and Catherine Woods. Long-term support of the BMSR Monitoring Team has kindly been provided by the Liz Claiborne and Art Ortenberg Foundation in collaboration with Yale University and the University of Cambridge. Numerous funding agencies have generously supported our on-going research at BMSR, notably National Science Foundation, World Wildlife Fund, Margot Marsh Biodiversity Foundation, Schwartz

Family Foundation Trust, St. Louis Zoo, Primate Conservation Inc., the International Primatological Society, the Indianapolis Zoo (Department of Science and Conservation), Tany Meva, MNP, the Wenner-Gran Foundation, National Sciences and Engineering Research Council of Canada, the University of North Dakota (SSAC; Faculty Research Seed Money Council; Arts, Humanities and Social Sciences Award Committee) and North Dakota EPSCoR, the American Society of Primatologists, the Lindbergh Fund, the John Ball Zoo Society, the National Geographic Society, the University of Colorado-Boulder (Council on Research and Creative Work, Innovative Grant Program), and Washington University. The electronic record of the primate specimens and their assigned Beza Mahafaly Osteological Collection (BMOC) numbers is available online through the University of Massachusetts, Amherst, Collections website (Brockman et al. 2008).
[http://www.umass.edu/anthmorphometricslab/BezaMahafalyOsteoCollection.htm]. Finally, we thank Peter Kappeler, David Watts, and an anonymous reviewer for their excellent comments on this paper.

References

Brockman DK (1994) Reproduction and mating system of Verreaux's sifaka, *Propithecus verreauxi*, at Beza Mahafaly, Madagascar. PhD thesis, Yale University, New Haven
Brockman DK (1999) Reproductive behavior of female *Propithecus verreauxi* at Beza Mahafaly, Madagascar. Int J Primatol 20:375–398
Brockman DK, van Schaik CP (2005) Seasonality and reproductive function. In: Brockman DK, van Schaik CP (eds) Seasonality in primates: studies of living and extinct human and nonhuman primates. Cambridge University Press, Cambridge, pp 269–306
Brockman DK, Whitten PL (1996) Reproduction in free-ranging *Propithecus verreauxi*: estrus and the relationship between multiple partner matings and fertilization. Am J Phys Anthropol 100:57–69
Brockman DK, Whitten PL, Richard AF, Schneider A (1998) Reproduction in free-ranging male *Propithecus verreauxi*: the hormonal correlates of mating and aggression. Am J Phys Anthropol 105:137–151
Brockman DK, Whitten PL, Richard AF, Benander B (2001) Birth season testosterone levels in male Verreaux's sifaka, *Propithecus verreauxi*: insights into socio-demographic factors mediating seasonal testicular function. Behav Ecol Sociobiol 49:117–127
Brockman DK, Godfrey LR, Dollar LJ, Ratsirarson J (2008) Evidence of invasive *Felis silvestris* predation on *Propithecus verreauxi* at Beza Mahafaly Special Reserve, Madagascar. Int J Primatol 29:135–152
Brockman DK, Cobden AK, Whitten PL (2009) Birth season glucocorticoids are related to the presence of infants in sifaka (*Propithecus verreauxi*). Proc R Soc Lond B 276:1855–1863
Crowley BE, Godfrey LR, Irwin MT (2011) A glance to the past: subfossils, stable isotopes, seed dispersal, and lemur species loss in southern Madagascar. Am J Primatol 73:25–37
Cuozzo FP, Sauther ML (2004) Tooth loss, survival, and resource use in wild ring-tailed lemurs (*Lemur catta*): implications for inferring conspecific care in fossil hominids. J Hum Evol 46:623–631
Cuozzo FP, Sauther ML (2006) Severe wear and tooth loss in wild ring-tailed lemurs (*Lemur catta*): a function of feeding ecology, dental structure, and individual life history. J Hum Evol 51:490–505
Dewar RE, Richard AF (2007) Evolution in the hypervariable environment of Madagascar. Proc Natl Acad Sci USA 104:13723–13727
Gentry AH (1993) Diversity and floristic composition of lowland tropical forest in Africa and South America. In: Goldblatt P (ed) Biological relationships between Africa and South America. Yale University Press, New Haven, pp 500–546

Gould L (1997a) Intermale affiliative behavior in ringtailed lemurs (*Lemur catta*) at the Beza Mahafaly Reserve, Madagascar. Primates 38:15–30

Gould L (1997b) Affiliative relationships between adult males and immature group members in naturally occurring ringtailed lemurs (*Lemur catta*). Am J Phys Anthropol 103:163–171

Gould L (2006) *Lemur catta* ecology: what we know and what we need to know. In: Gould L, Sauther ML (eds) Lemurs: ecology and adaptation. Springer, New York, pp 255–274

Gould L, Ziegler TE (2007) Variation in fecal testosterone levels, inter-male aggression, dominance rank and age during mating and post-mating periods in wild adult male ring-tailed lemurs (*Lemur catta*). Am J Primatol 69:1325–1339

Gould L, Sussman RW, Sauther ML (1999) Natural disasters and primate populations: the effects of a 2-year drought on a naturally occurring population of ring-tailed lemurs (*Lemur catta*) in southwestern Madagascar. Int J Primatol 20:69–84

Gould L, Sussman RW, Sauther ML (2003) Demographic and life-history patterns in a population of ring-tailed lemurs (*Lemur catta*) at Beza Mahafaly Reserve, Madagascar: a 15-year perspective. Am J Phys Anthropol 120:182–194

Gould L, Ziegler TE, Wittwer DJ (2005) Effects of reproductive and social variables on fecal glucocorticoid levels in a sample of adult male ring-tailed lemurs (*Lemur catta*) at the Beza Mahafaly Reserve, Madagascar. Am J Primatol 67:5–23

Heckman KL, Rasoazanabary E, Machlin E, Godfrey LR, Yoder AD (2006) Incongruence between genetic and morphological diversity in *Microcebus griseorufus* of Beza Mahafaly. BMC Evol Biol 6:e98. doi:10.1186/1471-2148-6-98

Kappeler PM, Fichtel C (2011) A 15-year perspective on the social organization and life history of sifaka in Kirindy Forest. In: Kappeler PM (ed) Long-term field studies of primates. Springer, Heidelberg

Lawler RR (2007) Fitness and extra-group reproduction in male Verreaux's sifaka: an analysis of reproductive success from 1989–1999. Am J Phys Anthropol 132:267–277

Lawler RR (2009) Monogamy, male-male competition, and mechanisms of sexual dimorphism. J Hum Evol 57:321–325

Lawler RR, Richard AF, Riley MA (2003) Genetic population structure of the white sifaka (*Propithecus verreauxi verreauxi*) at Beza Mahafaly Special Reserve, southwest Madagascar (1992–2001). Mol Ecol 12:2307–2317

Lawler RR, Richard AF, Riley MA (2005) Intrasexual selection in Verreaux's sifaka (*Propithecus verreauxi verreauxi*). J Hum Evol 48:259–277

Lawler RR, Caswell H, Richard AF, Ratsirarson J, Dewar RE, Schwartz M (2009) Demography of Verreaux's sifaka in a stochastic rainfall environment. Oecologia 161:491–504

Loudon JE (2009) The parasite ecology and socioecology of ring-tailed lemurs (*Lemur catta*) and Verreaux's sifaka (*Propithecus verreauxi*) inhabiting the Beza Mahafaly Special Reserve. PhD thesis, University of Colorado, Boulder

Loudon JE, Sauther ML, Fish KD, Hunter-Ishikawa M, Jack IAY (2006) One reserve, three primates: applying a holistic approach to understand the interconnections among ring-tailed lemurs (*Lemur catta*), Verreaux's sifaka (*Propithecus verreauxi*), and humans (*Homo sapiens*) at Beza Mahafaly Special Reserve, Madagascar. Ecol Environ Anthropol 2:54–74

Loudon JE, Sponheimer M, Sauther ML, Cuozzo FP (2007) Intraspecific variation in hair δ^{13}C and δ^{15}N values of ring-tailed lemurs (*Lemur catta*) with known individual histories, behavior, and feeding ecology. Am J Phys Anthropol 133:978–985

Millette JB, Sauther ML, Cuozzo FP (2009) Behavioral responses to tooth loss in wild ring-tailed lemurs (*Lemur catta*) at the Beza Mahafaly Special Reserve, Madagascar. Am J Phys Anthropol 140:120–134

Morris WF, Altmann J, Brockman DK, Cords M, Fedigan LM, Pusey AE, Stoinski TS, Bronikowski AM, Alberts SC, Strier KB (2011) Low demographic variability in wild primate populations: fitness impacts of variation, covariation, and serial correlation in vital rates. Am Nat 177:e14–28. doi:10.1086/657443

3 Beza Mahafaly Special Reserve: Long-Term Research

Rainwater TR, Sauther ML, Rainwater KAE, Mills RE, Cuozzo FP, Zhang B, McDaniel LN, Abel MT, Marsland EJ, Weber MA, Jack IAY, Platt SG, Cobb GP, Anderson TA (2009) Assessment of organochlorine pesticides and metals in ring-tailed lemurs (*Lemur catta*) at Beza Mahafaly Special Reserve, Madagascar. Am J Primatol 71:998–1010

Ratsirarson J (2003) Réserve Spéciale de Beza Mahafaly. In: Goodman SM, Benstead JP (eds) The natural history of Madagascar. University of Chicago Press, Chicago, pp 1520–1525

Ratsirarson J (2008) La Réserve Spéciale de Beza Mahafaly. In: Goodman SM (ed) Paysages naturels et biodiversité de Madagascar. Muséum National d'Histoire Naturelle, Paris, pp 615–626

Ratsirarson J, Randrianarisoa J, Ellis E, Emady RJ, Efitroarany RJ, Razanajaonarivalona EH, Richard AF (2001) Beza Mahafaly: écologie et réalités socio-économiques. Rech Dev B 18:1–104

Richard AF, Rakotomanga P, Schwartz M (1991) Demography of *Propithecus verreauxi* at Beza Mahafaly, Madagascar: sex ratio, survival, and fertility, 1984–1988. Am J Phys Anthropol 84:307–322

Richard AF, Rakotomanga P, Schwartz M (1993) Dispersal by *Propithecus verreauxi* at Beza Mahafaly, Madagascar: 1984–1991. Am J Primatol 30:1–20

Richard AF, Dewar RE, Schwartz M, Ratsirarson J (2000) Mass change, environmental variability and female fertility in wild *Propithecus verreauxi*. J Hum Evol 39:381–391

Richard AF, Dewar RE, Schwartz M, Ratsirarson J (2002) Life in the slow lane? Demography and life histories of male and female sifaka (*Propithecus verreauxi verreauxi*). J Zool Lond 256:421–436

Sauther ML, Cuozzo FP (2008) Somatic variation in living, wild ring-tailed lemurs (*Lemur catta*). Folia Primatol 79:55–78

Sauther ML, Cuozzo FP (2009) The impact of fallback foods on wild ring-tailed lemur biology: a comparison of intact and anthropogenically disturbed habitats. Am J Phys Anthropol 140:671–686

Sauther ML, Sussman RW, Cuozzo F (2002) Dental and general health in a population of wild ring-tailed lemurs: a life history approach. Am J Phys Anthropol 117:122–132

Sauther ML, Fish KD, Cuozzo FP, Miller DS, Hunter-Ishikawa M, Culbertson H (2006) Patterns of health, disease, and behavior among wild ringtailed lemurs, *Lemur catta*: effects of habitat and sex. In: Jolly A, Sussman RW, Koyama N, Rasamimanana H (eds) Ringtailed lemur biology. Springer, New York, pp 313–331

Sponheimer M, Codron D, Passey BH, de Ruiter DJ, Cerling TE, Lee-Thorp JA (2009) Using carbon isotopes to track dietary change in modern, historical, and ancient primates. Am J Phys Anthropol 140:661–670

Steele E (1975) Needed: virtue and money. Defenders Wildlife 50:90

Strier KB, Altmann J, Brockman DK, Bronikowski AM, Cords M, Fedigan LM, Lapp H, Liu X, Morris WF, Pusey AE, Stoinski TS, Alberts SC (2010) The Primate Life History Database: a unique shared ecological data resource. Methods Ecol Evol 1:199–211

Sussman RW (1991) Demography and social organization of free-ranging *Lemur catta* in the Beza Mahafaly Reserve, Madagascar. Am J Phys Anthropol 84:43–58

Sussman RW (1992) Male life history and intergroup mobility among ringtailed lemurs (*Lemur catta*). Int J Primatol 13:395–413

Sussman RW, Rakotozafy A (1994) Plant diversity and structural analysis of a tropical dry forest in southwestern Madagascar. Biotropica 26:241–254

Sussman RW, Ratsirarson J (2006) Beza Mahafaly Special Reserve: a research site in southwestern Madagascar. In: Jolly A, Sussman RW, Koyama N, Rasamimanana HR (eds) Ringtailed lemur biology. Springer, New York, pp 43–51

Sussman RW, Green GM, Porton I, Andrianasolondraibe OL, Ratsirarson J (2003) A survey of the habitat of *Lemur catta* in southwestern and southern Madagascar. Primate Conserv 19:32–57

Sussman RW, Sweeney S, Green GM, Porton I, Andrianasolondraibe OL, Ratsirarson J (2006) A preliminary estimate of *Lemur catta* population density using satellite imagery. In: Jolly A,

Sussman RW, Koyama N, Rasamimanana HR (eds) Ringtailed lemur biology: *Lemur catta* in Madagascar. Springer, New York, pp 16–31

White F (1983) The vegetation of Africa: a descriptive memoir to accompany the UNESCO/AETFAT/UNSO vegetation map of Africa. UNESCO (United Nations Educational, Scientific and Cultural Organization), Paris

Wingfield JC, Hegner RE, Dufty AM Jr, Ball GF (1990) The "Challenge Hypothesis": theoretical implications for patterns of testosterone secretion, mating systems, and breeding strategies. Am Nat 136:829–846

Chapter 4
Long-Term Lemur Research at Centre Valbio, Ranomafana National Park, Madagascar

Patricia C. Wright, Elizabeth M. Erhart, Stacey Tecot, Andrea L. Baden, Summer J. Arrigo-Nelson, James Herrera, Toni Lyn Morelli, Marina B. Blanco, Anja Deppe, Sylvia Atsalis, Steig Johnson, Felix Ratelolahy, Chia Tan, and Sarah Zohdy

Please note the erratum to this chapter at the end of the book.

Abstract We present findings from 25 years of studying 13 species of sympatric primates at Ranomafana National Park, Madagascar. Long-term studies have revealed that lemur demography at Ranomafana is impacted by climate change, predation from raptors, carnivores, and snakes, as well as habitat disturbance. Breeding is seasonal, and each species (except *Eulemur rubriventer*) gives birth synchronously to be able to wean before winter. Infant mortality is high (30–70%)

P.C. Wright (✉)
Institute for the Conservation of Tropical Environments, Stony Brook University, Stony Brook, NY, USA
e-mail: patchapplewright@gmail.com

E.M. Erhart
Department of Anthropology, Texas State University, San Marcos, TX, USA
e-mail: berhart@txstate.edu

S. Tecot
School of Anthropology, University of Arizona, Tucson, AZ, USA
e-mail: stecot@email.arizona.edu

A.L. Baden • J. Herrera • A. Deppe
IDPAS, Stony Brook University, Stony Brook, NY, USA
e-mail: andrea.baden@gmail.com; jherrera84@gmail.com; deppeam@hotmail.com

S.J. Arrigo-Nelson
Department of Biological and Environmental Sciences, University of California in Pennsylvania, California, PA, USA
e-mail: arrigonelson@calu.edu

T.L. Morelli
Museum of Vertebrate Zoology, Department of Environmental Sciences, Policy and Management, U.C. Berkeley, Berkeley, CA, USA
e-mail: morellitlm@gmail.com

M.B. Blanco
Department of Anthropology, University of Massachusetts, Amherst, MA, USA
e-mail: mbblanco@anthro.umass.edu

P.M. Kappeler and D.P. Watts (eds.), *Long-Term Field Studies of Primates*,
DOI 10.1007/978-3-642-22514-7_4, © Springer-Verlag Berlin Heidelberg 2012

and partly due to infanticide in *Propithecus edwardsi,* and perhaps *Varecia variegata.* Diurnal lemurs can live beyond 30 years in the wild and most females reproduce until death. Small-bodied *Microcebus rufus* live up to 9 years without signs of senescence. *Prolemur simus* migrates in search of new bamboo and mates, and related *V. variegata* mothers park their multiple offspring in "kindergartens," protected by others while mothers forage. Interference competition among sympatric lemurs occurs. Anthropogenic factors, such as past selective logging and climate change may influence the declining density of *E. rufifrons, P. simus,* and *P. edwardsi* while not affecting the density of pair-living species.

4.1 Introduction

Madagascar ranks as one of the world's top biodiversity hotspots because of its high endemism and high rate of habitat degradation (Myers et al. 2000; Ganzhorn et al. 2001). For primates, Madagascar has the highest conservation priority with 5 endemic families and 15 endemic genera (Mittermeier et al. 2010). Ninety-seven lemur species are now recognized of which 41% are threatened with extinction (8 critically endangered, 18 endangered, and 15 vulnerable), while 42 species remain data deficient (IUCN 2010; Mittermeier et al. 2010). Knowledge obtained from long-term field studies, such as the ones described here, is particularly valuable compared to short-term "snapshots" because long-term data can be especially useful to conservation management efforts. For effective management, park authorities can benefit from understanding the differences between normal fluctuations in population size and real trends over time, patterns which can only be detected with decades of data.

To date there are four long-term lemur research sites in Madagascar: Kirindy Forest in the dry deciduous forest of the west (Ganzhorn and Kappeler 1996; Kappeler and Fichtel 2012), the southern spiny desert reserve Berenty Private Reserve (Jolly et al. 2006; Jolly 2012), Beza Mahafaly Special Reserve (Richard

---

S. Atsalis • C. Tan
San Diego Zoological Society, The San Diego Zoo's Institute for Conservation Research, San Diego, CA, USA
e-mail: satsalis55@gmail.com; ctan@sandiegozoo.org

S. Johnson
Department of Anthropology, University of Calgary, Calgary, Canada
e-mail: steig.johnson@ucalgary.ca

F. Ratelolahy
Wildlife Conservation Society, Antananarivo, Madagascar
e-mail: Jeaf_ratel@yahoo.fr

S. Zohdy
University of Helsinki, Helsinki, Finland
e-mail: sarahzohdy@gmail.com

et al. 2002; Jolly and Sussman 2006; Sussman et al. 2012), and Ranomafana National Park (RNP) in the southeastern rainforest (Wright and Andriamihaja 2002; Wright 2004). At each site, lemurs have been studied for several decades (Jolly 1966; Sussman 1974; Richard 1978; Jolly and Sussman 2006; Sussman and Ratsirason 2006; Kappeler and Fichtel 2012). In this chapter, we discuss a compilation of the findings of long-term studies of the 13 lemur species found in RNP.

4.1.1 History of Ranomafana National Park

Ranomafana National Park (RNP) is a rainforest park located in southeastern Madagascar (21°16'S, 47°20'E). The landscape is dominated by submontane rainforest, which receives a mean of 3,000 mm of rain per year during the December through March rainy season. The RNP project was initiated in 1986 with the goal of protecting the habitat of a then newly discovered lemur species, the golden bamboo lemur (*Hapalemur aureus*), and the rediscovered greater bamboo lemur (*Prolemur simus*; Meier et al. 1987). From 1986 to 1989, logging concessions were granted by the forestry department and selective logging for valuable hardwood trees was intensive. In 1991, 41,000 ha of the montane rainforest were designated as a national park (Wright and Andriamihaja 2002). The Namorona River and a parallel paved road (Route 25) bisect the park into northern and southern parcels with a third parcel on the western boundary to the north. Patricia Wright, then at Duke University, spearheaded the initial park project, an integrated conservation and development project conducted simultaneously with research on lemurs and other aspects of biodiversity (Wright 1997; Wright and Andriamihaja 2002). Management transitioned to Stony Brook University (SBU) when Wright moved there in 1991. While retaining management of research in RNP, SBU handed park management over to the Association Nationale pour la Géstion des Aires Protegées (ANGAP), the national park system, in 1998. In 2006, the System of Protected Areas of Madagascar (SAPM) replaced ANGAP and incorporated sustainable practices of resource use into park creation and management. In 2009, SAPM became Madagascar National Parks (MNP).

4.1.2 Infrastructure at Ranomafana

The first research station was built in 1989 near the entrance to RNP. This structure was a small, one-story log cabin. In 2003, the station was upgraded to a three-story stone facility adjacent to the park and overlooking the rainforest. The new research station, named the International Centre for Research and Training for the Valorization of Biodiversity (Centre ValBio), is located on Route 25 near the park entrance (Wright 2004). This hub of scientific research and education is managed by a consortium of universities headed by SBU and the Institute for the Conservation of Tropical Environments (ICTE). Founding institutions include SBU and the Universities of Antananarivo, Fianarantsoa, and Helsinki. Currently, the main building houses administrative offices, a small laboratory, a library, and a dining hall that serves 65 people. A second four-story building (15,676 square feet) will open in 2012, and will be equipped with high speed internet, modern hormone, parasite, genetics and infectious disease laboratories, an audio/visual/computer center, and living accommodations for 54 students and researchers. The Centre ValBio has authorization to do research from the Ministry of Forests and Environment of the Government of Madagascar, and works closely with the MNP, especially on conservation management.

Centre ValBio's administration oversees the Departments of Research, Biodiversity Monitoring, Logistics and Management, and Community Outreach including Health and Education. Seventy-two local staff, many trained as lemur technicians and Malagasy biodiversity experts, work at Centre ValBio and live in the villages surrounding the park. The health and education team provides training and outreach programs to 22 nearby villages. Reforestation with native species and medicinal plant gardens managed by traditional healers are two important components of Centre ValBio's outreach efforts (Wright et al. 2005a). Twenty conservation clubs foster appreciation for conservation and a cooperative of artisanal women weavers is a sustainable contributor to village economics.

4.1.3   History of Ranomafana Lemur Research

Ranomafana contains 13 lemur species of which 8 have been subjects of long-term research (Table 4.1). Seven of the 13 species have been subject to taxonomic revision during the 20-year period (Table 4.2). Five species have been redesignated based on genetics and morphology. One new species has been described (*H. aureus*), one taxon has been raised to a new genus (*P. simus*), and one remains to be identified (*Cheirogaleus* sp.).

Twenty-six PhD dissertations and 26 Masters theses on lemurs have been completed since the park's initiation, and an additional eight dissertations are currently in progress. Major foci of these long-term studies have been diurnal lemurs, including behavioral ecology, demography, life history, reproductive biology, stress and reproductive hormones, parasites, feeding and nutritional ecology, morphometrics, predation, communication, and cognition (Table 4.1). Nocturnal lemurs have been studied less intensively, with emphases on reproduction, hibernation, parasites, and vocalizations (Table 4.1). Moreover, research in ecosystem dynamics and conservation with emphasis on lemur seed dispersal, climate, and tree phenology are also ongoing (Dew and Wright 1998; Wright et al. 2005b; Dunham 2008; Dunham et al. 2008, 2010; Ganzhorn et al. 2009).

Researchers have studied lemurs at four sites, each approximately 4 km^2, within the contiguous forest of the park (Fig. 4.1). Three sites (Vatoharanana, Valohoaka, and Mangevo), each with bush camp facilities, are in undisturbed or minimally

4 Long-Term Lemur Research at Centre Valbio, Ranomafana National Park, Madagascar 71

Table 4.1 List of RNP lemur species and references to research

Species	Common name	References
Avahi peyrierasi	Peyrierasi's woolly lemur	Harcourt (1987, 1988), Roth (1996), Andriantompohavana et al. (2007a)
Cheirogaleus crossleyi	Crossley's dwarf lemur	Wright and Martin (1995), Blanco et al. (2009), Groeneveld et al. (2011)
Cheirogaleus sp.		Not yet described
Daubentonia madagascariensis	Aye–aye	Sefczek (2009)
Eulemur rufifrons	Red-fronted brown lemur	Meyers et al. (1989), Overdorff (1991, 1993, 1996), Merenlender (1993), Johnson and Overdorff (1999), Overdorff et al. (1999), Johnson (2002), Johnson et al. (2005), Erhart and Overdorff (2008a)
Eulemur rubriventer	Red-bellied lemur	Overdorff (1991, 1993, 1996), Durham (2003), Overdorff and Tecot (2006), Tecot (2008, 2010), Wright et al. (2011), Tecot in press
Hapalemur aureus	Golden bamboo lemur	Meier et al. (1987), Wright et al. (1987), Glander et al. (1992), Tan (1999, 2007), Arrigo-Nelson and Wright (2004)
Hapalemur griseus	Gray gentle bamboo lemur	Meier et al. (1987), Tan (1999), Grassi (2002), Mutschler and Tan (2003), Arrigo-Nelson and Wright (2004), Herrera et al. in press
Lepilemur microdon	Small toothed sportive lemur	Porter (1998), Louis et al. (2006)
Microcebus rufus	Brown mouse lemur	Wright and Martin (1995), Atsalis et al. (1996), Atsalis (1999a, 1999, 1999b, 2000, 2008), Louis et al. (2006), Blanco (2008), Blanco and Meyer (2009), Durden et al. (2010), Deppe (2011)
Prolemur simus	Greater bamboo lemur	Meier et al. (1987), Wright et al. (1987), Tan (1999, 2007), Bergey and Patel (2008), Wright et al. (2008b), CVB census
Propithecus edwardsi	Milne Edwards' sifaka	Hemingway (1995, 1998), Wright (1995), Erhart and Overdorff (1998), Jernvall and Wright (1998), Pochron and Wright (2003), Arrigo-Nelson and Wright (2004), Mayor et al. (2004), Pochron et al. (2004), King et al. (2005, 2011), Pochron et al. (2004, 2005), Arrigo-Nelson (2006), Lehman et al. (2006), Irwin (2007, 2008), Morelli (2008), Bailey et al. (2009), Wright et al. (2009), Wright et al. (2011)
Varecia variegata editorium	Black-and-white ruffed lemur	White et al. (1995), Balko and Underwood (2005), Overdorff et al. (2005), Ratsimbazafy (2006), Baden et al. (2008)

Table 4.2 List of RNP lemurs including recent taxonomic changes with activity pattern: nocturnal (N), diurnal (D), and cathemeral (C); weight in grams; IUCN status

Current nomenclature	Previous	Activity	Status	Weight (g)	References
Avahi peyrierasi	*A. laniger*	N	DD	960	Zaramody et al. (2006)
Cheirogaleus crossleyi	*C. major*	N	DD	350	Groeneveld et al. (2011)
Daubentonia madagascariensis		N	NT	2,500	Feistner and Sterling (1995)
Eulemur rufifrons	*E. fulvus rufus*	C	NT	2,200	Mittermeier et al. (2010)
Eulemur rubriventer	*E. rubriventer*	C	VU	2,400	Mittermeier et al. (2010)
Hapalemur aureus	*New species*	D	EN	1,800	Meier et al. (1987)
Hapalemur griseus ranomafanensis	*H. griseus*	D	NE	990	Rabarivola et al. (2007)
Lepilemur microdon	*L. mustelinus*	N	DD	990	Louis et al. (2006)
Microcebus rufus	*M. rufus*	N	LC	45	Louis et al. (2006)
Prolemur simus	*Hapalemur simus*	D	CR	2,800	Groves (2001)
Propithecus edwardsi	*P. diadema edwardsi*	D	EN	5,800	Mayor et al. (2004)
Varecia variegata editorium	*V. variegata variegata*	D	CR	3,500	Groves (2001)

From IUCN 2010 RedList Guidelines: *NE* Not Evaluated, *DD* Data Deficient, *LC* Least Concern, *NT* Near Threatened, *VU* Vulnerable, *EN* Endangered, *CR* Critically Endangered, *EW* Extinct in the Wild, *EX* Extinct

disturbed rainforest. The fourth is Talatakely located near Route 25, selectively logged by hand from 1986 to 1989, and now accessible to tourists (Wright and Andriamihaja 2002).

Ad hoc transects and surveys have been conducted throughout the park since 1987. Since 2003, researchers conducted lemur surveys along eight 2 km transects, from the edges to the interior (Fig. 4.1). The surveys have led to the identification of new social groups and discovery of new species (Irwin et al. 2000, 2005; Arrigo-Nelson and Wright 2004; Andriantompohavana et al. 2007a, b; Wright et al. 2008b).

4.2 Long-Term Data Collection and Management

4.2.1 Long-Term Research on Focal Species

Most adults from 6 of the 7 diurnal lemur species at each of the main research sites have been marked with tags and collars for individual identification (*Propithecus edwardsi, Hapalemur aureus* and *Prolemur simus, Eulemur rubriventer,*

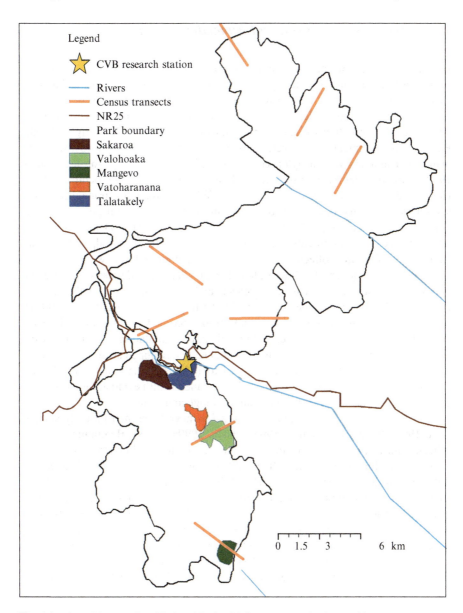

Fig. 4.1 Map of Ranomafana National Park with long-term study sites and long-term transects marked

E. rufifrons, and *Varecia variegata*). Many study groups have at least one member with a telemetry radio-collar for locating groups. Data collection on lemurs has been ongoing in Ranomafana for more than 24 years (Wright 2004; Table 4.1).

4.2.2 Weights, Measures, and Biomedical Data

Beginning in 1987, we established a protocol to obtain morphometric and health data on all seven diurnal lemur species. A trained team of Malagasy technicians capture individuals using remote injection techniques, whereby a Telinject blow gun or CO_2-powered rifle is used to tranquilize individuals with Telazol administered with lightweight darts (Glander et al. 1992; Wright 1995). Researchers, technicians, and veterinarians measure, collect samples from, and mark adult animals with a nylon collar and individual identification tags. Since 1999, captured animals also received a subcutaneous microchip (AVID, HomeAgain®) for permanent identification. While animals are tranquilized, the team uses a checklist developed to record information on general physical condition, body mass, and reproductive state (Glander et al. 1992). Dental molds are taken; physical measurements are recorded; and hair, fecal, blood, and external parasite samples are collected. We used data including body weights, canine eruption patterns, general tooth wear, female nipple length, and presence or absence of descended testes to assign age categories to all known study individuals (Johnson et al. 2005; Baden et al. 2008; Erhart and Overdorff 2008a; Wright et al. 2008a). Actual ages are known for most animals from recorded births. For individuals whose birth date is uncertain (i.e., individuals in the population since before 1986 or immigrants from other groups), Jernvall and King developed an accurate method to determine age by comparing year-to-year tooth wear (King et al. 2005). Beginning in 2003, veterinarians now also conduct detailed health evaluations, compiling biomedical profiles of all lemur individuals (Junge and Louis 2005).

Microcebus rufus and *Cheirogaleus crossleyi* have been intermittently studied since 1990 (Wright and Martin 1995; Atsalis 1999a, 2008). Beginning in 2003, subcutaneously placed microchips have allowed us to monitor individual mouse lemurs during long-term studies (Blanco 2008; Blanco and Meyer 2009; Blanco and Rahalinarivo 2010; Durden et al. 2010; Zohdy et al. 2010; Deppe 2011). Mouse lemurs are captured in Sherman traps baited with bananas, then weighed, measured, and released at the capture site. Over 300 individuals have now been marked. Dental tooth casts reveal ages, and repeated captures allow data collection on body mass fluctuations associated with torpor patterns (Atsalis et al. 1996; Atsalis 2008), reproductive status (Blanco 2008), parasite prevalence (Zohdy et al. 2010), and noninvasive behavioral experiments (Deppe and Wright 2009; Deppe 2011).

4.2.3 Long-Term Phenology, Climate, and Terrestrial Vertebrate Data

Tree phenology, daily rainfall, and temperature data have been recorded since 1987. Tree phenology began with monthly monitoring of 100 trees of 25 lemur fruit species and was expanded in 1995 with monthly monitoring of trees from more than 71 species, representing 26 genera and 19 families (Clark and Clark 2006, 2010). Initially, tree diameter measurements (diameter at breast height, DBH) were taken every 5 years, but beginning in 2004, DBH measurements are taken every 6 months (Clark et al. 2003) to correlate tree growth with rainfall. Maximum and minimum temperature and rainfall data are taken at 06:00 h each day with a team of technicians responsible for accuracy. Camera trap data are taken in distant regions of the park to monitor carnivores and terrestrial vertebrates (Gerber et al. 2010).

4.2.4 Database Management

Long-term data collection and management requires standardized protocols. The following are areas where protocols are in place: (1) permanent identification of individuals for behavioral observation and census, (2) lemur capture data including weights, measurements, and biomedical data, (3) focal lemur sampling, (4) tree phenological data, (5) daily temperature minima/maxima and rainfall, (6) ad libitum observations of predation, reproduction, and intergroup aggression, (7) GIS data for mapping, and (8) lemur fecal sample data. The phenology and climate data are archived and accessible to researchers at Centre ValBio and will soon be available on the ICTE/Centre ValBio website (http://icte.bio.sunysb.edu). Data from projects under the supervision of Dr. Wright are transcribed from field notebooks into Excel spreadsheets, by local research technicians with oversight by the CVB Chief Technical Advisor or US students at SBU with postdoctoral oversight. Data from other projects are the responsibility of the project's principal investigators. Three years after a researcher has terminated data collection, the data can be transferred into the Centre ValBio central database and made available to CVB researchers (with proper citation assured).

4.2.5 Research Highlights from Long-Term Data

Since 1986, the principal goal of primate research at Ranomafana has been to understand the factors driving the behavioral ecology of lemurs in a species-rich community. Our research has focused on particular species that have been the subjects of intensive study over 20+ years, as well as community level analyses to understand the effects of competition, predation, and habitat quality on species richness and on relative abundance. Further, the mixed history of anthropogenic disturbance has allowed comparative work to elucidate the effects of disturbance on lemur physiology, behavioral ecology, and community structure. However, for this chapter we will concentrate on population changes documented over decades of observation.

4.3 Population Ecology

4.3.1 Flexibility of Behaviors

Many social behaviors were not observed until after many years of study. Female takeovers of groups, male lethal aggression from other males, and simultaneous immigrations of related males into groups were not observed in the first 10 years of studying *P. edwardsi*, but were observed in the next decade (Morelli et al. 2009). After years of losing track of focal study groups, continuous long-term observations confirmed that *E. rufifrons* groups seasonally expand their range and travel over 4,000 m to find fruit (Erhart and Overdorff 2008a). Likewise, when *P. simus* group size began to decline in 2003, we began to observe patterns of male disappearance, as well as females' (and their offspring) migrations and wanderings for 3 months before returning to the original territory. This migration behavior was not seen in the first decade (CVB unpublished census data; Wright et al. 2008b). In addition, while breeding out of the birth peak had been observed early on in *E. rubriventer* (Overdorff unpublished data), it was not until almost 20 years of data were analyzed that it was determined that they have been observed to breed in eight different months of the year (Tecot 2010).

4.3.2 Lemur Group Size, Composition, and Social Organization

Social groupings have long been investigated in primates (Crook and Gartlan 1966; Eisenberg et al. 1972; Clutton-Brock and Harvey 1977) by correlating ecological factors to social organization (Janson 1992). For instance, it has been proposed that lemurs' relatively small group sizes may be due to nocturnal ancestry (van Schaik and Kappeler 1996) or smaller crowned fruit trees (Wright 1999). We initially reported that *P. edwardsi* have a multimale, multifemale social organization in groups of 3–9 individuals (Wright 1995). However, by 2003 we had observed every type of social grouping in this species with potential for all different types of mating systems. Moreover, we learned that group sizes can get as large as 11 (Pochron and Wright 2003). In contrast, we consistently found *E. rubriventer* in socially monogamous pairs accompanied by immature offspring (Merenlender 1993; Overdorff and Tecot 2006; Tecot 2008). Although previously described as pair-living (Tan 1999), *H. griseus* and *H. aureus* groups can contain two breeding females in habitats with abundant bamboo (Grassi 2002; Wright personal observation). Using the older mating system terminology did not reflect this flexibility and, as stated by Kappeler and van Schaik (2002), these groupings are really patterns of social organization. With that conceptual framework it makes sense that a high frequency of pair-living and small groups among lemurs may be a response to food resource scarcity and unpredictability (van Schaik and Kappeler 1996; Wright 1999), and may be linked to female dominance (Dunham 2008).

4.3.3 Dispersal

Understanding dispersal is critical to understanding population dynamics, but these data are difficult to collect. Long-term data on multiple groups make it possible to observe dispersal in long-lived primates. For *P. edwardsi,* we initially thought that only males emigrated from natal groups (Wright 1995). We have since observed emigration and immigration by both sexes (Morelli 2008), aggressive group takeovers by females (Morelli et al. 2009), and targeted aggression within and between groups (Wright 1995). Females disperse from their natal groups at a younger age than their male counterparts but male secondary dispersal is more common (Morelli et al. 2009). Male dispersal occurs in the 3–5 months before the breeding season. During this time, males visit other groups and male scent marking frequencies and testicular volume increase (Pochron et al. 2005; Pochron and Wright 2005). Testosterone levels also increase (Tecot et al. 2010). Depending on breeding opportunities, males undergo natal dispersal between 3 and 9 years of age. Most adult *P. edwardsi* males transfer at least three times in their lifetimes (Morelli et al. 2009).

In contrast, *E. rufifrons* males in Ranomafana transfer only once in their lifetimes, typically at 3–5 years old and just before the breeding season. Immigrant males are often accepted without aggression and become social partners of one adult female for 3–6 years, mating preferentially with her (Overdorff 1993, 1998; Erhart and Overdorff 1998). Female dispersal has not been seen in *E. rufifrons*, but groups may fission along matrilineal lines (Overdorff et al. 1999; Erhart and Overdorff 2008b), a behavior which differs from *Eulemur* groups in western Madagascar (Wimmer and Kappeler 2002; Ostner and Kappeler 2004).

Observations of *E. rubriventer* revealed that both sexes disperse from their natal group at 2.5–3 years of age (Overdorff and Tecot 2006). Hostile replacement of resident adult females by nongroup females has been observed, but males have not been aggressively replaced. One dispersing female was seen with a new male in an adjacent territory 15 years later (Overdorff and Tecot 2006). Our data on the genetics of *E. rubriventer* offspring revealed that the resident male consistently fathers the offspring, and there are no data that indicate extra-pair copulations (Merenlender 1993).

Molecular and behavioral data show that dispersal in *Varecia* is not sex-biased, as within-sex relatedness scores were similar for males and females (Baden 2011). Mark–recapture studies of brown mouse lemurs (*M. rufus*) have shown that male membership in the population changes at a higher rate than female membership (Atsalis 2000, 2008), and that males can disperse relatively long distances (Karanewski personal communication; Zohdy personal communication). In all species thus studied, dispersal patterns ensure heterozygosity, an advantage for conservation strategies (Merenlender 1993; Morelli et al. 2009, Bradley and Baden personal communication).

4.3.4 Reproductive Hormones

P. edwardsi, E. rubriventer, and *M. rufus* fecal hormone profiles have been developed (Tecot 2008, Tecot et al. 2009; Blanco and Meyer 2009). In *P. edwardsi* and *E. rubriventer,* estradiol, progesterone, testosterone, dihydrotestosterone, and cortisol are being measured to determine the ovarian steroid fecal metabolites that characterize reproduction. Progesterone levels can reliably indicate pregnancy in these species, and estradiol levels reliably indicate fetal sex (Tecot et al. 2009). As expected, patterns of change in fecally excreted steroid levels during the reproductive season in *M. rufus* showed estradiol (E-2) levels were elevated around estrus, whereas progesterone levels were highest during late pregnancy and around parturition (Blanco 2008; Blanco and Meyer 2009). Blanco (2008) documented moderate estrous synchrony among female mouse lemurs, with clusters of females showing strong estrous synchrony. Two females showing signs of abortion or perinatal death of offspring also showed renewed vaginal swelling in late December, suggesting that some form of polyestry (i.e., as reproductive compensation for fetal loss) exists at RNP (Blanco 2008). With these baselines and proof of concept established, we can now investigate questions associated with development, sexual relationships, and seasonal breeding.

Hormonal studies of dominance rarely investigate inter-sexual relationships. To determine whether female dominance might be mediated by hormonal levels, we investigated androgens (dihydrotestosterone and testosterone) in male and female *P. edwardsi* (Tecot et al. 2009). While DHT levels were higher in males than in females, there was no significant sex difference in testosterone levels. Similar testosterone results were found in *M. rufus*. These results differ from those found for all other mammals studied to date, in which male testosterone levels are consistently higher than female levels, with the exception of the female-dominant rock hyrax (Koren et al. 2006). In other masculinized mammals such as the hyena and ring-tailed lemur, androstenedione is elevated in females, but testosterone levels remain higher in males. This finding may have important consequences for understanding sex differences in lifespan and senescence. Maintaining high testosterone levels may explain why mammalian males frequently have shorter life expectancies than females, but testosterone burden may not explain the sex differences in lifespan in *P. edwardsi*. However, dispersal season testosterone levels increase significantly in both sexes, and if males continue to disperse throughout their lives and females do not, testosterone levels may still contribute to shorter male lifespan (Tecot et al. in prep.).

4.3.5 Reproductive Success

Obtaining lifetime reproductive success data from wild primates is possible for females who have been followed throughout their lifespans (Bronikowski et al. 2011).

4 Long-Term Lemur Research at Centre ValBio, Ranomafana National Park, Madagascar 79

By coupling life-long observations with genetic evidence, calculating male reproductive success is now possible. Currently, we have lifetime reproductive success for two females and one male *P. edwardsi*. One female, killed by a fossa at age 32, gave birth to 13 offspring (7 males and 6 females). Five of these animals lived to reproductive age with two males living to emigrate from their natal group (Pochron and Wright 2003; King et al. 2005, 2011). A second female who died at 16 had 7 offspring (3 males and 4 females). Three males and one female survived to reproductive age; the males migrated to breeding groups and the female reproduced in her natal group after her father transferred. Genetic evidence provides a measure of lifetime reproductive success for one male who produced offspring in two groups before he was killed by an immigrant male during a group takeover at age 19. He fathered 14 offspring, 9 females and 5 males, in one group with 4 male and 3 female offspring surviving to reproductive age. Following a second transfer, he sired a 15th offspring, which disappeared, a probable infanticide, after his father was killed. Without continuous long-term data collection we would not be able to have these data on lifetime reproductive success in even these few individuals.

In many primates, heavier females have more surviving offspring (Altmann 1980; Terborgh and Janson 1986). We also see this trend in lemurs. *P. edwardsi* females who were heavier during the previous mating season were significantly more likely to give birth in the following birth season than lighter females (Morelli et al. 2009). Habitat disturbance appears to have a disproportionate impact on the body mass of female *P. edwardsi*. A comparison between Talatakely (logged) and Valohoaka (unlogged) revealed that adult females but not males living within the unlogged forest weighed significantly more than those females living in the disturbed forest (Arrigo-Nelson 2006). When males and females were compared within sites, significant differences in body mass were found only at the disturbed forest site. Given the climatic and reproductive synchrony of the two study sites, and the fact that body mass is positively associated with reproductive success in some primates (Stevenson 2005), these data suggest that differences in *P. edwardsi* feeding behavior and nutrient intake may affect future reproductive success.

4.3.6 Health and Parasites

Although we follow many species daily, we rarely see signs of illness. Over the years, we have seen a wide range of effects of fighting and predation attempts. Wounds are relatively common during the breeding season and we have observed one or both testicles missing in individuals of *E. rufifrons* and *P. edwardsi*. In *P. edwardsi*, *E. rubriventer*, *E. rufifrons*, and *M. rufus* individuals have been found functioning with sight in only one eye (Erhart and Overdorff 2008a). Older individuals have worn teeth and in two individuals of *P. edwardsi* we have seen healed abscessed teeth (King et al. 2005; Wright et al. 2008a). In *E. rufifrons*,

E. rubriventer, and *V. variegata* we have captured very old individuals with teeth worn to the gums.

The diversity and prevalence of parasites has been found to influence health and fitness in other mammals (Hart 2007; Price and Kirkpatrick 2009). The variation among individuals and the transfer of parasites among lemur species is presently unknown, as is the incidence of disease that parasites cause. A variety of roundworms and pinworms have been observed in lemur intestines, but further study is necessary (Junge and Louis 2005; Junge and Sauther 2006; Irwin and Raharison 2009). Our initial studies suggest that one species of parasite may be found on many species of these sympatric lemurs, but there are differences in prevalence among species. For example, *Makialges* spp., a parasitic mite, was abundant on *P. edwardsi* (80%), *P. simus* (67%), and *H. aureus* (83%), yet rare on *E. rufifrons* (3%) and *V. variegata* (14%), and absent on *H. griseus* (Wright et al. 2009; Hogg et al. in press). Large group size has been proposed as a factor for higher parasitism (Freeland 1976), but the largest groups in *E. rufrifrons* had the lowest incidence. Large body size might be more attractive for parasites than smaller body size (Freeland 1976), and indeed larger species had the most ectoparasites.

Additionally, lemurs may have species-specific parasites. For example, *M. rufus* is ecto-parasitized by three tick and one louse species. This louse, *Lemurpediculus verruculosus*, is likely a brown mouse lemur-specific parasite (Durden et al. 2010). A new species of wingless, bloodsucking hippoboscid fly, *Allobosca crassipes,* was recently described as a parasite of *P. edwardsi* and *V. variegata* (Vaughn and McGee 2009).

Parasites may increase in primates living in forests with anthropogenic disturbance and be correlated with disease and decreased fitness (Dobson and May 1986; Chapman et al. 2009). We have some evidence that this trend holds true in the Malagasy rainforest. Wright et al. (2009) found that habitat disturbance may account for high ectoparasite loads in *P. edwardsi*. Endoparasite prevalence in *M. rufus* in 2007 was higher in more disturbed habitat than in the less disturbed habitat. In 2008 and 2009 this difference disappeared, and linking habitat disturbance with lemur parasites should be done with caution. Again, long-term studies allowed us to differentiate between minor fluctuations and the consistent correlations with factors such as climate, body size, group size, or habitat disturbance factors.

4.3.7 Mortality: Adults and Infants, and Infanticide

Adult mortality is generally caused by predation, rather than by illness, wounds, or infections (see Sect. 4.4.1). Adult mortality for the lemurs at Ranomafana is low, as would be expected for long-lived primates (Erhart and Overdorff 2008a; Pochron et al. 2004). In contrast, infant mortality is high (overall approximately 50%) and food stress due to environmental unpredictability may account for some mortality (Wright 1999; Richard et al. 2002).

Long-term observations have also allowed us to document infanticides and infanticide attempts. The killing of infants of up to 2 months of age has been observed in *P. edwardsi*, with both immigrant males and females as perpetrators (Wright 1995; Erhart and Overdorff 1998; Pochron et al. 2004; Morelli et al. 2009). Over 24 years, there have been 9 infanticides out of 60 births (15%) associated with immigration in 4 groups of *P. edwardsi* (Morelli et al. 2009). These infanticides have brought the mothers back into estrus a year earlier in a species that gives birth every other year, providing males with the opportunity to improve their reproductive success (Hrdy 1977; Erhart and Overdorff 1998). A potential infanticide attempt may have also been observed when a *V. variegata* male approached and then entered an unguarded nest, knocking two young infants to the ground over 10 m below; neither infant survived the fall. To date, we have not seen infanticide in either *E. rufifrons* or *E. rubriventer* (Durham 2003; Erhart and Overdorff 2008a; Tecot 2008).

4.3.8 Lifespan

Our long-term research has allowed us to document long lifespans (over 30 years old) in individuals of all the diurnal species of wild rainforest lemurs, regardless of body mass (Erhart and Overdorff 2008a; Baden 2011; King et al. 2011). For *P. edwardsi* we documented a dramatic difference in the maximum lifespans of males and females. Since 1986, few old males have been observed whereas three females are known to have lived beyond 30 years of age. Therefore, males cease contributing genetically to the population after about 20 years, whereas we have no evidence that old females cease reproducing (Wright et al. 2008a; King et al. 2011). In *E. rufifrons*, males over 10 years old were peripheralized and replaced in breeding position by younger nonnatal males (Overdorff et al. 1999). Aged males were burdened with handicaps; one had only one eye, another only partial use of the right hand, another had lost both testicles, and two had visible limps (Overdorff et al. 1999; Erhart and Overdorff 2008a).

In *E. rufifrons* older individuals have been seen, and one functioned with only one eye. These scars and wounds are male-biased and indicate violent male–male aggression, which may account for the shorter lifespan of males as observed in many primate species (Bronikowski et al. 2011). In *E. rubriventer* older individuals have been seen, with one female a minimum of 17 years of age (Tecot 2008), though scars and wounds in this pair-bonded species are generally not evident (Tecot personal observation). In *V. variegata*, both older males and older females were observed in the population at Mangevo, suggesting that male–male aggression may not be as pronounced in this species (Baden 2011). These individual life histories add up, over time, into a better understanding of the evolution of social organization in each species.

With new dental technology that has become available in the past 5 years, Zohdy has documented that wild brown mouse lemurs survive up to 9 years of age and do not experience any of the physical symptoms of senescence that are seen in captive congeners (Bons et al. 2006). On the basis of dental wear, we have found that many brown mouse lemurs survive past 5 years (the age of the onset of senescence in captivity) and these aged individuals represent 9% of those captured. It is possible that few mouse lemurs reach old age because of high predation rates (Goodman et al. 1993; Karpanty 2006; Karpanty and Wright 2007; Sefczek 2009; Deppe 2011).

4.3.9 Nutritional Ecology

In contrast with many sympatric monkey diets, rainforest lemur diets are very diverse (Terborgh 1983; Struhsaker 1997). Studying these species over the long term and in different environments revealed the flexibility in diet within certain constraints. Many lemurs have anatomical specializations such as a large cecum or a long foregut to better digest bamboo or leaves from other plants. For example, *Avahi* and *Lepilemur* both eat leaves, but *Avahi* chooses leaves with tannins, whereas *Lepilemur* chooses leaves with alkaloids (Ganzhorn et al. 1985). The three bamboo lemurs all consume nearly 95% bamboo, but two species can tolerate large amounts of cyanide in the shoots (Glander et al. 1992; Tan 1999; Ballhorn et al. 2009). Unlike any other lemur species, *P. simus* with its strong jaws and big teeth has physical capabilities to open the tough culm of the bamboo and eat the pith (Tan 1999; Vinyard et al. 2008; Yamashita et al. 2009), and *H. griseus* eats primarily bamboo leaf petioles. *Daubentonia* eats beetle larvae extracted from dead wood, a niche taken by woodpeckers in other continents (Cartmill 1974). Recently, a comparative study of the four diurnal frugivores revealed that there is much more specialization in fruit choice than previously thought, with the fruit of entire plant families exploited by only one diurnal lemur species in the forest (Wright et al. 2011).

Our most comprehensive dietary studies have been conducted on *P. edwardsi*. Early work by Hemingway (1998) on the Vatoharanana population revealed that they ate leaves, fruits, and seeds in nearly equal proportions. More recent work by Arrigo-Nelson (2006) has added comparative data on sifaka populations at Talatakely (disturbed forest) and Valohoaka (undisturbed forest). Selective logging has altered species composition in the disturbed forest; in response, sifakas have altered their diet by consuming plant taxa in disproportion to their abundance in the forest and by relying more heavily on food from plant life forms other than trees. Disturbance limits the ability of sifakas to consume fruit and seeds, their preferred food and, as they appear to consume leaves in an effort to replace these missing foods, this creates a discrepancy in the nutrient intake of sifakas living within this habitat (Arrigo-Nelson 2006). As fruit availability was found to be lowest during the most climatologically and reproductively harsh months of the year, we hypothesize that this discrepancy may severely impact infant survival and, with it, the reproductive success of sifakas living in disturbed forest habitats.

Habitat differences in diet are evident in *E. rubriventer* as well. In a 19-month study in Talatakely and Vatoharanana, Tecot (2008) found that during the scarce

4 Long-Term Lemur Research at Centre Valbio, Ranomafana National Park, Madagascar   83

season, dietary overlap decreased and the proportion of the diet composed of fruits, flowers, and leaves differed between the two sites. Seasonal changes in behavior and diet were greater in the undisturbed site, indicating more flexibility in that site. Most notably, during an entire month of the scarce season, animals in the disturbed site spent 100% of their time eating unripe fruit from the invasive Chinese guava (Tecot 2008).

Atsalis (1999, 2008) conducted a 17-month feeding study on *M. rufus* and found that this species fed on a wide variety of fruits, mistletoe berries, and insects, especially beetles. The seeds of *Bakerella* spp., epiphytic semiparasitic mistletoes high in lipids, were present in 42% of fecal samples that contained fruit and this food was consumed year-round irrespective of habitat-wide fruit availability (Atsalis 1999, 2008). This abundant mistletoe is eaten by many lemurs and has been documented to be a fallback food in both disturbed forest areas and forest fragments (Arrigo-Nelson 2006; Irwin 2006).

4.4 Community Ecology

4.4.1 Predators on Lemurs

Predation is a major selective factor in primates and major cause of mortality (Isbell 1994; van Schaik 1983; Zuberbühler 2007). Predators of Ranomafana lemurs include birds, mammalian carnivores, and snakes (Table 4.3). Two raptors,

Table 4.3 Known predation on lemurs in Ranomafana National Park

Lemur species	Carnivore	Boa constrictor	Raptor
Avahi peyrierasi	Not observed	Not observed	*Accipiter++* *Polyboroides ++*
Cheirogaleus crossleyi	*Galidia++*	Yes	*Accipiter ++*
Cheirogaleus sp. nov.	Unk.	Unk.	Unk.
Daubentonia madagascariensis	Not observed	Not observed	Not observed
Eulemur rufifrons	*Cryptoprocta+*	Not observed	*Accipiter++*
Eulemur rubriventer	*Cryptoprocta+*	Not observed	*Accipiter+*
Hapalemur aureus	Not observed	Not observed	Not observed
Hapalemur griseus	Not observed	Not observed	*Accipiter+++* *Polyboroides ++*
Lepilemur microdon	Not observed	Not observed	Not observed
Microcebus rufus	*Galidia++*	Yes	*Accipiter++* *Polyboroides++*
Prolemur simus	Not observed	Not observed	*Accipiter++*
Propithecus edwardsi	*Cryptoprocta+++*	Not observed	Not observed
Varecia variegata editorium	*Cryptoprocta++*	Not observed	*Accipiter+*

Wright et al. (1997), Wright and Martin (1995), Wright (1998); Karpanty and Wright (2006), Erhart and Overdorff (2008a), Baden (2011), Deppe (2011)
+ observed once, ++ 2–14 observations, +++ 15–30 observations

Table 4.4 Changes in *Propithecus edwardsi* population size over time in relation to the observed predation events by *Cryptoprocta ferox*

Time step	Population density	# Predation events	Comments
1990–1995	7 ind/km^2	3	Two adult and three immature members of three groups
1996–2000	5 → 12 ind/km^2	0	No known predation events
2001–2005	12 → 3–6 ind/km^2	5	Two immature, two old-age and one prime-age adults from two groups
2006–2010	6 → 3 ind/km^2	6	Two immature, three old-age and one prime-age adults from four groups; extinction of two study groups in 2007

The oscillation in population size in relation to known predation events indicates that fossa predation is a major cause of population size change for *P. edwardsi*. Data are generated from one study site, Talatakely over 20 years with observations of known groups, direct observations of corpses with indications of fossa predation and long-term census data. *Arrows* within population density column indicate trends in size change within the 5-year interval

Henstii's goshawk (*Accipiter hensteii*) and the Madagascar harrier hawk (*Polyboroides radiatus*) (Karpanty and Wright 2007) eat small-bodied nocturnal (*C. crossleyi, M. rufus,* and *A. peyrierasi*) and diurnal (*E. rufifrons* and *H. griseus*) lemurs. The lemurs preferred by raptors, *A. peyrierasi* and *H. griseus*, weigh approximately 1 kg (Karpanty 2006), suggesting that the body mass, not activity cycle, account for the preference.

A major predator of lemurs is the fossa (*Cryptoprocta ferox*), a viverrid carnivore that weighs between 8 and 12 kg, is agile in the trees, and hunts diurnally and nocturnally. The fossa has made a major impact on the population of the largest lemur in Ranomafana, *P. edwardsi* (Table 4.4; Wright 1995, 1998; Wright et al. 1997; Irwin et al. 2009) as well as on *V. variegata* in recent years (Baden personal communication). Over 20 years, it seems that predation events happen during temporally clumped periods that cut population densities by up to 50%. Fossa predation has caused the extinction of two long-term study groups and independent censuses of the study population support the tremendous impact on population density (Table 4.4; Irwin et al. 2009; see also Kappeler and Fichtel 2011). *Galidea elegans*, a diurnal mongoose weighing less than a kilogram, has been observed eating and stalking *M. rufus* and *C. crossleyi* (Wright and Martin 1995). Finally, boa constrictors (*Sanzinia madagascariensis*) have been mobbed by *Microcebus* and are known to eat *Cheirogaleus* (Wright and Martin 1995; Deppe 2011). Data on all of these rare events can only be accumulated during long-term studies.

4.4.2 Interspecific Aggression

Interference competition by close relatives has been cited as an important selective force in animal behavior (Case and Gilpin 1974; Terborgh 1983), and in a rainforest with 13 sympatric lemurs we would expect this to be the case. Results from observations over 25 years show that one strategy is for each lemur species to specialize on different fruit species (Wright et al. 2011). Yet interspecific aggression among lemurs exists, including agonistic vocalizations, chasing, and biting. These interactions occur during the season of ripe fruit availability (Overdorff and Tecot 2006; Wright personal observation). Since 88% of agonistic interactions among *E. rufifrons, E. rubriventer, P. edwardsi,* and *V. variegata* occurred over ripe fruit during the high season for fruit availability (Overdorff et al. 2005; Overdorff and Tecot 2006), interference competition for high quality fruit is the most likely driver of this agonism. *E. rubriventer* retreated in every case in the resource abundant season, but did not retreat in the few agonistic encounters which occurred with *P. edwardsi* and *E. rufifrons* during the scarce fruit season (Overdorff and Tecot 2006). *V. variegata* will often successfully defend fruit trees against all other lemur species (Balko 1998; Andrea Baden personal observation; Iris de Winter personal communication). The ruffed lemurs are most often the winner of any competition (the raucous barking of a whole group deters other species), even though the sifakas are twice their size. Congeners seem to compete most, and *E. rubriventer* and *E. rufifrons* have 0.04 aggressive interactions/hour, while *V. variegata* and *E. rubriventer* contest 0.006/h and *P. edwardsi* and *E. rubriventer* contest 0.003/h. The hierarchy established by % contests won is *V. variegata* (obligate frugivore), *E. rufifrons, P. edwardsi,* and *E. rubriventer* (Overdorff and Tecot 2006). During 3 months of fruit scarcity in 2010, *E. rufifrons* initiated and won 7 out of 7 contests with *E. rubriventer,* and 6 out of 7 aggressive interactions were adjacent to fruit trees (Iris de Winter personal communication).

Interference competition may also be playing a role in the population dynamics of *Eulemur* species. Across three long-term study sites, population density changes are inverse between *E. rufifrons* and *E. rubriventer;* when *E. rufifrons* population densities increase, *E. rubriventer* density decreases and vice versa (Fig. 4.2). Further evidence that competition for food resources is driving interspecific behavior, not predation, is that polyspecific associations are rare. This is contrary to observations of rainforest monkeys in Africa and South America where several species often feed together for protection against predators (Terborgh 1983; Holenweg et al. 1996).

4.4.3 Seasonal Breeding

Seasonal breeding is one female strategy to limit extra-pair copulation by males (Wright 1999). In the first decade, Ranomafana researchers suggested strict seasonality in lemur breeding (Wright 1999). Long-term results revealed that in *P. edwardsi, P. simus, V. variegata,* and *E. rufifrons* all females in a group and usually all those in the [study] population come into estrus within the same week for a day or two, with older females breeding first. In contrast, Tecot (2010) found that reproduction in *E. rubriventer* is not strictly tied to photoperiod, with births

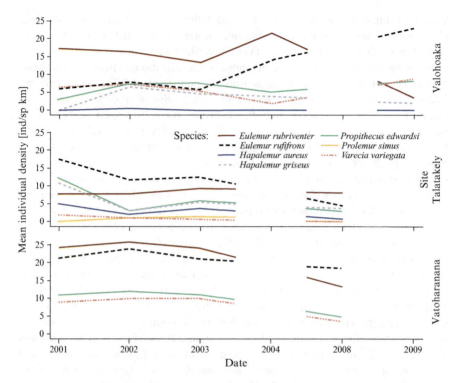

Fig. 4.2 Population densities of diurnal lemurs in Ranomafana National Park from 2001 to 2009 in three different study sites (Talatakely, Vatoharanana, and Valohoaka) within the park

occurring in eight different months. However, only infants born in the "seasonal breeding" window (with food abundance prebreeding) actually survived.

Another selective advantage to infants born at the same time is that synchronous births could be a successful strategy to satiate predators (Boinski 1987). *Varecia variegata* are strict seasonal breeders (Baden 2011) and use a boom or bust strategy (Ratsimbazafy 2006; Baden 2011). At Mangevo, a population of over 80 individuals, *V. variegata* reproduction was only observed once in 6 years of continuous observation. In 2007, 6 out of 7 adult females in the Mangevo population gave birth to twins or triplets (Baden 2011). This synchronous breeding at 6–8 year intervals has been observed in *Pongo pygmaeus* and was associated with fruit masting (Knott 1998). Preliminary analysis of plant phenological data from Ranomafana suggests that fruit availability is also the driver of *V. variegata* synchronous breeding (Wright unpublished data).

M. rufus has a breeding season from mid-October to mid-November, during which females have estrous periods of 5 days or more (Blanco 2008; Blanco and Rahalinarivo 2010), although some variation in the timing of breeding exists. Most females are gestating by mid- to late November (Atsalis 2008), although some are not gestating until December (Blanco 2008). Thus, the onset of estrus may not be completely controlled by photoperiod (Perret and Aujard 2001). There is no evidence for more than one litter a year (Atsalis 2008; Blanco 2008; Blanco and Rahalinarivo 2010).

4.4.4 Habitat Disturbance

To understand the effects of habitat disturbance on lemur demography, we compared life histories of lemurs at sites within the park with high and low levels of human disturbance. The high level disturbance site (Talatakely) was selectively logged for valuable hardwoods between 1986 and 1989, with the intrusion of invasive plant (Chinese guava) and animal (*Rattus rattus*) species (Arrigo-Nelson 2006; Brown and Gurevitch 2004; Laakkonen et al. 2003). The intermediate disturbance site (Vatoharanana) had less than 1% of the study area trees removed during the 1986–1989 period (approximately 1,000 trees; Balko 1998). The low level site (Valohoaka) is considered sacred by local villages and no known timber extraction has occurred (Herrera et al. 2009; Balko 1998). In contrast to the forests north of the park boundaries, there has been very little hunting within Ranomafana forest since the park was established (Lehman et al. 2006; Golden 2009).

The strictest frugivore, *V. variegata*, did not occur in the high disturbance level site (Talatakely) where the big canopied fruit trees were removed (Balko and Underwood 2005). Compared to forests in South America and Africa, Madagascar rainforest tree growth is slow (Struhsaker 1997; Ganzhorn et al. 1999; Clark and Clark 2010), and this delayed regeneration of the forest may impact lemur demography for decades. In the 20 years since selective logging, no one group of *V. variegata* has since arrived in the high disturbance site. Similarly, 10 years postselective logging, the effects on the physical structure of the forest, its species composition, and availability of *P. edwardsi* foods have continued at the highly disturbed site (Arrigo-Nelson 2006). Sifakas in the disturbed forest consumed foods from tree taxa in disproportion to their abundance and relied more on vines and epiphytes than counterparts at the unlogged site. As a consequence, intake of fats and sugars was lower for sifakas at the previously logged site. These differences in food availability and nutrient intake are reflected in significant differences in the body weights of female sifakas between sites and significant male–female differences in body mass in previously logged forest. Sifakas in the disturbed forest spent significantly less time interacting socially and significantly more time feeding and self-grooming than animals in the undisturbed forest. That all of these differences were greatest during lean season only makes their potential impact on the sifaka population of greater consequence. In the long term, these differences may lead to differences in group cohesion, survival, and/or reproductive success (Arrigo-Nelson 2006).

To investigate the impact of habitat disturbance on *E. rubriventer* demography, Tecot (2008, in press) collected data simultaneously on the red-bellied lemur populations in heavily logged and minimally logged sites. Results again indicate that logging has reduced the structure, species composition, abundance, and predictability of red-bellied lemur foods within the disturbed forest site. Additionally, red-bellied lemurs at the disturbed site were less active (Tecot 2008), bred out of peak season, and had higher infant mortality (Tecot 2010, in press).

High levels of cortisol have been implicated in reduced fitness (Bonier et al. 2009), and the effects of stress on lemur demography is being investigated (Tecot 2008). In a comparative study of stress hormones in adult *E. rubriventer* in selectively logged versus minimally logged sites, patterns of cortisol excretion were similar in both sites, but those in the undisturbed site showed little response to variation in food availability and rainfall. In contrast, at the disturbed site, fecal cortisol levels were significantly higher when fruit was scarce (parturition and early lactation) compared with when fruit was abundant (prebreeding season). Contrary to the Cortisol-Fitness Hypothesis (Bonier et al. 2009), cortisol levels were higher in the undisturbed site. Lower cortisol levels, minimal changes in hormones and behavior (Tecot 2008), and higher infant mortality (Tecot 2010) in the disturbed site indicate that there may be down-regulation of the cortisol stress response where environmental stress is prolonged (Tecot 2008, in press).

4.4.5 Trends in Population Densities of Lemurs

Over the past 20 years, we have documented population size changes across three sites with different histories of anthropogenic disturbance (Table 4.5; Fig. 4.2). These data allow us to determine changes, rather than trends, in population size over time. Our results indicate that changes in population size vary across sites, as

Table 4.5 Estimated population size of lemurs in Ranomafana National Park

Lemur species	Estimated density (ind/km^2) $\pm$ SE	Estimated population size (ind/330 km^2 of RNP forest)	Estimated biomass (kg/330 km^2)
Avahi peyrierasi	9.65 $\pm$ 1.92	3,185 $\pm$ 633	3,058 $\pm$ 607
Cheirogaleus crossleyi	13.48 $\pm$ 3.91	4,448 $\pm$ 1,291	1,557 $\pm$ 452
Daubentonia madagascariensis	rare	200 $\pm$ 50	600 $\pm$ 150
Eulemur rufifrons	6.75 $\pm$ 1.63	2,228 $\pm$ 537	4,902 $\pm$ 1,181
Eulemur rubriventer	5.46 $\pm$ 0.70	1,802 $\pm$ 231	4,325 $\pm$ 554
Hapalemur aureus	0.21 $\pm$ 0.14	69 $\pm$ 47	124 $\pm$ 85
Hapalemur griseus	2.48 $\pm$ 0.48	818 $\pm$ 159	614 $\pm$ 119
Prolemur simus	0.85–1.23 at one site in 2002–2003	20 $\pm$ 5	56 $\pm$ 15
Lepilemur microdon	0.99 $\pm$ 0.37	327 $\pm$ 128	327 $\pm$ 128
Microcebus rufus	23.52 $\pm$ 4.07	7,762 $\pm$ 1,343	233 $\pm$ 40
Propithecus edwardsi	4.73 $\pm$ 0.76	1,561 $\pm$ 251	9,366 $\pm$ 1,506
Varecia variegata	2.23 $\pm$ 0.81	736 $\pm$ 267	2,429 $\pm$ 1,082

These data are based on transect surveys from 2004 to 2009 (S.E. Johnson, F. Ratelolahy, P.C. Wright, J.P. Herrera). Species in bold are critically endangered (IUCN Redlist)

would be expected of meta-populations in a varying landscape (Hanski and Gilpin 1991; Table 4.5). Oscillating population densities of *E. rufifrons* and *E. rubriventer* suggest that these changes in population density may reflect some degree of congeneric competition (Overdorff et al. 1999; Erhart and Overdorff 2008a). In comparison, in Vatoharanana, the density of *E. rubriventer* and *V. variegata* increased over the study years from 15 to 25 individuals per km^2 and from 2 to 10 individuals per km^2, respectively, while the density of *P. edwardsi* remained at 11 individuals per km^2. Similar trends have been observed in the critically endangered *V. variegata*. In the low disturbance site, their population density has oscillated but increased slightly overall. Population density in the intermediate disturbance site increased gradually between 1990 and 2003, but seems to have declined again by 2008. Overall population densities are low (2–10 ind/km^2). In *P. edwardsi* and *V. variegata*, we have seen an increase in population densities at Valohoaka, but population densities appear to be declining at the intermediate and high disturbance sites, which can perhaps be partly attributed to predation events or fruit scarcity (Arrigo-Nelson 2006; Irwin et al. 2009). Long-term data show that the species with declining populations in normal circumstances have larger home ranges and larger group sizes (Morriss et al. 2009). Theoretically this suggests that food constraints are more important than predation, as larger group size provides more eyes and ears for detecting predators (Hamilton 1964; van Schaik 1983; van Schaik et al. 1983).

4.5 Conservation

4.5.1 Successes and Ongoing Problems

RNP has been designated as a conservation priority (Kremen et al. 2008) and in 2007 was declared a World Heritage Site. We have successfully integrated education, health, and economic assistance programs with biodiversity research and habitat protection goals (Wright 1997; Wright and Andriamihaja 2002; Lovejoy 2006). An evaluation of the educational impact has shown that local people have experienced a change in attitude (Korhonen 2006). Attempts to correlate human impact on lemur populations have shown that human immigration into the park's peripheral zone is correlated with increased deforestation rates (Brooks et al. 2009). Villages which are the farthest from the road tend to encroach further into the park than do the on-road villages (DeFries et al. 2010). Moreover, the park itself has maintained edges with minimum invasion by exotic plants (Brown et al. 2009). Eco-tourism to visit the park has been a great boon to the local economy. However, the 30,000 tourists do have negative consequences on habitat and breeding birds (Razafimahaimodison 2003). Better management of tourism is in process. Satellite photos suggest that there is minimum forest destruction since 1991, when the park was gazetted, but the recent political instability is worrisome.

4.5.2 Implications of Climate Change

Long-term data enable better understanding of the effects of climate change on rainforest ecology and lemur populations. Indeed, lemur observers with long-term research projects were among the first to gain evidence of the effect of climate fluctuations on mammal populations (Gould et al. 1999; Wright 1999, 2006). Using Madagascar climate data and the Centre ValBio long-term rainfall and temperature database, we discovered that dry seasons have become longer, and cyclones more frequent (Wright 2006). In particular, the November temperatures in 2007, 2008, and 2009 were over 30°C, much higher than ever previously recorded, and the gap between minimum and maximum temperatures has increased. But does this have an effect on lemur demographics?

Long-term demographic data show that older *P. edwardsi* mothers lose infants in years with extended hot, dry seasons (King et al. 2005). Further analysis shows that the average fecundity of lemurs was over 65% lower in El Niño years (Dunham et al. 2008). As El Niño years become more erratic and frequent (Fedorov and Philander 2000) this could lead to more extreme weather and increased impacts on biodiversity. The southern oscillation (ENSO) related to El Niño is known to cause drought (Thomson et al. 2003) and changes in vegetation indices (Ingram and Dawson 2005). Dunham et al. (2010) found that cyclones, ENSO phases, and rainfall levels affected the reproductive rates of *P. edwardsi*. Overall fecundity (defined as the number of offspring per female per year surviving to 1 year of age) was negatively associated with cyclone presence during gestation, and positively associated with colder ENSO phases during the second 6 months of life and during the period faced by mothers preceding conception. Wet season rainfall and intensity during gestation was negatively related to birth rates, and the number of drought months during lactation was negatively associated with first year survival. Finally, fluctuations in lemur stress hormones show an elevation of cortisol during the dry season (Tecot 2008; Tecot in press), which may ultimately impact reproduction. Longer dry seasons in consequent years could impact lemur populations negatively (Wright 2006; Dunham et al. 2008, 2010). The effects of ENSO events on population dynamics has also been seen in many species of New World monkeys (Wiederholt and Post 2010).

4.6 Summary and Conclusions

Most of the lemur species in Ranomafana were data deficient before we initiated our first studies in 1986. The virtue and vice of long-term research is that it is never complete. Here we have compiled a list of essential findings, the product of long-term research that has many times resulted in the reevaluation of earlier findings. Thus, over the years we established the number of species residing in RNP through the rediscovery of *P. simus* and the discovery of *H. aureus* and a potentially new species awaiting description: a high altitude species of *Cheirogaleus*. We have established demographic changes through time including life history events, mortality, lifespan, and dispersal patterns for *P. edwardsi*, *E. rufifrons*, *E. rubriventer*, *V. variegata*, *H. aureus*, *H. griseus*, *P. simus*, and to a lesser extent *M. rufus*. We have documented variability of social organization in each species, and we have described how populations recover after cyclones and droughts.

Furthermore, we have confirmed that lemur population densities vary over time, and that predation by raptors and mammalian carnivores can have a major impact on local lemur populations. We know that many lemur species are important to seed dispersal and thus to forest dynamics. Monitoring and measuring the long-term effects of habitat disturbance on lemur populations, we have evidence that selective logging may negatively impact population densities of *E. rufifrons*, *V. variegata*, and *P. edwardsi*, even a decade after the last logging disturbance. We have also determined that fertility of *P. edwardsi* females decreased during ENSO years and infants of older *P. edwardsi* females died in years with extended dry months.

Acknowledgments This research complied with all the laws of Madagascar and was authorized by Madagascar National Parks (MNP), CAFF/CORE, and the Madagascar Ministry of the Environment. The research was approved by the IACUC animal care committees of all research host institutions. The authors thank Benjamin Andriamihaja and the MICET and ICTE staffs who have expedited the research process and made our long-term work in Madagascar possible. For funding we acknowledge the Wenner-Gren Foundation (PCW), National Geographic Society (PCW), National Science Foundation USA (PCW BCS-0721233, PCW DBI-0829761, PCW DBI-0122357, SAN BCS-0333078, SRT BCS-0424234, EME SBS-0001351, ALB BCS-0725975), Earthwatch Institute (PCW and SAN), Primate Conservation Inc (PCW, SAN, ALB, MBB, AMD, RJ, TLM, SRT), The LSB Leakey Foundation (ALB), Primate Action Fund (PCW, SAN, SRT, ALB), The University of Texas at Austin (SRT), and SUNY-Stony Brook (PCW, SAN, SRT, SJK, ALB), Fulbright Foundation (SAN and ALB), American Society of Primatologists (SRT), PEO Foundation (SRT), AAUW (SRT), Saint Louis Zoo's FRC Committee (SAN), The University of Helsinki (PCW, SZ). We would also like to acknowledge our many research technicians: Georges Rakotonirina, Raymond Ratsimbazafy, and Remi Rakotovao, Georges René Randrianirina, Paul Rakotonirina, Georges Razafindrakoto, Dominique Razafindraibe, Armand Razafitsiafazato, Bernadette (Menja) Rabaovola, Laurent Randrianasolo, Rasendrinirina Victor, Rakotonirina A. Thierry Emile (Nirina), and Rakotoniaina Jean Félix (Rakoto), Razafindravelo Zafi, Solo Justin, and Telo Albert for their expert assistance following animals and collecting phenological data. Graduate students from the University of Antananarivo: Sahoby Ivy Randriamahaleo, Rakotonirina Laingonianina, Herifito Fidèle, Ralainasolo Fidy, and Andriamaharoa Hubert. P.C. Wright is grateful to Ingrid Daubechies for assistance with the figures and Patricia Paladines for her efforts and editorial assistance.

References

Altmann J (1980) Baboon mothers and infants. Harvard University Press, Cambridge, MA

Andriantompohavana R, Lei R, Zaonarivelo JR, Engberg SE, Nalanirina G, McGuire SM, Shore GD, Andrianasolo J, Herrington K, Brenneman RA, Louis EE Jr (2007a) Molecular phylogeny and taxonomic revision of the woolly lemurs, Genus *Avahi* (Primates: Lemuriformes). Spec Publ Mus Texas Tech Univ 51:1–59

Andriantompohavana R, Morelli TL, Behncke SM, Engberg SE, Brennneman RA, Louis EE Jr (2007b) Characterization of 20 microsatellite marker loci in the red-bellied brown lemur (*Eulemur rubriventer*). Mol Ecol Notes 7:1162–1165

Arrigo-Nelson SJ (2006) The impact of habitat disturbance on the feeding ecology of the Milne-Edwards' sifaka (*Propithecus edwardsi*) in Ranomafana National Park, Madagascar. PhD thesis, State University of New York, Stony Brook

Arrigo-Nelson SJ, Wright PC (2004) Survey results from Ranomafana National Park: new evidence for the effects of habitat preference and disturbance on the distribution of *Hapalemur*. Folia Primatol 75:331–334

Atsalis S (1999) Diet of the brown mouse lemur (*Microcebus rufus*) in Ranomafana National Park, Madagascar. Int J Primatol 20:193–229

Atsalis S (1999a) Feeding ecology and aspects of life history in *Microcebus rufus* (Family *Cheirogaleidae*, Order Primates). PhD thesis, City University of New York, New York

Atsalis S (1999b) Seasonal fluctuations in body fat and activity levels in a rainforest species of mouse lemur, *Microcebus rufus*. IJP 20(6):883–910

Atsalis S (2000) Spatial distribution and population composition of the brown mouse lemur (*Microcebus rufus*) in Ranomafana National Park, Madagascar, and its implications for social organization. Am J Primatol 51:61–78

Atsalis S (2008) A natural history of the brown mouse lemur. Pearson Prentice Hall Publishers, New Jersey

Atsalis S, Schmid J, Kappeler PM (1996) Metrical comparisons of three mouse lemur. J Hum Evol 31:61–68

Baden AL (2011) Communal breeding in ruffed lemurs. PhD thesis, Stony Brook University, Stony Brook

Baden AL, Brenneman RA, Louis EE Jr (2008) Morphometrics of wild black-and-white ruffed lemurs [*Varecia variegata*; Kerr, 1972]. Am J Primatol 70:913–926

Bailey CA, Lei R, Brenneman RA, Louis EE Jr (2009) Characterization of 21 microsatellite marker loci in the Milne-Edwards' sifaka (*Propithecus edwardsi*). Conserv Genet 10:1389–1392

Balko EA (1998) A behaviorally plastic response to forest composition and logging disturbance by *Varecia variegata* in Ranomafana National Park, Madagascar. PhD thesis, State University of New York, Syracuse

Balko EA, Underwood HB (2005) Effects of forest structure and composition on food availability for *Varecia variegata* at Ranomafana National Park, Madagascar. Am J Primatol 66:45–70

Ballhorn DJ, Kautz S, Rakotoarivelo FP (2009) Quantitative variability of cyanogenesis in *Cathariostachys madagascarensis* – the main food plant of bamboo lemurs in southeastern Madagascar. Am J Primatol 71:305–315

Bergey C, Patel ER (2008) A preliminary vocal repertoire of the greater bamboo lemur (*Prolemur simus*): classification and contexts. Nexus 1:69–84

Blanco MB (2008) Reproductive schedules of female *Microcebus rufus* at Ranomafana National Park, Madagascar. Int J Primatol 29:323–338

Blanco MB, Meyer JS (2009) Assessing reproductive profiles in female brown mouse lemurs (*Microcebus rufus*) from Ranomafana National Park, southeast Madagascar, using fecal hormone analysis. Am J Primatol 71:439–446

Blanco MB, Rahalinarivo V (2010) First direct evidence of hibernation in an eastern dwarf lemur species (*Cheirogaleus crossleyi*) from the high-altitude forest of Tsinjoarivo, central-eastern Madagascar. Naturwissenschaften 97:945–950

Blanco MB (2011) Timely estrus in wild brown lemur females at Ranomafana National Park, southeastern Madagascar. Am J Phys Anthropol 145:311–317

Blanco MB, Godfrey LR, Rakotondratsima M, Rahalinarivo V, Samonds KE, Raharison J-L, Irwin MT (2009) Discovery of sympatric dwarf lemur species in the high-altitude rain forest of Tsinjoarivo, eastern Madagascar: implications for biogeography and conservation. Folia Primatol 80:1–17

4 Long-Term Lemur Research at Centre Valbio, Ranomafana National Park, Madagascar 93

Boinski S (1987) Birth synchrony in squirrel monkeys (*Saimiri oerstedi*): a strategy to reduce neonatal predation. Behav Ecol Sociobiol 21:393–400

Bonier F, Martin PR, Moore IT, Wingfield JC (2009) Do baseline glucocorticoids predict fitness? Trends Ecol Evol 24:634–642

Bons N, Reiger F, Prudhomme D, Fisher A, Krause K-H (2006) *Microcebus murinus*: a useful primate model for human cerebral aging and Alzheimer's disease? Genes Brain Behav 5:120–130

Bronikowski AM, Altmann J, Brockman DK, Cords M, Fedigan LM, Pusey AE, Stoinski T, Morris WF, Strier KB, Alberts SC (2011) Aging in the natural world: comparative data reveal similar mortality patterns across primates. Science 331:1325–1328

Brooks CP, Holmes C, Kramer K, Barnett B, Keitt TH (2009) The role of demography and markets in determining deforestation rates near Ranomafana National Park Madagascar. PLoS One 4: e5783. doi:10.1371/journal.pone.0005783

Brown KA, Gurevitch J (2004) Long-term impacts of logging on forest diversity in Madagascar. Proc Natl Acad Sci USA 101:6045–6049

Brown KA, Ingram JC, Flynn DFB, Razafindrazaka R, Jeannoda V (2009) Protected area safeguard tree and shrub communities from degradation and invasion: a case study in eastern Madagascar. Environ Manage 44:136–148

Cartmill M (1974) *Daubentonia, Dactylopsila*, woodpeckers, and klinorhynchy. In: Martin RD, Doyle GA, Walker AC (eds) Prosimian biology. Duckworth, London, pp 655–670

Case TJ, Gilpin ME (1974) Interference competition and niche theory. Proc Natl Acad Sci USA 71:3073–3077

Chapman CA, Speirs ML, Hodder SAM, Rothman JM (2009) Colobus monkey parasite infections in wet and dry habitats: implications for climate change. Afr J Ecol 48:555–558

Clark DB, Clark DA (2006) Tree growth, mortality, physical condition, and microsite in an old-growth lowland tropical rain forest. Ecology 87:2132

Clark DA, Clark DB (2010) Assessing tropical forests' climatic sensitivities with long-term data. Biotropica 43:31–40

Clark DA, Piper SC, Keeling CD, Clark DB (2003) Tropical rain forest tree growth and atmospheric carbon dynamics linked to interannual temperature variation during 1984–2000. Proc Natl Acad Sci USA 100:5852–5857

Clutton-Brock TH, Harvey PH (1977) Primate ecology and social organization. J Zool Lond 183:1–39

Crook JH, Gartlan JS (1966) Evolution of primate societies. Nature 210:1200–1203

DeFries R, Rovero F, Wright PC, Ahumada J, Andelman S, Brandon K, Dempewolf J, Hansen A, Hewson J, Liu J (2010) From plot to landscape scale: linking tropical biodiversity measurements across spatial scales. Front Ecol Environ 8:153–160

Deppe AM (2011) Predator recognition in the brown mouse lemur (*Microcebus rufus*): experiments in Ranomafana National Park, Madagascar. PhD thesis, Stony Brook University, Stony Brook

Deppe AM, Wright PC (2009) Predator recognition in wild brown mouse lemurs (*Microcebus rufus*): field experiments in Ranomafana National Park Madagascar. Am J Primatol 71:67

Dew JL, Wright PC (1998) Frugivory and seed dispersal by four species of primates in Madagascar's eastern rain forest. Biotropica 30:425–437

Dobson AP, May RM (1986) Disease and conservation. In: Soulé ME (ed) Conservation biology: the science of scarcity and diversity. Sinauer Associates, Sunderland, MA, pp 345–365

Dunham AE (2008) Battle of the sexes: cost asymmetry explains female dominance in lemurs. Anim Behav 76:1435–1439

Dunham AE, Erhart EM, Overdorff DJ, Wright PC (2008) Evaluating effects of deforestation, hunting, and El Niño events on a threatened lemur. Biol Conserv 141:287–297

Dunham AE, Erhart EM, Wright PC (2010) Global climate cycles and cyclones: consequences for rainfall patterns and lemur reproduction in southeastern Madagascar. Global Change Biol 17:219–227

Durden LA, Zohdy S, Laakkonen J (2010) Lice and ticks of the eastern rufous mouse lemur, *Microcebus rufus*, with descriptions of the male and third instar nymph of *Lemurpediculus verruculosus* (Phthiraptera: Anoplura). J Parasitol 96:874–878

Durham DL (2003) Variation in responses to forest disturbance and the risk of local extinction: a comparative study of wild *Eulemurs* at Ranomafana National Park, Madagascar. PhD thesis, University of California, Davis

Eisenberg JF, Muckenhirn NA, Rudran R (1972) The relation between ecology and social structure in primates. Science 176:863–874

Erhart EM, Overdorff DJ (1998) Infanticide in *Propithecus diadema edwardsi*: an evaluation of the sexual selection hypothesis. Int J Primatol 19:73–81

Erhart EM, Overdorff DJ (2008a) Population demography and social structure changes in *Eulemur fulvus rufus* from 1988 to 2003. Am J Phys Anthropol 136:183–193

Erhart EM, Overdorff DJ (2008b) Rates of agonism by diurnal lemuroids: implications for female social relationships. Int J Primatol 29:1227–1247

Fedorov AV, Philander SG (2000) Is El Niño changing? Science 288:1997–2002

Feistner ATC, Sterling EJ (1995) Body mass and sexual dimorphism in the aye-aye *Daubentonia madagascariensis*. Dodo 31:73–76

Freeland WJ (1976) Pathogens and the evolution of primate sociality. Biotropica 8:12–24

Ganzhorn JU, Kappeler PM (1996) Lemurs of the Kirindy Forest. Primate Report 46–1:257–274

Ganzhorn JU, Abraham JP, Razanahoera-Rakotomalala M (1985) Some aspects of the natural history and food selection of *Avahi laniger*. Primates 26:452–463

Ganzhorn JU, Fietz J, Rakotovao E, Schwab D, Zinner DP (1999) Lemurs and the regeneration of dry deciduous forest in Madagascar. Conserv Biol 13:794–804

Ganzhorn JU, Lowry PP, Schatz GE, Sommer S (2001) The biodiversity of Madagascar: one of the world's hottest hotspots on its way out. Oryx 35:346–348

Ganzhorn JU, Arrigo-Nelson SJ, Boinski S, Bollen A, Carrai V, Derby A, Donati G, Koenig A, Kowalewski M, Lahann P, Norscia I, Polowinsky SY, Schwitzer C, Stevenson PR, Talebi MG, Tan CL, Vogel ER, Wright PC (2009) Possible fruit protein effects on primate communities in Madagascar and the Neotropics. PLoS One 4:e8253. doi:10.1371/journal.pone.0008253

Gerber B, Karpanty SM, Crawford C, Kotschwar M, Randrianantenaina J (2010) An assessment of carnivore relative abundance and density in the eastern rainforests of Madagascar using remotely-triggered camera traps. Oryx 44:219–222

Glander KE, Wright PC, Daniels PS, Merenlender AM (1992) Morphometrics and testicle size of rain forest lemur species from southeastern Madagascar. J Hum Evol 22:1–17

Golden CD (2009) Bushmeat hunting and use in the Makira Forest, north-eastern Madagascar: a conservation and livelihoods issue. Oryx 43:386–392

Goodman SM, O'Connor S, Langrand O (1993) A review of predation on lemurs: implications for the evolution of social behavior in small, nocturnal primates. In: Kappeler PM, Ganzhorn JU (eds) Lemur social systems and their ecological basis. Plenum, New York, pp 51–66

Gould L, Sussman RW, Sauther ML (1999) Natural disasters and primate populations: the effects of a 2-year drought on a naturally occuring population of ring-tailed lemurs (Lemur catta) in southwestern Madagascar. Int J Primatol 20:69–84

Grassi C (2002) Sex differences in feeding, height, and space use in *Hapalemur griseus*. Int J Primatol 23:677–693

Groeneveld LF, Rasoloarison RM, Kappeler PM (2011) Morphometrics confirm taxonomic deflation in dwarf lemurs (Primates: *Cheirogaleidae*), as suggested by genetics. Zool J Linn Soc Lond 161:229–244

Groves CP (2001) Primate taxonomy. Smithsonian Institution Press, Washington, DC

Hamilton WD (1964) The genetical evolution of social behaviour, I and II. J Theor Biol 7:1–52

Hanski I, Gilpin M (1991) Metapopulation dynamics: brief history and conceptual domain. Biol J Linn Soc 42:3–16

Harcourt CS (1987) Ecology and behaviour of *Avahi laniger*. Int J Primatol 8:501

Harcourt CS (1988) *Avahi laniger*: a study in inactivity. Primate Eye 35:9

4 Long-Term Lemur Research at Centre Valbio, Ranomafana National Park, Madagascar 95

Hart D (2007) Predation on primates: a biogeographical analysis. In: Gursky SL, Nekaris KAI (eds) Primate anti-predator strategies. Springer, Chicago, pp 27–59

Hemingway CA (1995) Feeding and reproductive strategies of the Milne-Edwards' sifaka, *Propithecus diadema edwardsi*. PhD thesis, Duke University, Durham

Hemingway CA (1998) Selectivity and variability in the diet of Milne-Edwards' sifakas (*Propithecus diadema edwardsi*): implications for folivory and seed-eating. Int J Primatol 19:355–377

Herrera J, Lauterbur E, Wright PC, Ratovonjanahary L, Taylor LL (2009) Rapid assessment of lemurs in disturbed and undisturbed habitats in southeastern Madagascar 71: 90 DOI 10, 10021 Am J Primatol. 20733

Herrera JP, Wright PC, Lauterbur E, Ratovonjanahary L, Taylor LL (in press) The effects of habitat disturbance on lemurs at Ranomafana National Park, Madagascar. Int J Primatol DOI 10.1007/s10764-011-9525-8

Hogg KL, Wade S, Wright PC (in press) Parasites of nine lemur species from Ranomafana National Park, Madagascar. J Zoo Wildlife Med

Holenweg A-K, Noë R, Schabel M (1996) Waser's gas model applied to associations between red colobus and Diana monkeys in the Taï National Park, Ivory Coast. Folia Primatol 67:125–136

Hrdy SB (1977) Infanticide as a primate reproductive strategy. Am Sci 65:40–49

Ingram JC, Dawson TP (2005) Climate change impacts and vegetation response on the island of Madagascar. Phil Trans R Soc Lond A 363:55–59

Irwin MT (2006) Ecological impacts of forest fragmentation on diademed sifakas (*Propithecus diadema*) at Tsinjoarivo, eastern Madagascar: implications for conservation in fragmented landscapes. PhD thesis, Stony Brook University, Stony Brook

Irwin MT (2007) Living in forest fragments reduces group cohesion in diademed sifakas (*Propithecus diadema*) in eastern Madagascar by reducing food patch size. Am J Primatol 69:434–447

Irwin MT (2008) Feeding ecology of *Propithecus diadema* in forest fragments and continuous forest. Int J Primatol 29:95–115

Irwin MT, Raharison J-L (2009) A review of the endoparasites of the lemurs of Madagascar Malagasy. Nature 2:66–93

Irwin MT, Smith TM, Wright PC (2000) Census of three eastern rainforest sites north of Ranomafana National Park: preliminary results and implication for lemur conservation. Lemur News 5:20–22

Irwin MT, Johnson SE, Wright PC (2005) The state of lemur conservation in south-eastern Madagascar: population and habitat assessments for diurnal and cathemeral lemurs using surveys, satellite imagery and GIS. Oryx 39:204–218

Irwin MT, Raharison J-L, Wright PC (2009) Spatial and temporal variability in predation on rainforest primates: do forest fragmentation and predation act synergistically? Anim Conserv 12:220–230

Isbell LA (1994) Predation on primates: ecological patterns and evolutionary consequences. Evol Anthropol 3:61–71

IUCN (2010) IUCN red list of threatened species. Species Survival Commission

Janson CH (1992) Evolutionary ecology of primate social structure. In: Smith EA, Winterhalder B (eds) Evolutionary ecology and human behavior. Aldine de Gruyter, New York, pp 95–130

Jernvall J, Wright PC (1998) Diversity components of impending primate extinctions. Proc Natl Acad Sci USA 95:11279–11283

Johnson SE (2002) Ecology and speciation in brown lemurs: white-collared lemurs (*Eulemur albocollaris*) and hybrids (*Eulemur albocollaris* x *Eulemur fulvus rufus*) in southeastern Madagascar. PhD thesis, University of Texas, Austin

Johnson SE, Overdorff DJ (1999) Census of brown lemurs (*Eulemur fulvus* sspp.) in southeastern Madagascar: methods-testing and conservation implications. Am J Primatol 47:51–60

Johnson SE, Gordon AD, Stumpf RM, Overdorff DJ, Wright PC (2005) Morphological variation in populations of *Eulemur albocollaris* and *E. fulvus rufus*. Int J Primatol 26:1399–1416

Jolly A (1966) Lemur behavior: a Madagascar field study. University of Chicago Press, Chicago

Jolly A (2012) Berenty reserve, Madagascar: a long time in a small space. In: Kappeler PM (ed) Long-term field studies of primates. Springer, Heidelberg

Jolly A, Sussman RW (2006) Notes on the history of ecological studies of Malagasy lemurs. In: Gould L, Sauther ML (eds) Lemurs: ecology and adaptation. Springer, New York, pp 19–39

Jolly A, Koyama N, Rasamimanana HR, Crowley H, Williams GW (2006) Berenty Reserve: a research site in southern Madagascar. In: Jolly A, Sussman RW, Koyama N, Rasamimanana HR (eds) Ringtailed lemur biology: Lemur catta in Madagascar. Springer, New York, pp 32–42

Junge RE, Louis EE Jr (2005) Preliminary biomedical evaluation of wild ruffed lemurs (*Varecia variegata and V. rubra*). Am J Primatol 66:85–94

Junge RE, Sauther ML (2006) Overview on the health and disease ecology of wild lemurs: conservation implications. In: Gould L, Sauther ML (eds) Lemurs: ecology and adaptation. Springer, New York, pp 423–440

Kappeler PM, Fichtel C (2012) A 15-year perspective on the social organization and life history of Sifaka in Kirindy Forest. In: Kappeler PM (ed) Long-term field studies of primates. Springer, Heidelberg

Kappeler PM, van Schaik CP (2002) Evolution of primate social systems. Int J Primatol 23:707–740

Karpanty SM (2006) Direct and indirect impacts of raptor predation on lemurs in southeastern Madagascar. Int J Primatol 27:239–261

Karpanty SM, Wright PC (2007) Predation on lemurs in the rainforest of Madagascar by multiple predator species: observations and experiments. In: Gursky SL, Nekaris KAI (eds) Primate anti-predator strategies. Springer, New York, pp 77–99

King SJ, Arrigo-Nelson SJ, Pochron ST, Semprebon GM, Godfrey LR, Wright PC, Jernvall J (2005) Dental senescence in a long-lived primate links infant survival to rainfall. Proc Natl Acad Sci USA 102:16579–16583

King SJ, Morelli TL, Arrigo-Nelson SJ, Ratelolahy FJ, Godfrey LR, Wyatt J, Tecot S, Jernvall J, Wright PC (2011) Morphometrics and pattern of growth in wild sifakas (*Propithecus edwardsi*) at Ranomafana National Park, Madagascar. Am J Primatol 73:155–172

Knott CD (1998) Changes in orangutan caloric intake, energy balance, and ketones in response to fluctuating fruit availability. Int J Primatol 19:1061–1079

Koren L, Mokady O, Geffen E (2006) Elevated testosterone levels and social ranks in the female rock hyrax. Horm Behav 49:470–477

Korhonen K (2006) The rocky road of social sustainability: the impact of integrated biodiversity conservation and development on the local realities in Ranomafana National Park, Madagascar. PhD thesis, University of Helsinki, Finland

Kremen C, Cameron A, Moilanen A, Phillips SJ, Thomas CD, Beentje H, Dransfield J, Fisher BL, Glaw F, Good TC, Harper GJ, Hijmans RJ, Lees DC, Louis EE, Nussbaum RA, Raxworthy CJ, Razafimpahanana A, Schatz GE, Vences M, Vieites DR, Wright PC, Zjhra ML (2008) Aligning conservation priorities across taxa in Madagascar with high-resolution planning tools. Science 320:222–226

Laakkonen JT, Lehtonen JT, Ramiarinjanahary H, Wright PC (2003) Trypanosome parasites in the invading *Rattus rattus* and endemic rodents in Madagascar. In: Singleton GR, Hinds LA, Krebs CJ, Spratt DM (eds) Rats, mice and people: rodent biology and management. Australian Centre for International Agricultural Research, Canberra, pp 37–39

Lehman SM, Ratsimbazafy J, Rajaonson A, Day S (2006) Decline of *Propithecus diadema edwardsi* and *Varecia variegata variegata* (Primates: Lemuridae) in south-east Madagascar. Oryx 40:108–111

Louis EE Jr, Engberg SE, Lei R, Geng H, Sommer JA, Randriamampionona R, Randriamanana JC, Zaonarivelo JR, Andriantompohavana R, Randria G, Prosper RB, Rakotoarisoa G, Rooney A, Brenneman RA (2006) Molecular and morphological analyses of the sportive lemurs (Family Megaladapidae: Genus *Lepilemur*) reveals 11 previously unrecognized species. Spec Publ Mus Texas Tech Univ 49:1–47

4 Long-Term Lemur Research at Centre Valbio, Ranomafana National Park, Madagascar 97

Lovejoy TE (2006) Protected areas: a prism for a changing world. Trends Ecol Evol 21:329–333

Mayor MI, Sommer JA, Houck ML, Zaonarivelo JR, Wright PC, Ingram C, Engel SR, Louis EE Jr (2004) Specific status of *Propithecus* spp. Int J Primatol 25:875–900

Meier B, Albignac R, Peyriéas A, Rumpler Y, Wright PC (1987) A new species of *Hapalemur* (Primates) from southeast Madagascar. Folia Primatol 48:211–215

Merenlender AM (1993) The effects of sociality on the demography and genetic structure of *Eulemur fulvus rufus* (polygamous) and *Eulemur rubriventer* (monogamous) and the conservation implications. PhD thesis, University of Rochester, Rochester

Meyers DM, Rabarivola C, Rumpler Y (1989) Distribution and conservation of Sclater's lemur: implications of a morphological cline. Prim Conserv 10:77–81

Mittermeier RA, Louis EE Jr, Richardson M, Schwitzer C, Langrand O, Rylands AB, Hawkins F, Rajaobelina S, Ratsimbazafy J, Rasoloarison RM, Roos C, Kappeler PM, Mackinnon J (2010) Lemurs of Madagascar 3 rd edn. Tropical field guide series. Conservation International, Arlington/VA

Morelli TL (2008) Dispersal, kinship, and genetic structure of an endangered Madagascar primate, *Propithecus edwardsi*. PhD thesis, Stony Brook University, Stony Brook

Morelli TL, King SJ, Pochron ST, Wright PC (2009) The rules of disengagement: takeovers, infanticide, and dispersal in a rainforest lemur, *Propithecus edwardsi*. Behaviour 146:499–523

Morriss DH, Arrigo-Nelson SJ, Karpanty SM, Gerber BD, Wright PC (2009) Ranging behavior flexibility in response to habitat disturbance by Milne-Edwards' sifakas (*Propithecus edwardsi*) in Ranomafana National Park, Madagascar. Am J Phys Anthropol 138(48):194

Mutschler T, Tan CL (2003) *Hapalemur*, bamboo or gentle lemurs. In: Goodman SM, Benstead JP (eds) The natural history of Madagascar. University of Chicago Press, Chicago, pp 1324–1329

Myers N, Mittermeier RA, Mittermeier CG, da Fonseca GAB, Kent J (2000) Biodiversity hotspots for conservation priorities. Nature 403:853–858

Ostner J, Kappeler PM (2004) Male life history and the unusual adult sex ratios of red-fronted lemur, *Eulemur fulvus rufus*, groups. Anim Behav 67:249–259

Overdorff DJ (1991) Ecological correlates to social structure in two prosimian primates, *Eulemur fulvus rufus* and *Eulemur rubriventer* in Madagascar. PhD thesis, Duke University, Durham

Overdorff DJ (1993) Similarities, differences, and seasonal patterns in the diets of *Eulemur rubriventer* and *Eulemur fulvus rufus* in the Ranomafana National Park, Madagascar. Int J Primatol 14:721–753

Overdorff DJ (1996) Ecological correlates to activity and habitat use of two prosimian primates: *Eulemur rubriventer* and *Eulemur fulvus rufus* in Madagascar. Am J Primatol 40:327–342

Overdorff DJ (1998) Are *Eulemur* species pair-bonded? Social organization and mating strategies in *Eulemur fulvus rufus* from 1988–1995 in southeast Madagascar. Am J Phys Anthropol 105:153–166

Overdorff DJ, Tecot SR (2006) Social pair-bonding and resource defense in wild red-bellied lemurs (*Eulemur rubriventer*). In: Gould L, Sauther ML (eds) Lemurs: ecology and adaptation. Springer, New York, pp 235–245

Overdorff DJ, Merenlender AM, Talata P, Telo A, Forward ZA (1999) Life history of *Eulemur fulvus rufus* from 1988–1998 in southeastern Madagascar. Am J Phys Anthropol 108:295–310

Overdorff DJ, Erhart EM, Mutschler T (2005) Does female dominance facilitate feeding priority in black-and-white ruffed lemurs (*Varecia variegata*) in southeastern Madagascar? Am J Primatol 66:7–22

Perret M, Aujard F (2001) Regulation by photoperiod of seasonal changes in body mass and reproductive function in gray mouse lemurs (*Microcebus murinus*): differential responses by sex. Int J Primatol 22:5–24

Pochron ST, Wright PC (2003) Variability in adult group compositions of a prosimian primate. Behav Ecol Sociobiol 54:285–293

Pochron ST, Wright PC (2005) Testes size and body weight in the Milne-Edwards' sifaka (*Propithecus edwardsi*) of Ranomafana National Park, Madagascar, relative to other strepsirhine primates. Folia Primatol 76:37–41

Pochron ST, Tucker WT, Wright PC (2004) Demography, life history, and social structure in *Propithecus diadema edwardsi* from 1986–2000 in Ranomafana National Park, Madagascar. Am J Phys Anthropol 125:61–72

Pochron ST, Morelli TL, Scirbona J, Wright PC (2005) Sex differences in scent-marking in *Propithecus edwardsi* in Ranomafana National Park, Madagascar. Am J Primatol 66:97–110

Porter LM (1998) Influences on the distribution of *Lepilemur microdon* in the Ranomafana National Park, Madagascar. Folia Primatol 69:172–176

Price TD, Kirkpatrick M (2009) Evolutionarily stable range limits set by interspecific competition. Proc R Soc Lond B 276:1429–1434

Rabarivola C, Prosper P, Zaramody A, Andriaholinirina N, Hauwy M (2007) Cytogenetics and taxonomy of the genus *Hapalemur*. Lemur News 12:46–49

Ratsimbazafy J (2006) Diet composition, foraging, and feeding behavior in relation to habitat disturbance: implications for the adaptability of ruffed lemurs (*Varecia variegata editorium*) in Manombo Forest, Madagascar. In: Gould L, Sauther ML (eds) Lemurs: ecology and adaptation. Springer, New York, pp 403–422

Razafimahaimodison JC (2003) Biodiversity and ecotourism: impacts of habitat disturbance on an endangered bird species in Madagascar. Trop Biodiv 4:12–23

Richard AF (1978) Behavioral variation: case study of a Malagasy lemur. Bucknell University Press, Lewisburg, PA

Richard AF, Dewar RE, Schwartz M, Ratsirarson J (2002) Life in the slow lane? Demography and life histories of male and female sifakas (*Propithecus verreauxi verreauxi*). J Zool Lond 256:421–436

Roth O (1996) Ecology and social behaviour of the woolly lemur (*Avahi laniger*), a nocturnal Malagasy prosimian. Master thesis, University of Basel, Basel

Sefczek TM (2009) Diurnal evidence of a nocturnal feeder: using feeding traces to understand aye-ayes feeding strategy in Ranomafana National Park, Madagascar. PhD thesis, San Diego State University, San Diego

Stevenson PR (2005) Potential keystone plant species for the frugivore community at Tinigua Park, Colombia. In: Dew JL, Boubli JP (eds) Tropical fruits and frugivores: the search for strong interactors. Springer, Netherlands, pp 37–57

Struhsaker TT (1997) Ecology of an African rain forest: logging in Kibale and the conflict between conservation and exploitation. University of Florida Press, Gainsville

Sussman RW (1974) Ecological distinction in sympatric species of lemur. In: Martin RD, Doyle GA, Walker AC (eds) Prosimian biology. Duckworth, London, pp 75–108

Sussman RW, Ratsirason J (2006) Beza Mahafaly Special Reserve: a research site in southwestern Madagascar. In: Jolly A, Sussman RW, Koyama N, Rasamimanana HR (eds) Ringtailed lemur biology: *Lemur catta* in Madagascar. Springer, New York, pp 43–51

Sussman RW, Richard AF, Ratsirarson J, Sauther ML, Brockman DK, Gould L, Lawler R, Cuozzo FP, Mahafaly B (2012) Special reserve: long-term research on Lemurs in Southwestern Madagascar. In: Kappeler PM (ed) Long-term field studies of primates. Springer, Heidelberg

Tan CL (1999) Group composition, home range size, and diet of three sympatric bamboo lemur species (Genus *Hapalemur*) in Ranomafana National Park, Madagascar. Int J Primatol 20:547–566

Tan CL (2007) Behavior and ecology of gentle lemurs (Genus *Hapalemur*). In: Gould L, Sauther ML (eds) Lemurs: ecology and adaptation. Springer, New York, pp 369–381

Tecot SR (2008) Seasonality and predictability: the hormonal and behavioral responses of the red-bellied lemur, *Eulemur rubriventer*, in Ranomafana National Park in southeastern Madagascar. PhD thesis, University of Texas, Austin

Tecot SR (2010) It's all in the timing: birth seasonality and infant survival in *Eulemur rubriventer*. Int J Primatol 31:715–735

Tecot SR (in press) Variable energetic strategies in disturbed and undisturbed rain forest habitats: fecal cortisol levels in southeastern Madagascar. In: Leaping Ahead: Advances in Prosimian

4 Long-Term Lemur Research at Centre Valbio, Ranomafana National Park, Madagascar 99

Biology. Developments in Primatology series. Master J, Gamba M, Genein F (eds) Springer, New York

Tecot S, King SJ, Jernvall J, Wright PC (2009) Lemur pregnancy in the wild: Noninvasive monitoring of reproductive function in Milne-Edwards' sifaka, *Propithecus edwardsi*, in Ranomafana National Park, Madagascar. Am J Phys Anthropol S44:393

Tecot S, Zohdy S, King S, Wright PC, Jernvall J (2010) Wimpy males and formidable females: Testosterone levels in *Propithecus edwardsi*. Am J Phys Anthropol S50:228

Tecot SR, Wright PC (2010) Primate conservation efforts. In: Hill McGraw (ed) Yearbook of science and technology. McGraw Hill, New York, pp 310–315

Terborgh J (1983) Five New World primates: a study in comparative ecology. Princeton University Press, Princeton

Terborgh J, Janson CH (1986) The socioecology of primate groups. Annu Rev Ecol Syst 17:111–135

Thomson MC, Abayomi K, Barnston AG, Levy M, Dilley M (2003) El Niño and drought in South Africa. Lancet 361:437–438

van Schaik CP (1983) Why are diurnal primates living in groups? Behaviour 87:120–144

van Schaik CP, Kappeler PM (1996) The social systems of gregarious lemurs: lack of convergence with anthropoids due to evolutionary disequilibrium? Ethology 102:915–941

van Schaik CP, van Noordwijk MA, Warsano B, Satriono E (1983) Party size early detection of predators in Sumatran forest primates. Primates 24:211–221

Vaughn SE, McGee EM (2009) Association of *Allobosca crassipes* (Diptera: *Hippoboscidae*) with the black and white ruffed lemur (*Varecia variegata variegata*) and Milne-Edwards' sifaka (*Propithecus edwardsi)* in southeastern Madagascar. Pan Pac Entomol 85:162–166

Vinyard CJ, Yamashita N, Tan CL (2008) Linking laboratory and field approaches in studying the evolutionary physiology of biting in bamboo lemurs. Int J Primatol 29:1421–1439

White FJ, Overdorff DJ, Balko EA, Wright PC (1995) Distribution of ruffed lemurs (*Varecia variegata*) in Ranomafana National Park, Madagascar. Folia Primatol 64:124–131

Wiederholt R, Post E (2010) Tropical warming and the dynamics of endangered primates. Biol Lett 6:257–260

Wimmer B, Kappeler PM (2002) The effects of sexual selection and life history on the genetic structure of redfronted lemur, *Eulemur fulvus rufus*, groups. Anim Behav 64:557–568

Wright PC (1995) Demography and life history of free-ranging *Propithecus diadema edwardsi* in Ranomafana National Park, Madagascar. Int J Primatol 16:835–854

Wright PC (1997) The future of biodiversity in Madagascar: a view from Ranomafana National Park. In: Goodman SM, Patterson BD (eds) Natural change and human impact in Madagascar. Smithsonian University Press, Washington, DC, pp 381–405

Wright PC (1998) Impact of predation risk on the behaviour of *Propithecus diadema edwardsi* in the rain forest of Madagascar. Behaviour 135:483–512

Wright PC (1999) Lemur traits and Madagascar ecology: coping with an island environment. Yearb Phys Anthropol 42:31–72

Wright PC (2004) Centre ValBio: long-term research commitment in Madagascar. Evol Anthropol 13:1–2

Wright PC (2006) Considering climate change effects in lemur ecology and conservation. In: Gould L, Sauther ML (eds) Lemurs: ecology and adaptation. Springer, New York, pp 385–401

Wright PC, Andriamihaja BR (2002) Making a rainforest national park work in Madagascar: Ranomafana National Park and its long-term research commitment. In: Terborgh J, van Schaik CP, Davenport L, Rao M (eds) Making parks work: strategies for preserving tropical nature. Island Press, Washington, DC, pp 112–136

Wright PC, Martin LB (1995) Predation, pollination and torpor in two nocturnal prosimians: Cheirogaleus major and Microcebus rufus in the rainforest of Madagascar. In: Alterman L, Doyle GA, Izard MK (eds) Creatures of the dark: the nocturnal prosimians. Plenum Publishing, New York, pp 45–60

Wright PC, Daniels PS, Meyers DM, Overdorff DJ, Rabesoa J (1987) A census and study of *Hapalemur* and *Propithecus* in southeastern Madagascar. Primate Conserv 8:84–88

Wright PC, Heckscher SK, Dunham AE (1997) Predation on Milne-Edward's sifaka (*Propithecus diadema edwardsi*) by the fossa (*Cryptoprocta ferox*) in the rain forest of southeastern Madagascar. Folia Primatol 68:34–43

Wright PC, Andriamihaja BR, Raharimiandra SA (2005a) Tanala synecological relations with lemurs in southeastern Madagascar. In: Paterson JD, Wallis J (eds) Commensalism and conflict: the human-primate interface. Kluwer Press, New York, pp 118–145

Wright PC, Razafindratsita VR, Pochron ST, Jernvall J (2005b) The key to Madagascar frugivores. In: Dew JL, Boubli JP (eds) Tropical fruits and frugivores. Springer, Netherlands, pp 121–138

Wright PC, King SJ, Baden A, Jernvall J (2008a) Aging in wild female lemurs: sustained fertility with increased infant mortality. In: Atsalis S, Margulis SW, Hof PR (eds) Primate reproductive aging: cross-taxon perspectives. Karger, Basel, pp 17–28

Wright PC, Johnson SE, Irwin MT, Jacobs R, Schlichting P, Lehman SM, Louis EE Jr, Arrigo-Nelson SJ, Raharison J-L, Rafalirarison RR, Razafindrasita V, Ratsimbazafy J, Ratelolahy FJ, Dolch R, Tan CL (2008b) The crisis of the critically greater endangered bamboo lemur (*Prolemur simus*). Primate Conserv 23:5–17

Wright PC, Arrigo-Nelson SJ, Hogg KL, Bannon B, Morelli TL, Wyatt J, Harivelo AL, Ratelolahy F (2009) Habitat disturbance and seasonal fluctuations of lemur parasites in the rain forest of Ranomafana National Park, Madagascar. In: Chapman C, Huffman M (eds) Primate parasite ecology: the dynamics and study of host-parasite relationships. Cambridge University Press, Cambridge, pp 311–330

Wright PC, Tecot SR, Erhart EM, Baden AL, King SJ, Grassi C (2011) Frugivory in four sympatric lemurs: implications for the future of Madagascar's forests. Am J Primatol 73:585–602

Yamashita N, Vinyard CJ, Tan CL (2009) Food mechanical properties in three sympatric species of *Hapalemur* in Ranomafana National Park, Madagascar. Am J Phys Anthropol 139:368–381

Zaramody A, Fausser J-L, Roos C, Zinner DP, Andriaholinirina N, Rabarivola C, Norscia I, Tattersall I, Rumpler Y (2006) Molecular phylogeny and taxonomic revision of the eastern woolly lemurs (*Avahi laniger*). Primate Report 74:9–23

Zohdy S, Tecot S, Rakotoarinivo TH, Carag J, King SJ, Jernvall J, Wright PC (2010) Wild brown mouse lemurs live long and prosper. Am J Phys Anthropol 141(suppl 50):251

Zuberbühler K (2007) Predation and primate cognitive evolution. In: Gursky SL, Nekaris KAI (eds) Primate anti-predator strategies. Springer, Chicago, pp 3–26

Chapter 5
A 15-Year Perspective on the Social Organization and Life History of Sifaka in Kirindy Forest

Peter M. Kappeler and Claudia Fichtel

Abstract In this chapter, we summarize some fundamental demographic and morphometric data from the first 15 years of a long-term study of Verreaux's sifaka (*Propithecus verreauxi*) at Kirindy Forest in Western Madagascar. We first describe this research site, its history, and infrastructure, as well as the methods employed to study a local sifaka population. Regular censuses, behavioral observations, and systematic captures of members of up to 11 groups began in 1995 and yielded a data set on demography and life history that can contribute comparative insights about sifaka life history. Our analyses revealed that average group size fluctuated very little around a mean of six individuals across years. Group composition was modified by dispersal (mostly male transfers) or disappearances, births, and deaths. Predation and female transfer were the main mechanisms triggering group extinctions and foundation of new groups ($N = 5$ cases in 149 group years). These exceptional cases of female transfer were most likely motivated by female competition or inbreeding avoidance. One female was a member of at least four different groups. Median age at first birth was 5 years. All females gave birth to single infants, but the proportion of adult females reproducing varied between 25 and 85% across years. The mean interval between 112 births was 15.1 months. Loss of an infant before weaning reduced the subsequent inter-birth interval only by about 1 month. The probability that individual females reproduced successfully decreased as the number of adult females per group increased, implying that subtle forms of female competition limit group size. Mortality is especially high (62%) in

---

P.M. Kappeler (✉)
Department of Sociobiology and Anthropology, CRC Evolution of Social Behavior, University of Göttingen, Göttingen, Germany

Behavioral Ecology & Sociobiology Unit, German Primate Center, Göttingen, Germany
e-mail: pkappel@gwdg.de

C. Fichtel
Behavioral Ecology & Sociobiology Unit, German Primate Center, Göttingen, Germany
e-mail: claudia.fichtel@gwdg.de

P.M. Kappeler and D.P. Watts (eds.), *Long-Term Field Studies of Primates*,
DOI 10.1007/978-3-642-22514-7_5, © Springer-Verlag Berlin Heidelberg 2012

the first 2 years of life. Predation by the fossa (*Cryptoprocta ferox*) is the main cause of death. Maximum female reproductive lifespan is at least 15 years, but longevity is still impossible to estimate. These analyses revealed new insights into female reproductive strategies and their interaction with social organization that were only possible because of the long-term nature of the study, but problems of small sample size still limit the analysis of many vital statistics.

5.1 Introduction

Sifakas (*Propithecus* spp.) represent a genus of lemurs that has played an important role in the history of primatology in Madagascar for at least two reasons. First, sifakas include some of the largest extant lemurs, and they are diurnal and group-living. Because all these traits facilitate behavioral observations, sifakas were among the first lemurs to be studied in the wild (Jolly 1966; Richard 1974a,b; Albignac et al. 1988), and some of the most detailed and long-term lemur data sets are available from several sifaka populations, notably from Beza Mahafaly (*P. verreauxi*: Richard et al. 1991, 1993, 2002; Sussman et al. 2012) and Ranomafana (*P. edwardsi*: Wright 1995; Pochron and Wright 2003; Pochron et al. 2004; Morelli et al. 2009; Wright et al. 2012). [With long-term, we refer not only to periods that far extend typical Ph.D. field projects of 1 or 2 years but also to periods exceeding species-typical ages of first reproduction or dispersal]. Second, recent phylogenetic analyses indicated that diurnality and group-living in sifakas evolved independently from other lemurs and primates (Horvath et al. 2008). Sifakas therefore present a valuable opportunity to study fundamental adaptations to primate sociality from a comparative perspective. Interesting levels of comparison include analysis of variation among groups within populations, among populations of the same species, among different sifaka species, and ultimately between sifaka and other group-living lemurs in the family Lemuridae, and between sifakas and ecologically similar anthropoids. The latter comparative perspectives are beyond the scope of the present chapter, however.

At present, nine species of sifaka are recognized (Mittermeier et al. 2010). They inhabit most remaining dry and rain forests around the island where they feed on flowers, leaves, and fruits (Richard 2003). Sifakas are strictly diurnal (Erkert and Kappeler 2004). At night, they often retreat into emergent trees. During austral winters, they may begin the subsequent day with a sunbath because they lower their body temperatures overnight to conserve energy (Richard and Nicoll 1987). Sifakas range in body mass from about 3 to 9 kg and locomote mostly by vertical leaping. Their main predators include the fossa (*Cryptoprocta ferox*), Harrier hawk (*Polyboroides radiatus*), and boas (*Acrantophis* spp.; Rasoloarison et al. 1995; Wright 1998; Burney 2002; Karpanty 2006), to which they represent some of the most profitable prey because of their size and density. They live in groups of 2–12 individuals that typically contain multiple adult males and females. Females tend to be slightly larger than males, and they dominate them socially (Richard 1974a;

Kappeler 1991; Pochron et al. 2003). Home ranges vary between just a few to about 200 ha among study sites. Mating is confined to a few days within a brief annual season of a few weeks (Richard 1974b; Brockman and Whitten 1996). The single infants require 3–5 years to attain sexual maturity (Richard et al. 2002). Dispersal is primarily by males (Richard et al. 1993), but female dispersal has also been documented (Morelli et al. 2009). Infant mortality is high, but longevity has been projected to exceed 30 years (Richard et al. 2002; Wright et al. 2012). Sifakas communicate with scent marks and several vocalizations (Fichtel and Kappeler 2002; Lewis 2005; Pochron et al. 2005; Fichtel 2008), one of which is responsible for their onomatopoetic name.

Verreaux's sifakas (*P. verreauxi*) have been studied at Beza Mahafaly Special Reserve for more than 3 decades (Richard et al. 2002; Sussman et al. 2012). Results of this long-term study of marked individuals were instrumental in characterizing the ecology, demography, and social behavior of this species. Because *P. verreauxi* has the largest distribution of all sifakas, ranging from the dry spiny forests of the far south to the deciduous baobab forests of the central west, opportunities to identify intraspecific behavioral flexibility and fine-grained adaptations across habitat gradients exist (Richard 1978; Fichtel and Kappeler 2011). In this chapter, we summarize the first 15 years of research on a *P. verreauxi* population at Kirindy Forest to contribute to the comparative approach outlined above.

5.2 The Study Site: Kirindy Forest

Kirindy Forest is one of the largest remaining tracts of dry deciduous forest in Madagascar. The infrastructure for long-term studies of lemur ecology and behavior was established there by the German Primate Center (DPZ) in 1993. In this section, we briefly characterize the study site, outline its history as a lemur study site, and describe the current research infrastructure and methods.

5.2.1 Forest Characteristics

The DPZ research station is located at 44°39′ E 20°03′ S near the center of Kirindy Forest in the central Menabe region of western Madagascar, about 20 km inland from the Mozambique Channel. A 12,500-ha forest concession forms the core of Kirindy Forest. It is (still) connected by a narrow forest corridor to Ambadira Forest to the north (Fig. 5.1). These two forest blocks form the heart of a future protected area (Aire Protégée Menabe Antimena), but it remains officially unprotected as of 2011 (N.B.: Kirindy is sometimes confused with Kirindy-Mitea National Park, which is located south of Morondava). This area is characterized by a hot wet season from November to March and a dry season that can last up to 9 months (Fig. 5.2). The mean annual maximum temperature is 35.8°C and mean annual

Fig. 5.1 Map of Kirindy Forest (© Google maps)

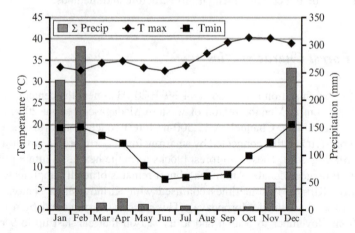

Fig. 5.2 Climate at Kirindy. Mean monthly maximum (*diamonds*) and minimum (*squares*) temperature as well as monthly precipitation (*bars*)

5 A 15-Year Perspective on the Social Organization and Life History of Sifaka

rainfall is 900 mm. Mean annual minimum temperature is 13.5°C, but individual nights during the dry season can get very cold (recorded minimum, 3°C).

Kirindy forest grows on predominantly sandy red and yellow soils just above sea level. It is home to more than 200 species of trees (Rakotonirina 1996). With the exception of a few emergent trees (10–50 trees/ha with >25 cm DBH) along riverbeds or in other humid sites, the vast majority of trees at Kirindy do not exceed 25 m in height. In fact, mean canopy height is 12–18 m. Most canopy layer trees are deciduous, and other plant adaptations to water stress, such as small leaves, spines or thickened stems are common. Visually, three species of baobab (*Adansonia* spp.) as well as *Commiphora*, *Poupartia*, *Colvillea*, and *Terminalia* trees stick out. Near the ground, the forest is very dense with 5,000–19,000 stems/ha with <10 cm DBH and 300–400 trees/ha with a DBH between 12 and 25 cm (Sorg et al. 2003). Depending on the locally prevalent soil type (yellow, red, brown, black, rock), forest structure and composition are very heterogeneous and can change dramatically over distances of a few dozen meters. The phenology of 55 tree species, including all commercially valuable ones, has been studied in great detail (Sorg and Rohner 1996).

Various surveys have revealed the local presence of 15 species of amphibians, 45 species of reptiles, 82 species of birds, and 35 species of mammals, 8 of which are lemurs (Ganzhorn and Sorg 1996). This vertebrate community contains several endangered and locally endemic species, such as the Madagascar jumping frog (*Aglyptodactylus laticeps*), the flat-shelled tortoise (*Pyxis planicauda*), the white-breated mesite (*Mesitornis variegata*), the giant jumping rat (*Hypogeomys antimena*), and Madame Berthe's mouse lemur (*Microcebus berthae*).

5.2.2 History

The Centre de Formation Professionnelle Forestière de Morondava (CFPF) was established by the Malagasy government in 1978 to develop sustainable silvicultural practices for the selective logging of the forests in the Menabe region. To this end, the CFPF installed a forestry concession in Kirindy Forest. The center's activities were coordinated with Swiss development (Coopération Suisse) and research (ETH Zürich) activities. Apart from the training of forestry personnel, the initial period of semi-mechanized logging (1978–1984) was characterized by research projects focusing on forest ecology, silviculture, and methods of forest exploitation and reforestation. In a second phase (1984–1992), silvicultural research was extended to agroforestry experiments outside the forest. After 1992, more emphasis was put on developing ways to use the land and wood resources around the forest area sustainably. In 1995, forestry-related activities inside the forest were reduced, and the CFPF established an ecotourism site that relied to a large extent on the initial forest camp infrastructure. In 2008, the CFPF was formally replaced by the Centre National de Formation, d'Etudes et de Recherche en Environnement et Foresterie (CNFERF), which has a national mission in environmental and forestry training and research and which manages the concession today, including a popular ecotourism site with bungalows and a restaurant.

Research on the fauna of Kirindy began in 1987, when Jörg Ganzhorn (then at the University of Tübingen, Germany) first visited Kirindy Forest. In the following years, he returned annually with some students to explore the effects of selective logging on lemurs and the small mammal fauna (Ganzhorn et al. 1990, 1999; Ganzhorn 1995). Construction of a simple research camp next to the CFPF field camp began in early 1993, after Jörg Ganzhorn had taken up a position at the German Primate Center (DPZ) in Göttingen as the head of a newly created research unit on primate behavior and ecology. Jutta Schmid was the first Ph.D. student (studying mouse lemur energetics: Schmid 2000; Schmid et al. 2000), and Peter Kappeler took up a postdoc position in that group. In 1997, Jörg Ganzhorn left the DPZ for a professorship in Hamburg, and Peter Kappeler took over the management of the research station. Rodin Rasoloarison subsequently joined the team to support the project administration at the national level while also conducting lemur biodiversity research (Rasoloarison et al. 2000; Yoder et al. 2000, 2005; Mittermeier et al. 2008; Groeneveld et al. 2009; Weisrock et al. 2010). Léonard Razafimanantsoa became the station manager in 1998, when the first four permanent local field assistants were also employed and trained. Today, the project employs 15 assistants, including cooks, guardians, and a driver. By 2011, field work for 28 Ph.D. and 75 Master's projects by students from nine countries has been conducted at Kirindy, and regular field courses for students from Göttingen and the University of Antananarivo, as well as the Tropical Biology Association (Cambridge) take place there.

5.2.3   Research Infrastructure

The DPZ research camp at Kirindy was renovated and enlarged in 2004. Today, it offers room for eight researchers, who live in tents on platforms in the forest surrounding the central camp facilities. The core station facilities include brick buildings for the five permanent field assistants, associated staff, a kitchen, several storage buildings for research materials, and a simple laboratory. Electrical power is provided by solar panels; all water and food has to be brought in with a project vehicle from a nearby village (Beroboka) or the provincial town (Morondava), respectively. A base camp ("Villa Mirza") in Morondava provides facilities for the storage of equipment, data management, and recreation. The long-term maintenance and perspective of this site are due to financial support by the DPZ, who provides an annual budget for all local salaries, infrastructure, and basic operating costs.

Animal research at Kirindy has focused on the behavior and ecology of its eight sympatric lemur species (*Microcebus berthae, M. murinus, Cheirogaleus medius, Mirza coquereli, Phaner pallescens, Lepilemur ruficaudatus, Eulemur rufifrons,* and *Propithecus verreauxi*). Lemur research in this dense forest was facilitated by

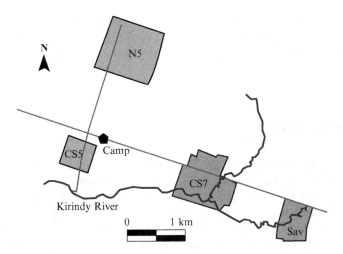

Fig 5.3 Map of local study areas at Kirindy Forest

establishing several trail systems. Checkerboard-like systems of foot trails were established in four areas between 1993 and 2010 (Fig. 5.3). In most cases, trails intersect every 25 m. In total, about 128 km of trails provide access to about 181 ha of forest. All trails are mapped, and all intersections are marked with a small plastic tag. The two oldest grid systems (locally known as N5 and CS7, respectively) also include phenology transects that include several hundred individually marked trees that are checked twice a month.

The Verreaux's sifakas inhabiting the core area (about 50 ha) of the grid system CS7 situated along the Kirindy river were first captured and marked in March 1995. Enafa, an adroit field assistant of the Beza Mahafaly project (Sussman et al. 2012), has been darting the Kirindy sifakas with a blowgun ever since 1995. The anesthetized animals are weighed, permanently marked with a subdermally implanted transponder and subjected to standard morphometric field measurements. A small skin biopsy from the ear is stored in ethanol and used for later DNA extraction. Mitochondrial DNA has been sequenced to study group structure and histories (Kappeler et al. in prep), and variation in nuclear microsatellites has been used to determine paternities (Kappeler and Schäffler 2008). Finally, all individuals are fitted with small nylon collars, each with a unique pendant. One individual per group is fitted with a radio collar. Juveniles, who are born in July/August, are captured in March/April of the subsequent year together with any immigrants. Radio collars (Holohill, Canada) were initially replaced annually; nowadays, battery life is 3 years. The composition and location of all marked sifaka groups are monitored by Tiana Andrianjanahary several times a week, so that all important demographic events, including births, predation events, and disappearances or dispersal events can be dated to within several days. Several Master's (Hussmann, Rümenap, Carrai, Kraus, Randriamanalina, Dill, Rakotondravony, Dirac, Trillmich, Scholz, Benadi) and Ph.D. studies (Carrai, Fichtel, Lewis, Mass, Koch) on sifaka ecology or behavior have yielded additional behavioral data on most groups over the years. Demographic, morphometric, genetic, and behavioral data have been centralized and are currently being transferred from Excel files into a data bank.

5.3 Kirindy Sifaka

During the initial capture in 1995, 24 individuals from six groups (A–F) were marked within a 25-ha grid. With the eventual extension of the grid system, additional groups were added to the study population over the years, and a total of 213 individuals were captured at least once. After several group extinctions, fissions and fusions, there are currently 9 groups with 52 marked individuals. In the intervening 149 group years, 141 infants were born into these study groups and 154 animals disappeared or died. In this section, we summarize some of the basic demographic events of these first 15 years of our study, focusing on some rare events as well as on births and deaths.

5.3.1 Group Size and Composition

The average group of Kirindy sifaka contains 1.6 ± 0.8 (mean $\pm$ SD) adult females, 1.9 ± 0.8 adult males, 0.7 ± 0.8 juvenile females, 0.8 ± 0.9 juvenile males, 0.1 ± 0.3 unsexed juveniles, 0.3 ± 0.5 female infants, and 0.4 ± 0.6 male infants (unsexed infants 0.3 ± 0.5). Group size fluctuated very little around a grand mean of 6.1 (± 1.8) over 149 group years. For the present analyses, we used group composition in April of every year to determine the number of adults, juveniles, and infants present. Infants were captured for the first time in March or April, i.e., when they were still within their first year of life and about 8 months old. Infants who disappeared before this first capture could not be sexed, and typically, no material for paternity analyses was available. Individuals in their second and third year were classified as juveniles because (female) sexual maturity and the earliest age of first reproduction were observed at age 4 (see below). Characteristics of the average growth pattern could be used to age unknown individuals below 3 years of age reliably (see Sect. 5.3.5). All other animals were classified as adults.

5.3.2 Group Histories

The histories of some groups were characterized by extinctions or fusions, but we also witnessed the establishment of new groups. Predation and female transfer were the key mechanisms in these events. We operationally define groups as bisexual units, i.e., they consisted of at least one adult male and female. In 149 group years, only five events that resulted in changes in the number of study groups occurred. Below, we describe the circumstances of these rare but important events in detail.

Two groups essentially went extinct following confirmed fossa predation. First, group D was among the groups captured in 1995. It consisted of an adult pair (DFRom and DMMil) and their juvenile son (DMNap; acronyms denote group (in which an animal was born or first captured)* sex* individual ID). They were also regularly visited by a roaming male (AMPin). In July 1995, a female infant (DFPis) was born. On July 22, 1996, DMMil was killed by a fossa. The 1996 infant of DFRom disappeared within a week after its birth on July 24. Its disappearance was likely due to infanticide as it coincided with the immigration of a new adult male (DMFir) between the 25th and 30th of July 1996. On August 14, 1996, DMNap was killed by a fossa. For most of the following year, group D consisted of DFRom, DFPis, DMFir, and AMPin. On August 10, 1997, DFRom gave birth to another infant. On August 27, 1997, she disappeared together with her new infant; circumstances also imply fossa predation as the cause of their disappearance. In October 1997, the 2-year-old DFPis emigrated (together with AMPin) into the neighboring group H, where she stayed until March 2001. DMFir was left behind by himself. He paired up with another solitary adult male (BMBer) between February and June 1998, but then became solitary again and stayed alone until he was last seen in October 1998. Thus, group D dissolved following emigration of a juvenile female after all of her relatives were killed by fossa within a year.

Second, group K was also exterminated by fossa predation. In 2006, it consisted of two adult females (KFAlm and KFJal), an adult male (KMChe), a juvenile male (FMGor) who disappeared on 8 June, and KFAlm's infant KMCha. On July 29, 2007, KFAlm, and her infant (born on 16 July) were killed by a fossa. KFJal was last seen on the same day. KMCha was last seen together with his father KMChe on August 4, 2007. After a last solitary sighting on August 6, the yearling disappeared. KMChe was still solitary in August 2010. Thus, presumably, a single fossa attack ultimately led to the extinction of this group.

In three other cases, female transfer significantly modified group compositions and ultimately changed the number of groups. In one example, the female members of group A voluntarily joined one of their neighboring groups. In 2006, group A consisted of two closely related adult females (AFSil and AFSis), an immigrant adult male (FMPho), a 3-year-old natal male (AMAnt), a same-aged juvenile female (AFRos), and the 2006 infant of AFSil. In early January 2007, two males from neighboring group F, 9-year-old FMChi and 4-year-old FMFra, immigrated into group A. On January 8, 2007, AFSil was attacked by a fossa and disappeared shortly thereafter. Her 2006 infant disappeared on the same day. FMFra went back to his natal group in April 2007. FMPho, the half brother of FMChi and father of AFRos, disappeared on June 18, 2007. On September 29, 2007, EFSis emigrated into neighboring group E, which had lost its sole adult female EFAli together with her 2007 infant in a fossa attack on September 16, 2007, leaving 4 males (EMDar and his sons, 7-year-old EMHar, 4-year-old EMMel, and 3-year-old EMAlb) behind. AFRos followed 2 days later, leaving behind AMAnt and FMChi, who stayed together until they immigrated into FMChi's natal group in March 2008. Group A therefore ceased to exist because its two females emigrated. Inbreeding avoidance (AFSis and AFRos) and, perhaps, other reproductive interests led to voluntary female emigration that was facilitated in this case because the target group contained no other females at the time of fusion.

Female transfers also led to massive changes in the composition of group F and ultimately to the formation of the new group F1. In 2005, group F consisted of two adult females (FFDal and her daughter FFSav), immigrant adult male FMJun, natal adult male FMChi, three juvenile males (FMFra, FMDet, and FMRal), and juvenile female FFTam. FFDal was the mother of all natal group members, except for FMDet, who was FFSav's son. On October 18, 2005, FFSav and FFDal were first seen with a group of four unmarked males about 150 m away from the rest of group F. On October 25, they were back with FMJun and their offspring. On October 28 and 30, the two females were again with the new males. On 31 October and 1 November, FFSav was with the new males, whereas FFDal was back with FMJun and her sons. On November 2, both females were with the new males; on November 9, both of them were back in group F. On November 13, FFSav was again with the new males; on the next day, her mother had followed. On November 19, both were again in group F; on the 22nd both were with the new males. On November 23, FFDal and FFSav were seen with the males of group F for the last time. The four unmarked animals were captured on April 4, 2006. FFDal and FFSav both gave birth on July 7, 2006. FFDal and her infant disappeared on July 17, 2006. The new group remained in a distinct area to the east of group F's range and was named F1. [FFSav had given birth to her first infant in 2003. It was one of the very few infants sired by a nonresident male. She had another infant in 2004 (FMDet) and none in 2005]. As all males in her natal group were close relatives, inbreeding avoidance might have triggered her emigration; why her mother followed remains unknown, however. Throughout 2006, FMJun and his two sons FMChi and FMFra stayed together in their habitual home range. On October 17, 2007, they were joined by an unmarked adult female (FFOma), who gave birth to an infant in July 2008.

Groups I and C represent two other related cases in which the same adult female (IFCal) formed a new group. She appeared in the study area in February 1998 as a fully adult female and was first captured on September 8. She had been joined by two adult males from different groups, EMSyd (on February 20, 1998) and BMBer (on June 7, 1998) to form group I. In January 1999, BMBer left group I and led a solitary life until he was last sighted on May 28, 1999. On that day, IFCal and EMSyd immigrated together into group C. Group C's matriarch CFAnt (together with her infant of 1998) had been killed by a fossa on October 11, 1998. The only other resident female, CFTam, left her group on the next day, leaving two natal juvenile males (CMMaf and CMBel) and three adult immigrant males (CMTul, CMAnt, and CMDau) behind. Beginning on October 26, 1998, one of two adult females of group B (BFFul), one of group C's neighbors, associated with the five males of group C. She was last seen in group C on April 27, 1999 before returning to group B. CMAnt and CMDau emigrated into group E in March 1999. Thus, group C consisted of two juvenile and one adult male when IFCal and EMSyd joined them.

EMSyd left group C again on July 5, 1999 and was by himself until he was last seen on July 17, 1999.

More than a year later, on September 16, 2000, IFCal left group C and joined five unmarked animals. She did not have an infant in 2000. It turned out that her new group (L) consisted of one adult (LFPat) female and two juvenile females (LFDel and LFMad) living with two adult males (LMBha and LMBom). The three initial resident females of group L disappeared in August 2001, July 2002, and October 2003, respectively. IFCal had her first infant in 2002 and subsequently five more infants before she disappeared on April 7, 2009. Thus, the circumstances triggering this exceptional female's dispersals remain unclear, but her case shows that females can initiate new groups by attracting males, and that females who have left their natal groups can disperse repeatedly – in this case three times.

5.3.3 Reproduction

Mating in Kirindy sifaka is limited to a few weeks in January and February. Direct observations of matings are extremely rare, despite considerable effort by some students (e.g., Kraus et al. 1999; Lewis 2004; Mass et al. 2009). For one thing, they occur at the peak of the rainy season, when observation conditions are very difficult. Moreover, they tend to be very uneventful and short. A study relying on fecal hormone analyses to determine the timing of female reproductive periods confirmed that fertile periods of individual females are limited to 2–4 days and that the fertile periods of co-resident females rarely overlap (Mass et al. 2009).

Births occur about 6 months later, i.e., in July and August. Only single infants have ever been born. Between 1995 and 2009, a total of 29 different females gave birth to a total of 141 infants. The sex ratio of infants that survived long enough to be captured and sexed ($N = 102$) did not differ from unity (1.37; 59 males, 43 females, $X^2 = 0.96$, $df = 1$, ns). However, the proportion of females giving birth differed notably among years, varying between 25 and 85% (Fig. 5.4). This proportion is not significantly correlated with total annual rainfall ($r = 0.35$, $N = 14$, ns) or other climatic variables.

Eleven females born into the study groups had reached reproductive age by 2009. Their median age at first birth was 5 years (range 4–6). Twenty-four females gave birth at least twice, so that 112 inter-birth intervals were available for analysis. The mean inter-birth interval was 15.1 months (± 5.4). If the infant survived to post-weaning (i.e., the following April; $N = 93$), mean inter-birth intervals were only slightly longer (15.3 months ± 5.7) than if it died during its first 9 months of life (14.7 months ± 4.9). Early loss of an infant thus did not increase the probability of subsequent conception (Fisher's exact test, $p = 0.47$).

The number of adult females per group varied between 1 and 4 (Fig. 5.5). In 64% of 64 group years in which only one adult female was present, this female reproduced. When a group contained two adult females (66 group years), the probability that both of them reproduced was reduced to 39%; in 41% of these

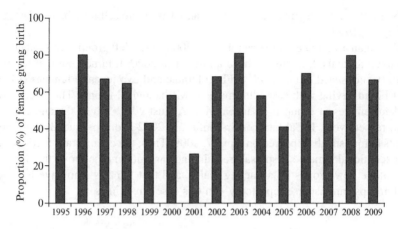

Fig. 5.4 Proportion of sifaka females giving birth between 1995 and 2009

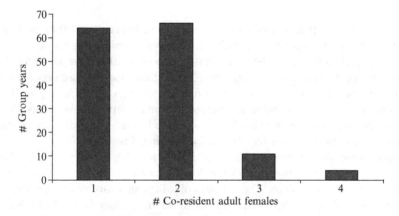

Fig. 5.5 Number of co-resident sifaka females per group over 149 group years

cases, at least one female reproduced. In none of 11 group years with three adult females present did all three produce an infant; in 55% of these years, there were at least two infants. Finally, in the two group years with four adult females, a maximum of two females had an infant in one of those years (50%). Assuming that no other factors play a role, the probability of individual reproduction therefore decreased with the number of females per group from 64% ($N = 1$ female) to 40% ($N = 2$ females), to 37 % ($N = 3$ females), to 25% ($N = 4$ females). Thus, competition among females appears to limit reproductive opportunities in these Verreaux's sifakas.

Paternities were determined for 54 surviving infants born between 1995 and 2003 by comparing patterns of individual variation at 15 microsatellite loci (Kappeler and Schäffler 2008). More than 91% of all infants were sired by the dominant adult male of a group when a group contained two or more non-natal adult males. Only one infant was the result of an extra-group mating. The youngest confirmed male fathering offspring was 4 years and 7 months old at the time of conception.

5.3.4 Disappearances

Between 1995 and 2010, 154 individuals disappeared. Animals that disappeared could have emigrated or died. Evidence for mortality ($N = 31$) was most often due to confirmed cases of fossa predation (Fig. 5.6). Kirindy harbors a dense population of fossas that heavily prey on sifakas, especially in the second half of the dry season (June–October), when most alternative prey hibernate (see also Rasoloarison et al. 1995). In some cases, hunts were directly observed. Some of the fossa hunts were cooperatively (Lührs and Dammhahn 2010). In all other confirmed cases, corpses or at least the animal's nylon collars together with some fur and blood were found. Predation by the Harrier hawk, other raptors, or snakes has not been observed or inferred (cf. Karpanty and Goodman 1999). Infanticide by strange males has been observed or suspected as the cause of some infant deaths (see above; Lewis et al. 2003). Several infants have died within the first weeks of life, presumably from maternal neglect (observed in several cases) or failure to produce enough milk. Juvenile or adult sifakas dying from disease or other causes were never discovered.

About a third of all newborns did not survive their first year of life, and 62.4% of individuals died within the first 2 years (Fig. 5.7). Female dispersal was extremely rare (12 cases in 149 group years). The majority of disappeared females and juvenile males are therefore most likely dead and did not emigrate. Thus, a

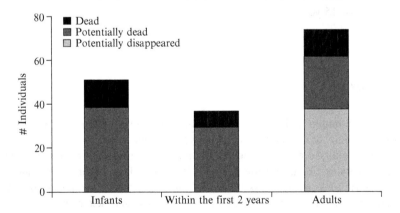

Fig. 5.6 Number of infants, juveniles within the first 2 years, and adults that either died are potentially dead or potentially disappeared. Individuals that disappeared at an age under the average age of natal dispersal (males median 60 months (IQR 12); females median 54 months (IQR 18)) were considered as potentially dead; all other individuals were classified as potentially disappeared

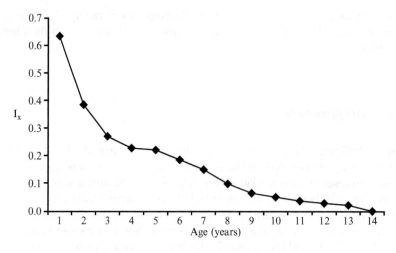

Fig. 5.7 Age-dependent survival of sifakas at Kirindy Forest

minimum of 38 and a maximum of 54 of the disappeared individuals presumably died.

Longevity is still difficult to estimate 15 years into the study. Of the 11 oldest animals with known ages (born in 1994 and 1995), none are still alive today. The oldest ones disappeared at age 14. Currently, the oldest female alive with known age is 13, and three males are of the same age. However, one female first captured in 1995 as a young adult (BFCol) had an infant in 1996 and was killed by a fossa in September 2010. If the 1996 infant was her first one, she was most likely born in 1991 (±1 year), and was thus in her 20th year. She had an infant (her 13th!) on July 29, 2010, but abandoned it on 13 August. Thus, maximum female reproductive life span is at least 15 years.

5.3.5 Growth

Newborn and immigrant individuals were captured at least once a year. In addition, dead radio collars or broken nylon collars were replaced whenever possible (i.e., if the bearer was not a pregnant or lactating female). As a result, we have accumulated a mix of cross-sectional and longitudinal data on body mass and other standard morphometric measurements (skull length and width, canine and testes size, body and tail length) for 184 individuals of different age and sex classes whom we captured a total of 384 times across the years. Here, we present some growth data that have helped us to age unknown individuals. Because of the seasonality of reproduction, identifying an individual's year of birth is equivalent to reconstructing its age to within a month.

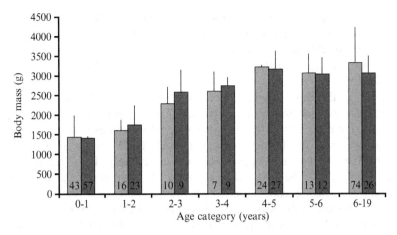

Fig. 5.8 Mean (±SD) body mass of female (*light*) and male (*dark*) sifakas per age category. Numbers inside bars denote sample size. Please note that most individuals were weighed repeatedly at different ages

The lightest individual ever weighed was a 15-day-old infant that was abandoned by its mother when it weighed 58 g. Of the individuals with exactly known ages, the heaviest (3,700 g) was a 6-year-old female. The heaviest female ever captured weighed 4,400 g; the heaviest male weighed 3,850 g. At first capture (mean age 241 days), recently weaned individuals had a mean weight of 1,381 ± 28 g (both N's = 89; Fig. 5.8). During their second year of life (mean age 441 days), juveniles weighed on average 1,601 ± 530 g (N = 32). During their third year (mean age 855 days), 11 juveniles had a mean weight of 2,331 ± 594 g. Between ages 3 and 4, the mean weight increased to 2,626 ± 221 g (N = 8). Adults between 4 and 13 years old weighed an average of 2,973 ± 124 g (N = 34). Animals without known ages (immigrants, first capture of a group) with a body mass of >3,000 g were therefore assigned a minimum age of 5 years, so that their subsequently obtained data could be included into future analyses.

5.4 Comparative Aspects Based on Long-Term Data

This first summary of the social organization and life history of the Kirindy sifaka provides a basis for comparisons with other studies of sifaka, in particular, that of the population at Beza Mahafaly. At Beza, groups also contain about 6 individuals on average (Fig. 2 in Richard et al. 1993), with a slight average preponderance of males (Fig. 4 in Richard et al. 1993). Dispersal is also male biased, with all 17 males aged 3–6 years of one particular study cohort emigrating from their natal groups; in most cases, into neighboring groups (Richard et al. 1993). As at Kirindy, female transfer at Beza was rare (Sussman et al. 2012). During a 7-year period, only two neighboring females switched home ranges and males. In six additional cases, an adult pair formed a new group, but the circumstances of the departures of these adult females from their original groups were unknown (Richard et al. 1993). Population genetic analyses of the Beza population reflected the genetic consequences of this sex bias in dispersal (Lawler et al. 2003; Sussman et al. 2012). Thus, the overall pattern of social organization of sifakas at both sites is rather similar.

In contrast, the demographic structures of the two populations appear very different, despite great similarities in female life history parameters. The youngest females to give birth at Beza were 3 years old, but no infants of females younger than 5 years survived to 12 months (Richard et al. 2002). Five years is the age at which Kirindy females also begin to reproduce regularly. Even though female fertility declined after the age of 18 at Beza, females continued to give birth until at least the age of 28 (Richard et al. 2002), whereas we have so far only evidence that a single female in the Kirindy population is even close to being 20 years old. Close to 60% of females at Beza reproduced successfully within 1 year of a surviving birth, indicating that average inter-birth intervals are similar at the two sites. Infant survival during the first year is slightly lower at Beza (51 vs 55%), but later mortality is substantially lower, with many animals surviving well into their twenties (Richard et al. 2002). Infanticide is similarly rare at both study sites. However, during the 15 years on which they based their analysis, Richard and colleagues (1992) found 28 decomposing sifaka corpses (compared to zero at Kirindy), and report only a single probable episode of fossa predation and no attacks by the Harrier hawk. Predation by introduced wild cats (*Felis sylvestris*) is rare and apparently ecologically insignificant (Brockman et al. 2008). Thus, it appears likely that different predation rates, particularly by fossa, underlie a major component of the observed differences in demography between the two study sites.

Propithecus edwardsi at Ranomafana National Park exhibit some interesting contrasts and similarities with the two *P. verreauxi* populations. Groups of this rainforest species consist of 4.6 individuals on average, including 1.44 adult females and 1.46 adult males (Pochron et al. 2004). Female dispersal is much more common in *P. edwardsi*, however, and even equaled dispersal rates among males in a small sample (Morelli et al. 2009). Inbreeding avoidance appears to be a major cause of female dispersal, but dispersing females also commit infanticide in their new groups and evict the corresponding mothers (Morelli et al. 2009). Thus, reproductive opportunities for females appear to be limited, and dispersal may represent the main behavioral mechanism that females use to gain breeding positions (see also Wright et al. 2012).

Female *P. edwardsi* also begin reproducing as early as age 4, but substantial fertility is only achieved after age 6. Reproduction continues until age 18, with an average inter-birth interval of 1.56 years, but early infant death results in a significant reduction of the average inter-birth interval to 1.1 years (Pochron et al. 2004). Life expectancy at Ranomafana also declines at age 15 and ends before age 20. As in *P. verreauxi*, half of the newborn (female) infants fail to survive their first year of life. Infanticide is a more common cause of infant death (12% of young females' deaths; see also Morelli et al. 2009) than in either *P. verreauxi* population, but, importantly, as in Kirindy, predation by fossa is responsible for a large proportion (64%) of deaths for which the causes are known (Wright et al. 1997; Pochron et al. 2004). Thus, high predation pressure by fossa has resulted in similar demographic structures of the sifaka populations at Kirindy and Ranomafana, both of which clearly differ from that at Beza Mahafaly.

5.5 Conclusions: Limitations and Highlights of a Young Long-Term Study

This summary of some aspects of our sifaka study both reveals some of the major limitations of our research thus far and highlights a few findings that were only possible because of the long-term nature of the study.

A 3-kg sifaka is a small-to-medium-sized mammal. However, to describe even the cornerstones of its social organization and life history, studying >150 individuals over 15 years turned out to be far from sufficient. Relatively low birth and developmental rates characteristic of most primates, and sifakas in particular (Richard et al. 2002), limit sample size for other analyses considerably. Most importantly, even after 10 or more years of study, small sample size hampers many aspects of life table analyses (see Alberts and Altmann 2003). For example, the sample size for female age at first reproduction – the key functional life history variable – is still so small that it is statistically not justified to calculate an arithmetic mean. Thus, studying multiple groups in species with such small average group sizes as sifaka is essential for capturing natural variation across groups and time, and a long-term approach is the only way to reduce these natural limitations.

The specific highlights of our insights into sifakas sociality due to the long-term nature of our study mainly concern rare but important events, of which female transfer is clearly the most important one. Female transfer happens so rarely, quickly, and uneventfully that it is extremely easy to miss in short-term studies. Whereas genetic analyses allowed us to infer its occurrence (see also Di Fiore et al. 2009), only regular observations have permitted us to identify some of the potential ultimate causes motivating it (see also Morelli et al. 2009). Because it has been known for some years that females of group-living Lemuridae evict other females from their group, even though they are close relatives by anthropoid standards (Vick and Pereira 1989), we initially suspected that these rare dispersing females might also be victims of targeting aggression in their natal groups. However, as our long-term analyses of female reproductive success as a function of natural variation in female group size revealed, female competition in sifakas appears to be ubiquitous despite peaceful coexistence of two or more adult females. Unlike those species of Lemuridae that form stable social groups, active expulsion of female competitors has never been observed among the Kirindy sifaka. Exploiting opportunities for reproduction in groups that have lost adult females to predation or other events may instead trigger many of these unusual female dispersals.

A second, related point concerns the importance of knowing the genealogical relationships among all group members, especially because these cannot be determined with precision from genetic analyses alone (Harris et al. 2009). In some of the documented cases of female dispersal in this study, female transfer can be reasonably interpreted as an attempt to avoid incestuous matings with closely related males.

Given the low reproductive rates of sifakas in combination with high rates of early infant mortality, a long-term approach to studying the distribution of paternities among males was also necessary because the sample sizes required for meaningful analyses accumulated very slowly. Only after more than 10 years did we have a sufficient number of infants who survived long enough to be captured to obtain a tissue sample for genetic analyses (Kappeler and Schäffler 2008). The results of this study contributed to the illumination of a long-standing problem in the study of lemur sociality related to the male-biased adult sex ratios (Kappeler et al. 2009). The relative reproductive success of males pursuing apparently different reproductive tactics could also be only determined after witnessing a sufficient number of group take-overs or knowing individual dispersal strategies (Kappeler and Schäffler 2008). These data now provide an opportunity to examine interesting intraspecific variation in male reproductive strategies because paternity analyses of the Beza sifaka revealed that the higher population density there, combined with more strongly overlapping home ranges during an equally short mating season, has resulted in greater opportunities for male extra-group mating opportunities (Lawler 2007; Sussman et al. 2012) and presumably smaller male reproductive skew, suggesting that sifaka males at Kirindy appear to be better able to mate-guard receptive females (Mass et al. 2009).

Finally, the combination of results from long-term studies at different sites made it possible to begin explaining other differences among populations. A qualitative comparison of sifakas demography showed Beza Mahafaly to be unusual, especially with respect to its age structure and mortality schedules, compared to Ranomafana and Kirindy. The absence of fossa predation at Beza can account for this demographic difference, something that provides a natural experiment to study the behavioral and long-term demographic consequences of the changes in one key ecological variable.

Acknowledgments Our foremost thanks go to our field assistant Tiana Andrianjanahary for his reliable and enthusiastic collection of census data. Enafa, Mily, Elysée, Léonard Razafimanantsoa, Rodin Rasoloarison, Mamitiana Razafindrasamba, Jean-Pierre Ratolojanahary, Nielsen Rabarijaona, Remy Ampataka, the late Jean-Claude Beroboka, Dietmar Zinner, Rebecca Lewis, and numerous student baby-sitters helped with the capture of sifakas. Dr. Joelison Ratsirarson supported our sifaka study in numerous ways. The CFPF/CNFEREF, Département Biologie Animale Université d'Antananarivo, and the Ministère Forêts et Environnement de Madagascar continue to authorize and support our research. The Bundesamt für Naturschutz provided the permits for tissue imports, and Heike Klensang was instrumental in obtaining them. Ulrike Walbaum managed the sifaka data base for many years, and Katharina Peters helped in extracting relevant information. Without the ground-breaking activities of Jörg Ganzhorn and the continuous financial support of the Deutsches Primatenzentrum and the Deutsche Forschungsgemeinschaft, this ongoing long-term project would have been impossible. We thank all these individuals and institutions for their support, and David Watts for helpful comments on this chapter.

References

Alberts SC, Altmann J (2003) Matrix models for primate life history analysis. In: Kappeler PM, Pereira ME (eds) Primate life histories and socioecology. University of Chicago Press, Chicago, IL, pp 66–102

Albignac R, Fontenille D, Maleyran D, Duvernoy F (1988) Evolution de l'organisation sociale et territoriale de *Propithecus verreauxi coquereli* pendant 6 ans, dans les forêts sèches du nord-ouest de Madagascar (Ankarafantsika). In: Rakotovato L, Barre V, Sayer J (eds) L'équilibre des ecosystèmes forestiers à Madagascar: actes d'un séminaire international. IUCN, Gland and Cambridge, pp 90–94

Brockman DK, Whitten PL (1996) Reproduction in free-ranging *Propithecus verreauxi*: estrus and the relationship between multiple partner matings and fertilization. Am J Phys Anthropol 100:57–69

Brockman DK, Godfrey LR, Dollar LJ, Ratsirarson J (2008) Evidence of invasive *Felis silvestris* predation on *Propithecus verreauxi* at Beza Mahafaly Special Reserve, Madagascar. Int J Primatol 29:135–152

Burney DA (2002) Sifaka predation by a large boa. Folia Primatol 73:144–145

Di Fiore A, Link A, Schmitt CA, Spehar SN (2009) Dispersal patterns in sympatric woolly and spider monkeys: integrating molecular and observational data. Behaviour 146:437–470

Erkert HG, Kappeler PM (2004) Arrived in the light: diel and seasonal activity patterns in wild Verreaux's sifakas (*Propithecus v. verreauxi*; Primates: Indriidae). Behav Ecol Sociobiol 57:174–186

Fichtel C (2008) Ontogeny of conspecific and heterospecific alarm call recognition in wild Verreaux's sifakas (*Propithecus verreauxi verreauxi*). Am J Primatol 70:127–135

Fichtel C, Kappeler PM (2002) Anti-predator behavior of group-living Malagasy primates: mixed evidence for a referential alarm call system. Behav Ecol Sociobiol 51:262–275

Fichtel C, Kappeler PM (2011) Variation in the meaning of alarm calls in Coquerel's and Verreaux's sifakas (*Propithecus coquereli, P. verreauxi*). Int J Primatol 32:346–361

Ganzhorn JU (1995) Low-level forest disturbance effects on primary production, leaf chemistry, and lemur population. Ecology 76:2084–2096

Ganzhorn JU, Sorg J-P (1996) Ecology and economy of a tropical dry forest in Madagascar. Primate Rep 46–1:1–382

Ganzhorn JU, Ganzhorn AW, Abraham J-P, Andriamanarivo L, Ramananjatovo A (1990) The impact of selective logging on forest structure and tenrec populations in western Madagascar. Oecologia 84:126–133

Ganzhorn JU, Fietz J, Rakotovao E, Schwab D, Zinner DP (1999) Lemurs and the regeneration of dry deciduous forest in Madagascar. Conserv Biol 13:794–804

Groeneveld LF, Weisrock DW, Rasoloarison RM, Yoder AD, Kappeler PM (2009) Species delimitation in lemurs: multiple genetic loci reveal low levels of species diversity in the genus *Cheirogaleus*. BMC Evol Biol 9:30. doi:10.1186/1471-2148-9-30

Harris TR, Caillaud D, Chapman CA, Vigilant L (2009) Neither genetic nor observational data alone are sufficient for understanding sex-biased dispersal in a social-group-living species. Mol Ecol 18:1777–1790

Horvath JE, Weisrock DW, Embry SL, Fiorentino I, Balhoff JP, Kappeler PM, Wray GA, Willard HF, Yoder AD (2008) Development and application of a phylogenomic toolkit: resolving the evolutionary history of Madagascar's lemurs. Genome Res 18:489–499

Jolly A (1966) Lemur behavior: a Madagascar field study. University of Chicago Press, Chicago, IL

Kappeler PM (1991) Patterns of sexual dimorphism in body weight among prosimian primates. Folia Primatol 57:132–146

Kappeler PM, Schäffler L (2008) The lemur syndrome unresolved: extreme male reproductive skew in sifakas (*Propithecus verreauxi*), a sexually monomorphic primate with female dominance. Behav Ecol Sociobiol 62:1007–1015

Kappeler PM, Mass V, Port M (2009) Even adult sex ratios in lemurs: potential costs and benefits of subordinate males in Verreaux's sifaka (*Propithecus verreauxi*) in the Kirindy Forest CFPF, Madagascar. Am J Phys Anthropol 140:487–497

Karpanty SM (2006) Direct and indirect impacts of raptor predation on lemurs in southeastern Madagascar. Int J Primatol 27:239–261

Karpanty SM, Goodman SM (1999) Diet of the Madagascar harrier-hawk, *Polyboroides radiatus*, in southeastern Madagascar. J Rap Res 33:313–316

Kraus C, Heistermann M, Kappeler PM (1999) Physiological suppression of sexual function of subordinate males: a subtle form of intrasexual competition among male sifaka (*Propithecus verreauxi*)? Physiol Behav 66:855–861

Lawler RR (2007) Fitness and extra-group reproduction in male Verreaux's sifaka: an analysis of reproductive success from 1989–1999. Am J Phys Anthropol 132:267–277

Lawler RR, Richard AF, Riley MA (2003) Genetic population structure of the white sifaka (*Propithecus verreauxi verreauxi*) at Beza Mahafaly Special Reserve, southwest Madagascar (1992–2001). Mol Ecol 12:2307–2317

Lewis RJ (2004) Male-female relationships in sifaka (*Propithecus verreauxi verreauxi*): power, conflict and cooperation. PhD thesis, Duke University, Durham

Lewis RJ (2005) Sex differences in scent-marking in sifaka: mating conflict or male services? Am J Phys Anthropol 128:389–398

Lewis RJ, Razafindrasamba SM, Tolojanahary JP (2003) Observed infanticide in a seasonal breeding prosimian (*Propithecus verreauxi verreauxi*) in Kirindy Forest, Madagascar. Folia Primatol 74:101–103

Lührs M-L, Dammhahn M (2010) An unusual case of cooperative hunting in a solitary carnivore. J Ethol 28:379–383

Mass V, Heistermann M, Kappeler PM (2009) Mate-guarding as a male reproductive tactic in *Propithecus verreauxi*. Int J Primatol 30:389–409

Mittermeier RA, Ganzhorn JU, Konstant WR, Glander K, Tattersall I, Groves CP, Rylands AB, Hapke A, Ratsimbazafy J, Mayor MI, Louis EE Jr, Rumpler Y, Schwitzer C, Rasoloarison RM (2008) Lemur diversity in Madagascar. Int J Primatol 29:1607–1656

Mittermeier RA, Louis EE Jr, Richardson M, Schwitzer C, Langrand O, Rylands AB, Hawkins F, Rajaobelina S, Ratsimbazafy J, Rasoloarison RM, Roos C, Kappeler PM, Mackinnon J (2010) Lemurs of Madagascar, 3rd edn, Tropical Field Guide Series. Conservation International, Arlington, VA

Morelli TL, King SJ, Pochron ST, Wright PC (2009) The rules of disengagement: takeovers, infanticide, and dispersal in a rainforest lemur, *Propithecus edwardsi*. Behaviour 146:499–523

Pochron ST, Wright PC (2003) Variability in adult group compositions of a prosimian primate. Behav Ecol Sociobiol 54:285–293

Pochron ST, Fitzgerald J, Gilbert CC, Lawrence D, Grgas M, Rakotonirina G, Ratsimbazafy R, Rakotosoa R, Wright PC (2003) Patterns of female dominance in *Propithecus diadema edwardsi* of Ranomafana National Park, Madagascar. Am J Primatol 61:173–185

Pochron ST, Tucker WT, Wright PC (2004) Demography, life history, and social structure in *Propithecus diadema edwardsi* from 1986–2000 in Ranomafana National Park, Madagascar. Am J Phys Anthropol 125:61–72

Pochron ST, Morelli TL, Scirbona J, Wright PC (2005) Sex differences in scent-marking in *Propithecus edwardsi* of Ranomafana National Park, Madagascar. Am J Primatol 66:97–110

Rakotonirina (1996) Composition and structure of a dry forest on sandy soils near Morondava. Primate Rep 46–1:81–87

Rasoloarison RM, Rasolonandrasana BPN, Ganzhorn JU, Goodman SM (1995) Predation on vertebrates in the Kirindy Forest, western Madagascar. Ecotropica 1:59–65

Rasoloarison RM, Goodman SM, Ganzhorn JU (2000) Taxonomic revision of mouse lemurs (*Microcebus*) in the western portions of Madagascar. Int J Primatol 21:963–1019

Richard AF (1974a) Intra-specific variation in the social organization and ecology of *Propithecus verreauxi*. Folia Primatol 22:178–207

5 A 15-Year Perspective on the Social Organization and Life History of Sifaka

Richard AF (1974b) Patterns of mating in *Propithecus verreauxi verreauxi*. In: Martin RD, Doyle GA, Walker AC (eds) Prosimian biology. Duckworth, London, pp 49–74

Richard AF (1978) Behavioral variation: case study of a Malagasy lemur. Bucknell University Press, Lewisburg

Richard AF (2003) *Propithecus*, sifakas. In: Goodman SM, Benstead JP (eds) The natural history of Madagascar. University of Chicago Press, Chicago, IL, pp 1345–1348

Richard AF, Nicoll ME (1987) Female social dominance and basal metabolism in a Malagasy primate, *Propithecus verreauxi*. Am J Primatol 12:309–314

Richard AF, Rakotomanga P, Schwartz M (1991) Demography of *Propithecus verreauxi* at Beza Mahafaly, Madagascar: sex ratio, survival and fertility, 1984–1988. Am J Phys Anthropol 84:307–322

Richard AF, Rakotomanga P, Schwartz M (1993) Dispersal by *Propithecus verreauxi* at Beza Mahafaly, Madagascar: 1984–1991. Am J Primatol 30:1–20

Richard AF, Dewar RE, Schwartz M, Ratsirarson J (2002) Life in the slow lane? Demography and life histories of male and female sifaka (*Propithecus verreauxi verreauxi*). J Zool Lond 256:421–436

Schmid J (2000) Daily torpor in the gray mouse lemur (*Microcebus murinus*) in Madagascar: energetic consequences and biological significance. Oecologia 123:175–183

Schmid J, Ruf T, Heldmaier G (2000) Metabolism and temperature regulation during daily torpor in the smallest primate, the pygmy mouse lemur (*Microcebus myosin's*) in Madagascar. J Comp Physiol B 170:59–68

Sorg JP, Rohner U (1996) Climate and tree phenology of the dry deciduous forest of the Kirindy Forest. Primate Rep 46–1:57–80

Sorg JP, Ganzhorn JU, Kappeler PM (2003) Forestry and research in the Kirindy Forest/Centre de Formation Professionnelle Forestière. In: Goodman SM, Benstead JP (eds) The natural history of Madagascar. University of Chicago Press, Chicago, IL, pp 1512–1519

Vick LG, Pereira ME (1989) Episodic targeting aggression and the histories of *Lemur* social groups. Behav Ecol Sociobiol 25:3–12

Weisrock DW, Rasoloarison RM, Fiorentino I, Ralison JM, Goodman SM, Kappeler PM, Yoder AD (2010) Delimiting species without nuclear monophyly in Madagascar's mouse lemurs. PLoS One 5:e9883. doi:10.1371/journal.pone.0009883

Wright PC (1995) Demography and life history of free-ranging *Propithecus diadema edwardsi* in Ranomafana National Park, Madagascar. Int J Primatol 16:835–854

Wright PC (1998) Impact of predation risk on the behaviour of *Propithecus diadema edwardsi* in the rain forest of Madagascar. Behaviour 135:483–512

Wright PC, Heckscher SK, Dunham AE (1997) Predation on Milne-Edward's sifaka (*Propithecus diadema edwardsi*) by the fossa (*Cryptoprocta ferox*) in the rain forest of southeastern Madagascar. Folia Primatol 68:34–43

Yoder AD, Rasoloarison RM, Goodman SM, Irwin JA, Atsalis S, Ravosa MJ, Ganzhorn JU (2000) Remarkable species diversity in Malagasy mouse lemurs (Primates, *Microcebus*). Proc Natl Acad Sci USA 97:11325–11330

Yoder AD, Olson LE, Hanley C, Heckman KL, Rasoloarison RM, Russell AL, Ranivo J, Soarimalala V, Karanth KP, Raselimanana AP, Goodman SM (2005) A multidimensional approach for detecting species patterns in Malagasy vertebrates. Proc Natl Acad Sci USA 102:6587–6594

Sussman R et al (2012) Beza Mahafaly Special Reserve long-term research on lemurs in southwestern Madagascar. In: Kappeler PM, Watts DP (eds) Long-term field studies of primates. Springer, Berlin

Wright PC et al (2012) Long-term lemur research at Centre Valbio, Ranomafana National Park, Madagascar. In: Kappeler PM, Watts DP (eds) Long-term field studies of primates. Springer, Berlin

Part III
America

Chapter 6
The Northern Muriqui (*Brachyteles hypoxanthus*): Lessons on Behavioral Plasticity and Population Dynamics from a Critically Endangered Species

Karen B. Strier and Sérgio L. Mendes

Abstract Since its onset in the early 1980s, our ongoing field study of the northern muriqui in southeastern Brazil has yielded original data on the behavioral ecology, reproductive biology, and life histories of one of the most critically endangered primates in the world. At the same time, a sixfold expansion in the size of our study population has provided insights into the plasticity of behavior and life history patterns that have important implications for muriqui conservation as well as for comparative models of primate socioecology. In this review of the history, growth, and diversification of our long-term study, we describe the transformation of our field site into a federally protected private reserve, the progression of the research questions as our knowledge has increased, and our predictions about the effects of increased population density on key demographic and life history variables. We also reiterate the need for more comparative studies of other muriqui populations, and reflect on the essential role that long-term, international collaborations have played in advancing the scientific and conservation agendas we have pursued from the start.

6.1 Introduction

Like most other members of the Atelinae, the northern muriqui (*Brachyteles hypoxanthus*) is a large-bodied New World monkey characterized by slow, ape-like life histories and a social system that includes male philopatry and female-biased dispersal (Strier 1992a, 1999a; Nishimura 2003; Di Fiore et al. 2011;

K.B. Strier (✉)
Department of Anthropology, University of Wisconsin-Madison, Madison, WI, USA
e-mail: kbstrier@wisc.edu

S.L. Mendes
Departamento de Ciências Biológicas, Universidade Federal do Espírito Santo, Vitória, ES, Brazil
e-mail: slmendes1@gmail.com

P.M. Kappeler and D.P. Watts (eds.), *Long-Term Field Studies of Primates*,
DOI 10.1007/978-3-642-22514-7_6, © Springer-Verlag Berlin Heidelberg 2012

Fig. 6.1 Male northern muriquis (*Brachyteles hypoxanthus*) at the RPPN-Feliciano Miguel Abdala, Minas Gerais, Brazil. Photo © Carla B. Possamai

Fig. 6.1). Unlike the other three genera of Atelinae (*Ateles, Lagothrix,* and *Oreonax*), muriquis are endemic to the Atlantic Forest of southeastern Brazil, with the southern muriqui (*Brachyteles arachnoides*) found in the states of Rio de Janeiro and São Paulo and in forest fragments in Paraná (Aguirre 1971; Koehler et al. 2002), and the northern muriqui, restricted to only a dozen of the remaining forest fragments in the states of Minas Gerais and Espírito Santo (Mendes et al. 2005a). With a known population of less than 1,000 individuals, the northern muriqui is one of the most critically endangered primates in the world (Mittermeier et al. 2006).

Concern for the muriqui's conservation status dates back more than 40 years, when it was still considered to be a monotypic genus and virtually nothing about its behavior and ecology was known (Aguirre 1971; Coimbra-Filho 1972). Subsequent analyses of the morphological and genetic differences between northern and southern populations led to their reclassification as separate species (Rylands et al. 1995; Groves 2001; Rylands and Mittermeier 2009). The two muriqui species face different kinds of primary threats from habitat loss and hunting, respectively (Strier and Fonseca 1996/1997).

The elevation of the northern muriqui to separate species status is only one of the many changes we have witnessed since 1982, when unbeknownst to us and by independent paths, we had separately visited the same small patch of privately owned Atlantic Forest, located on Fazenda Montes Claros in the municipality of Caratinga, Minas Gerais, and had each caught our first glimpses of a wild muriqui. One of us (K.B. Strier) had gone there with the explicit goal of assessing the feasibility of studying muriquis; the other (S.L. Mendes) went to evaluate the prospects for a study of the sympatric brown howler monkey (*Alouatta guariba*; previously, *Alouatta fusca*), another species endemic to the Atlantic Forest and just as poorly known as the muriqui. We both opted to return to the forest to pursue our respective studies in 1983, which thus marks the beginning of what has developed into the longest-running field study on the northern muriqui, as well as our mutual enduring interest in the future of this forest and the endangered primates it supports.

We begin this chapter by reviewing the history of our long-term study site and the administrative and ecological changes it has undergone as a result of international and Brazilian conservation efforts mediated through some key nongovernmental organizations (NGOs) and the conservation initiatives taken by members of the family that owns the forest. We then summarize some of the main findings that have emerged from the nearly 30 years that this muriqui population has been systematically monitored. As is true for most long-term field studies, the focus of our research has shifted from initial quantitative depictions of basic behavior and ecology to analyses of reproductive patterns and life history strategies, which are only possible to investigate with data and perspectives obtained over the course of multiple generations in the muriquis' lives (Strier 2003a, 2009; Strier and Mendes 2009). Our key findings thus include the insights into the behavioral consequences of demographic changes that could not have been made during a shorter study period. We also discuss some of our predictions about the effects of demography on fundamental variables such as female dispersal and male maturation. We conclude by reflecting on the synergy between the two driving forces – conservation and research – that have fueled the long-term study from its inception.

6.2 History and Ecology of the Study Site

In 1944, Sr. Feliciano Miguel Abdala purchased a plot of land about 60 km south of Caratinga, Minas Gerais, a town that sits at the crossroads between the transnational highways that connect São Paulo in the south with Bahia in the north, and Belo Horizonte in the west with Vitória in the east. The land, known as Fazenda Montes Claros, was a productive coffee plantation and cattle ranch, but also included what has remained one of the largest privately owned tracts of Atlantic Forest in the region. According to local history, Sr. Feliciano promised the fazenda's previous owner that he would preserve the forest and protect its inhabitants (Abdala Passos 2003). The area is known today as the Reserva Particular do Patrimônio Natural (RPPN)-Feliciano Miguel Abdala; its persistence as a federally recognized "Private Natural Heritage Reserve" is a testimony to Sr. Feliciano's honored promise (Castro 2001).

The presence of muriquis at this site was first reported to the scientific community by zoologist Álvaro Aguirre in 1971. By the late 1970s, zoologists from the Universidade Federal de Minas Gerais (UFMG), led by the indefatigable Professor Célio Valle, had initiated a campaign for the protection and scientific study of this area. Japanese primatologist Akisato Nishimura (1979) visited the site and provided some of the first observations of muriqui behavior. Other primatologists and conservationists, stimulated by Russell A. Mittermeier, Adelmar F. Coimbra Filho, and Ibsen G. Câmara, joined Célio Valle's group from UFMG in an international collaborative effort to study and preserve the local biodiversity (Mittermeier et al. 1982). Besides the presence of brown howler monkeys and muriquis, the discoveries of another endangered primate, the buffy-headed marmoset (*Callithrix flaviceps*), and of a large population of tufted capuchin monkeys (*Cebus nigritus*) signaled the forest's unusual potential to become a major site for field research on wild primates in the Atlantic Forest of Brazil. To help realize this potential, Sr. Feliciano donated a small house at the edge of the forest, and in collaboration with the UFMG, the World Wildlife Fund (WWF), and the Brazilian Foundation for the Conservation of Nature (FBCN), the Estação Biológica de Caratinga (EBC) was inaugurated in May 1983.

The EBC provided essential infrastructure for researchers and was thus critical for establishing our long-term study (Strier and Mendes 2003). Although the accommodations are simple, the house has undergone many improvements over the years, including the acquisition of electricity from nearby power lines, the expansion of living space for researchers and visitors, and the construction of a laboratory for storing and processing materials such as plants (Boubli et al. in press) and feces used for the noninvasive analyses of muriqui gastrointestinal parasites (Stuart et al. 1993; Santos et al. 2004), steroid hormones (e.g., Strier and Ziegler 1997, 2000), and genetics (Fagundes et al. 2008). The EBC has also undergone transitions in its administration, which passed from the FBCN to Fundação Biodiversitas, then to Conservation International-Brasil, and, in 2001, to the Sociedade para a Preservação do Muriqui, or SPM, established by the Abdala family to administer the activities in their Reserve.

Consistent with Sr. Feliciano's conservation vision for his forest, the Reserve's most important function continues to be the sanctuary it offers to its endangered flora and fauna. The natural regeneration of nearly 100 ha of pasture and agricultural land within and surrounding the Reserve has provided additional habitat that all four species of primates increasingly exploit (Strier and Boubli 2006). In addition, plans to create corridors to link the Reserve with some of the smaller forest fragments that remain on other privately owned properties surrounding it are underway; once established, the corridors will increase the available habitat for muriquis and thus should permit the population's continued expansion (Strier et al. 2005; Tabacow et al. 2009b).

The local predator community has increased in diversity along with the increased habitat protection and expansion over the past three decades. Suspected muriqui predators include tayra and large hawks (Printes et al. 1996), and possibly semi-feral dogs (Mourthé et al. 2007). The first ocelot since the onset of the study was sighted in 1990 (Strier 1999b), and scat analyses have since confirmed that muriquis in our study population are among their prey (Bianchi and Mendes 2007).

Over the past three decades, many of the plants and animals, and all four species of primates, have been the targets of systematic field studies. The integration of research with conservation efforts, like our own work with the muriquis, has characterized many of these other projects and benefitted from the strong international collaborations that were initiated decades ago when the value of the forest for conservation and for science was first recognized. These collaborations, along with our mutual commitment to capacity building, have been critically important to the continuity of our long-term study. Since 1983, some 45 Brazilian students have participated in the muriqui project alone, and many of these students have gone on to pursue scientific and conservation-oriented careers (Strier and Mendes 2003, 2009; Strier and Boubli 2006).

6.3 Demography, Group Dynamics, and Life Histories

Systematic studies were initiated on one of the two original muriqui groups (Matão group) present in the forest in 1982, yielding detailed individual life history data since July 1983. Systematic studies on the other groups in this population were initiated in 2002, yielding demographic and life history data on the entire population from 2003 through the present. One of these groups (Jaó) was present in 1982; the other two groups were established when the Jaó group fissioned in 1988 (M2 group) and 2002 (Nadir group) (Strier et al. 1993, 2006). Thus, as of June 2010, our accumulated demographic and life history data span a 28-year period on the Matão group and the last 7 years on the entire population.

All individuals in this isolated population can be identified by their natural markings and are monitored by teams of trained Brazilian students who are in the forest on a near-daily basis. This has made it possible to follow known individuals from their birth through maturity, and to track females after they disperse from their natal groups (Strier and Mendes 2009).

6.3.1 Population Expansion

The Matão group has increased steadily from 22 to 107 individuals between July 1982 and June 2010 (Fig. 6.2). Some of its original members are still alive, including five adult females estimated to be >35 years old. The increase in this group's size and in the number of muriqui groups (from 2 to 4) can be attributed to the increase in the population's size and density during its recovery from past disturbances that included the forest's initial fragmentation, selective logging, and fire (Strier 1999b; Strier and Boubli 2006). In the late 1960s, the muriqui population was estimated at 20–25 individuals (Aguirre 1971) and by the early 1980s, it was estimated at 40–50 individuals (Valle et al. 1984); as of June, 2010, the population included 288 individuals. This sixfold size increase in less than 30 years has offered us an unanticipated opportunity to document the muriquis' behavioral responses to demographic changes (e.g., Dias and Strier 2003; Strier et al. 2006; Tabacow et al. 2009a; Strier 2011).

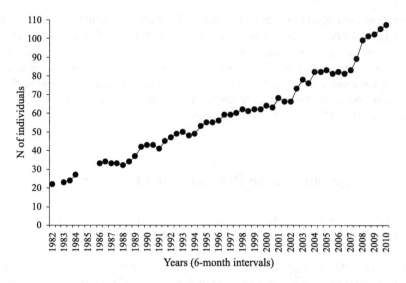

Fig. 6.2 Changes in the size of the Matão group. Group counts are shown at 6-month intervals, beginning in July 1982; gaps indicate missed counts. Updated from Strier (2005)

6.3.2 Behavioral Stability and Plasticity

Despite the group and population expansion, there have been no changes in the unusually peaceful, egalitarian relationships that distinguish the societies of northern muriquis from those of other primates (Strier 1990, 1994; also reviewed in Di Fiore et al. 2011). There is no overt evidence of dominance relationships among or between males and females, and both the rates and intensity of intragroup aggression have remained relatively low (Strier 1992b; Strier et al. 2000). Philopatric males maintain particularly close spatial associations and do not interfere with one another's access to fertile females, who routinely mate with multiple partners at times in their cycles when they are most likely to conceive (Strier 1997a; Strier and Ziegler 1997; Strier et al. 2002; Possamai et al. 2007). Despite more than 28 years of near-continuous field observations in our main study group, there has never been any indication of the kind of lethal aggression that has been observed in spider monkeys, which also show male philopatry (Campbell 2006), and in one population of southern muriquis living under different ecological conditions than those at our study site (Talebi et al. 2009).

Nonetheless, other behavioral changes can be directly attributed to the dramatic demographic changes the population has undergone. For example, within 15 years after the systematic monitoring had begun, the size of both the Matão group and its home range had roughly doubled. Although there was no corresponding increase in their daily travel distances, the group's previously cohesive pattern of association shifted to more fluid associations involving variably sized parties (Dias and Strier 2003) like those seen in other primates in which fission–fusion sociality is associated with the avoidance of direct feeding competition (Aureli et al. 2008). Indeed, shifts from cohesive to fluid grouping patterns appear to be a consistent response to increasing group sizes that distinguishes primates living in patrilocal societies from those living in matrilocal societies (Strier 2009).

The fluid grouping patterns that now characterize the Matão group may be responsible, at least in part, for the persistence of this group despite its continued (and ongoing) growth. In contrast to the Jaó group, which has undergone two prior fissioning events since the onset of the population monitoring in 1982 (Strier et al. 2006), the Matão group has maintained its integrity as a group, as evidenced by the routine reuniting of subgroups as well as by its fidelity to a common home range. Previous predictions about the Matão group's fate have focused on the effects of group size, age and sex composition, and intra- and intergroup male relationships (Strier et al. 1993; Strier 2011). For example, increasingly male-biased adult sex ratios might lead to higher rates of male–male aggression or even favor male dispersal if the sex ratios in other groups are more favorable than those in their natal groups (Strier et al. 2006). Evaluating these predictions is an empirical question that only ongoing observations over the long term will be able to resolve.

With roughly 0.30 muriquis per hectare, our study population's density is now one of the highest known for this species (Mendes et al. 2005a). Extensive home range overlap among the four groups that now occupy the forest (Boubli et al. 2005) and a decline in the density of sympatric brown howler monkeys, whose diets overlap with those of muriquis (Almeida-Silva et al. 2005), provide indirect indications that our study population might be approaching the carrying capacity of this forest. In addition, the documented increase in the use of terrestrial substrates by members of the Matão group may reflect an expansion in their vertical niches (Mourthé et al. 2007; Tabacow et al. 2009a). Moreover, recent sightings of nulliparous females from our study population moving between the Reserve and some neighboring forest fragments located adjacent to the Reserve suggest that they may be seeking new habitat to colonize as the population density of muriquis inside the Reserve has increased (Tabacow et al. 2009b).

Although much of our ongoing research is now focused explicitly on understanding the effects of habitat saturation on the muriquis' behavior and ecology, there are also some intriguing illustrations of the synergistic interactions between behavior and population dynamics that we are simultaneously pursuing. For example, the increase in terrestrial behavior in our main study group appears to have occurred along two dimensions: The first involved the expansion from essential activities, such as feeding and drinking, to include nonessential activities, such as resting and socializing; the second involved the spread of the behavior along male social networks and subsequently to females including recent female immigrants to the group (Tabacow et al. 2009a). The terrestrial behavior of the Matão group can be considered a new local tradition according to the criteria applied to other types of traditions in other organisms (see Perry et al. 2011): it was socially transmitted and adopted by the all or most group members, and it has endured over time (Tabacow et al. 2009a). Consequently, we are now investigating whether females who have acquired this terrestrial tradition in their natal Matão group import it into the groups into which they disperse, and thereby effectively diffuse the custom of engaging in nonessential terrestrial activities throughout the population.

6.3.3 Habitat Saturation and Female Dispersal

We can expect flexibility in life history components in response to density dependent effects associated with habitat saturation. For example, the effects of high levels of intragroup competition in large groups might include delays in the maturation of philopatric individuals and earlier dispersal ages in the dispersing sex, whereas high levels of intergroup competition in saturated habitats can result in delays in dispersal age and in extreme cases may lead to the permanent retention of both sexes in their natal groups (e.g., Alberts and Altmann 1995; Ferrari and Digby 1996; Altmann and Alberts 2003; Charpentier et al. 2008).

Plasticity in dispersal provides a mechanism for adjusting the size and sex ratio of breeding groups relative to those of other groups in the population (Moore 1992; Strier 2003b). Dispersal costs previously documented for female muriquis include reduced survivorship, with 28.4% mortality estimated for 38 females of dispersal age (5–7 years) from 2002 to 2007 compared to 4.55% mortality for philopatric males in the same age class. There was also significantly later ages at first reproduction compared to females that remained and reproduced in their natal group. Specifically, natal Matão females have dispersed at 5.25–7.85 years of age (mean $= 6.15 \pm 0.60$ years, $N = 34$; updated from Strier et al. 2006), prior to the onset of sexual activity or hormonal evidence of puberty (Strier and Ziegler 2000). Age at first reproduction was significantly earlier in the Matão females that reproduced in their natal group (7.77 ± 0.72 years, $N = 3$) compared to females that dispersed from the Matão group and whose first reproductions in their new groups have been documented (9.58 ± 0.87 years, $N = 9$; $z = 2.40$, $p < 0.02$); this difference can be attributed to the earlier onset of puberty in the non-dispersing females instead of shorter cycling-to-first conception delays (Martins and Strier 2004).

Delayed reproduction associated with dispersal has been documented in red howler monkeys (*Alouatta seniculus*), where it may be associated with eviction from their natal groups (Crockett and Pope 1993), but not in mountain gorillas (*Gorilla beringei*), where the voluntary dispersal of nulliparous females from their natal groups (Robbins et al. 2009) more closely resembles that of muriquis to date (Printes and Strier 1999). Nonetheless, fluctuations in group and population conditions in our study population could alter both the contexts under which females disperse and the consequences of dispersal for reproduction. For example, if females are sensitive to levels of intragroup competition, then we might expect the age at dispersal of female muriquis to decline as the size of their natal group increases. However, there is no evidence that this has occurred over the decades during which the Matão group has been monitored (Fig. 6.3).

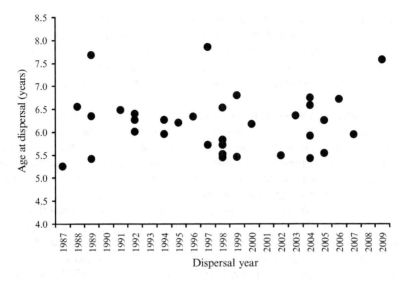

Fig. 6.3 Female dispersal age over time. Natal Matão group females ($N = 34$) have dispersed between 5.25 and 7.85 years of age, and there is no evidence for changes in female dispersal age despite increases in group size, population size, and population density over time

Alternatively, if females are sensitive to intergroup competition associated with population density, then they might postpone dispersal from their natal groups, and experience higher levels of aggression from females in the groups they try to join and higher mortality costs during dispersal as population density increases. Although high population density should facilitate encounter rates with other groups (Strier 2000a), increased competition and aggression should make joining these groups more difficult now than it has been in the past. However, although we found early on that the rate at which females received displacements from other female was higher for new immigrants than for long-term residents (Printes and Strier 1999), this difference has not increased, nor have we seen increases in the rate of aggression directed at immigrants.

6.3.4 Male Maturation and Philopatry

Changes in female dispersal and maturation can have significant implications for breeding sex ratios as population density has increased, and thus affect levels of male competition within and between groups, particularly in species in which males remain in their natal groups (Strier 2000b, 2011). To date, males in the Matão group become sexually active between 4.10 and 8.27 years of age (mean $= 6.18 \pm 1.04$, $N = 26$) and when they have reached sexual maturity, defined behaviorally by their first observed copulation that culminates with ejaculation, between 5.21 and 8.36 years of age (mean $= 6.81 \pm 0.86$, $N = 25$). Males in the Matão group have thus entered the potential breeding population about 0.3 years (4 months) younger, on average, than the age at which the females that have (atypically) remained in their natal groups have conceived (7.10 years; calculated from subtracting the hormonally determined mean gestation length of 216 days, or 0.59 years; Strier and Ziegler 1997), and more than 2.2 years earlier than the average age at which dispersing females have conceived.

Despite more than a 3-year range of variation in age of male sexual maturity, there has been no pattern of change in the variation over time (Fig. 6.4). This consistency in the age of sexual maturation in male muriquis might reflect a balance between demographic pressures that should delay maturation on the one hand, and the effects of increased competition on survivorship that should favor faster maturation on the other hand (Janson and van Schaik 1993). Alternatively, it could reflect a lag between the effects of group size and density effects, such that social pressures have not yet begun to exert the expected delaying effect on male life histories (Joffe 1997). As operational sex ratio in the Matão group becomes increasingly male-biased (Strier et al. 2006) and scramble competition for access to females correspondingly increases (Strier et al. 2002; Strier 2003b), we expect increasing delays in the onset of male sexual maturity and possibly dispersal to groups with more favorable breeding sex ratios. Extraordinary demographic conditions may underlie other examples of deviations from male philopatry, such as in an isolated chimpanzee community at Bossou, New Guinea (Sugiyama 2004), in the case of male bonobos thought to have transferred into a community with more favorable sex ratios (Hohmann 2001), and in one group of wooly monkeys in a population that might have been affected by past hunting (Di Fiore and Fleischer

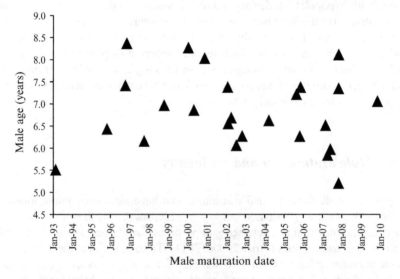

Fig. 6.4 Male age at sexual maturity over time. Matão males ranged from 5.21 to 8.36 years ($N = 25$) and there was no obvious change over time despite increases in group size, population size, and population density

6 The Northern Muriqui (*Brachyteles hypoxanthus*)

2005; Di Fiore et al. 2009). Indeed, the social flexibility seen in both woolly monkeys and northern muriquis may extend to their dispersal patterns (Di Fiore and Strier 2004), and thus provide a mechanism to offset the predicted effects of increased social and ecological competition on male maturational stages.

6.4 Past Perspectives and Prospectives

After nearly 30 years of field research on the northern muriqui, we have gained some clear insights into features of the behavioral ecology, reproductive biology, and life histories of our study population. Although our understanding of other muriqui populations is still limited, both the changes and the equally important lack of changes we have documented in our study population provide a crucial comparative context for research on these populations. Neither the extent of the population expansion nor the corresponding behavioral changes that our study has tracked were anticipated aims of the project from the onset. Yet, in many respects it has been this first-hand glimpse into the dynamics of the muriquis' lives that has captured – and continues to hold – our attention.

Reflecting on the history of our long-term field study, five particular points come to mind, and all are connected in one way or another to the northern muriquis' critically endangered status. Perhaps the most obvious of these has been the methodological choices we have made to minimize the impact of our research on the animals. Thus, we have restricted ourselves and our students to observational studies instead of field experiments, even though the careful deployment of feeding platforms or of controlled playback techniques might have permitted us to discriminate between correlated and causal responses. We have also opted to restrict investigations of muriqui parasites, hormones, and genetics to what could be gleaned from analyses of their dung. This has sometimes meant greater costs and longer delays than we might have incurred, for example, if we had captured the muriquis to obtain samples of their blood for genetic analyses before the techniques for extracting and amplifying DNA from dung were developed. Yet, even with these self-imposed constraints, we are cautious about the possible impacts of our long-term research presence (Strier 2010).

A second major consideration pertains to the demographic fluctuations and behavioral plasticity we have documented. The increases in the size of our main study group and the entire study population have clearly shown that primate groups and populations are not static entities with fixed properties, but instead, that they have group and population histories of their own (Strier 1997b). This realization raises all kinds of questions about how much we can conclude about the behavioral adaptations of northern muriquis – or any other primates – that are currently living under what are likely to be highly altered ecological and demographic conditions (Strier 2003b, 2009). Indeed, after all of these years, we cannot determine what group size or population density would be typical for this species or the extent to which these vary among populations. Nonetheless, we think that questions about how much of the behavior we observe in our study subjects today reflect adaptations as a result of past selection pressures versus mismatches between past and current conditions that could negatively impact their survival are important and interesting, not only for northern muriqiuis, but for all other primates whose futures are threatened (Strier and Mendes 2009).

Demographic changes in our main study group were responsible for a third significant shift in our perspective, for as the number of natal female emigrants began to exceed the number of female immigrants, our focus expanded from studying a single group to studying the dynamics of the entire population (Strier 2005; Strier et al. 2006). This catapulted the importance of this particular population – and of this particular forest, which supports nearly 30% of the entire species – into a broader perspective, stimulating our interest in facilitating new research initiatives on the vegetation and ecology of the forest (Strier and Boubli 2006; Boubli et al. in press) and our ongoing collaboration with the NGO that now administers the protected reserve.

The expansion and diversification of research contributed to a fourth major advance, which involved the launching of comparative field studies of other populations of northern muriquis. This includes one at the municipality of Santa Maria de Jetibá (SMJ), Espírito Santo State (Mendes et al. 2005b) that is particularly valuable because the demography of that population differs greatly from that at the RPPN-FMA. Whereas our study population includes multiple large social groups in a single isolated forest of nearly 1,000 ha, the SMJ population is represented by at least 13 single small groups of 2–20 individuals occupying forest fragments of 60–350 ha, where opportunities for dispersal are limited (updated from Mendes et al. 2005b). Although the behavioral and ecological comparisons between these populations hold great theoretical potential, our greatest concern at the moment is how to apply what we have learned about northern muriquis so far to insure the survival of both of these – and other – remaining populations.

A final reflection that we are in a privileged position to make involves our ongoing appreciation for the synergy that can come from a true international collaboration (Strier and Mendes 2009). The continuity in our long-term study of the northern muriqui would not have been possible without it.

Acknowledgments We thank CNPq for permission for us to conduct research in Brazil and for the support of part of this work, the Abdala family for permission to conduct this research at the RPPN Feliciano Miguel Abdala, and the Sociedade para a Preservação do Muriqui (Preserve Muriqui), Conservation International (CI) and CI-Brasil for their help with logistics and long-term collaboration. We thank the many people who have contributed to the long-term demographic data records (in alphabetical order): L. Arnedo, M.L. Assunção, N. Bejar, J.P. Boubli, A. Carvalho, D. Carvalho, C. Cäsar, A.Z. Coli, C.G. Costa, P. Coutinho, L. Dib, Leonardo G. Dias, Luiz G. Dias, D.S. Ferraz, J. Fidelis, J. Gomes, D. Guedes, V.O. Guimarães, R. Hack, M.F. Iurck, M. Kaizer, M. Maciel, W.P. Martins, F.D.C. Mendes, I.M. Mourthé, F. Neri, M. Nery, S. Neto, C.P. Nogueira, A. Odalia Rímoli, A. Oliva, L. Oliveira, F.P. Paim, C.B. Possamai, R.C. Printes, J. Rímoli, S.S. Rocha, R.C. Romanini, R.R. dos Santos, B.G.M. da Silva, J.C. da Silva, V. Souza, D.V. Slomp, F.P. Tabacow, W. Teixeira, M. Tokudo, K. Tolentino, and E.M. Veado. We especially thank Carla de Borba Possamai and Fernanda Pedreira Tabacow for their commitment to the long-term demographic data. The field study has been supported by a variety of sources, including the National

6 The Northern Muriqui (*Brachyteles hypoxanthus*)

Science Foundation (BNS 8305322, BCS 8619442, BCS 8958298, BCS 9414129, BCS 0621788, BCS 0921013), National Geographic Society, the Liz Claiborne and Art Ortenberg Foundation, Fulbright Foundation, Sigma Xi Grants-in-Aid, Grant #213 from the Joseph Henry Fund of the NAS, World Wildlife Fund, L.S.B. Leakey Foundation, Chicago Zoological Society, Lincoln Park Zoo Neotropic Fund, Center for Research on Endangered Species (CRES), Margot Marsh Biodiversity Foundation, Conservation International, the University of Wisconsin-Madison, and CNPq – Brazilian National Research Council. This research has complied with all U.S. and Brazilian regulations. We thank Peter Kappeler for inviting us to participate in the conference that led to this volume, and for including our contribution here despite our inability to attend. We appreciate the comments that he and David Watts provided on an earlier version of this chapter.

References

Abdala Passos R (2003) A legacy of love for nature. In: da Fonseca MT (ed) 20 anos da Estação Biológica de Caratinga: reserva privada do patrimônio natural. Conservation International do Brasil, Belo Horizonte, pp 6–9

Aguirre AC (1971) O mono *Brachyteles arachnoides* (E. Geoffroy): situação atual da espécie no Brasil. Academia Brasileira de Ciências, Rio de Janeiro

Alberts SC, Altmann J (1995) Balancing costs and opportunities: dispersal in male baboons. Am Nat 145:279–306

Almeida-Silva B, Cunha AA, Boubli JP, Mendes SL, Strier KB (2005) Population density and vertical stratification of four primate species at the Estação Biológica de Caratinga/RPPN-FMA, Minas Gerais, Brazil. Neotrop Primates 13(suppl):25–29

Altmann J, Alberts SC (2003) Variability in reproductive success viewed from a life-history perspective in baboons. Am J Hum Biol 15:401–409

Aureli F, Schaffner CM, Boesch C, Bearder SK, Call J, Chapman CA, Connor R, Di Fiore A, Dunbar RIM, Henzi SP, Holekamp K, Korstjens AH, Layton R, Lee PC, Lehmann J, Manson JH, Ramos-Fernandez G, Strier KB, van Schaik CP (2008) Fission-fusion dynamics: new research frameworks. Curr Anthropol 49:627–654

Bianchi RC, Mendes SL (2007) Ocelot (*Leopardus pardalis*) predation on primates in Caratinga Biological Station, Southeast Brazil. Am J Primatol 69:1173–1178

Boubli JP, Couto-Santos F, Strier KB (in press) Structure and floristic composition of a semideciduous Atlantic Forest fragment in Minas Gerais, Brazil: implications for the conservation of the critically endangered northern muriqui, *Brachyteles hypoxanthus*. Ecotropica

Boubli JP, Tokuda M, Possamai CB, Fidelis J, Guedes D, Strier KB (2005) Dinâmica intergrupal de muriquis-do-norte, Brachyteles hypoxanthus, na Estação Biológica de Caratinga, MG: o comportamento de uma unidade de machos (all male band) no vale do Jaó. Paper presented at the XI Congresso Brasileiro de Primatologia, Porto Alegre, Brasil

Campbell CJ (2006) Lethal intragroup aggression by adult male spider monkeys (*Ateles geoffroyi*). Am J Primatol 68:1197–1201

Castro MI (2001) RPPN Feliciano Miguel Abdala – a protected area for the northern muriqui. Neotrop Primates 9:128–129

Charpentier MJE, Tung J, Altmann J, Alberts SC (2008) Age at maturity in wild baboons: genetic, environmental and demographic influences. Mol Ecol 17:2026–2040

Coimbra-Filho AF (1972) Mamíferos ameaçados de extinção no Brasil. In: Academia Brasileira de Ciências (ed) Espécies da Fauna Brasileira Ameaçadas de Extinção. Academia Brasileira de Ciências, Rio de Janeiro, pp 13–98

Crockett CM, Pope TR (1993) Consequences of sex differences in dispersal for juvenile red howler monkeys. In: Pereira ME, Fairbanks LA (eds) Juvenile primates: life history, development, and behavior. Oxford University Press, New York, pp 104–118

Di Fiore A, Fleischer RG (2005) Social behavior, reproductive strategies, and population genetic structure of *Lagothrix poeppigii*. Int J Primatol 26:1137–1173

Di Fiore A, Strier KB (2004) Flexibility in social organization in atelin primates. Folia Primatol 75(suppl 1):140–141

Di Fiore A, Link A, Schmitt CA, Spehar SN (2009) Dispersal patterns in sympatric woolly and spider monkeys: integrating molecular and observational data. Behaviour 146:437–470

Di Fiore A, Link A, Campbell CJ (2011) The atelines: behavioral and socioecological diversity in a New World radiation. In: Campbell CJ, Fuentes A, MacKinnon KC, Bearder SK, Stumpf RM (eds) Primates in perspectives, 2nd edn. Oxford University Press, New York, pp 155–188

Dias LG, Strier KB (2003) Effects of group size on ranging patterns in *Brachyteles arachnoids hypoxanthus*. Int J Primatol 24:209–221

Fagundes V, Paes MF, Chaves PB, Mendes SL, Possamai CB, Boubli JP, Strier KB (2008) Genetic structure in two northern muriqui populations (*Brachyteles hypoxanthus*, Primates, Atelidae) as inferred from fecal DNA. Genet Mol Biol 31:166–171

Ferrari SF, Digby LJ (1996) Wild *Callithrix* groups: stable extended families? Am J Primatol 38:19–27

Groves CP (2001) Primate taxonomy. Smithsonian Institution Press, Washington, DC

Hohmann G (2001) Association and social interactions between strangers and residents in bonobos (*Pan paniscus*). Primates 42:91–99

Janson CH, van Schaik CP (1993) Ecological risk aversion in juvenile primates: slow and steady wins the race. In: Pereira ME, Fairbanks LA (eds) Juvenile primates: life history, development, and behavior. Oxford University Press, New York, pp 57–74

Joffe TH (1997) Social pressures have selected for an extended juvenile period in primates. J Hum Evol 32:593–605

Koehler A, Pereira LCM, Nicola PA (2002) New locality for the woolly spider monkey, *Brachyteles arachnoides* (E. Geoffroy, 1806) in Paraná state, and the urgency of strategies for conservation. Estudos de Biologia 24:25–28

Martins WP, Strier KB (2004) Age at first reproduction in philopatric female muriquis (*Brachyteles arachnoides hypoxanthus*). Primates 45:63–67

Mendes SL, de Melo FR, Boubli JP, Dias LG, Strier KB, Pinto LPS, Fagundes V, Cosenza B, De Marco Jr P (2005a) Directives for the conservation of the northern muriqui, *Brachyteles hypoxanthus* (Primates, Atelidae). Neotrop Primates 13(suppl):7–18

Mendes SL, Santos RR, Carmo LP (2005b) Conserving the northern muriqui in Santa Maria de Jetibá, Espírito Santo. Neotrop Primates 13(suppl):31–35

Mittermeier RA, Coimbra-Filho AF, Constable ID, Rylands AB, Valle C (1982) Conservation of primates in the Atlantic forest region of eastern Brazil. Int Zoo Yrbk 22:2–17

Mittermeier RA, Valladares-Pádua C, Rylands AB, Eudey AA, Butynski TM, Ganzhorn JU, Kormos R, Aguiar JM, Walker S (2006) Primates in Peril: the world's 25 most endangered primates, 2004–2006. Primate Conserv 20:1–28

Moore J (1992) Dispersal, nepotism, and primate social behavior. Int J Primatol 13:361–378

Mourthé IMC, Guedes D, Fidelis J, Boubli JP, Mendes SL, Strier KB (2007) Ground use by northern muriquis (*Brachyteles hypoxanthus*). Am J Primatol 69:706–712

Nishimura A (1979) In search of woolly spider monkey. Kyoto Univ Overseas Res Rep New World Monkeys 1:21–37

Nishimura A (2003) Reproductive parameters of wild female *Lagothrix lagotricha*. Int J Primatol 24:707–722

Perry S, Godoy I, Lammers W (2011) The Lomas Barbudal Monkey Project: Two Decades of Research on *Cebus capucinus*. In: Kappeler PM (ed) Long-term field studies of primates. Springer, Heidelberg

Possamai CB, Young RJ, Mendes SL, Strier KB (2007) Socio-sexual behavior of female northern muriquis (*Brachyteles hypoxanthus*). Am J Primatol 69:766–776

Printes RC, Strier KB (1999) Behavioral correlates of dispersal in female muriquis (*Brachyteles arachnoides*). Int J Primatol 20:941–960

6 The Northern Muriqui (*Brachyteles hypoxanthus*)

Printes RC, Costa CG, Strier KB (1996) Possible predation on two infant muriquis, *Brachyteles arachnoides*, at the Estação Biologica de Caratinga, Minas Gerais, Brasil. Neotrop Primates 4:85–86

Robbins AM, Stoinski TS, Fawcett KA, Robbins MM (2009) Does dispersal cause reproductive delays in female mountain gorillas? Behaviour 146:525–549

Rylands AB, Mittermeier RA (2009) The diversity of the New World primates (Platyrrhini): an annotated taxonomy. In: Garber PA, Estrada A, Bicca-Marques JC, Heymann EW, Strier KB (eds) South American primates: comparative perspectives in the study of behavior, ecology, and conservation. Springer, New York, pp 23–54

Rylands AB, Mittermeier RA, Rodriguez Luna E (1995) A species list for the New World primates (Platyrrhini): distribution by country, endemism, and conservation status according to the Mace-Land system. Neotrop Primates 3(suppl):113–160

Santos SMC, Nogueira CP, Carvalho ARD, Strier KB (2004) Levantamento coproparasitológico em muriqui (*Brachyteles arachnoides hypoxanthus*). In: Mendes SL, Chiarello AG (eds) A primatologia no Brasil-8. IPEMA/Socieda de Brasileira de Primatologia, Vitória, Espírito Santo, pp 327–332

Strier KB (1990) New World primates, new frontiers: insights from the woolly spider monkey, or muriqui (*Brachyteles arachnnoides*). Int J Primatol 11:7–19

Strier KB (1992a) Atelinae adaptations: behavioral strategies and ecological constraints. Am J Phys Anthropol 88:515–524

Strier KB (1992b) Causes and consequences of nonaggression in woolly spider monkeys, or muriqui (*Brachyteles arachnnoides*). In: Silverberg J, Gray JP (eds) Aggression and peacefulness in humans and other primates. Oxford University Press, New York, pp 100–116

Strier KB (1994) Brotherhoods among atelins: kinship, affiliation, and competition. Behaviour 130:151–167

Strier KB (1997a) Mate preferences of wild muriqui monkeys (*Brachyteles arachnoides*): reproductive and social correlates. Folia Primatol 68:120–133

Strier KB (1997b) Behavioral ecology and conservation biology of primates and other animals. Adv Stud Behav 26:101–158

Strier KB (1999a) The atelines. In: Dolhinow P, Fuentes A (eds) The nonhuman primates. McGraw Hill, New York, pp 109–114

Strier KB (1999b) Faces in the forest: the endangered muriqui monkeys of Brazil. Harvard University Press, Cambridge, MA

Strier KB (2000a) Population viabilities and conservation implications for muriquis (*Brachyteles arachnoides*) in Brazil's Atlantic forest. Biotropica 32:903–913

Strier KB (2000b) From binding brotherhoods to short-term sovereignty: the dilemma of male Cebidae. In: Kappeler PM (ed) Primate males: causes and consequences of variation in group composition. Cambridge University Press, Cambridge, pp 72–83

Strier KB (2003a) Primatology comes of age: 2002 AAPA luncheon address. Yrbk Phys Anthropol 122:2–13

Strier KB (2003b) Demography and the temporal scale of sexual selection. In: Jones CB (ed) Sexual selection and reproductive competition in primates: new perspectives and directions. American Society of Primatologists, Norman, OK, pp 45–63

Strier KB (2005) Reproductive biology and conservation of muriquis. Neotrop Primates 13 (suppl):41–46

Strier KB (2009) Seeing the forest through the seeds: mechanisms of primate behavioral diversity from individuals to populations and beyond. Curr Anthropol 50:213–228

Strier KB (2010) Long-term field studies: positive impacts and unintended consequences. Am J Primatol 72:772–778

Strier KB (2011) Social plasticity and demographic variation in primates. In: Sussman RW, Cloninger CR (eds) Origins of altruism and cooperation. Springer, New York

Strier KB, Boubli JP (2006) A history of long-term research and conservation of northern muriquis (*Brachyteles hypoxanthus*) at the Estação Biológica de Caratinga/RPPN-FMA. Primate Conserv 20:53–63

Strier KB, Fonseca GAB (1996/1997) The endangered muriquis of Brazil's Atlantic forest. Primate Conserv 17:131–137

Strier KB, Mendes SL (2003) Research center. In: Fonseca MT (ed) 20 anos da Estação Biológica de Caratinga: reserva privada do patrimônio natural. Conservation International do Brasil, Belo Horizonte, pp 18–22

Strier KB, Mendes SL (2009) Long-term field studies of South American primates. In: Garber PA, Estrada A, Bicca-Marques JC, Heymann EW, Strier KB (eds) South American primates: comparative perspectives in the study of behavior, ecology, and conservation. Springer, New York, pp 139–155

Strier KB, Ziegler TE (1997) Behavioral and endocrine characteristics of the reproductive cycle in wild muriqui monkeys, *Brachyteles arachnoides*. Am J Primatol 42:299–310

Strier KB, Ziegler TE (2000) Lack of pubertal influences on female dispersal in muriqui monkeys, *Brachyteles arachnoides*. Anim Behav 59:849–860

Strier KB, Mendes FDC, Rímoli J, Rímoli AO (1993) Demography and social structure of one group of muriquis (*Brachyteles arachnoides*). Int J Primatol 14:513–526

Strier KB, Carvalho DS, Bejar NO (2000) Prescription for peacefulness. In: Aureli F, de Waal FBM (eds) Natural conflict resolution. University of California Press, Berkeley, pp 315–317

Strier KB, Dib LT, Figueira JEC (2002) Social dynamics of male muriquis (*Brachyteles arachnoides hypoxanthus*). Behaviour 139:315–342

Strier KB, Pinto LPS, Paglia A, Boubli JP, Mendes SL, Marini-Filho OJ, Rylands AB (2005) The ecology and conservation of the muriqui (*Brachyteles*): reports from 2002–2005. Introduction. Neotrop Primates 13(suppl):3–5

Strier KB, Boubli JP, Possamai CB, Mendes SL (2006) Population demography of northern muriquis (*Brachyteles hypoxanthus*) at the Estação Biológica de Caratinga/Reserva particular do Patrimônio Natural-Feliciano Miguel Abdala, Minas Gerais, Brazil. Am J Phys Anthropol 130:227–237

Stuart MD, Strier KB, Pierberg SM (1993) A coprological survey of parasites of wild muriquis, *Brachyteles arachnoides*, and brown howling monkeys, *Alouatta fusca*. J Helminthol Soc Wash 60:111–115

Sugiyama Y (2004) Demographic parameters and life history of chimpanzees at Bossou, Guinea. Am J Phys Anthropol 124:154–165

Tabacow FP, Mendes SL, Strier KB (2009a) Spread of a terrestrial tradition in an arboreal primate. Am Anthropol 111:238–249

Tabacow FP, Possamai CB, de Melo FR, Mendes SL, Strier KB (2009b) New sightings of northern muriqui (*Brachyteles hypoxanthus*) females in forest fragments surrounding the Estação Biológica de Caratinga-RPPN Feliciano Miguel Abdala, Minas Gerais, Brasil. Neotrop Primates 16:67–69

Talebi MG, Beltrão-Mendes R, Lee PC (2009) Intra-community coalitionary lethal attack of an adult male southern muriqui (*Brachyteles arachnoides*). Am J Primatol 71:860–867

Valle CMC, Santos IB, Alves MC, Pinto CA, Mittermeier RA (1984) Preliminary observations on the behavior of the monkey (*Brachyteles arachnoides*) in a natural environment (Fazenda Montes Claros, Município de Caratinga, Minas Gerais, Brasil). In: Thiago de Mello M (ed) A primatologia no Brasil. Sociedade Brasileira de Primatologia, Belo Horizonte, pp 271–283

Chapter 7
The Lomas Barbudal Monkey Project: Two Decades of Research on *Cebus capucinus*

Susan Perry, Irene Godoy, and Wiebke Lammers

Abstract The Lomas Barbudal Monkey Project began in 1990 with the study of a single white-faced capuchin monkey (*Cebus capucinus*) group, and has since expanded to 11 groups. Social behavior has always been the primary focus of our research, with emphasis on communication, social learning, and life history strategies. Genetic analyses in the context of this long-term study have enabled research of many standard behavioral ecology topics such as kin-based altruism, reproductive skew, and inbreeding avoidance. Long-term research on numerous groups, and collaboration with researchers at other *C. capucinus* sites, has permitted the documentation of social traditions regarding both communicative rituals and foraging techniques.

7.1 The History and Infrastructure of the Study Site

7.1.1 History

The Lomas Barbudal Monkey Project began with my (SP's) dissertation work on the evolution of intelligence. I was looking for a stable country where I could begin a long-term project investigating social relationships and social intelligence in capuchin monkeys, which were known for their large relative brain sizes. Costa Rica, with its friendly policy toward researchers, excellent environmental record, superb medical system and lack of an army, was the obvious choice. Following the

---

S. Perry (✉) • I. Godoy
Department of Anthropology, Center for Behavior, Evolution and Culture, University of California, Los Angeles, CA, USA
e-mail: sperry@anthro.ucla.edu; godoy@ucla.edu

W. Lammers
Proyecto de Monos, Bagaces, Costa Rica
e-mail: wiebkelammers@gmail.com

P.M. Kappeler and D.P. Watts (eds.), *Long-Term Field Studies of Primates*,
DOI 10.1007/978-3-642-22514-7_7, © Springer-Verlag Berlin Heidelberg 2012

advice of Colin Chapman, who had censused monkeys in Guanacaste province, I settled on Lomas Barbudal as a study site and conducted a pilot study in 1990.

Lomas Barbudal Biological Reserve is a tropical dry forest site located in Guanacaste Province, northwestern Costa Rica. It was established as a reserve by UC-Berkeley entomologist Gordon Frankie, who has described the vegetation structure (Frankie et al. 1988). The monkeys range well outside the reserve into other public and private lands. This highly disturbed forest includes riparian forest, dry deciduous forest, mesic forest, and regenerative forest. Lomas receives 1,000–2,200 mm of rain annually between the months of May–November (Frankie et al. 1988), and fires are common in the dry season. Lomas Barbudal is approximately 55 km away from the better-known Santa Rosa National Park, which also hosts a long-term study of white-faced capuchins (see Fedigan and Jack 2012). The two sites have similar ecologies and plant lists (Panger et al. 2002); however, the Lomas monkeys have plenty of access to fresh running water, whereas the Santa Rosa monkeys rely heavily on waterholes in the dry season.

In 1991, the year following my pilot study, Joseph Manson and Julie Gros-Louis joined me as field assistants for my thesis research and became co-founders of the site, assisting me in the set-up of the site and co-managing it with me until the end of 2001. For the first 4 years, we documented the social behavior of a single group, Abby's group (AA).

Following my thesis research, Joe Manson and I received half-time tenure track jobs at UCLA, and began developing plans to make Lomas a long-term project. Our UCLA startup funds allowed for the purchase of a 1977 LandCruiser and a tent, and funds from my postdoctoral fellowships purchased Psion handheld computers, to streamline data collection. From 1994 to 1998, we pitched our tents on a rice farm owned by some *campesino* friends (the Rosales family) and ran our laptop computer on a truck battery powered by solar panels. In the meantime, Julie Gros-Louis began her dissertation research and was primarily responsible for the habituation of a second study group (Rambo's, RR). Because all of us had commitments teaching or taking classes at our respective universities, there were periods of time when no one could be present at the site. We initially attempted to solve this problem by inviting graduate students from other universities to work there in exchange for contributing to the demographic database. This, however, was a failure in terms of the long-term goals of the project; visiting students were not willing to invest the time to learn the identities of young monkeys who were not their focal subjects, or to track the movements of migrating males. And it was difficult to persuade anyone to continue data collection through the worst of the rainy season. It became clear that obtaining an accurate demographic database required employing someone whose sole responsibility was to collect these data for the project during our absence.

By 2001, things were not looking good for the future of the field site. The rice farm where we lived had been sold. Julie was finishing her final field season for her doctoral research. To ensure proper management of the site and the long-term database, it was imperative that I maintain a continuous presence at the site, and this was not possible while I was employed by UCLA. At this critical juncture,

I was fortunate to be appointed director of the Cultural Phylogeny research group at the Max Planck Institute for Evolutionary Anthropology (MPI-EVAN) in Leipzig. Germany. This job allowed me to devote all of my time to research and to hire a large crew to run the project when I was in Germany for 6 months of each year. The project rented two adjacent houses in the nearby town of Bagaces (35 min from the forest where we worked), which had electricity, running water, and even internet access, so that I could supervise my crew's work remotely. The modernization of the site made it possible to attract field assistants who would stay for a year or more.

During the MPI phase of the project, we employed 6–9 interns per year, plus a permanent staff of seven people. The MPI-EVAN job ended in 2006, at which time I returned to my half-time job at UCLA. Thus far we have succeeded in keeping the project running on far less funding than was available at MPI, via short-term grants from NSF, the Leakey Foundation, and the National Geographic Society. We still maintain a staff of six interns, two full-time permanent staff and two part-time permanent staff. But now that we lack the security of long-term funding from MPI, we are extremely vulnerable to funding crises that could suddenly end the project at any time. This insecurity makes it harder to retain valuable staff members.

Beginning in January 2002, we completed habituation of a third study group (Pelon, FF) and were able to monitor all of our groups for up to 25 days/month year-round. These groups grew and fissioned (see Fig. 7.1), and males migrated to new groups. Currently, we have 11 study groups, nine of which we follow for several days per month, and two of which we visit more sporadically. Since 1994, we have collected fecal samples for genetic analysis from virtually all members of the study groups (aside from infants who die before we can obtain samples), and since 2006 we have been collecting fecal samples for hormonal analysis (fecal corticosteroids and testosterone). Our genetic and hormonal work has been carried out by graduate

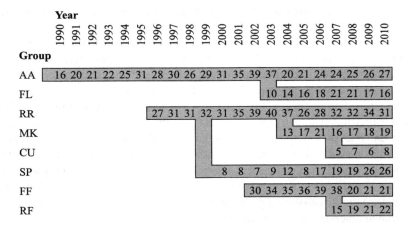

Fig. 7.1 Maximum group sizes during each year of observation, and timing of fissions. For fission years, we report the size of the group right before the fission for the larger fission product, and the size of the group right after the fission for the smaller fission product

students working in the MPI-EVAN laboratories of Linda Vigilant and Tobias Deschner.

The topics of investigation have always centered on social behavior. The first few years of the project were devoted to documenting basic natural history of social communication and the structure of social relationships (Perry 1996a, b; 1997; 1998a, b; 2003; Manson et al. 1999; Gros-Louis et al. 2003; Perry et al. 2004; Perry and Manson 2008). Between 1995 and 2001, we investigated communication (vocal, gestural and olfactory) in greater detail and became particularly interested in the way in which individuals negotiate their social relationships (Manson et al. 1997, 2004a, 2005; Manson 1999; Manson and Perry 2000; Gros-Louis 2002, 2004a, b, 2006; Fichtel et al. 2005; Campos et al. 2007; Gros-Louis et al. 2008; Perry and Manson 2008). Our discovery of group-specific communicative rituals launched me on a multi-year study of social learning and social traditions (Panger et al. 2002; Perry et al. 2003a, b; Rose et al. 2003; Perry and Ordoñez Jiménez 2006; Perry and Manson 2008; Perry 2009). The interest in social learning spurred me to begin a developmental study of 46 infants in 2001 that continues to the present (Perry 2009). The availability of genetic data has enabled us to answer basic questions in behavioral ecology about issues such as kin-based altruism, inbreeding avoidance, and mating systems (Muniz et al. 2006, 2010; Perry and Manson 2008; Perry et al. 2008). This knowledge contributes to our long-term study of life history strategies.

7.1.2 Data Collection Methods

In the early years of the study, a team of observers (a narrator and a spotter, to check ID's) collected data, and data were narrated directly onto microcassette recorders. This method was essential in the early days, due to the richness of the behavioral repertoire, which had not yet been thoroughly documented. In 1997, we switched to using Psion palmtop computers, and I developed an elaborate system for rapidly coding behaviors in fine detail. We still switch to microcassette recorder when the action is extremely rapid or complex (e.g., when there are coalitionary fights or multi-party play bouts, or when the animals innovate and we have no codes to describe their actions accurately). These inserts are transcribed into the data after they are transferred into Excel spreadsheets at the end of the day. During data collection, one person types and double-checks IDs while the second person narrates the action and constantly watches the focal animal. Focal follows are the top priority in data collection, and these vary in length from 10 min to 12 h, depending on the topic of investigation. During these follows, we collect activity data and proximity data during point samples every 2.5 min, and continuous data on all social interactions, vocalizations, object handling and foraging events. We also conduct group scans approximately every 30 min, in which activities, food type (for foraging activities) and proximity to other animals are recorded. *Ad libitum* data on predator encounters, fights, agonistic interactions, sex, grooming, and innovative or traditional behaviors are also collected. Census sheets are filled out daily, reporting who was in each group, health and wounding status, and reproductive states. At the end of each year of employment with the project, each observer fills out a 26-item personality questionnaire on the animals they studied. During some years, there are special protocols in addition to these core protocols, which can include recording (audio or visual), field experiments of various types, and food processing protocols that note processing techniques, proximity between foragers, and gaze directed toward other foragers. Fecal samples are collected for both genetic and hormonal analysis. We have used a variety of methods for collection of genetic samples (Muniz and Vigilant 2008), all of which have worked: collection in (a) 96% ethanol, (b) silica, (c) RNA-later, and (d) a two-step process involving first ethanol and then silica. We dry and grind feces in the field for hormone analysis and mail them to a lab for extraction.

7.1.3 *Current Infrastructure and Logistics of the Site*

The bulk of the core data collection is performed by a crew of six or more interns. Most are people who have just finished a bachelor's or master's degree and are seeking fieldwork experience before continuing to a Ph.D. program or a career in conservation. Two Costa Ricans (Alex Fuentes Jiménez and Juan Carlos Ordoñez Jiménez) have been employed by the project for a number of years to help with data collection, logistics, and plant identification. Two managers aid me in the running of the site. One lives in the project house and is responsible for the training of assistants, monthly inter-observer reliability checks, scheduling, maintenance of the house and project equipment, and the running of the fecal sample laboratory. The other (Wiebke Lammers) is in charge of data organization, project purchasing and accounting, and the environmental education program. This second manager and the Costa Ricans are invaluable in coordinating conservation efforts and meeting with government officials when I am out of the country. The managers and I remain quite active in data collection, so that we can reliably identify all animals and maintain continuity in the demographic database. In addition to this core staff, who are primarily responsible for the long-term database, there are typically one or two graduate students or postdoctoral scholars who are doing independent research, and they also have field assistants. Most of the graduate students have previously served as field assistants.

In a project of this size that involves a collective attempt to produce a complex data set, inter-observer reliability is always a concern. Before anyone is cleared to collect usable data, they must pass code and syntax tests, speed typing tests on the Psion Workabout handheld computers, shadow follows (in which a trainee follows a trained team of observers and types along, and the two sets of records are compared), monkey ID tests, and vocalization recognition tests. It typically takes 6 weeks to 3 months for an intern to be considered fully trained enough to be the senior member of a two-person data collection team. Every line of data is tagged by spotter and typist ID so that if any errors are discovered later, the relevant data can be fixed or discarded. Each month, all observers are retested on their knowledge of codes, syntax, and vocalization recognition. If there is reason to suspect that someone's typing speed has declined since the initial training, then typing speed tests are retaken as well. Because we rotate work partners regularly, and work partners constantly double-check one another's monkey IDs, there is not much chance for errors to creep into monkey IDs. Whenever there is lack of agreement about monkey IDs or about what behaviors occurred, the follow is aborted and discarded. Whenever there is doubt about plant identification, a sample is collected and brought back to the Costa Rican botanist. We have a short staff meeting every night during dinner to discuss any doubts about protocol that have arisen during the day's data collection. Vocalization tests are always the hardest for achieving inter-observer reliability, so we just keep training records regarding which observers reliably recognize each vocalization and only analyze vocal data from those observers who are reliable for that call type.

Currently the project has amassed approximately 70,600 h of behavioral data, stored in the form of Excel spreadsheets. With this amount of data, the analysis process has become unwieldy, and so we are in the process of constructing a MySQL relational database to aid in the analysis process, modeled roughly after Babase, the database created for the Amboseli Baboon Project (Alberts and Altmann 2012).

7.2 Demography

7.2.1 Dispersal Patterns

Female capuchins at Lomas Barbudal are philopatric, as at other sites where this species and genus have been studied (Fragaszy et al. 2004; Perry et al. 2008; Jack and Fedigan 2009). Although we have directly observed immigrations or immigration attempts by 85 males (many of these migrating multiple times), we have still never witnessed a migration by a female (but see Jack and Fedigan (2009) and Fedigan and Jack (2012) for a few exceptions to this pattern at Santa Rosa). There have been 90 adult females above age 5 observed in our study for a period of up to 20 years. We have witnessed females interact non-aggressively with females from other groups only three times in 20 years and have never seen females seriously attempt to join other groups. In two cases, the females were separated from their own groups briefly following an intergroup encounter, and in the third case, it appeared that an estrous female and a male wandered away from their group and temporarily joined a neighboring group for half a day, until there was an intergroup encounter between the two groups and they returned to their normal group.

Typically females only remove themselves from close proximity with female kin when their group becomes so large that coordination of group movement becomes difficult (i.e., group size of >30). In these cases, groups fission fairly neatly along matrilines, so that the average relatedness among females is higher in the fission products than in the original group (Muniz 2008). The fission process may take up to 3 months, during which the two subgroups apparently try to stay together, but finally they decide who will be in which subgroups and adopt hostile relations toward the members of the other subgroup. For the first several months after the fission, intergroup relations are actually more hostile than is typical in intergroup encounters, and even females physically attack one another (whereas in typical intergroup encounters, active participation is almost exclusively by males). Figure 7.1 shows the history of group fissions at Lomas, with the number of monkeys that was present in each group at the time the fission occurred.

Males can either migrate singly or in groups of 2–8. Often co-migrant males are kin, and co-migration with kin can occasionally result in higher relatedness among adult males than among adult females. However, in 10 out of 11 demographic situations analyzed from three social groups at Lomas at different time periods, adult females had higher average relatedness than adult males (Muniz 2008). Males born into our three primary study groups, (AA, RR and FF) tend to make their first migration at age 92 months, regardless of whether we use a sample of 21 males from our developmental study whose birthdates and migration dates are fairly accurately known, or a sample of 44 males from the broader study whose birthdates and migration dates are somewhat less accurately estimated. The three youngest males to migrate were 4 years old, and they migrated in the context of a fission and/or an alpha male takeover in which a large all-male group composed of their male relatives formed. It is likely that they would not have migrated so early if the fission had not occurred. The seven oldest males to leave their natal groups were 11 years old. Additionally, one 11-year-old remains in a fission product of his natal group (but does not reside with either parent). This male is missing a hand, which may inhibit him from migrating, though he did accompany some of his brothers on a visit to a neighboring group once.

This mean age of migration (7.6 years) is considerably higher than that seen at the nearby site of Santa Rosa National Park, where males tend to make their first migration at a mean age of 4.17 and never remain in natal groups past age 8 (Jack and Fedigan 2004; Fedigan and Jack 2012). It is not clear why this difference between the two nearby, ecologically similar sites exists. Possibly the apparently higher rate of lethal coalitionary aggression at Lomas Barbudal (Gros-Louis et al. 2003) makes migration more dangerous, but the exact rate of such killings at Santa Rosa has not yet been reported. Males at Lomas tend to co-migrate more often than Santa Rosa males do, and so many males may spend extra years in their natal groups waiting for close kin to mature to migration age so that they can move together. Every one of the 44 males born into our study groups whom we have seen to migrate has migrated with other natal males, at least initially (though sometimes the co-migrants do not remain with their co-migration partners after one male has claimed the alpha position). In contrast, only 71% of males at Santa Rosa co-migrate (Jack and Fedigan 2004). We do see occasional solo migrations at Lomas, but thus far it has always been males from unhabituated groups who have migrated alone into our study groups; therefore, we know nothing about the migration options for these males.

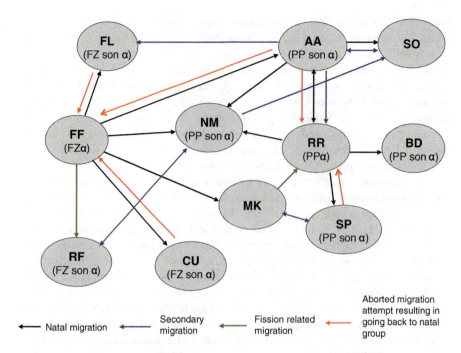

Fig. 7.2 Diagram of male migrations. Each oval represents a social group. Two long-term alpha males are noted, as are their alpha male sons. See text for the four types of migrations, which are denoted by different types of *arrows*

Figure 7.2 demonstrates the patterning of male migration. There are four basic types of migration or transfer: (a) natal transfer (i.e., transfer from the natal group to another group), (b) secondary transfer (transfer from one non-natal group to another non-natal group), (c) "fission transfer" in which males initially go with their mothers during a fission, but later move to the other fission product that has less closely related adult females, and (d) "returning home" transfers, in which males first transfer to a non-natal group and then return to the natal group.

Some males make many false starts; that is, they transfer to non-natal groups, but return to the natal group for extended periods between each transfer attempt; in these cases, we counted all emigrations out of the natal group as natal migrations for the purpose of Fig. 7.2. We only included multi-male/multi-female groups in this diagram, but there are also many all-male groups in the study area that shift composition at high rates. Males often spend prolonged periods in these groups in between periods of residency in groups containing females. Some of the secondary migrations reported in Fig. 7.2 might not have been direct migrations, but rather involved visits to all-male groups and perhaps even other multi-male/multi-female groups in between. Males typically transfer to an adjacent group with females or to an all-male group, at least initially.

It has been proposed (Schoof et al., 2009) that one of the reasons males engage in parallel dispersal is because it gives them access to willing coalition partners who can enhance their competitive ability in within-group and between-group competition. One of the arguments for why males are expected to engage in parallel dispersal is to keep male kin together so that they can engage in kin-selected altruism (Schoof et al. 2009). Indeed, at Lomas there is a tendency for male co-migrant dyads to be more closely related than randomly selected male–male dyads from a given group (Wilcoxon signed ranks test: $Z^+ = -2.132$, $p = 0.016$, $N = 13$ migration events; Muniz 2008). At Lomas, it is definitely true that males aid one another in defending the group from intruders (Perry 1996b; Perry and Manson 2008), and that they regularly aid one another in takeovers of other groups (Perry and Manson 2008, unpublished data). It is less certain whether co-migrants aid one another in within-group coalitionary aggression in their new groups more often than do pairs of males who are not co-migrants; this topic will require further research. Certainly we do see many cases in which co-migrants side against one another in within-group coalitionary aggression. Once a takeover has been achieved, we sometimes see cases in which co-migrant males engage in fierce battles over the alpha position with their own brothers, inflicting severe wounds. Strong circumstantial evidence from Lomas indicates that males sometimes kill their brothers and their brothers' infant offspring in disputes over the alpha position, despite their close genetic relatedness (unpublished data).

7.2.2 Genetic Structure of the Population

The Lomas Barbudal capuchins are noteworthy for often having quite long alpha male tenures, lasting up to 18 years. These long tenures, combined with high reproductive skew, mean that natal individuals are often related both through the maternal and the paternal line. Out of 2,111 dyads analyzed from five social groups, 4% were full siblings, 5% were maternal half siblings, and 24% were paternal half siblings (Muniz 2008). Another unusual feature of capuchin social structure imposed by long alpha male tenures is the high frequency with which alpha males co-reside with their daughters and grand-daughters. In most mammalian populations, father–daughter inbreeding is avoided by having one sex disperse. But capuchins avoid father–daughter inbreeding even while co-residing (Muniz et al. 2006). Paternity data on 97 infants born during 15 stable alpha male tenures show that in the years before an alpha male's daughters have started to mature, the alpha males have sired 96% of all offspring (data from Godoy 2010; Muniz et al. 2010). However, subordinate males who have been helping an alpha male defend his group's females from extra-group males experience greatly increased breeding opportunities when the alpha male's female descendents finally mature. Subordinate males have sired 94% (31/33) of offspring produced by females that were the daughters or granddaughters of their groups' current alpha males (Godoy 2010; Muniz et al. 2010). Thus, the longer an alpha male remains in place, the lower the degree of reproductive skew, due to the increasing numbers of his female descendants.

Figure 7.3 shows the breeding histories for the males in Rambo's group from 1991 to 2008. This pattern is similar to what we observed during two other long-term tenures. At the start of the study, alpha male Pablo (PP) co-resided with two other immigrant

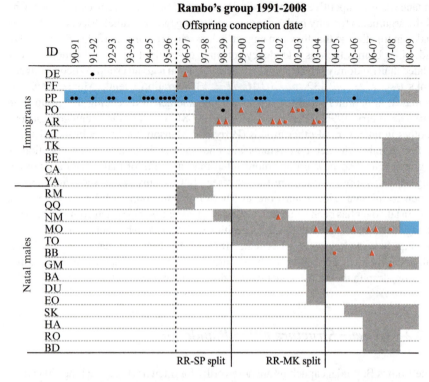

Fig. 7.3 Distribution of paternity over time in Rambo's group. All males residing in a group during a particular year are shaded in grey, with the alpha male having *blue* shading. Offspring of the current alpha males' daughters are represented by *red triangles*, and offspring of the alpha male's granddaughters are represented by *red circles*, while offspring of all other females are represented by *black circles*. Vertical dashed lines indicate the beginning of direct observations

males with whom he had non-antagonistic relationships. In 1997, a trio of migrant males immigrated relatively peacefully and did not attempt to overthrow PP. Again in 2008, four migrant males who had known one another in AA group immigrated peacefully. At the end of 2008, PP was overthrown by MO, who was the offspring of PP's former ally DE. Aside from MO, RM and QQ, all other natal males in the group were PP's sons or grandsons. PP remained in RR group after losing alpha status.

PP sired 25 out of 30 offspring (83% of offspring) born to females who were not his direct descendants; the remainder were sired by immigrant males. Of those offspring born to PP's daughters and granddaughters, 12 were sired by immigrant males, 4 were sired by PP's sons (i.e., his sons bred with their paternal half sisters or nieces), and 7 were sired by MO, a natal male who was unrelated to PP and only distantly related to most of the females with whom he bred. Thus, PP did not breed with any of his daughters or granddaughters.

Long-term alpha males have far more reproductive influence in the population than is apparent just by examining the patterns of reproduction in the groups in which they reside. This is because long-term alpha males have ample opportunity to produce sons,

7 The Lomas Barbudal Monkey Project

who then migrate and often become alpha males themselves, monopolizing reproduction in other groups. Figure 7.2 shows which adjacent groups have alpha males who are sons of the long-term alpha males PP (of RR group) and FZ (of FF group). These males may have additional sons who are alpha males in groups outside our study area. To the best of our knowledge, PP has sired at least 25 offspring, 83 grandoffspring, and nine great-grandoffspring at the present time (and of course the numbers of his second and third generation descendents will continue to increase after his death).

7.2.3 Sources of Mortality and Ages at Mortality

Although white-faced capuchins can live to 55 years of age in captivity (Hakeem et al. 1996), maximum life span is in the wild is unknown because no field study has come close to spanning that amount of time. Certainly, however, it is rare for capuchins to live that long. Based on genealogical data, we estimate that our oldest living monkey is 36 years old. Of the 24 monkeys who were members of AA group during the first 3 years of our 20-year study, only 2 females are still living.

Mortality rates at Lomas Barbudal were calculated using individuals with birthdates known to an accuracy of plus or minus 3 months ($N = 262$). We assigned two values for age at death to those individuals presumed dead ($N = 108$); these incorporated inaccuracy in both birth and death dates. We used these values to generate two curves for mortality rates across time (Fig. 7.4a), one using the lowest possible value for age at death and second curve using the highest possible value. Since birth dates were relatively accurate, the two curves produced similar results. We also included in this analysis stillborn infants and obvious miscarriages (when it was clear that a female was pregnant based on the size and shape of her abdomen, and then she suddenly decreased in size). Because mortality was highest in the first year (26–30%), we generated a separate graph to describe mortality rates within that year in four 3-month increments (Fig. 7.4b).

The principle cause of mortality for infants under 1 year of age seems to be infanticide. The mean interbirth interval for those cases in which the first infant lives until the next infant is born is 749 days $\pm$ 145 ($N = 41$ IBI's for which birthdates were known to an accuracy of 1 week). This is slightly shorter than the 2.25-year (~821-day) interbirth intervals reported for Santa Rosa (Fedigan and Jack 2011). However, when the previous infant dies or is abandoned before the second is conceived, the mean interbirth interval is 444 days $\pm$ 168 ($N = 21$), i.e., significantly shorter than when the previous infant lives (two-sample t-test with unequal variance, $t = 7.09$ (37), $p < 0.0001$). The mean time to conception following an infanticide was 478 days $\pm$ 212 ($N = 6$), or 271 days shorter than in cases where infanticide was not committed.

We analyzed the effects of adult male migrations and alpha male turnovers on infant mortality rates using a sample of 210 births for which we had accurate birthdates. For each birth, we scored whether a male migration or a turnover in the alpha male position occurred within 6 months before or 1 year after the infant's

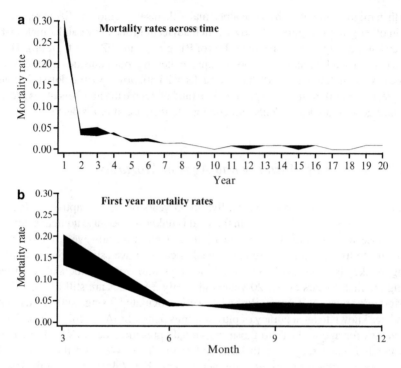

Fig. 7.4 Mortality rates over the life span. (**a**) Shows the mortality rates over the first 20 years of life; (**b**) shows the timing of deaths during the first year of life. *Shaded areas* in both graphs denote the upper and lower bounds of mortality rates

birth. Only 18% of infants died in periods characterized by stable alpha males, whereas 49% of those infants born in alpha male turnover periods died before reaching 1 year of age. This is a significant difference (Pearson's $\chi^2 = 23.38$, 1df, 1-tailed $p < 0.001$), and replicates similar findings from the Santa Rosa capuchins (Fedigan 2003; Fedigan et al. 2008). We witnessed 20 peaceful immigrations, i.e., migrations in which no alpha male turnover occurred. The mortality rate for infants born during times of peaceful male immigration (25%) did not differ significantly from that for infants born during periods in which no male immigrated (23%) (Pearson's $\chi^2 = 0.038$, 1df, 1-tailed $p = 0.845$). We do not always witness infant deaths, so we cannot be certain what proportion are due to infanticide. We directly observed six infanticides, and in five other cases we witnessed the alpha male stalking the infant for days previously, and/or the mother alarm-calling at the alpha male right after the infant was killed (Manson et al. 2004b; Perry and Manson 2008). In another four cases, infants disappeared in the middle of an infanticide spree when a new alpha was seen to kill other same-aged infants. In another 11 cases, a female was pregnant on the last day of observation during one month, but when next seen was no longer pregnant and had no infant; it was impossible to know whether there was a miscarriage or an infanticide. Thus, although we cannot give a precise rate, infanticide is clearly the biggest source of infant mortality at our site.

7.2.4 Age at First Reproduction for Each Sex

Females at Lomas Barbudal give birth for the first time at a mean age 6.22 years (SD = 0.58). This sample is based on 30 females for whom the age of their own birth and the age of their first infant's birth is known to an accuracy of ±3 months. It includes three miscarriages. If we look only at the age of first live birth, then the mean age of first birth increases to 6.30 (SD 0.62). The youngest female to give birth was 5.50 years old.

Determining the age of first reproduction for males is much more difficult because there is such high reproductive skew that very few males actually reproduce. Only 27 males in our population are known to have sired offspring, and we only had reasonably accurate birthdates (+3 months) for two of these, since it is primarily the older males who reproduce. Of these two males, one conceived his first offspring at 7.3 years of age, and the other conceived his first offspring at 9.0 years. Based on our much rougher age estimates for other sires, we suspect that the typical age at which males first conceive an offspring is much later, however.

7.2.5 Birth Seasonality

67.1% of births at Lomas Barbudal ($N = 173$ births, during years for which we have continuous demographic monitoring) occurred between April and July, though some births occurred during all months but October. Thus, the birth season straddles the end of the dry season and the start of the rainy season, which typically begins in mid-May. Figure 7.5 shows the distribution of births across years, using

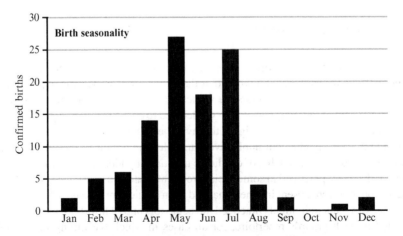

Fig. 7.5 Number of births in each month in a sample of 106 births for which birth date was known accurately enough to assign it to a particular month

only those births for which we knew the precise month ($N = 106$). The timing of this birth season corresponds roughly to that of Santa Rosa (see Fig. 8.2 in Fedigan and Jack 2011), where 40% of births occur in April–June.

7.3 Topics We Could Not Have Studied Without Long-Term Research: Social Learning and Traditions

Lack of temporal depth in datasets from wild animal populations usually prevents researchers from identifying *traditions*, defined as "enduring behavior patterns shared among members of a group that depend to a measurable degree on social contributions to individual learning, resulting in shared practices among members of a group" (Fragaszy and Perry 2003). Simply knowing that a behavior pattern varies across sites in cross-sectional samples is insufficient to conclude that the behavior is a tradition, because many factors, including between-site genetic and ecological variation, can affect behavioral variation. To determine whether a behavior pattern is a true tradition, it is useful to know whether and how it varies over time. Two forms of long-term temporal patterning strongly suggest that social learning affects the acquisition of a behavior. First, a behavior may appear suddenly in an individual (i.e., as an innovation) and then spread, in a transmission chain, through that individual's social network. Second, a general behavior pattern may occur universally in a population, but exhibit discrete variant forms, and immature individuals may preferentially acquire the variants favored by their close spatial and social associates. We have documented both patterns over multi-year periods at Lomas Barbudal, the first with respect to dyadic social rituals and the second with respect to foraging techniques.

7.3.1 Social Conventions

Early in the history of the project, SP began visiting other field sites and communicating with the researchers who worked at these sites. It soon became apparent that some of the more bizarre behaviors we witnessed were not common features of the repertoire of capuchins everywhere, and we decided to coordinate our methods so that we could more rigorously investigate the hypothesis that capuchins exhibit social conventions – i.e., dyadic communicative rituals that are unique to particular cliques or social groups. We made a list of candidate traditions by identifying forms of social behavior that were practiced in some social groups at a rate of at least once per 100 h of observation, but were absent in others that had been studied for at least 250 h. We also kept track of which individuals practiced the behaviors and of how long the behaviors persisted in each group's repertoire, for all cases in which the timing of the field seasons permitted documentation of the innovation event and included the periods during which it could have spread and, where relevant, when it became extinct.

In our original study, which included 19,000 h of behavioral observations on 13 social groups from four sites in Costa Rica over a 13-year period, five behaviors qualified as traditions according to our operational criteria (Perry et al. 2003a). One of these, handsniffing, involved the insertion of a finger into the nose of another monkey; this behavior was often mutual and exhibited many minor variations in form. Handsniffing was common for a period of 1–7 years in five out of the 13 groups we studied and at three of the four study sites. The sucking of body parts (ears, tails or fingers) was common in two of our study groups for periods of 6 months or more. Three "games" were invented by a single monkey (Guapo of Abby's group at Lomas Barbudal) and had roughly the same format, in which one monkey has something in its mouth that the partner tries to remove from the mouth (a finger or a tuft of hair from the partner, or an inanimate inedible object such as bark or a stick). The mouth is pried open using hands, feet, and perhaps the mouth, and the object is passed back and forth from mouth to mouth.

In the case of the three games ("finger-in-mouth", "hair" and "toy"), we could construct social transmission chains. The "toy" game was invented in 1991, and the other two were invented in 1992. The "toy" game had 13 practitioners and three links in the transmission chain and lasted 9 years. The "finger-in-mouth" game and hair-passing games both lasted for 10 years and had 11 and 14 practitioners, respectively. The finger-in-mouth game had two links in its transmission chain, whereas the hair-passing game had three. In the games, adult male-juvenile males were the primary participants. The games persisted in group repertoires for 2–3 years after the innovator (a subordinate male) became alpha male and stopped participating. Some of the main practitioners continued to play these games together even after co-emigrating.

In subsequent years, we have continued to document the innovation and spread of social conventions. The larger the data set grows, the more often we find that similar innovations occur. For example, the males in Flakes group have independently invented all three of the games that were previously played in Abby's group. Despite the fact that Flakes is a fission product of Abby's group, none of the current game players were resident in the group at the time these games were played in Abby's group, so we believe them to be independent inventions. We have also observed the invention of a rare, new variant of handsniffing that includes the insertion of the partner's finger deep into the eye socket, up to the first knuckle in some cases. This variant has been slower to spread than most traditions and also not quite common enough to meet all of our operational criteria for tradition in the original study, but it nonetheless appears to be socially learned.

As the study continues and the cumulative number of observation hours per group increases, we are discovering that (a) there are more observations of groups that exhibit temporal variation in their behavioral repertoires, and (b) many groups that formerly exhibited a total absence of particular traditional behaviors now have an occasional observation of a behavior thought to be unique to another group, even if it does not spread to other group members or become common in the repertoire. The longest-studied group, Abby's group, has exhibited particularly interesting variation in the expression of handsniffing over the past 20 years. This behavior was not observed in 1990. From 1991 to 1997, it was commonly practiced by up to

12 individuals and 13 dyads in any given year. From 1998 to 2001, it was never seen, but females often sniffed their own hands. During 2002–2003, when there were frequent migrations, dyadic handsniffing was practiced at low frequency by 14–16 individuals and 15–16 dyads per year. The practice dwindled until by 2008, dyadic handsniffing had once again vanished from the group's repertoire, though solo handsniffing remained. The solo handsniffing in Abby's group and its fission product (Flakes) is never seen in three of our other study groups (MK, CU, NM) and occurs roughly an order of magnitude less often in the neighboring RR group.

7.3.2 Social Learning About Foraging Techniques

Short-term studies of *C. capucinus* have demonstrated between-site differences in the ways foods are processed (Panger et al. 2002). Providing conclusive evidence for a substantial role of social learning is impossible in such cross-site studies, although the fact that all of the sites in this study were tropical dry forest sites in northwestern Costa Rica makes a major role for ecological or genetic variation unlikely. Within-site studies are highly suggestive of a role of social learning of food-processing techniques. White-faced capuchin monkeys are more prone to observe group-mates foraging at close range when they are foraging on foods that require multiple steps to process before ingestion (Perry and Ordoñez Jiménez 2006). However, showing that monkeys observe one another does not mean that they necessarily learn from the observations. At Lomas Barbudal (Perry and Ordoñez Jiménez 2006), Palo Verde (Panger et al. 2002) and Santa Rosa (O'Malley and Fedigan 2005), pairs of monkeys who frequently associate are more prone to share the same food-processing techniques than pairs who associate less often, but the results of these comparisons are generally only marginally significant.

A longitudinal approach is preferable to a cross-sectional approach to this issue because it allows researchers to document the association patterns at the time when young animals are acquiring their food-processing techniques and are most subject to social influence, rather than simply measuring association patterns in adulthood and assuming that these represent the patterns that held when the monkeys acquired their current techniques. It is time-consuming and costly to conduct longitudinal studies of the development of large numbers of juvenile primates in the wild, so such studies are rare. We have conducted the largest such study to date, focusing on the acquisition of techniques for processing *Luehea candida* fruits, a staple item that contributes up to 15.4% of the diet during the peak fruiting period (Perry and Ordoñez Jiménez 2006). This fruit consists of a woody capsule with five cracks, from which small but nutritious wind-dispersed seeds can be extracted either by pounding or scrubbing the fruit. Both techniques are approximately equally efficient (Perry 2009). The lack of difference in efficiency means that individuals are probably less likely to select a technique on the basis of individual trial and error learning, as opposed to social learning.

All of the social groups included in the study (three groups and three of their fission products) had some group members who were primarily pounders and others who were primarily scrubbers. However, some between-group differences in the tendency to pound or scrub existed, particularly for the philopatric sex. For example, currently nine of the 10 adult females in AA group are scrubbing specialists (the youngest female still combines pounding and scrubbing), whereas in RR group, all of the six adult females are pounders.

During the first developmental year (i.e., the first year in which the infants have exposure to *Luehea* fruits and are off the mother's back during part of *Luehea* season), most infants do not regularly process *Luehea* fruits, although they handle them and eat the seeds protruding from the ends. In their second year of exposure to the fruits, they employ a variety of techniques (about four per monkey), including both pounding and scrubbing (Perry 2009). As juveniles mature, they select a preferred technique and use it increasingly until they reach adulthood (Fig. 7.6). By 3 years, inefficient techniques have largely been eliminated from individual repertoires, and individuals use their dominant technique 82% of the time ($N = 49$ individuals), and by age 7 years, they use the dominant technique 95% of the time ($N = 27$).

Infants remain on their mothers' backs for most of the first 3 months of life. During months 4–6, they spend increasing amounts of time with alloparents, and by 6–9 months of age, they are within 40 cm of their mothers only 10% of the time (Perry 2009). Even after becoming largely independent of their mothers, they spend much time within observation range of other monkeys, and by age 5 years, they still spend 73% of their time within 4 m of at least one other monkey. Juvenile males and females do not differ significantly with regard to the amount of time they spend in proximity to their mothers, time spent alone, or time spent observing other foragers (Perry 2009).

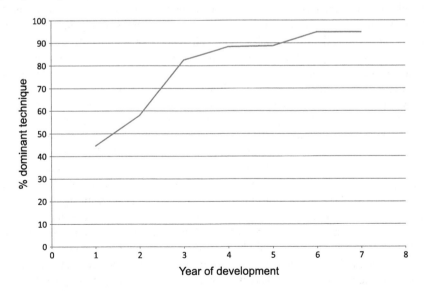

Fig 7.6 Percentage of Luehea processing incidents in which the individual's dominant (favored) processing technique is used during each year of development

We collected data on the *Luehea* processing techniques of all members of our study groups over a period of 7 years, which allowed us to follow 48 immature monkeys (21 females and 27 males) during the first 5 years of their lives. In many cases we had to drop particular monkey-years from the data set due to inadequate sample sizes for (a) number of fruits processed by the focal monkey, or (b) opportunities for the focal animal to observe others' foraging techniques. Once these cases were eliminated, we had 79 cases (i.e., monkey years) in the sample (see Perry 2009 for further details). Whenever a monkey entered a *Luehea* patch, we recorded the technique(s) used to process each fruit, the distance between the focal animal and other individuals who were processing *Luehea*, the gaze directions between animals, and the techniques being used by other animals in the same patch.

I analyzed the data using a Poisson regression model, adjusting the standard errors for within-subject correlation (see Perry 2009 for details of measurement and analysis). The primary predictor variable was the technique to which the focal subject was exposed, which is a measure of the relative exposure to pounding as opposed to scrubbing in foraging neighbors (see Perry 2009 for details of measurement and analysis). Sex was also a predictor variable. The outcome variable was the proportion of *Luehea* fruits that the focal monkeys processed by pounding them (i.e., number of fruits pounded, divided by the sum of number of fruits pounded and number of fruits scrubbed). Developmental years were control variables.

Figure 7.7 shows the effect of observed technique on the practiced technique for males vs. females across the first five developmental years. In Fig. 7.7a, only observations of maternal foraging are included in the independent variable. Figure 7.7b includes only observation of non-maternal foraging in the independent variable. In both cases, the impact of observed techniques on practiced techniques was greatest in the second year of development. Females were more strongly influenced by observed technique than were males, across all years of development. Non-maternal influence is slightly greater than maternal influence for both males and females.

In a separate analysis (Perry 2009), I used a broader data set ($N = 106$ monkeys) in which I also included adults and included all cases for which we had adequate behavioral data to characterize individuals' techniques during the most recently available processing season and also knew the mothers of the subjects. Females were significantly likely to use the same technique as their mothers (Fisher's exact $p = 0.002, N = 48$), but males were not ($p = 0.18, N = 58$). This result parallels findings by Lonsdorf et al. (2004) at Gombe, in which the acquisition of termiting techniques was studied in 14 immature chimpanzees over a period of 4 years. In their study, daughters were more likely than sons to adopt a termiting tool length similar to the mother's; daughters acquired the technique earlier than sons and were more efficient termite-fishers. Lonsdorf et al. found that this sex difference could be explained by the fact that sons paid less attention to the mothers than the daughters did when the mother was termiting. However, in the Lomas capuchins, there are no differences between sons and daughters with regard to time spent in proximity to the mother (or other group members) or in the tendency to visually focus on the model during demonstrations of the *Luehea* processing technique. It may be that

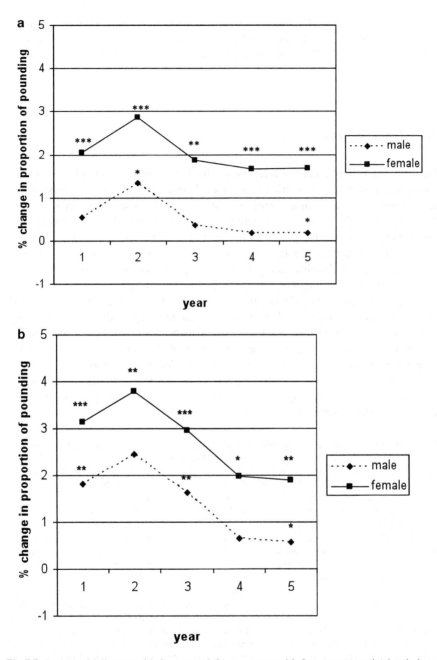

Fig 7.7 Impact of (**a**) maternal influence and (**b**) non-maternal influence on practiced technique for processing *Luehea* fruits. Y-axis is "% change in proportion of pounding practiced" resulting from a 1% change in observed technique. These two figures are slight modifications of Fig. 3b, c in Perry (2009) *=$p \leq 0.05$; **=$p \leq 0.01$; ***$p \leq 0.001$

female capuchins, being the philopatric sex, are more inclined to identify with their closest associates in the group, and hence conform to their techniques, in accordance with de Waal's "Bonding and identification-based learning model" (de Waal 2001).

7.4 Conclusions

The most interesting pieces of information that have emerged from our project are findings that were only possible to obtain via long-term study and extensive inter-site collaboration. To gain a deeper understanding of the social strategies of these long-lived animals, we needed to integrate data from multiple social groups and sites for hundreds of individuals over a period of 20 years, and even now, we are far from having complete life histories for many individuals in our sample. The discovery of father–daughter and grandfather–granddaughter inbreeding avoidance was possible only by obtaining genealogical data for multiple generations. Migration strategies can only be documented by knowing the natal groups of males and following them for multiple decades as they migrate repeatedly under changing demographic circumstances. The role of social learning in acquiring foraging skills was documented by collecting extraordinarily detailed observations at high density for a 7-year period in multiple groups. Finally, the documentation of group-specific communicative rituals was possible only by collecting long-term data on transmission chains, changes in groups' behavioral repertoires over multi-year periods, and comparative data from multiple sites.

Acknowledgments The following field assistants contributed a year or more of data to this data set, in addition to data collected by the authors: B. Barrett, L. Beaudrot, M. Bergstrom, R. Berl, A. Bjorkman, L. Blankenship, J. Broesch, J. Butler, F. Campos, C. Carlson, M. Corrales, N. Donati, C. Dillis, G. Dower, R. Dower, K. Feilen, A. Fuentes J., M. Fuentes A., C. Gault, H. Gilkenson, I. Gottlieb, J. Gros-Louis, S. Herbert, S. Hyde, L. Johnson, S. Leinwand, T. Lord, M. Kay, E. Kennedy, D. Kerhoas-Essens, E. Johnson, S. Kessler, J. Manson, W. Meno, C. Mitchell, A. Neyer, C. O'Connell, J.C.Ordoñez Jiménez, N. Parker, B. Pav, K. Potter, K. Ratliff, H. Ruffler, M. Saul, I. Schamberg, C. Schmitt, J. Verge, A. Walker-Bolton, E. Wikberg, and E. Williams. We are particularly grateful to H. Gilkenson, W. Lammers, C. Dillis and M. Corrales for managing the field site. E. Wikberg and W. Lammers contributed a year or more of effort to organizing the dataset. The genetic analysis was conducted by L. Muniz and I. Godoy in Linda Vigilant's lab. This project is based on work supported by the funding provided to SEP by the Max Planck Institute for Evolutionary Anthropology, the National Science Foundation (grants No. SBR-9870429, 0613226 and 6848360, a graduate fellowship, and an NSF-NATO postdoctoral fellowship), five grants from the L.S.B. Leakey Foundation, three grants from the National Geographic Society, The Wenner-Gren Foundation, Sigma Xi, an I.W. Killam postdoctoral fellowship, and several faculty development or student grants and fellowships from University of California-Los Angeles and The University of Michigan. IG was supported by the following fellowships: Eugene Cota-Robles, Ford Predoctoral Diversity, NSF graduate research, UCLA NSF AGEP competitive edge, UCLA IRSP, UC DIGSSS summer research mentorship, and two UCLA anthropology research grants. Any opinions, findings, and conclusions or recommendations expressed in this material are those of the author(s) and do not necessarily reflect the views of the National Science Foundation or other funding agencies. I thank the Costa Rican park service (MINAET and SINAC, currently), Hacienda Pelon de la Bajura, Hacienda Brin D'Amour, and the residents of San Ramon de Bagaces for permission to work on their land. This research was performed in compliance with the laws of Costa Rica, and the protocol was approved by the University of Michigan IACUC (protocol #3081) and the UCLA animal care committee (ARC #1996–122 and 2005–084 plus various renewals). J. Manson provided comments on the manuscript.

References

Alberts SC, Altmann J (2012) The Amboseli Baboon Research Project: Forty Years of Continuity and Change. In: Kappeler PM (ed) Long-term field studies of primates. Springer, Heidelberg

Campos F, Manson JH, Perry S (2007) Urine washing and sniffing in wild white-faced capuchins (*Cebus capucinus*): testing functional hypotheses. Int J Primatol 28:55–72

de Waal FBM (2001) The ape and the sushi master: cultural reflections of a primatologist. Harvard University Press, Cambridge, MA

Fedigan LM (2003) Impact of male takeovers on infant deaths, births and conceptions in *Cebus capucinus* at Santa Rosa, Costa Rica. Int J Primatol 24:723–741

Fedigan LM, Jack KM (2012) Tracking Neotropical Monkeys in Santa Rosa: Lessons from a Regenerating Costa Rican Dry Forest. In: Kappeler PM (ed) Long-term field studies of primates. Springer, Heidelberg

Fedigan LM, Carnegie SD, Jack KM (2008) Predictors of reproductive success in female white-faced capuchins (*Cebus capucinus*). Am J Phys Anthropol 137:82–90

Fichtel C, Perry S, Gros-Louis J (2005) Alarm calls of white-faced capuchin monkeys: an acoustic analysis. Anim Behav 70:165–176

Fragaszy DM, Perry S (2003) The biology of traditions: models and evidence. Cambridge University Press, Cambridge

Fragaszy DM, Visalberghi E, Fedigan LM (2004) The complete capuchin: the biology of the genus *Cebus*. Cambridge University Press, Cambridge

Frankie GW, Vinson SB, Newstrom LE, Barthell JF (1988) Nest site and habitat preferences of *Centris* bees in the Costa Rican dry forest. Biotropica 20:301–310

Godoy I (2010) Testing Westermarck's hypothesis in a wild primate population: proximity during early development as a mechanism of inbreeding avoidance in white-faced capuchin monkeys (*Cebus capucinus*). Master thesis, University of California, Los Angeles

Gros-Louis J (2002) Contexts and behavioral correlates of trill vocalizations in wild white-faced capuchin monkeys (*Cebus capucinus*). Am J Primatol 57:189–202

Gros-Louis J (2004a) The function of food-associated calls in white-faced capuchin monkeys, *Cebus capucinus*, from the perspective of the signaller. Anim Behav 67:431–440

Gros-Louis J (2004b) Responses of white-faced capuchins (*Cebus capucinus*) to naturalistic and experimentally presented food-associated calls. J Comp Psychol 118:396–402

Gros-Louis J (2006) Acoustic analysis and contextual description of food-associated calls in white-faced capuchin monkeys (*Cebus capucinus*). Int J Primatol 27:273–294

Gros-Louis J, Perry S, Manson JH (2003) Violent coalitionary attacks and intraspecific killing in wild white-faced capuchin monkeys (*Cebus capucinus*). Primates 44:341–346

Gros-Louis J, Perry S, Fichtel C, Wikberg E, Gilkenson H, Wofsy S, Fuentes A (2008) Vocal repertoire of *Cebus capucinus*: acoustic structure, context and usage. Int J Primatol 29:641–670

Hakeem A, Sandoval GR, Jones M, Allman J (1996) Brain and life span in primates. In: Birren JE, Schaie KW (eds) Handbook of the psychology of aging, 4th edn. Academic, San Diego, pp 78–104

Jack KM, Fedigan LM (2004) Male dispersal patterns in white-faced capuchins, *Cebus capucinus*. Part 1: Patterns and causes of natal emigration. Anim Behav 67:761–769

Jack KM, Fedigan LM (2009) Female dispersal in a female-philopatric species, *Cebus capucinus*. Behaviour 146:471–497

Lonsdorf EV, Eberly LE, Pusey AE (2004) Sex differences in learning in chimpanzees. Nature 428:715–716

Manson JH (1999) Infant handling in wild *Cebus capucinus*: testing bonds between females? Anim Behav 57:911–921

Manson JH, Perry S (2000) Correlates of self-directed behaviour in wild white-faced capuchins. Ethology 106:301–317

Manson JH, Perry S, Parish AR (1997) Nonconceptive sexual behavior in bonobos and capuchins. Int J Primatol 18:767–786

Manson JH, Rose LM, Perry S, Gros-Louis J (1999) Dynamics of female-female relationships in wild *Cebus capucinus*: data from two Costa Rican sites. Int J Primatol 20:679–706

Manson JH, Navarrete CD, Silk JB, Perry S (2004a) Time-matched grooming in female primates? New analyses from two species. Anim Behav 67:493–500

Manson JH, Gros-Louis J, Perry S (2004b) Three apparent cases of infanticide by males in wild white-faced capuchins (*Cebus capucinus*). Folia Primatol 75:104–106

Manson JH, Perry S, Stahl D (2005) Reconciliation in wild white-faced capuchins (*Cebus capucinus*). Am J Primatol 65:205–219

Muniz L (2008) Genetic analyses of wild white-faced capuchins (*Cebus capucinus*). PhD thesis, Universität Leipzig, D

Muniz L, Vigilant L (2008) Isolation and characterization of microsatellite markers in the white-faced capuchin monkey (*Cebus capucinus*) and cross-species amplification in other New World monkeys. Mol Ecol Res 8:402–405

Muniz L, Perry S, Manson JH, Gilkenson H, Gros-Louis J, Vigilant L (2006) Father-daughter inbreeding avoidance in a wild primate population. Curr Biol 16:R156–R157

Muniz L, Perry S, Manson JH, Gilkenson H, Gros-Louis J, Vigilant L (2010) Male dominance and reproductive success in wild white-faced capuchins (*Cebus capucinus*) at Lomas Barbudal, Costa Rica. Am J Primatol 72:1118–1130

O'Malley RC, Fedigan LM (2005) Evaluating social influences on food-processing in white-faced capuchins (*Cebus capucinus*). Am J Phys Anthropol 127:481–491

Panger MA, Perry S, Rose L, Gros-Louis J, Vogel E, MacKinnon KC, Baker M (2002) Cross-site differences in foraging behavior of white-faced capuchins (*Cebus capucinus*). Am J Phys Anthropol 119:52–66

Perry S (1996a) Female-female social relationships in wild white-faced capuchin monkeys, *Cebus capucinus*. Am J Primatol 40:167–182

Perry S (1996b) Intergroup encounters in wild white-faced capuchins (*Cebus capucinus*). Int J Primatol 17:309–330

Perry S (1997) Male-female social relationships in wild white-faced capuchins (*Cebus capucinus*). Behaviour 134:477–510

Perry S (1998a) Male-male social relationships in wild white-faced capuchins, *Cebus capucinus*. Behaviour 135:139–172

Perry S (1998b) A case report of a male rank reversal in a group of wild white-faced capuchins (*Cebus capucinus*). Primates 39:51–70

Perry S (2003) Coalitionary aggression in white-faced capuchins, *Cebus capucinus*. In: de Waal FBM, Tyack P (eds) Animal social complexity: intelligence, culture and individualized societies. Harvard University Press, Cambridge, MA, pp 111–114

Perry S (2009) Conformism in the food processing techniques of white-faced capuchin monkeys (*Cebus capucinus*). Anim Cogn 12:705–716

Perry S, Manson J (2008) Manipulative monkeys: the capuchins of Lomas Barbudal. Harvard University Press, Cambridge, MA

7 The Lomas Barbudal Monkey Project

Perry S, Ordoñez Jiménez JC (2006) The effects of food size, rarity, and processing complexity on white-faced capuchins' visual attention to foraging conspecifics. In: Hohmann G, Robbins MM, Boesch C (eds) Feeding ecology in apes and other primates. Cambridge University Press, Cambridge, pp 203–234

Perry S, Baker M, Fedigan L, Gros-Louis J, Jack K, MacKinnon KC, Manson JH, Panger M, Pyle K, Rose L (2003a) Social conventions in wild white-faced capuchin monkeys: evidence for traditions in a neotropical primate. Curr Anthropol 44:241–268

Perry S, Panger M, Rose L, Baker M, Gros-Louis J, Jack K, MacKinnon KC, Manson JH, Fedigan L, Pyle K (2003b) Traditions in wild white-faced capuchin monkeys. In: Fragaszy DM, Perry S (eds) The biology of traditions: models and evidence. Cambridge University Press, Cambridge, pp 391–425

Perry S, Barrett HC, Manson JH (2004) White-faced capuchin monkeys show triadic awareness in their choice of allies. Anim Behav 67:165–170

Perry S, Manson JH, Muniz L, Gros-Louis J, Vigilant L (2008) Kin-biased social behaviour in wild adult female white-faced capuchins, *Cebus capucinus*. Anim Behav 76:187–199

Rose LM, Perry S, Panger MA, Jack K, Manson JH, Gros-Louis J, MacKinnon KC, Vogel E (2003) Interspecific interactions between *Cebus capucinus* and other species: preliminary data from three Costa Rican sites. Int J Primatol 24:759–796

Schoof VAM, Jack KM, Isbell LA (2009) What traits promote male parallel dispersal in primates? Behaviour 146:701–726

Chapter 8
Tracking Neotropical Monkeys in Santa Rosa: Lessons from a Regenerating Costa Rican Dry Forest

Linda M. Fedigan and Katharine M. Jack

Abstract The Santa Rosa primate project began in 1983 and we have studied the behavioral ecology of the resident primate species (*Cebus capucinus, Alouatta palliata* and *Ateles geoffroyi*) continuously since then. Most of our research has concentrated on the behavior, ecology, and life history of multiple groups of capuchins and on documenting the effects of forest protection and regeneration on the howler and capuchin populations. Our examination of capuchin life histories has shown that they lead complex and intriguing lives, many aspects of which are affected by the frequent movement of adult males between social groups throughout the course of their lives. Over the past 28 years, we have documented increases in both the capuchin and howler populations. However, the howler population apparently reached carrying capacity in 1999, whereas the capuchin population continues to grow, probably because of their ability to occupy early-regeneration habitats. Our long-term examination of the population structure and life history of these two species clearly demonstrate that many species-specific aspects of biology and behavioral ecology differentially influence patterns of primate population recovery. It is only after decades of research that we can begin to understand the underlying constraints and variability in the lives of these animals.

L.M. Fedigan (✉)
Department of Anthropology, University of Calgary, Calgary, Canada
e-mail: fedigan@ucalgary.ca

K.M. Jack
Department of Anthropology, Tulane University, New Orleans, LA, USA
e-mail: kjack@tulane.edu

P.M. Kappeler and D.P. Watts (eds.), *Long-Term Field Studies of Primates*,
DOI 10.1007/978-3-642-22514-7_8, © Springer-Verlag Berlin Heidelberg 2012

8.1 History and Infrastructure of the Santa Rosa Monkey Project

In the early 1980s, I (L.M. Fedigan) began searching for a site to conduct long-term studies of primates. My wish list of ideal field site characteristics included the following: the site would be in a stable sociopolitical setting where I could safely bring students; the habitat would contain multiple primate species, at least one of which was relatively unstudied; the flora and fauna would be protected; the terrain would not be too formidable and the primates would be accessible for study and amenable to habituation. After a false start in one country where the monkeys were accessible and protected but the political situation was not stable, and in another where the government was stable but the primates were poorly protected, I made a trip to Costa Rica in 1982. After checking out many Costa Rican reserves and parks, I visited Santa Rosa National Park, home to white-faced capuchins (*Cebus capucinus*), mantled howlers (*Alouatta palliata*) and black-handed spider monkeys (*Ateles geoffroyi*). Not only was Santa Rosa an ideal location, but the three primate species, which all form multi-male, multi-female groups, display a range of dispersal patterns, dietary specializations, and social systems exhibited by polygynandrous primates, making them ideal for comparative studies. I knew immediately that I had found a place that met my criteria and where I could establish a stable long-term primate field site.

Santa Rosa National Park (SRNP) was created in 1971. The history of the park is a fascinating story in itself, far too long for this chapter, but well described in Evans (1999) and Allen (2001). SRNP was one of the first parks established by the fledgling Costa Rican National Park Service. It constitutes about 100 km^2 of tropical dry forest in Guanacaste Province and was chosen for early park designation mainly because its large central ranch house was the site of a famous 1857 battle in which a volunteer army of Costa Ricans repelled an invasion by American mercenaries. At the time of the park's establishment, the area was overrun with squatters, shacks, and roaming cattle owned by absentee landlords who planted African grasses that they burned annually to create pastureland. Soon after the park was established, the squatters were removed, hunting and logging were banned, a small complex of cement block buildings was constructed for park personnel, and the guards were given guns and horses (but no uniforms or vehicles!) to drive the poachers and the neighboring ranchers' cattle out of the park. Throughout the 1970s and 1980s, efforts were made on a shoestring budget to attract visitors and control the major forms of anthropogenic disturbance: poaching, logging, grass fires and grazing by cattle.

In the mid-1980s, Dan Janzen, a renowned tropical ecologist who first visited Santa Rosa before it became a park and who has made it his lifetime study site, convinced the park service and Costa Rican government to establish a large mega-park, now called Área de Conservación Guanacaste (ACG; Janzen 1988, 2000, 2002, 2004). Dr. Janzen created a non-profit conservation organization (Guanacaste Dry Forest Conservation Fund) that initially raised funds to purchase the properties

8 Tracking Neotropical Monkeys in Santa Rosa

surrounding Santa Rosa, in order to connect it to the small nearby parks and reserves. More recently, ACG has expanded further, as the Conservation Fund purchases cloud forest and Atlantic rainforest habitat on the eastern slopes of the mountains. Santa Rosa has now become a "sector" in the mega-park and is the core of a much larger protected zone (163,000 ha and counting) that covers nine Life Zones from the Atlantic rainforests of the east, to the volcanoes at the tops of the mountain range that runs along Guanacaste like a backbone, and into the Pacific Ocean on the west. ACG was declared a UNESCO World Heritage Site in 1999. The park now has ten sectors, most with their own buildings and staff. ACG also has a professionally-trained fire fighting team, locally-trained park researchers, police protection service, ecotourism office and outreach educational programs. Although far from wealthy and still dependent on donations from conservationists, the park has come a long way from the days when a small and poorly funded group of rangers tried to protect and maintain the land. Under Janzen's visionary guidance, the goal of ACG is not only to protect the old growth rainforest and cloud forest that remain in the mountainous regions of Guanacaste Province, but also to engage the local populace in efforts to regenerate the severely endangered tropical dry forest that was the original habitat type of the western lowland areas. Over the past two decades, major fires have become a thing of the past in Santa Rosa, the introduced African grasses have mostly died out, and the pastures are transforming into newly regenerating dry forest (Janzen 2002, 2004).

When I first requested a research permit from the Costa Rican National Park Service in 1983, the administrators made it clear they wanted me to monitor how the monkey populations were faring in the park. No counts of the primate populations had occurred in Santa Rosa since Curtis Freese, a Peace Corps volunteer had censused the monkeys in 1972, 1 year after the area came under protection (Freese 1976). Therefore, along with my plans to study the life histories and behavioral ecology of the monkeys, I agreed to monitor the park-wide populations. Beginning in 1983, my students, field assistants and I conducted annual censuses in May/June of every year for 6 years, after which we switched to less frequent park-wide censuses with a goal of at least one census every 4 years.

In 1984, I also selected three groups of capuchins, four groups of howlers, and one community of spider monkeys to be habituated and closely tracked as our "study groups." We found it difficult to distinguish individual howler monkeys and to track the rapidly-moving spider monkeys around their large home ranges. Therefore, with the help of Glander et al. (1991) we captured, marked, measured, and released many individuals in our howler study groups and several in our spider monkey community (Fedigan et al. 1988). In 1985, we started to systematically observe recognizable individuals in all three species in order to record births, deaths, disappearances, and dispersal, as well as foraging and social behaviors. The official "start" date for our life history data is June 1986, because it took us nearly a year to develop standardized and efficient methods for reliable data collection.

In the mid-1990s, I made the decision to concentrate my research on five groups of capuchins (Fig. 8.1), except for the park-wide censuses of monkey populations in

Fig. 8.1 A subadult female from one of our study groups carrying her infant sister (© Fernando Campos). Alloparenting, including extensive allonursing is very common in this species

May/June of designated census years. I maintained my focus on long-term life history data and started to direct and oversee many shorter-term (6–12 month) behavioral ecology projects carried out by graduate advisees.

Graduate students were actively involved in the Santa Rosa monkey project from the beginning. Dr. Colin Chapman was part of our original census team in 1983 and continued research in our park until 1989, focusing on the spider monkeys latterly (e.g., Chapman 1989, 1990; Chapman et al. 1988, 1989a, b). A young Costa Rican biologist, Rodrigo Morera Avila, worked as my local project manager for 10 years, until he completed his master's degree in wildlife management at the National University of Costa Rica and took up a post at the University of Costa Rica in Heredia. Since 1984, 30 graduate students from 12 universities and 6 countries have pursued and/or completed thesis projects on the monkeys of Santa Rosa (see partial list at http://people.ucalgary.ca/~fedigan/fedigan.htm). In 2000, Drs. Filippo Aureli and Colleen Schaffner initiated a long-term project on the spider monkeys of Santa Rosa. In 2004, I invited a former graduate advisee, Dr. Kathy Jack, now of Tulane University, to join me as co-director of the Santa Rosa capuchin project. Dr. Jack and her graduate students focus on capuchin behavioral ecology and life histories from the male perspective whereas my team addresses similar questions from the female perspective. We conduct many behavioral projects and all project data collection collaboratively.

Another major event in the history of our project was the creation of the Santa Rosa Database in June 2001 by Dr. John Addicott. This research tool is essential to our overall endeavor, although it will forever be a work-in-progress. It includes ~33 linked tables covering not only census and life history information, but also data on genetics (e.g., opsin genes, microsatellite DNA), climate (temperature, rainfall), phenology, food lists, dominance hierarchies, researcher names and other data

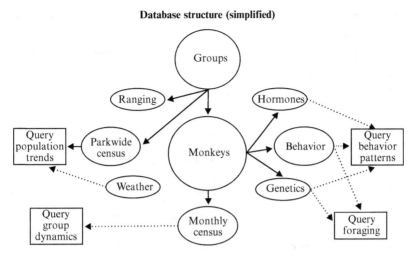

Fig. 8.2 Schematic diagram of selected linked tables that make up the Santa Rosa database. For additional information on the data included in each table, please contact the authors

(Fig. 8.2). We are putting this database on a server that will be remotely accessible by any user with appropriate permission.

Over the 28 years of this project, we have been variously housed: in tents, in an original 1971 park administration building, in a house outside of Santa Rosa that played a role in the "Iran-Contra" scandal, and in a more modern "dormitory" built in 1994. As of March 2011, we live in a designated "casa carablanca" that we built in the park with our own funds and grants from the Zemurray Foundation. This multi-roomed building is owned by the park but largely managed and maintained by our monkey research team.

8.2 Life History Parameters, Dispersal and Reproductive Success in the Santa Rosa Capuchins

Here we briefly review a few of the many aspects of capuchin life history, reproductive patterns, and socio-ecology that we could not have documented or understood without longitudinal data collection (see also Perry and Manson 2008; Perry et al. 2011). Table 8.1 summarizes capuchin reproductive and life history parameters at Santa Rosa. It did not take us long to discern that white-faced capuchins conformed to the *Cebus* pattern of male dispersal and female philopatry. However, only after many years could we determine the average age at which natal dispersal occurs, discover that males continue to change groups throughout their lives (Jack and Fedigan 2004b), and realize that adult females also disperse under some circumstances (Jack and Fedigan 2009). Although white-faced capuchins are best described as residing in multi-male/multi-female groups (current average

Table 8.1 Summary of life history parameters for white-faced capuchin monkeys in Santa Rosa National Park, Costa Rica between 1986 and 2010

		Value	Range	N
Gestation (days)		158 ± 8^a		6
Sex ratio at birth (proportion male)		0.605^b		119
Infant survival to age 1 (proportion)	All infants of known sex	0.823		119
	Females	0.914		47
	Males	0.763		72
	All infants (incl unknown sex)	0.680		144
Mean age at first birth (years)	Females	6.50	5.80–7.90	19
Mean interbirth interval (years)	First infant dies before age 1	1.05	0.67–1.75	22
	First infant survives to age 1	2.25	1.48–5.58	52
Mean age at 1st dispersal (years)	Females	$\geq 7^c$		5
	Males	4.5		30
Mean age at death (years) for individuals surviving to at least age 1	Females	9.41		18
	Males	2.96^d		14
Oldest age at death (years)	Females	27^e		
	Males	$24.70^{d,\ e}$		

[a]Sarah Carnegie unpublished data, based on ovarian hormone analyses
[b]Of the 144 infants born in our study groups, 25 (17%) were not sexed prior to their neonatal death
[c]All five females that immigrated into our study groups were considered adults (all but one was parous)
[d]For males, death is not easily distinguished from dispersal
[e]Birth dates of monkeys alive at the start of the study or immigrating into study groups are estimated

group size: 15.2, Table 8.2), their social structure varies and we have documented small uni-male/multi-female groups, and in one case, a group that had no resident adult, or even subadult, male for 10 months.

Even though the number of adult males per group varies across our five study groups and 48 census groups, all groups contain multiple, usually closely related, females and their immature offspring. Ovarian hormone analyses from fecal samples show that gestation length in our females is approximately 5.5 months, and that females typically experience only one or two cycles before they conceive (Carnegie 2011). Females appear to provide no visible cues to ovulation and often approach and direct proceptive signals to subordinate males when they are not fecund and even when they are pregnant (Carnegie et al. 2005). The occurrence of female-initiated, non-conceptive mating with subordinate males may explain why we see all resident adult males copulate, but genetic data show that the alpha males father most of the infants (Jack and Fedigan 2006).

Age at first birth for a female is usually around 6.5 years (Table 8.1), reflecting the slow life history of this species relative to its body size ($\male = 3.7$ kg, $\female = 2.7$ kg; Ford and Davis 1992). In captivity, capuchins may live into their fifth decade (Fragaszy et al. 2004), but in the wild, mortality rates at our site are such that female

8 Tracking Neotropical Monkeys in Santa Rosa   171

Table 8.2 Census data for (a) Capuchin monkeys and (b) Howler monkeys in Santa Rosa National Park, Costa Rica between 1972 and 2007

Year	# Monkeys counted	# Groups counted	Average group size	Estimated # groups	Estimated population size[a]
(a) Capuchins					
1972	?[b]	1	17.5	17	297
1983	226	20	11.5	28	318
1984	338	25	13.6	28	378
1985	175	13	14.8	28	397
1986	284	18	16.4	28	448
1987	173	10	16.7	28	474
1988	140	8	16.4	28	468
1990	314	18	17.7	28	491
1992	541	30	18.0	30	541
1999	521	31	16.8	35	588
2003	655	49	13.4	49	655
2007	594	39	15.2	48	716
(b) Howlers					
1972	65	8	8.1	10	85
1983	217	19	11.4	24	274
1984	295	23	12.8	25	321
1985	262	19	13.8	26	359
1986	315	19	16.6	28	464
1987	181	16	11.3	30	339
1988	212	12	17.7	31	548
1990	432	27	16.0	33	528
1992	563	35	16.1	34	547
1999	545	46	11.8	50	592
2003	529	44	12.0	49	589
2007	463	41	11.3	54	610

[a]Estimated population size was computed as: # Monkeys Counted + ((Estimated # Groups − # Groups Counted) × Average Group Size))

[b] Freese (1976) did not state how many capuchins he counted. Instead he estimated 15–20 groups of capuchins in Santa Rosa. The one group he tracked over time had on average 17.5 members

life expectancy at age one is only around 9 years whereas male life expectancy at age one is much lower, around 3 years (Table 8.1; Bronikowski et al. 2011). Santa Rosa has a large and intact predator community (cats, canids, mustelids, raptors, snakes; Janzen 1988) and we have seen capuchins killed by a *Boa constrictor* (Chapman 1986) and a puma (McCabe unpublished data), as well as observing predators stalking the monkeys. Other documented and inferred sources of mortality are intra-specific aggression, parasites, and contagious diseases (Fedigan 2003; Parr 2011). Obviously, some individuals live much longer than 3–9 years and we estimate that the oldest females in our study groups are about 27 years of age and the oldest males approximately 25 years. Estimating the life expectancy of capuchin males is problematic as it is often difficult to distinguish death from dispersal in this sex. At Santa Rosa, males disperse from their natal groups at around 4.5 years of age (Jack et al. 2011, but see Perry et al. (2012) for an older average male age at natal dispersal, 7.6 years) and they continue change groups at approximately 4-year intervals throughout their lives (Jack and Fedigan 2004b).

Infant mortality in the first year of life is quite variable from year to year, but averages around 30%. If a female's infant dies, she is likely to produce another infant in about a year, whereas if her infant survives, her interbirth interval averages 2.25 years. We examined the factors that may affect length of interbirth interval and infant survival, using 21 years of data, 24 adult females, and 74 completed interbirth intervals (Fedigan et al. 2008). In brief, we found that the pace of a female's reproduction can be predicted by the number of matrilineal kin in her group (sources of supportive allomothering and coalitions) and by the availability of resources, as inferred from the amount of rainfall that occurred in the 12 months subsequent to each infant's birth (see justification for use of rainfall as a proxy for food availability in Murphy and Lugo 1986; Fedigan et al. 2008). But the strongest predictor of the length of interbirth intervals is whether or not the first infant in the interval survives. In turn, the survival of infants is best predicted by whether male membership in the group is stable at the time of birth or whether there is a take-over in the first year of the infant's life. Infant survival is also affected by sex – a higher proportion of female than male infants make it through their first year (Table 8.1). Somewhat surprisingly, the mother's dominance rank at the time of the infant's birth affects neither its survival nor the length of interbirth intervals. Thus, dominance rank does not influence female reproductive success, whereas we have found such a relationship for males (Jack and Fedigan 2006).

Santa Rosa is a highly seasonal tropical dry forest and white-faced capuchins prefer to drink water every day. Rainfall amounts vary greatly across the months of the year as well as across years (range: 818–4,012 mm per annum). The rainy season typically occurs between mid-May and mid-November, during which an average of 1,792 mm of rain falls. Between mid-November and mid-May, especially from January through April, virtually no rainfall occurs, most trees drop their leaves, all the streams dry up, and most sources of standing water disappear. In tropical dry forests, rainfall is a major influence on plant and insect productivity and thus available food energy (Murphy and Lugo 1986; Fedigan et al. 2008). Therefore, we have been interested in the extent of seasonality in capuchin conceptions and births. Analyses of 144 birthdates for capuchin infants in our study groups over many years show that births are significantly more likely to occur in the late dry/ early wet season (Fig. 8.3; Fedigan 2003; Fedigan and Jack 2004). Given a 5.5-month gestation period, this indicates that conceptions are clustered in the second half of the wet and early part of the dry season (Carnegie et al. 2011). With 44% of infants born within the 3-month period of May through July (and therefore conceived within a 3-month period), the Santa Rosa white-faced capuchins display moderate breeding seasonality (see van Schaik et al. 1999 for classification overview). Although white-faced capuchins do not show strict birth seasonality, as has been demonstrated for the black capuchins (*Cebus nigritus*) of Iguazu National Park, Argentina, studied by Di Bitetti and Janson (2000), (see also Janson and

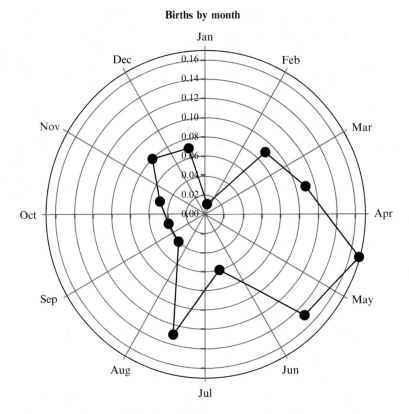

Fig. 8.3 The monthly pattern of births for 144 *Cebus capucinus* infants born in our study groups between 1986 and 2010. The dry season occurs from mid-December through mid-May

Verdolin 2005, Janson et al. 2012) they are similar to black capuchins and other more seasonal neotropical primates in that their birth peak coincides with, or slightly precedes, the seasonal peak in resource availability (Carnegie et al. 2011).

Interestingly, the white-faced capuchins studied at Lomas Barbudal (Perry et al. 2012) show even stronger reproductive seasonality than the Santa Rosa capuchins with 66.5% of all infants born between April and July. Lomas Barbudal and Santa Rosa are merely 55 km apart and are ecologically very similar so the reasons for this difference in birth seasonality are not easily explained. Variation in strength of reproductive seasonality may, perhaps, be tied to the fact that the capuchins at Lomas have year round access to fresh water, whereas the Santa Rosa capuchins drink from standing water holes, most of which have completely vanished by the end of the dry season. Indeed this single ecological difference may explain several of the demographic and life history differences that have been reported from these two nearby long-term study sites (e.g. group size, age at natal dispersal, and adult male tenure length). We plan to collaboratively address some of these inter-site differences with the Perry team of investigators over the next few years.

In some primate species, male dispersal is intimately linked to/timed with the breeding season (e.g. *Propithecus edwarsi*: Morelli et al. 2009; *Semnopithecus entellus*: Borries 2000), but the moderate seasonality displayed by our population of white-faced capuchins does not appear to directly influence male mobility. In our long-term analysis of natal and secondary dispersal, we did not find a significant relationship between the timing of male dispersal and the conception peaks (Fedigan and Jack 2004; Jack and Fedigan 2004a). Most dispersal events occur during the dry season months of January through April, a full 3–6 months prior to the conception peak. During the dry season, however, intergroup encounters are more likely to occur, usually around shared water holes, and males may use these encounters to appraise reproductive opportunities in neighboring groups and to assess the strengths of resident males in other groups. Assessment of the strength of groups targeted by would-be immigrants could be particularly important for dispersing adult males who typically fight their way into groups, although a few manage to peacefully join groups previously abandoned by resident males, a pattern we refer to as "waltz-ins". Males usually disperse in cohorts of two or more ($77\%, N = 74$; Jack and Fedigan 2004a, b), and in almost all of the successful takeovers that we have observed to date ($N = 20$), coalitions of invading males outnumber those of resident males. This pattern of male dispersal leads to the complete replacement of resident males approximately every 4 years.

Over the years, we have observed 25 male replacement events in our five study groups and they are usually associated with the wounding, deaths, and disappearances of individuals from all age-sex classes (Fedigan 2003; Fedigan and Jack 2004; Jack and Fedigan 2009). Indeed male replacement events strongly influence many life history patterns. For example, the occurrence of a group takeover is the most significant factor explaining the highly variable age at which male natal dispersal occurs (Jack et al. 2010) and infant deaths are significantly more common, and females are more likely to disperse, during years with male replacements than during peaceful years (Fedigan 2003; Jack and Fedigan 2009). Given that white-faced capuchins show very high reproductive skew (Jack and Fedigan 2006; Muniz et al. 2010), with alpha males siring most infants, it is no surprise that males frequently change groups – likely in an attempt to maximize their reproductive opportunities (reviewed by Jack 2003). By changing groups, formerly subordinate males experience rank gains, which should also confer reproductive benefits (Jack and Fedigan 2004b). Even alpha males have been observed to abandon groups to join others, taking risks that seem counterintuitive at first. However, they target groups that proffer a more favorable sex ratio (i.e., more potential mates). While secondary/breeding dispersal by adult males is frequently suggested as a mechanism for the avoidance of breeding with maturing daughters (see Smith 1982), this does not appear to be the main proximate cause for the frequent dispersal of male white-faced capuchins (Jack and Fedigan 2004b). The average tenure length for an adult male in our study groups is 4 years (alphas and subordinates do not differ in this regard), which is well under the age that females first conceive and give birth (Table 8.1). Indeed in many primate species secondary dispersal by adult males is better explained by intrasexual mating competition rather than inbreeding avoidance (reviewed by Jack 2003). Genetic studies of the Lomas Barbudal population of white-faced capuchins (Perry et al. 2012), where alpha males have much longer tenures (e.g. up to 18 years) than those in Santa Rosa, have shown that there are behavioral mechanisms in place to ensure inbreeding avoidance; namely, the daughters of alpha males reproduce with subordinate males rather than with their fathers (Muniz et al. 2006).

8.3 Population Recovery in a Regenerating Tropical Dry Forest

Our demographic research began in 1983 and 1984 with extensive attempts to count all the monkeys located throughout SRNP, and since that time we have conducted 11 park-wide censuses of capuchins and howlers (Table 8.2). Except for one census conducted in August–November of 1992, these have all been carried out in April through July and mainly in May and June. For the capuchin and howler groups, we use a modified quadrat ("complete count") technique that has proven useful in fragmented forest patches (see Fedigan et al. 1996, 1998; Fedigan and Jack 2001 for details). However, unlike capuchins and howlers, spider monkeys have fission–fusion societies and individuals range over very large areas in "parties" of frequently changing composition. Therefore, we have had to assess spider monkey densities from transect studies, which have been conducted much less frequently than our censuses of capuchin and howler populations (see Chapman et al. 1988, 1989b; Sorensen and Fedigan 2000; DeGama-Blanchet and Fedigan 2006).

We choose one area of the park at a time (usually a large forest fragment) and walk all known trails and dry stream beds there to locate monkeys. We consider any individual monkey within 100–300 m of the group and consistently traveling in the same direction as the group (even if in a peripheral position) to be part of that group. We use unique markings, known individuals and distinctive age/sex compositions to identify the same group on successive days for repeat counts. Multiple observers repeatedly count a group until achieving a stable count and composition, and plot its location on a map. After establishing a stable count on one group, we locate its nearest neighbor group. Whenever possible, with the aid of two-way radios, we use simultaneous contact with neighboring groups by different observers to establish their independence. With many years of practice, it has become increasingly easy to relocate our census groups in successive years and to determine when new groups have appeared or former groups have become extinct.

Table 8.2 shows the number of monkeys and groups we counted and the estimated population sizes of capuchins and howlers in Santa Rosa between 1972 (Freese's original count) and 2007, a 35-year period subsequent to the establishment of the park. Whether we begin with Freese's 1972 census or our first census in 1983, the number of capuchins in SRNP has more than doubled (e.g., from 318 in 1983 to 716 in 2007) and it gives every indication of continuing to grow (Fig. 8.4).

Howler population dynamics are different. Freese concluded that there were only ten howler groups in Santa Rosa in 1972 with a total population of only

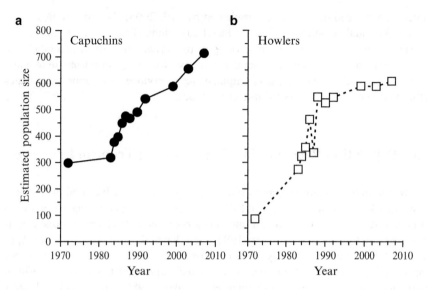

Fig. 8.4 Estimated population sizes of (**a**) capuchins and (**b**) howlers in Sector Santa Rosa, based on 12 park-wide censuses between 1972 and 2007

85 monkeys and he said that howlers were located only on the lower plateau, near the Pacific Ocean. The howler population apparently crashed shortly before the establishment of the park (see Fedigan et al. 1998; Fedigan and Jack 2001). However, at the time of our first census in 1983, we estimated there were 24 howler groups spread throughout the park, with a population of about 281 monkeys. That would indicate the howler population more than tripled in the 11-year time period between 1972 and 1983. Although howler females can first give birth at 3.5 years and can thereafter produce an infant every 20 months (Fedigan and Rose 1995), this would still be a remarkable rate of population growth. More conservatively, if we begin with our first census in 1983, the number of howlers more than doubled in a 14-year period (from 281 in 1983 to 620 in 2007). However, the Santa Rosa population of howlers began to level out in 1999 and since that time has experienced a near zero growth rate (Table 8.2).

Whether we begin our comparison of capuchins and howlers with Freese's 1972 census or our own first census in 1983, there were clearly more capuchins than howlers in Santa Rosa in the early days of the park's existence. The howler population then grew at a faster rate than the capuchins during the 1980s and 1990s, probably because howlers have a faster pace of reproduction and greater intrinsic rate of increase (earlier age at first birth, shorter interbirth intervals, Fedigan and Rose 1995). However, howlers are differentially found in the older (>60 years) evergreen forest that includes many large trees (DBH >63 cm, which is the smallest size tree in which Santa Rosa howlers rest and forage, probably to accommodate their larger body size and possibly also their folivorous diet). In the first two decades of our study, we observed several new groups of howlers colonize strips of old growth riverine forest and small patches of secondary forest that were transitioning into primary forest. But the stabilization of their population size may indicate that, at least for now, howlers have run out of suitable habitat into which they can expand as the forest fragments of Santa Rosa slowly regenerate. The smaller bodied, omnivorous capuchins, in contrast, can occupy newly regenerating forests and, as of our most recent censuses in 2003 and 2007, they once again outnumber the howlers (Table 8.2).

Figure 8.5 shows important differences in how the two populations grew in the 1980s and 1990s versus the most recent decade. Between 1983 and 1999, the capuchin population grew mainly via increases in group size (average group size went from 11 to 17) while the number of groups in the park only increased from 28 to 35. In contrast, the average howler group size fluctuated between census years, but showed no steady increase, whereas the number of howler groups doubled from 24 to 50 between 1983 and 1999 (Fig. 8.5b). During that time period, we repeatedly saw lone howler males move into unoccupied forest fragments and howl until they were joined by females and thus we surmise that the howler population initially increased by budding off small new groups. This pattern of creating new groups is feasible for howlers because both males and females regularly disperse. In contrast, the mechanism of capuchin population growth was via an increase in group size (Fig. 8.4b). Unlike the lone howler males who moved into previously unoccupied forest fragments, entire groups of capuchins began to range into new areas and because we often saw them in young forest, we infer that they accommodated their

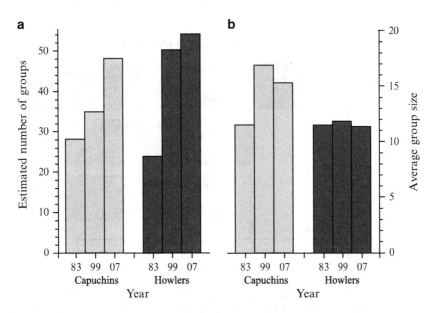

Fig. 8.5 (a) The estimated number of groups and (b) the mean group sizes of capuchin and howlers in Sector Santa Rosa in 1983, 1999 and 2007

increasing group size by expanding their home ranges into newly regenerating patches of forest.

However, beginning with our 2003 census, we started to see more new capuchin groups, their average group size stabilized and we ceased to locate new howler groups. The earlier species difference in population growth might have broken down because the howler population reached carrying capacity and because a large group size imposes reproductive costs on female capuchins – larger groups have lower female reproductive success (Fedigan and Jack 2011).

DeGama-Blanchet and Fedigan (2006) showed that the age of forest fragments significantly affects the densities of howlers and capuchins throughout ACG and that the availability of a dry season water source significantly predicts higher densities of capuchins. As pointed out by Altmann (1974) in his examination of baboon resources and home range sizes, water is the essential limiting resource for obligate drinkers living in arid regions. Sorensen and Fedigan (2000) found that capuchins can make use of forest patches as young as 25 years (Fig. 8.6), but whenever possible, they range in such a way as to have a source of dry season water within their home range from which they drink at least once daily. Capuchin densities follow a fairly linear pattern of increase from newly regenerating forest patches up to 180-year old (primary or evergreen) forest. Howlers do not usually appear in forest fragments under 60 years old and only become common in forests of 100–150 years old. Spider monkeys in Santa Rosa are only rarely seen in forest fragments less than 100–200 years old and they prefer larger patches, presumably to accommodate their trap-line pattern of foraging on fruit (Fig. 8.6). As described in Fedigan and Jack (2001), our long-term examination of the population dynamics of

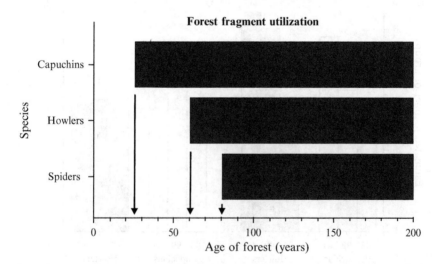

Fig. 8.6 The age of forest fragments that are utilized by capuchin, howler, and spider monkeys in Sector Santa Rosa. Capuchins can be found ranging, resting, and foraging in forest fragments ≥25 years of age, howlers in forest fragments ≥60 years and spider monkeys in forest fragments ≥80 years of age

these three species in a tropical dry forest habitat indicates that the fundamental requirement for capuchins is a year-round water source, whereas howlers (which seldom drink water) need large trees with leaves and fruit that have low levels of secondary compounds, and spider monkeys require large tracts of old growth forest to maintain their fission–fusion social system and frugivorous diet.

Another important difference in how the capuchin and howler populations have grown over the past 25 years concerns changes in age–sex compositions (Fig. 8.7). The age/sex composition of the howler population has fluctuated but not varied substantially between 1983 and 2007. However, adult males have accounted for an increasing proportion of the capuchin population. Furthermore, the ratio of adult males to females has gradually shifted from 0.47 in 1983 to 1.1 in 2007. Table 8.3 shows that while the numbers of infants, juveniles, and adult females doubled over the 25-year period, the number of adult males increased fivefold.

Why would there be a differential increase in the numbers of adult male capuchins? We offer three suggestions. First, the removal of hunters from the park has probably allowed more adult males to survive. Adult capuchin males assume a highly visible and audible protective role in their social groups and during encounters with humans they are always at the forefront of the group and are much more likely to be harmed or wounded than are other group members. Second, capuchins are a male-dispersal species and males are thus better able than are females to take advantage of protection in the park by immigrating into the more secure forests within the park boundaries. Finally, the infant sex ratios (Table 8.1) we recorded in our study groups between 1984 and 2000 were highly biased toward males in our groups and may reflect a population-wide bias toward male infants

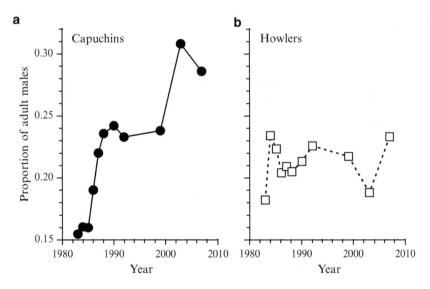

Fig. 8.7 Proportion of adult males for (**a**) capuchins and (**b**) howlers in the Santa Rosa population as counted during park-wide censuses between 1983 and 2007

Table 8.3 Age–sex composition of populations of (a) Capuchin monkeys and (b) Howler monkeys in Santa Rosa National Park, Costa Rica between 1983 and 2007

Year	# Counted	Prop. adult males	Prop. adult females	Prop. juveniles	Prop. infants	Prop. unknown	Adult males per adult female
(a) Capuchins							
1983	226	0.155	0.332	0.323	0.137	0.053	0.467
1984	338	0.160	0.337	0.349	0.109	0.044	0.474
1985	175	0.160	0.314	0.400	0.120	0.006	0.509
1986	284	0.190	0.313	0.345	0.109	0.042	0.607
1987	173	0.220	0.266	0.324	0.156	0.035	0.826
1988	123	0.236	0.276	0.382	0.098	0.008	0.853
1990	314	0.242	0.293	0.331	0.108	0.025	0.826
1992	541	0.233	0.298	0.237	0.176	0.057	0.783
1999	521	0.238	0.345	0.257	0.106	0.054	0.689
2003	655	0.308	0.276	0.342	0.072	0.002	1.116
2007	594	0.286	0.264	0.278	0.125	0.047	1.083
(b) Howlers							
1983	231	0.182	0.468	0.173	0.169	0.009	0.389
1984	299	0.234	0.425	0.217	0.124	0.000	0.551
1985	273	0.223	0.359	0.260	0.158	0.000	0.622
1986	313	0.204	0.403	0.201	0.176	0.016	0.508
1987	187	0.209	0.364	0.235	0.193	0.000	0.574
1988	220	0.205	0.382	0.205	0.209	0.000	0.536
1990	431	0.213	0.390	0.239	0.158	0.000	0.548
1992	579	0.226	0.408	0.185	0.169	0.012	0.555
1999	548	0.217	0.391	0.232	0.146	0.015	0.556
2003	538	0.188	0.411	0.227	0.169	0.006	0.457
2007	463	0.233	0.443	0.173	0.145	0.006	0.527

during this time period. If the infant sex ratios we observed in our study groups during this time period are reflective of the capuchin population as a whole, then the current population-wide male-biased adult sex ratio is not unexpected. Interestingly, the infant sex ratios in our study groups has changed to a nearly 1:1 ratio over the past decade and we are interested to see how this change will be reflected in the population-wide adult sex ratios in the years to come. Subsequent to the Trivers–Willard hypothesis (Trivers and Willard 1973), much has been published on factors affecting primate sex ratios at birth (see review and meta-analyses in Brown and Silk 2002). However, we agree with Strier (2009) that the sample sizes common in studies such as our own are too small to test whether the observed changes over time are truly adaptive or are random deviations from what will eventually prove to be a 50:50 ratio between sons and daughters at birth. Only the continuation of our already long-term study will tell.

8.4 Conclusions

Under the protection of the ACG park service, the typical tree species of a neotropical dry forest are slowly re-establishing themselves in former pasture land, creating arboreal corridors between fragments and growing into secondary forests that can be exploited by capuchins. Also, the forest fragments of Santa Rosa that were not cleared for agriculture but were selectively logged and subjected to other forms of human exploitation are now protected. Our long-term research has clearly demonstrated that monkey populations increase under these conditions. However, we can also draw the wider lesson that because the life history pattern and interaction of each species with its environment is unique, the successful restoration of primate populations is more complicated than simply removing the disturbances that humans introduced in the first place (hunters, loggers, cattle, non-native plants, anthropogenic fires).

It is essential to understand the species-specific aspects of vulnerability and potential for recovery. As pointed out by Chapman et al. (2010), a complex set of interactions govern changes in habitat composition and structure and the associated changes in animal populations. Chapman and colleagues found from repeated transect censuses in Kibale National Park between 1970 and 2006, that two of five primate species (mangabeys, black-and-white colobus) had increased in density over the study period, two (redtails, red colobus) were stable, and one (blue monkeys) had declined. In contrast, our long-term research indicates that population sizes of all three monkey species we study have increased since 1971, but differentially so. We attribute their differential patterns of population recovery to the many species-specific aspects of their biology and behavioral ecology (e.g., differences in body size, diet, life history pace, social organization and behavioral flexibility in response to change). It is encouraging that the monkey populations in the protected forests of Santa Rosa have grown substantially in the 40 years since the park was established. Although it requires great energy, optimism, and patience to work toward the regeneration of a forest, our research shows that humans have successfully created the conditions to re-grow the monkey populations of Santa Rosa.

It also takes effort, diligence and a long view to maintain a field study over a nearly 30-year period. But without a multi-year study, there is much vital information about these monkeys we could not know. We hope that this brief overview of some of the findings from our Santa Rosa project demonstrate that the rewards are well worth the effort.

Acknowledgments We thank the Costa Rican National Park Service for permission to work in SRNP from 1983 to 1989 and the administrators of the Area de Conservación Guanacaste (especially Roger Blanco Segura) for allowing us to continue research in the park through the present day. Many people contributed to the census and life history database on the Santa Rosa monkeys and we are grateful to all of them. John Addicott developed the database and helped with the figures. Greg Bridgett maintains the database and helped with editorial matters. Research protocols reported in this paper complied with all institutional and government regulations regarding ethical treatment of our study subjects. L.M. Fedigan's research is supported by NSERC and the Canada Research Chairs Program. K.M. Jack's research is supported by grants from Tulane Unversity's Research Enhancement Fund and Committee on Research. We also thank the Zemurray Foundation for support of our project.

References

Allen W (2001) Green phoenix: restoring the tropical forests of Guanacaste. Oxford University Press, New York

Altmann SA (1974) Baboons, space, time, and energy. Am Zool 14:221–248

Borries C (2000) Male dispersal and mating season influxes in Hanuman langurs living in multi-male groups. In: Kappeler PM (ed) Primate males: causes and consequences of variation in group composition. Cambridge University Press, Cambridge, pp 146–158

Bronikowski AM, Altmann J, Brockman DK, Cords M, Fedigan LM, Pusey AE, Stoinski T, Morris WF, Strier KB, Alberts SC (2011) Aging in the natural world: comparative data reveal similar mortality patterns across primates. Science 331:1325–1328

Brown GR, Silk JB (2002) Reconsidering the null hypothesis: is maternal rank associated with birth sex ratios in primate groups? Proc Natl Acad Sci USA 99:11252–11255

Carnegie SD (2011) Reproductive behavour and endocrinology of female white-faced capuchins (*Cebus capucinus*). PhD Dissertation, University of Calgary

Carnegie SD, Fedigan LM, Ziegler TE (2005) Behavioral indicators of ovarian phase in white-faced capuchins (*Cebus capucinus*). Am J Primatol 67:51–68

Carnegie SD, Fedigan LM, Melin AD (2011) Reproductive seasonality in female capuchins (*Cebus capucinus*) in Santa Rosa (Area de Conservación Guanacaste), Costa Rica. Int J Primatol. doi:10.1007/s10764-011-9523-x

Chapman CA (1986) *Boa constrictor* predation and group response in white-faced cebus monkeys. Biotropica 18:171–171

Chapman CA (1989) Spider monkey sleeping sites: use and availability. Am J Primatol 18:53–60

Chapman CA (1990) Ecological constraints on group size in three species of neotropical primates. Folia Primatol 55:1–9

Chapman CA, Fedigan LM, Fedigan L (1988) A comparison of transect methods of estimating population densities of Costa Rican primates. Brenesia 30:67–80

8 Tracking Neotropical Monkeys in Santa Rosa

Chapman CA, Fedigan LM, Fedigan L, Chapman LJ (1989a) Post-weaning resource competition and sex ratios in spider monkeys. Oikos 54:315–319

Chapman CA, Chapman L, Glander KE (1989b) Primate populations in northwestern Costa Rica: potential for recovery. Primate Conserv 10:37–44

Chapman CA, Struhsaker TT, Skorupa JP, Snaith TV, Rothman JM (2010) Understanding long-term primate community dynamics: implications of forest change. Ecol Appl 20:179–191

DeGama-Blanchet HN, Fedigan LM (2006) The effects of forest fragment age, isolation, size, habitat type, and water availability on monkey density in a tropical dry forest. In: Estrada A, Garber PA, Pavelka MSM, Luecke L (eds) New perspectives in the study of Mesoamerican primates: distribution, ecology, behavior, and conservation. Springer, New York, pp 165–188

Di Bitetti MS, Janson CH (2000) When will the stork arrive? Patterns of birth seasonality in neotropical primates. Am J Primatol 50:109–130

Evans S (1999) The green republic: a conservation history of Costa Rica. University of Texas Press, Austin

Fedigan LM (2003) Impact of male takeovers on infant deaths, births and conceptions in *Cebus capucinus* at Santa Rosa, Costa Rica. Int J Primatol 24:723–741

Fedigan LM, Jack KM (2001) Neotropical primates in a regenerating Costa Rican dry forest: a comparison of howler and capuchin population patterns. Int J Primatol 22:689–713

Fedigan LM, Jack KM (2004) The demographic and reproductive context of male replacements in *Cebus capucinus*. Behaviour 141:755–775

Fedigan LM, Jack KM (2011) Two girls for every boy: the effects of group size and composition on the reproductive success of male and female white-faced capuchins. Am J Phys Anthropol 144:317–326

Fedigan LM, Rose LM (1995) Interbirth interval variation in three sympatric species of neotropical monkey. Am J Primatol 37:9–24

Fedigan LM, Fedigan L, Chapman CA, Glander KE (1988) Spider monkey home ranges: a comparison of radio telemetry and direct observation. Am J Primatol 16:19–29

Fedigan LM, Rose LM, Morera Avila R (1996) See how they grow: tracking capuchin monkey populations in a regenerating Costa Rican dry forest. In: Norconk MA, Rosenberger AL, Garber PA (eds) Adaptive radiations of neotropical primates. Plenum Press, New York, pp 289–307

Fedigan LM, Rose LM, Morera Avila R (1998) Growth of mantled howler groups in a regenerating Costa Rican dry forest. Int J Primatol 19:405–432

Fedigan LM, Carnegie SD, Jack KM (2008) Predictors of reproductive success in female white-faced capuchins (*Cebus capucinus*). Am J Phys Anthropol 137:82–90

Ford SM, Davis LC (1992) Systematics and body size: implications for feeding adaptations in New World monkeys. Am J Phys Anthropol 88:415–468

Fragaszy DM, Visalberghi E, Fedigan LM (2004) The complete capuchin: the biology of the genus *Cebus*. Cambridge University Press, Cambridge

Freese CH (1976) Censusing *Alouatta palliata*, *Ateles geoffroyi*, and *Cebus capucinus* in the Costa Rican dry forest. In: Thorington RW, Heltne PG (eds) Neotropical primates: field studies and conservation. National Academy of Science, Washington, DC, pp 4–9

Glander KE, Fedigan LM, Fedigan L, Chapman CA (1991) Field methods for capture and measurement of three monkey species in Costa Rica. Folia Primatol 57:70–82

Jack KM (2003) Males on the move: evolutionary significance of secondary dispersal in male primates. Primate Rep 67:61–83

Jack KM, Fedigan LM (2004a) Male dispersal patterns in white-faced capuchins, *Cebus capucinus*. Part 1: patterns and causes of natal emigration. Anim Behav 67:761–769

Jack KM, Fedigan LM (2004b) Male dispersal patterns in white-faced capuchins, *Cebus capucinus*. Part 2: patterns and causes of secondary dispersal. Anim Behav 67:771–782

Jack KM, Fedigan LM (2006) Why be alpha male? Dominance and reproductive success in wild white-faced capuchins (*Cebus capucinus*). In: Estrada A, Garber PA, Pavelka MSM, Luecke L (eds) New perspectives in the study of Mesoamerican primates: distribution, ecology, behavior, and conservation. Springer, New York, pp 367–386

Jack KM, Fedigan LM (2009) Female dispersal in a female-philopatric species, *Cebus capucinus*. Behaviour 146:471–497

Jack KM, Sheller C, Fedigan L (2010) Predicting natal dispersal in male white-faced capuchins (*Cebus capucinus*). Am J Phys Anthropol 141(suppl S50):133

Jack KM, Sheller C, Fedigan LM (2011) Social factors influencing natal dispersal in male white-faced capuchins (*Cebus capucinus*). Am J Primatol 73: 1–7

Janson CH, Verdolin JL (2005) Seasonality of primate births in relation to climate. In: Brockman DK, van Schaik CP (eds) Seasonality in primates: studies of living and extinct human and non-human primates. Cambridge University Press, Cambridge, pp 307–350

Janson C, Baldovino MC, Di Bitetti M (2012) The Group Life Cycle and Demography of Brown Capuchin Monkeys (*Cebus [apella] nigritus*) in Iguazú National Park, Argentina. In: Kappeler PM (ed) Long-term field studies of primates. Springer, Heidelberg

Janzen DH (1988) Guanacaste National Park: tropical ecological and biocultural restoration. In: Cairns J Jr (ed) Rehabilitating damaged ecosystems. CRC Press, Baton Rouge/FL, pp 143–192

Janzen DH (2000) Costa Rica's Area de Conservación Guanacaste: a long march to survival through non-damaging biodevelopment. Biodiversity 1:7–20

Janzen DH (2002) Tropical dry forest: Area de Conservación Guanacaste, northwestern Costa Rica. In: Perrow MR, Davy AJ (eds) Handbook of ecological restoration, vol 2, Restoration in practice. Cambridge University Press, Cambridge, pp 559–583

Janzen DH (2004) Ecology of dry forest wildland insects in the Area de Conservación Guanacaste. In: Frankie GW, Mata A, Vinson SB (eds) Biodiversity in Costa Rica: learning the lessons in a seasonal dry forest. University of California Press, Berkeley, pp 80–96

Morelli TL, King SJ, Pochron ST, Wright PC (2009) The rules of disengagement: takeovers, infanticide, and dispersal in a rainforest lemur, *Propithecus edwarsi*. Behaviour 146:499–523

Muniz L, Perry S, Manson JH, Gilkenson H, Gros-Louis J, Vigilant L (2006) Father-daughter inbreeding avoidance in a wild primate population. Curr Biol 16:R156–R157

Muniz L, Perry S, Manson JH, Gilkenson H, Gros-Louis J, Vigilant L (2010) Male dominance and reproductive success in wild white-faced capuchins (*Cebus capucinus*) at Lomas Barbudal, Costa Rica. Am J Primatol 72:1118–1130

Murphy PG, Lugo AE (1986) Ecology of tropical dry forest. Annu Rev Ecol Syst 17:67–88

Parr N (2011) Predictors of parasitism in wild white-faced capuchins. MA thesis, University of Calgary

Perry S, Manson JH (2008) Manipulative monkeys: the capuchins of Lomas Barbudal. Harvard University Press, Cambridge, MA

Perry S, Godoy I, Lammers W (2012) The Lomas Barbudal Monkey Project: Two Decades of Research on Cebus capucinus. In: Kappeler PM (ed) Long-term field studies of primates. Springer, Heidelberg

Smith DG (1982) Inbreeding in three captive groups of rhesus monkeys. Am J Phys Anthropol 58:447–451

Sorensen TC, Fedigan LM (2000) Distribution of three monkey species along a gradient of regenerating tropical dry forest. Biol Conserv 92:227–240

Strier KB (2009) Seeing the forest through the seeds: mechanisms of primate behavioral diversity from individuals to populations and beyond. Curr Anthropol 50:213–228

Trivers RL, Willard DE (1973) Natural selection of parental ability to vary the sex ratio of offspring. Science 179:90–92

van Schaik CP, van Noordwijk MA, Nunn CL (1999) Sex and social evolution in primates. In: Lee PC (ed) Comparative primate socioecology. Cambridge University Press, Cambridge, pp 204–240

Chapter 9
The Group Life Cycle and Demography of Brown Capuchin Monkeys (*Cebus [apella] nigritus*) in Iguazú National Park, Argentina

Charles Janson, Maria Celia Baldovino, and Mario Di Bitetti

Abstract This study reports demographic and social changes across 20 years in a population of brown capuchin monkeys living in Iguazú National Park in northeastern Argentina. Three sets of results emerge that are critical to understanding the evolution of social behavior in this population. First, patterns of age-related mortality clearly highlight certain periods of increased mortality (postnatal 6 months, onset of reproduction, late senescence) and near-perfect survival (2–6-year-old juveniles, young adult females). Second, tracking the migrations and rank-related reproductive strategies of males helps to uncover the causes and consequences of long male reproductive tenures that average 5 years. Finally, observations of relatively rare male takeovers of the alpha breeding position reveal a predictable sequence of stages in a group's life cycle that tie together female fecundity, infanticide, group size, and kinship-based group fissions. These coordinated aspects of demography and kinship in different stages set the context for understanding differences between groups in social structure and organization.

9.1 Introduction

The population of *Cebus [apella] nigritus* in the Macuco trail area of Iguazú National Park, Argentina, has been studied since 1988 (Brown and Zunino 1990). This work has included censuses at least twice a year since 1991 (Di Bitetti and Janson 2001a; Ramírez-Llorens et al. 2008), and intensive behavioral and

---

C. Janson (✉)
Division of Biological Sciences, University of Montana, Missoula, MT, USA
e-mail: charles.janson@mso.umt.edu

M.C. Baldovino • M. Di Bitetti
CONICET, Instituto de Biología Subtropical, Universidad Nacional de Misiones, Puerto Iguazú, Argentina
e-mail: macelia@arnet.com.ar; dibitetti@yahoo.com.ar

P.M. Kappeler and D.P. Watts (eds.), *Long-Term Field Studies of Primates*,
DOI 10.1007/978-3-642-22514-7_9, © Springer-Verlag Berlin Heidelberg 2012

ecological studies spanning over 80 months (e.g., Janson 1996, 1998, 2007; Di Bitetti and Janson 2001a; Di Bitetti 2005; Baldovino and Di Bitetti 2008; Wheeler 2009; Baldovino 2010). During this period, groups have grown, split, joined, experienced male replacements, changes in matriline dominance, dispersal of subadult males, and complete replacement of the original cohort of females by their daughters. This report summarizes data on the rates of such events and on their demographic and behavioral consequences.

Both kinds of information are needed to understand the importance of a given selective force in a particular study population (Koenig et al. 2006; Kappeler and Fichtel 2012). Short-term studies can provide excellent data on the frequencies and consequences of common sources of natural selection but may miss rare albeit important sources of selection or may estimate their impact poorly. For instance, during the 30 years prior to 2004, several groups of researchers had studied various populations of *Cebus "apella"* superspecies (using the old taxonomy; these are now more properly referred to as species in the genus *Sapajus*: Lynch Alfaro et al. in press). Despite such effort, none had reported direct evidence of infanticide by males (although Izawa (1980) presented suggestive evidence). In 2005, we saw a male who attacked and killed an infant, and recorded two additional infant disappearances after a male takeover of our main study group (Ramírez-Llorens et al. 2008; see also Izar et al. 2007). Retrospective analysis of long-term demographic data suggests that infanticide has been the most important single source of mortality of unweaned infants in this study population, despite the rarity of male takeovers and the difficulty in documenting the behavior (Ramírez-Llorens et al. 2008). Thus, long-term studies can provide crucial context to interpret the social structure and organization of a species (e.g., Mitani et al. 2002). An important insight emerging from our analysis is that groups may show predictable "life cycles" – coordinated repetitive temporal patterns of group size, composition, and social change – that set the context for understanding rare demographic events in our study population.

9.2 Description of the Study Population

The Macuco trail area of Iguazú National Park, Argentina, is accessed from the Centro de Investigaciones Ecologicas Subtropicales (25°40′43″ S, 54°26′57″ W), just north of the main tourist attraction of Iguazú Falls. Floristically, the study area belongs to the Upper Paraná Atlantic Forest (Giraudo et al. 2003). The climate is humid and subtropical, with marked seasonality in temperature and day length, but little seasonality in rainfall. Fleshy fruits and arthropods, the most important resources for capuchin monkeys, are scarce in winter (July–August) and most abundant during spring and early summer (October–December; Placci et al. 1994; Di Bitetti and Janson 2001a; Di Bitetti 2009).

The study area contains forest in various stages of recovery from anthropogenic disturbance. First, the entire area to the north and west of the falls was a logging concession up until 1934, when the area was designated a national park by the

9 The Group Life Cycle and Demography of Brown Capuchin Monkeys

federal government (Ministerio del Medio Ambiente 2005). Timber harvesting was highly selective but had considerable impact on the natural vegetation, removed the largest trees and created logging roads to bring the timber to the Iguazú River; two of these logging roads, Yacaratia and Macuco, remain as tourist trails within the study site. A dirt airstrip was created in 1937. It was abandoned in 1971, but signs of former human habituation are still evident near its remnants, along the southwestern part of the Macuco trail. An important legacy of this period was the introduction of several nonnative fruit species, primarily citrus varieties and *Hovenia dulcis* (Rhamnaceae), which have escaped cultivated areas and to greater or lesser degrees become naturalized in the forest.

Current protection efforts by the Argentine National Park service are largely effective at preventing illegal logging and poaching (Di Bitetti et al. 2008; Paviolo et al. 2009). The study area boasts a fairly complete set of likely and actual monkey predators: black hawk-eagles (*Spizaetus tyrannus*), ornate hawk-eagles (*Spizaetus ornatus*), jaguarundis (*Puma yagouaroundi*), oncillas (*Leopardus tigrinus*), margays (*Leopardus wiedii*), ocelots (*Leopardus pardalis*), pumas (*Puma concolor*), jaguars (*Panthera onca*), and tayras (*Eira barbara*). Two species of large raptors, crested eagles (*Morphnus guianensis*) and harpy eagles (*Harpia harpyja*), formerly bred in the area, but the removal of the large emergent trees that they favor for nesting, in addition to general habitat loss, has made them no more than occasional visitors. Most of the five pit viper species (Crotalinae) that are expected for the region are seen routinely in the summer months; the medium-sized rainbow boa (*Epicrates cenchria*) is rare. Current anthropogenic disturbances inside the study area are routine use of the Macuco trail by tourists, noise from nearby roads, and helicopter flights relatively low above the canopy. Since 2007, increasingly frequent interactions (mostly begging for or stealing food) between individuals of one of the study groups and tourists near the falls have become a source of concern.

The study animals live in multi-male, multi-female groups of between 7 and 44 individuals (Di Bitetti and Janson 2001a; Agostini and Visalberghi 2005). With rare exceptions, all individuals of both sexes can be arranged into a linear dominance hierarchy based on decided agonistic and approach-avoid interactions. As in other *apella* populations, each study group has a clear alpha male who tends to be centrally positioned within the group spread and has feeding and mating priority (Janson 1984, 1985). The Iguazú capuchins often feed on multiple food sources at a time, thereby reducing aggression and monopolization of food by dominants (cf. Janson 1996; Di Bitetti and Janson 2001b). Females are philopatric and form alliances based on kinship and reciprocal social grooming (Di Bitetti 1997).

The data for this study come from nine social groups in the vicinity of the Macuco trail, inhabiting an area of over 6 km^2 (Ramírez-Llorens et al. 2008). The early (1991–1992) groups were Macuco (MAC), Yacaratia (YAC), Barrio (BAR), Rubias (RUB), Silver (SIL), and Laboratorio (LAB). In 2005, MAC split, giving rise to two additional daughter groups, Rita (RI) and Gundolf (GUN); MAC split again in 2009, producing an additional daughter group Spot (SP). Individual animals were recognizable through a combination of relatively stable patterns of fur color, tuft shape, size, healed injuries, as well as sex and group membership; altogether, 289 distinct individuals are in the database, although not all of them survived long enough to be named or even sexed. We did not capture and permanently mark the study animals, so the displacements of emigrating animals (mostly males) were recorded only opportunistically and for at most a year after their departure from one of the main study groups. The only group for which we have consistent census information throughout the 20-year study period is MAC and its daughter groups; for these groups, it is possible to state mother–infant relationships for the 147 descendants of the initial cohort of 5 adult females in 1991. Many researchers have contributed to the demographic database over the years, but nearly all of the consistent census data come from the authors for the following periods (CHJ: 1991; MSD: 1992–1996; MCB: 1997–2010). The study groups were either the subject of intensive behavioral study or were censused thoroughly at least twice per year, once late in the year (late November or December) after the major period of births and once in May–June after the period of high infant mortality.

Like other capuchin monkeys, the study population is omnivorous, with a diet consisting mainly of ripe fruits and arthropods (Robinson and Janson 1987). Seasonal trends in diet reflect the varying abundance of these resources. In the austral winter, when neither ripe fruit nor insects are readily available, the capuchins consume large amounts of vegetation, primarily leaf bases of bamboo and bromeliads (Brown and Zunino 1990). Births are highly seasonal; nearly all occur between October and February (Di Bitetti and Janson 2001a). Because of the scarcity of ripe fruit during June through August, Janson has used this population (primarily the MAC study group) for a series of large-scale provisioning experiments using up to 27 feeding platforms to test several hypotheses about foraging cognition and feeding competition (Janson 1996, 1998, 2007; Janson and Di Bitetti 1997; Di Bitetti and Janson 2001b). Smaller-scale use of feeding platforms has enabled studies on other topics in this population (e.g., Visalberghi et al. 2003; Di Bitetti 2005; Wheeler 2009). Use of the feeding platforms has not had a detectable effect in increasing birth rates (Di Bitetti and Janson 2001a) or infant survival rates (Ramírez-Llorens et al. 2008) in MAC relative to other groups. Some nonprovisioned groups in the population experienced similar population growth during parts of this period (e.g., SIL).

9.3 Demography of the Iguazú Capuchin Study Groups, 1991–2010

Nine study groups have been followed or censused with sufficient continuity to allow tracking of life events of identified individuals. The length of study in years for each group is 1 (SP), 2.5 (BAR), 4 (RUB), 5 (RI), 5.5 (YAC, LAB, GUN), 6 (SIL), and 20 (MAC); for further details from 1991 to 2006, see Ramírez-Llorens et al. (2008). Shorter periods have provided data for demographic rates and rare

9 The Group Life Cycle and Demography of Brown Capuchin Monkeys

events, while longer-term observations have permitted analyses of key events in the life cycle of groups (see Sect. 9.6 below). For rare events such as adult male immigration or male replacements, the sample size is the total observation years across all groups (55 group-years). In all cases, when an event happened between two censuses, the event was attributed to the date midway between the census dates, unless other information (e.g., infant size or behavior) suggested that the true date was closer to the first or second census date.

9.3.1 Fecundity

Female capuchin monkeys give birth on average every 19.4 months, without strong effects of group size or offspring sex (Di Bitetti and Janson 2001a). Birth intervals are significantly shorter if the previous infant dies before 8 months of age than if it survives its first year (14.1 versus 20.4 months; Ramírez-Llorens et al. 2008). Because births are highly seasonal, actual birth intervals cluster around 1 year or 2 years (rarely 3 or more years). If a female's prior offspring survives past 8 months, the modal birth interval is 2 or more years (52 of 80 cases), whereas if the prior infant does not survive to 8 months, the modal interval is 1 year (27 of 34 cases).

Female fecundity increases rapidly between the ages of 5 and 8, followed by a period of highest fecundity for females between 9 and 16 years of age, ending with a period of decreasing fecundity in females above 16 years of age (Fig. 9.1). There is clear statistical evidence for reproductive senescence in the declining birth rates of older females (Fig. 9.1) but not for menopause. Three of the four females estimated to have survived beyond 22 years continued to reproduce (one had an unusually long interbirth interval of 4 years); the fourth one (GU) moved with her daughter RI to form a new group in early 2005 and has had no offspring since late 2005 (ages 24–28).

The age at first birth of 27 known-age females ranged from 47.5 to 96.2 months, with a median of 71.64 months or about 6 years. Age at first birth is significantly earlier for females of alpha matrilines than for those in lower-ranking matrilines (6 of 8 first births at about 5 years of age in alpha matrilines versus 1 of 12 first births in non-alpha matrilines, Fisher exact test, two-tailed $p = 0.0044$). The likelihood that a female gave birth in a given season, after controlling for maternal age and survival of the prior infant, did not depend significantly on matriline rank measured as modal rank for foundress females during the first decade of the study and for their daughters as their mother's rank at the time of the daughter's first parturition (logistic regression, effect of rank: $p = 0.66$).

9.3.2 Survival

Survivorship is calculated based on two distinct data sets: (1) the histories of all individuals observed to be born into study groups and followed until disappearance,

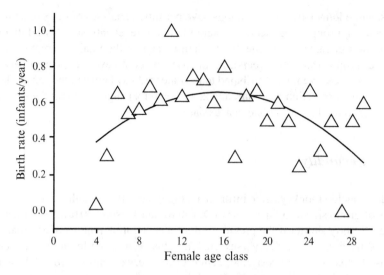

Fig. 9.1 Birth rate as a function of female age. There is a significant quadratic relationship between the fraction of females bearing offspring in a given year and the females' ages (% born = $0.73116065 - 0.0045456 \times$ age $- 0.0021112 \times$ (age-16)^2; ANOVA $F_{2,23} = 4.0$, $p = 0.032$). These cross-sectional values are validated by a more appropriate longitudinal analysis. Using a logistic General Linear Mixed Model of the probability of giving birth in a given year within individual females, using female identity as a random effect, the effect of having a prior surviving infant is negative ($p < 0.0001$), and the effect of maternal age has a positive linear ($p = 0.006$) and a negative quadratic coefficient ($p = 0.006$)

death, dispersal, or the end of the present study period, and (2) individuals already present at the start of the study and similarly followed. Histories of individuals still alive at the end of the present study are right-censored, whereas histories of individuals alive for an unknown period prior to the start of observations are left-censored. However, because of predictable age-related changes in size, shape, and fur patterning from birth through early adulthood (10–15 years of age), we estimated the ages of all individuals present at the start of the study for purposes of estimating age-specific survivorship.

The survivorship curve for individuals in MAC shows several periods of distinct mortality rates (Fig. 9.2). First is a period of high postnatal mortality (48%/year) lasting until roughly 8 months of age. About 40% of this early mortality occurs within the first 2–3 weeks after birth and is likely due to infant weakness; there is some clustering of these deaths among females, as 9 of the 12 early infant deaths occurred in only 3 of the 12 females that bore at least six infants each (contingency table, G with Williams' correction $= 25.8$, $df = 11$, $p = 0.007$). The remaining early mortality is evenly spread across the first 6 months of life and includes infanticide, which accounts for at least 27.3% of all infant mortality (Ramírez-Llorens et al. 2008). Second is a period of lower mortality (10.1%/year) between about 8 and 18 months of age, during which weaning generally occurs. Third is a period of remarkably low mortality (2.75%/year) between 18 months and about

9 The Group Life Cycle and Demography of Brown Capuchin Monkeys

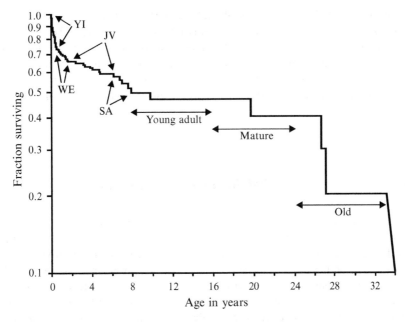

Fig. 9.2 Age-dependent survival. Survivorship graph for all individuals (male and female) born into MAC and its daughter groups in 1991–2010, combined with data on adult females categorized as young adult (YA; 8–16 years), mature (M; 16–24 years), and old (O; over 24 years) at the start of the study, for a total of 715 individual-years. Based on differences in the (log)linear slopes of contiguous regions, four periods of relatively uniform mortality can be recognized prior to adulthood (see text): YI = young infants (0–8 months old, sample size: $N = 146$), WE = weanlings (8–18 months, $N = 100$), JV = juveniles (18 months–6 years old, $N = 84$), SA = subadult (6–8 years old, $N = 37$). Sample sizes for the older female age classes are YA = 20, M = 9, O = 5. Expected lifespan at birth is 8 years, but females that survive to 8 months can expect to live another approximately 24 years on average

6 years, during which juveniles are sheltered from both competition for food (by the dominant male and their mothers) and from predation risk (by being able to occupy central positions in the group spread; Janson 1990; Di Bitetti and Janson 2001b).

Up to 6 years of age, the data include roughly equal numbers of males and females; there is no sex difference in survivorship for known-sex individuals (typically those that survive past 4 months of age; proportional hazards model, effect of individual's sex: $p = 0.50$). From 6 to 8 years, mortality increases (to about 8.6%/year) among females (most males have dispersed by the end of this age range; see below). This period includes the first birth and nursing for most females, and the higher mortality may be associated with the stresses of starting reproduction. After 8 years is a long period of negligible mortality among females (0.7%/year; Fig. 9.2). This halcyon period of near-perfect survival and high fecundity (Fig. 9.1) lasts roughly from 8 to 16 years of age, a period we refer to as "young adulthood." There follows a period of about 8 years during which "mature" females (ages 16–24) retain high fecundity but begin to experience modest mortality

(2.5%/year, Fig. 9.2). After the age of 24, "old" females experience a reduction in fecundity (Fig. 9.1) and increase in mortality (to 14.0%/year, Fig. 9.2).

Maternal rank (see Sect. 9.3.1 for definition) did not significantly affect juvenile survival for any interval of infant and juvenile ages (up to 1 month of age, 1–8 months, 8–18 months, and 18–60 months) or for cumulative survival from birth to 60 months (logistic regressions of infant survival, effects of rank: all $p > 0.34$).

9.3.3 Matriline Rank Stability and Growth Rates

Once a daughter reached adulthood, her dominance rank usually closely approximated that of her mother (Table 9.1). However, while she was still a juvenile, her rank appeared to depend more on her age and size relative to other juveniles than on her mother's rank, although the protection and intervention of a high-ranking mother could allow a daughter preferred access to contested food sources. Juvenile females nearly always rank below any adult female in access to contested food sources, but sometimes, subadult females of high-ranking lineages could displace low-ranking adult females.

Matrilines maintained stable ranks as long as the matriarch was alive. Notable changes in matriline rank were observed only twice in the 15 study years that MAC did not divide (1991–2005), but these corresponded to the only two instances of mature or old female mortality in this period. The first case was that of OL and her presumed daughter SP (born prior to the start of the study, but associated with OL spatially and socially). OL was very old but still very aggressive, often initiating threats toward other females or soliciting coalitions to threaten other group members. Although an exact nondependent rank for OL is hard to estimate (due to the high frequency of coalitionary aggression), she was not a peripheral group member and fed in many contested food sources along with SP. Following the death of OL in 1992, SP was still a juvenile and almost immediately became one of the group's most peripheral members, rarely seen feeding in the main group and mostly avoiding agonism in contested food trees throughout the subsequent 14 years; she split from her natal group in 2010, following a change in the dominant male, and formed a new group with one adult male and her three youngest offspring.

The second case was that of F2. She was the clear alpha female from 1991 to her disappearance and presumed death in mid-2001. During this period, both her male and female offspring benefitted from F2's agonistic protection when entering and feeding in contested food sources. Her oldest daughter (MF) that survived to reproduce clearly ranked closely below F2 at the top of the female hierarchy but disappeared (and is presumed to have died) a few months after her mother's disappearance. F2's younger surviving daughter (UR) was still a juvenile when F2 disappeared; UR survived another 5 years yet never attained high rank and was among the lowest-ranked adult females at the time she first gave birth.

Table 9.1 Changes in matriline dominance across years of the study

Rank	1991–1993	1996	1998	2000	2003	2005–2006	2006–2007
1	F2	F2	F2	F2	YO(DO)	TH(DO)	TH(DO)
2	GR = OL	GR	GR	MF(F2)	TH(DO)	YO(DO)	EST(TH < DO)
3		WC(GR)	WC(GR)	GR	GU	GR	YO(DO)
4	GU = DO	GU	LU(GR)	LU(GR)	RI(GU)	CHI(GO < DO)	CHI(GO < DO)
5		DO	mf(F2)	MG(GR)	SP(OL)	SP(OL)	GR
6	sp(OL)	lu(GR)	GU	WC(GR)	CL(SP < OL)	EST(TH < DO)	CL(SP < OL)
7	wc(GR)	th(DO)	DO	GU	WC(GR)	CL(SP < OL)	OLI(SP < OL)
8		SP(OL)	TH(DO)	RI(GU)	GV(WC < GR)	OLI(SP < OL)	SP(OL)
9		ri(GU)	SP(OL)	DO	WE(WC < GR)	UR(F2)	eva(CL < SP < OL)
10		yo(DO)	ri(GU)	TH(DO)	GR	eva(CL < SP < OL)	bia(SP < OL)
11		mf(F2)	yo(DO)	YO(DO)	LU(GR)	jos(SP < OL)	jos(SP < OL)
12		mg(GR)	mg(GR)	GO(DO)	MG(GR)	bia(SP < OL)	maw(UR < F2)
13		go(DO)	go(DO)	SP(OL)	GRE(GR)		
14		cl(SP < OL)	cl(SP < OL)	CL(SP < OL)	ur(F2)		
15		we(GR)	we(WC < GR)	we(WC < GR)	est(TH < DO)		
16			gre(GR)	gre(GR)	chi(GO < DO)		
17				gv(WC < GR)			
18				mi(TH < DO)			
19				ur(F2)			

Sufficient dyadic agonistic data to produce female dominance hierarchies were obtained only by certain researchers, so sampling is not as uniform or complete as desirable. Early hierarchies were arranged by eye, but later ones were calculated using MatMan. The adult females present at the start of the study are considered the matriarchs and are given with no abbreviation in parentheses. Each offspring's abbreviation is followed by her genealogy in parentheses, with "<" meaning "daughter of." Juveniles are given in lowercase, and adults in uppercase; infants are not included. In 2004–2005, GR's matriline (except GR) fissioned from MAC to form the GUN group, and GU's matriline split off to form RI group. Unrelated females linked by '=' were not distinguishable in rank.

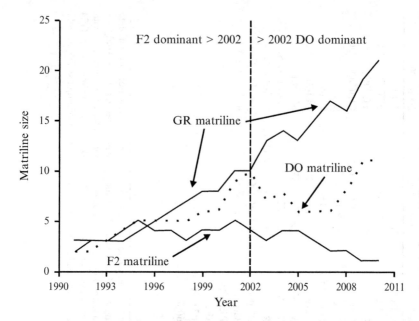

Fig. 9.3 Matriline dynamics. Growth of lineages of three of the five adult females present in the Macuco group in 1991; kinship among these females is not known. Female F2 was the alpha female during 1991–2001, GR was the beta, and DO was either delta or epsilon (depending on the year). F2 died in mid-2001, and by mid-2002, DO's matriline had moved from near the bottom of the female hierarchy to become alpha. For clarity, the growth rates of two other matrilines are not shown, but they were intermediate between GR and F2 and had net losses in only 1 of 20 observation years on average. The two dominant matrilines had losses in 4 of 20 observation years

Interestingly, after the death of the alpha female and her adult daughter, the beta female GR did not become the new alpha nor did her matriline become top-ranking, although she had three adult living daughters. Instead, the new alpha position was attained by DO's two oldest daughters of the fourth-ranking matriline (the two daughters exchanged the alpha position early in this period).

Despite the lack of significant maternal rank effects on fecundity or infant survival, the long-term success of different matrilines in MAC and its daughter groups differed markedly (Fig. 9.3). Rank effects were absent largely because alpha matrilines had the lowest recruitment of all matrilines (Fig. 9.3; F2 1991–2001, DO from 2002 onwards), whereas the beta matriline (GR) had the most consistent and highest growth rate. The low effective recruitment success of the alpha female F2's matriline was due to overproduction of male versus female offspring (ratio of 5:2 compared to 44:47 for the remaining females in MAC), along with the deaths of both of her subadult (6–8 years old) daughters. After it shifted from subordinate to alpha status in 2002, the low recruitment success of DO's matriline was due mostly to low juvenile survival (survival to age 2 of 0.4 versus 0.70 for other females). The different causes for low recruitment of these two dominant matrilines do not suggest any simple cause, but the overall pattern implies that alpha status confers no strong fecundity or survival benefits for females in this population.

9.4 Male Life Histories and Strategies

9.4.1 Natal Dispersal

We documented the emigration of 27 natal males during 55 group-years of observation. Because we do not mark individuals with permanent tags, we cannot follow the fates of dispersing males beyond about 1 year after they leave their natal or current group, and even this ability is restricted to a subset of males that we are lucky enough to find again. Thus, we cannot extend the survivorship graphs of juvenile males beyond the age of dispersal (cf. Fig. 9.3 for females), although we know the fates of a few individuals. Unlike in *Cebus capucinus* (Jack and Fedigan 2004), natal dispersals in our population were isolated events. We documented only a single clear codispersal, out of 21 natal male dispersal events in well-studied groups (MAC, GUN, RI), when two males disappeared on the same day in 2006. They were found together about 16 km from their natal group in the local town (Puerto Iguazú) a few days later. The very long dispersal distance and the unusual coincidence of their simultaneous departure gave rise to the suspicion that they had been captured by local tourists and released in the town.

Dispersal events are not clustered in time. Although in five cases, two natal males dispersed from the same group within 3 months of each other, this frequency of clustered dispersals does not differ from random (given a rate of 0.688 natal male dispersals per group-year of study, we should expect 3.32 of 21 dispersals to occur at random within 3 months of a previous dispersal, $G = 0.9$, $df = 1$, $p > 0.5$). In only one of these cases of closely spaced dispersal were the two males known to have dispersed together to the same group, which was their mother's natal group. In addition to the case of simultaneous long-distance dispersal, the fates of eight dispersing natal males were known: two were seen alone in the natal territory several weeks to months after emigration and did not appear in neighboring groups, and in six cases, the males were found in a neighboring group several months after emigration. Given that we frequently observed neighboring groups, the above data suggest that male dispersal is usually (21/27 cases, across all study groups) further than one group away.

Natal males disperse across a range of ages, but most dispersal events occur between the ages of 5 and 7 years, with a mean of 5.7 (Fig. 9.4). Both cases of infant dispersal occurred when their mothers apparently left their natal group. Three cases of male dispersal at younger than 4 years of age were all from a daughter group (RI) back to the mother's natal group (MAC); none of the mother's (known) matrilineal relatives lived in the latter. Excluding these five cases raises the mean age of dispersal to 6.5 years (median of 6.1 years). At the time of dispersal, they are clearly smaller than fully adult males of 9–10 years of age.

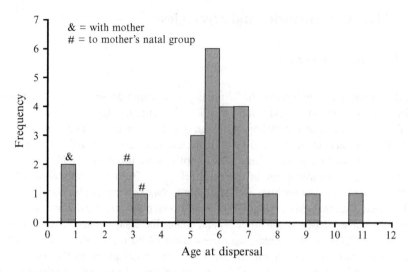

Fig. 9.4 Distribution of the ages of dispersal for natal males in all study groups 1991–2010. The two cases of dispersal at less than 1 year of age were infant males that disappeared along with the mothers and are presumed to have codispersed with them (but the possibility of coincident death cannot be excluded). The three unusual cases of dispersal between 2.5 and 3.5 years of age were from a recent fission group of Macuco (Rita) back to the Macuco group, and all occurred within 1 year

9.4.2 Immigrant Males and Their Fates

We documented the immigration of 39 subadult (5–8 years old) and young adult (9–11 years old) males into focal groups in 55 group-years; there were also 12 presumed immigrant adult males present in various study groups when we started sampling, including five that were alpha males (treated separately below). Some of the immigrant males had been seen previously alone in the home ranges of the groups that they joined, but the prior histories of most were unknown, suggesting that they did not move in from immediately neighboring study groups. Immigrant males stayed in a group until they either emigrated, challenged the dominant male for alpha status, or achieved alpha status some other way. We can assign "fates" to each of 46 non-alpha males; 17 of these were still subordinates at the end of the samples on their groups, and the remaining 29 males were involved in 35 changes in status (Table 9.2). The most striking aspect of these changes is the existence of three distinct "pathways to power": a subordinate male can become alpha by challenging the existing alpha, inheriting the alpha position if both the challenger and the alpha male are badly injured in a fight, or taking over as the alpha male of a daughter or splinter group of females that usually occurs shortly after a new male becomes established (see Sect. 9.5.1). The frequency of nonaggressive paths to power (inheritance or new group creation) is nearly as high as aggressive ones (9 versus 11 cases).

Table 9.2 "Fates" (changes in status) of 47 subordinate adult males following immigration to one of the study groups, and the mean and range (in parentheses) of values for each variable

	Challenge and replace prior alpha male	Inherit alpha role without fight	Leave as alpha male in new daughter group	Emigrate without obvious cause	Emigrate during or soon after alpha male change	Still subordinate at end of sample
N males (only complete tenures)	11 (3)	2 (1)	7 (4)	8 (3)	7 (6)	17
Mean number of years of residence prior to status change, all data	3.84 (0.02–8.49)	2.15 (1.64–2.67)	3.07 (0.41–5.0)	2.0 (0.54–5.5)	0.89 (0.22–2.36)	1.46 (0–4.64)
Mean number of years of residence prior to status change, complete tenures	3.95 (0.02–6.55)	2.67	3.92 (3.01–5.0)	2.83 (0.54–5.5)	0.85 (0.22–2.36)	NA
Mean age at status change	14.25 (10.5–21.7)	8.65 (7.6–9.6)	14.15 (10.1–18.4)	10.37 (7.8–14.5)	8.6 (6.2–12.4)	NA

Several males were assigned more than one fate, as they either changed status within a group or were tracked through more than one study group. Many of the residence durations are censored because males were already present at the start of study on each group; results are provided both for all data and only complete tenures.

Replacement of the alpha male position is associated with predictable emigration of the group's remaining non-natal males. In 6 of the 7 well-observed replacements of the alpha male in groups with three or more non-natal males, most or all of the group's non-natal males who did not obtain the alpha position left the group within days to 8 months, accounting for 10 of the 15 total individual emigrations recorded in the whole study. Even counting each bout of postreplacement emigrations as a single event, the frequency of such emigration bouts within 8 months of a male change is much higher than expected by chance (6 versus 1.33, out of 11 total bouts of emigration, $G = 11.9$, $df = 1$, $p < 0.001$).

Bouts of immigration are also more likely following alpha male replacements. In 13 of the 33 cases that could be scored, the arrival of new males to a group occurred within 6 months after the replacement of the group's alpha male. This frequency is much greater than the 3.6 immigration cases expected at random, given only 12 alpha male replacements in the same 55 group-years ($G = 17.97$, $df = 1$, $p < 0.001$). Thus, it appears that newly vacated "slots" for subordinate males are rapidly filled. In most (10/13) cases, the new males came into large groups with over eight adult females, suggesting a preference to join large female groups. Potential immigrants may take into account the number of males already in a group before moving: in the most closely monitored study group, the number of males lost and gained per year was well correlated (excluding the 11 years with neither losses nor gains, $N = 9$ years, $r = 0.95$, $p < 0.0001$).

Once a male joins a new group, his subsequent fate (Table 9.2) depends on his age and chance. Males less than 9 years old are not of full adult size and do not challenge adult males for high rank. They typically remain for shorter periods, emigrating predictably after 2–3 years of residence or when the alpha male is replaced, unless they happen to inherit the alpha status, a rare event over which they have little control. Fully adult male immigrants of at least 9 years of age take positions in the higher ranks of the male hierarchy, sometimes (three cases) challenging the alpha male within 5 months of immigrating but more commonly waiting long periods (an average of 5.2 years, $N = 8$) for successful opportunities to challenge the alpha or to become the alpha male of daughter or splinter groups of females at group fission (associated with alpha male replacement – see below). Males that took the alpha position in fission groups were of the same age on average as males that challenged the alpha (14.15 versus 14.25 years old), and spent nearly the same average time in residence before assuming an alpha role (3.92 versus 3.95 years, Table 9.2). Beta males can be very "patient" waiting for an opportunity to defeat the alpha male, which may take over 8 years to arrive (the case of GE in MAC, 1991–1999). These long waits for a beta male to take over the dominant position may provide opportunities for the beta male to reproduce, as the beta male had preferential access to females that happened to be in estrus at the same time as a more dominant female (Janson 1984) and to the daughters of the dominant male, who did not solicit their fathers as mating partners (Di Bitetti and Janson 2001a).

9.4.3 Tenure and Replacements of Alpha Males

Unlike many other primates in large multi-female groups in which the dominant male of a group may change frequently, sometimes more than once per year (e.g., Palombit et al. 2000), the alpha males in all known populations of *Cebus apella* superspecies tend to have long tenures. Figure 9.5 shows the survival plot of all tenure lengths (including censored durations) documented in the alpha males of the Iguazú population. Very short tenure lengths (less than 4 months) occurred when a male challenged the alpha successfully but the challenger was seriously injured (one case), failed to kill the former alpha male who returned to claim the alpha role (one case), returned to beta status following a group fusion (one case), or inherited alpha status and was subsequently challenged for the alpha position by an older male (one case). Tenure lengths for males that acquire alpha status by challenge versus group fission did not differ in our study (ANOVA, $F_{1,14} = 0.001, p = 0.98$). If a male retained alpha status past the first 4 months, tenure lengths were typically long, with a median duration of 5.01 years and a 0.76 chance of lasting at least 3 years. One male, Pecoso, was documented to be the alpha (in Silver group) for at least 11.6 years.

One case of failed male replacement revealed a likely cause of the relatively long tenures of dominant males in this population. Dedos, who had been the dominant male of MAC for over 6 years, was attacked and injured, but not killed, by Gendarme on about September 8, 1997. Dedos withdrew from MAC but remained

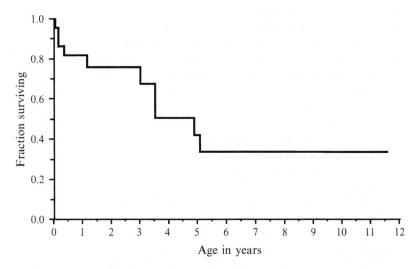

Fig. 9.5 Alpha male tenure duration across all study groups ($N = 23$ males). A tenure length was calculated as the number of days from the date of a male's first appearance as alpha to midway between his last known date as alpha and the date of the next census on his group. About 18% of alpha males held their rank for less than 4 months, but if they made it past this initial period, their expected tenure was 5 years. All four tenure lengths above 5.1 years (5.3, 5.9, 8.5, 11.6) were censored and thus represent minimal estimates of true tenure lengths

in the group's home range for over a month following his injury. Dedos then returned to MAC in late October 1997, and the two males faced off against each other. Dedos was backed up by a cohort of females, including all but one of the group's adult females, whereas Gendarme was backed up only by one female, WC, a putative daughter of Dedos. Dedos and his supporters won this encounter, whereas Gendarme suffered additional wounds during a series of fights. Dedos returned as the dominant male for another nearly 2 years until being badly injured by Gendarme in July of 1999. That the support of Dedos by other group members was more than symbolic is suggested by the fact that in the 1997 series of fights, both the dominant female, F2, and an older natal male, DF, were injured; similar injuries of natal males and adult females occurred during some other male takeover periods. Thus, the dominant male appears to enjoy support by most existing adult females and his sons in conflicts against male challengers, perhaps allowing him to maintain his breeding position in the group far longer than would be the case without this support.

9.5 Changes in Group Size, Group Fission, and Group Fusion

We have detailed data on composition, births, deaths, and migrations for all individuals in MAC and its descendant groups GUN, RI, and SP, from February 1991 through the end of 2010. During this 20-year period, numbers of individuals in these groups increased at a steady rate of about 4% per year (Fig. 9.6), regardless of alpha male changes, infanticides, and two group fissions.

Group counts from other groups span at most six continuous years and are less informative about long-term growth rates. Most groups grew in numbers over the period of observation, but four groups either went extinct or moved entirely out of the known study area and were not seen again. Two of these were small splinter groups (one from the breakup of YAC in 2009; one from the fission of MAC in 2009); in both cases, the group composition was one adult male, one adult female, and one or two juveniles. Two other groups (BAR, RUB) were not seen after 1996 and are presumed to have gone extinct or moved out of the study area; the BAR group partially occupied an area of regenerating forest that was cut down to provide new tourist facilities starting in 1998.

Temporary changes in group size occurred (see also Lynch Alfaro 2007). Very large groups (over 30 animals) tended to show signs of subgrouping behavior, which is not reflected in the total counts, but could lead to subgroups of defined composition ranging independently of other subgroups for several hours up to several days, even on occasion sleeping in different areas. Such subgrouping behavior appeared especially in the Argentine winter, the period of greatest food scarcity (Di Bitetti and Janson 2001a). When these large groups eventually split, the resulting descendant groups were not necessarily well-predicted by the composition of the previous subgroups, except that one fully adult male from the original large group tended to join each descendant group as the alpha (see Sect. 9.4).

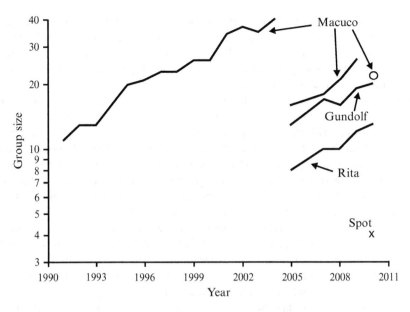

Fig. 9.6 Group sizes (excluding non-natal males) of MAC group and its daughter groups during the study period. The y-axis uses a log scale, so a straight line of numbers versus time represents a constant rate of growth. For the entire period, the best estimate of the growth rate is 0.04, and it does not differ before and after the group fission, nor among the daughter groups. The numbers for Rita group include the three juvenile males that emigrated back to their mother's natal group at unusually young ages (see Fig. 9.4)

9.5.1 Processes of Group Fission

We observed the detailed process of group fission in two study groups (three fission events: YAC 1991; MAC 2004–2005; MAC 2009–2010) and monitored the results of fission in a third group (SIL 1994–1995) via periodic censuses. In the first three cases, group splitting occurred in the largest groups known to us in the population at the time: YAC had 34 individuals (9 adult females) just prior to the split, and 2004–2005 MAC had 48 (14 adult females), and 2009–2010 MAC had 33 (10 adult females). In all four cases, splits occurred in association with a replacement of the dominant male in the original groups. However, not every male replacement led to group changes (see Sect. 9.4). In all of the well-observed cases, the clear establishment of the new alpha male was associated quickly with the disappearance of a majority of infants of less than 6 months of age, most likely by infanticide (Ramírez-Llorens et al. 2008). At least one group split (Rita from MAC 2004–2005) may have been the direct consequence of an adult female seeking to avoid infanticidal behavior by the new dominant male (Ramírez-Llorens et al. 2008); the same may be true of OLI's departure with her infant from MAC in late 2009, which preceded the eventual wave of infant disappearances in February–April 2010.

The fourth case of group fission occurred about 1 year following the fusion of the neighboring RUB and SIL groups, producing an enlarged SIL group in 1993. Between mid-1994 and mid-1995, the alpha male in the SIL group (LF) was replaced by a new immigrant adult male (PEC). During the same period, the two subordinate adult males in the SIL group (IN, BI) left the group with two of the three females from the original RUB group, plus one of the females formerly resident in the SIL group. One of the original RUB group females remained in the SIL group. This new RUB group remained as a separate social group, with IN as the alpha male, at least through the end of 1996. In neither SIL nor RUB did any of the infants of the 1994–1995 birth season survive to the mid-1995 census.

Group splits usually produced daughter groups of quite different sizes. In three of the four cases of group splits, the majority of the adult females remained in the original home range whereas the two new daughter groups were in each case composed of a few adult females (numbers of females for YAC 1991 = 6:2:1, MAC 2004–2005 = 8:4:2, and MAC 2009–2010 = 7:2:1) and their immature offspring. In the fourth case (Silver 1994), three females went with each daughter group.

In two well-observed group splits, smaller descendant groups were not at a demographic disadvantage compared to larger ones. There was no marked difference between the per capita growth rates of small versus large descendant groups in the 3 years following a group split, either for MAC (Fig. 9.6) or the YAC group (e.g., the Silver group increased from an initial size of about 9 animals with only two adult females in late 1991 to at least 23 individuals by 2000, an average annual increase of 10%). Because group splits were associated with a change in the dominant male and attendant infanticide, the year after a split was a year of high birth rates, as nearly all the females in a group produced offspring in the year following a male replacement.

When genealogies were known, daughter groups were found to be composed of individuals from single matrilines but did not contain complete matrilines. In the 2004–2005 MAC breakup, GUN contained all the daughters of GR and their offspring but not GR herself, who remained in MAC. Likewise, when Rita left with her offspring, her mother, GU, did not leave with her, although she joined her daughter within a couple of days. The 2009–2010 MAC split produced two splinter groups composed of portions of the matriline of SP (1 = OLI and infant; 2 = SP, BIA, DAN, FRA), yet three other adult daughters of SP (CL, EVA, JOS) stayed in MAC with their offspring, as did one of OLI's subadult daughters (OFE). Thus, the process of group fission in this population appears to produce new groups in which females are likely to be very closely related yet might well have some equally close kin in adjoining groups.

The process of group fission could take any period from 1 day to 6 months. In the case of YAC, it took about 3 months following the change in the dominant male for subgroup composition to stabilize into a close resemblance of the resulting daughter groups, during which time the group went from foraging as a single unit nearly all the time to increasing periods of 2–3 subgroups foraging separately from each other. After the fourth month, the subgroups (ST, SIL) were not seen to rejoin YAC to form the entire original group, and by the fifth month, they occupied largely distinct but adjacent core areas. In the case of MAC 2004–2005, one of the group's adult males (GUN) formed part of several spatially distinct subgroups before the alpha male (GEN) was replaced, but after GEN was removed, GUN became more independent, and within 4 months, he and a subgroup moved to a new home range adjacent to, but not overlapping with, the original range. In contrast, the second split (RI) in MAC 2004–2005 occurred over 1 day with the dispersal of Rita and her daughters to a peripheral area in MAC's home range, where she was later joined by her mother and two of MAC's former subordinate males. In the MAC 2009–2010 split, one female (OLI) and her new infant started to stay away from MAC after the first intense male–male fights in early November, and she was last seen about 1 month later in the company of one of the adult males of the group (JE). Following the death of MAC's alpha male on November 20th, 2009, continued aggression and unstable ranks among MAC males lasted until late January 2010; during December, the group showed indeterminate cohesion, some days moving as a single group and on other days dividing into at least two distinct foraging parties that sometimes even slept in different areas. After one male (ERN) assumed the alpha role starting in February, 2010, 4 of the 7 infants in the group disappeared within 2 months; by May 2010, another subgroup (SP with BIA, DAN, FRA) had left MAC.

9.5.2 Group Fusion

One case of group fusion was documented. In early 1993, RUB was composed of two adult males (NA, SO), three adult females, and several immatures. At this time, the neighboring SIL group had two adult males (LF = alpha, IN = beta), one subadult male (BI), three adult females, and several juveniles. On the 5th of June, 1993, MD discovered SO alone and with deep wounds to his hands, elbow, and shoulder. On the 14th of June, 1993, NA was also no longer with RUB, and instead, the group had only the two subordinate males from SIL, IN, and BI. These two males stayed with RUB until at least the 28th of June as the only adult males in that group. Meanwhile, on the 21st of June, SO was seen as a subordinate male in SIL, but on the 27th of August, he was back with RUB, along with IN and BI. SO had fresh wounds on his face and at the base of his tail. That same day, RUB had an intergroup encounter with SIL, during which MD saw LF (the alpha male of SIL) with deep wounds on the face. After more chases and a fight during this encounter, SO followed and gave submissive facial gestures to LF. From that date until the middle of 1994, covering seven separate censuses of the group, SIL and RUB comprised a single unit, with LF as the alpha male, IN, SO, and BI as subordinate males, and all the females and juveniles of both groups. NA, the former alpha male of RUB, was not seen again. Thus, it appears that he was displaced as the alpha male of RUB by the two subordinate males of SIL, who then rejoined SIL with the RUB females. SO and all the other known individuals of RUB stayed with SIL for about 1 year. After mid-1994, RUB and SIL fissioned (see above).

9.6 Discussion

9.6.1 Slow Life Histories

This population of *apella*-like capuchins illustrates one of the slower life histories known among monkeys. With an age at first reproduction of 6 years for females, this population would not persist without high juvenile and adult survival, which leads to a relatively high expected (at birth) lifespan of 8 years. There is some evidence that starting reproduction is stressful for females: their survival declines at age 6, and female fecundity is lower at age 5 than later (Fig. 9.1). Once a female reaches full maturity at 8 years, she has a long future life expectancy of 19 years. Although fecundity and survival begin to decline notably as a female surpasses the age of 24, there is no evidence for menopause, and even the oldest females in our study population continued to come into estrus.

Most likely because of slow infant development (Charnov and Berrigan 1993), mothers typically produce an offspring only once every other year, if the previous infant survives. This alternation of birth years sets up the conditions for sexually selected infanticide by males to be adaptive (Ramírez-Llorens et al. 2008), despite the fact that the Iguazú capuchin population is highly seasonal in its breeding (Di Bitetti and Janson 2000), which would make infanticide of little benefit to a newly dominant male in an annual breeder (van Schaik 2000). Although the great longevity and slow life history found in this population is surely assisted by the lack of large eagles preying upon adult capuchins, the reproductive parameters of Iguazú (high age at first birth, alternation of birth years) are commonly found in other capuchins (e.g., Robinson 1988; Fedigan and Rose 1995), even in areas with relatively high adult predation risk (Janson 1984). If the current notion is correct that these slow life histories are a direct result of the relatively large brain size of capuchins (Barton and Capellini 2011), then the cost of large brain size is extreme – the theoretical intrinsic rate of population growth (a rough measure of maximum fitness) in our population (from Figs. 9.1 and 9.2) is 0.1/year, a full order of magnitude less than the r_{max} of about 1 expected of mammals of equivalent body mass (Calder 1996). It is difficult to imagine what benefits of larger brains might repay such a tenfold fitness cost!

The survival of males, far more so than of females, is dictated by events related directly to their reproductive strategy, principally dispersal from their natal group and challenging an established alpha male for the dominant position in a social group. Natal dispersal is an apparent consequence of generalized incest avoidance (Di Bitetti and Janson 2001a). Prior to dispersal (typically by age 6), there is no statistical difference in survival between the sexes in natal animals in MAC. As we rarely can track dispersing males after they leave their natal group, their mortality during dispersal is not easily measured. However, we can estimate the extent of such mortality by examining group adult sex ratios, knowing the survivorship of adult females and the sources and extent of male mortality once they join established groups. Assuming the population as a whole is not growing and using the results from Fig. 9.1 and Table 9.2, the difference between the adult sex ratio expected based on postimmigration patterns of male mortality and that actually observed implies that at least 38% of dispersing males do not survive to show up in social groups. This mortality is likely spread across 2 or more years, reflecting the difference in average age between the ages of natal dispersers (6 years old; Fig. 9.4) and the estimated ages of immigrant males (typically 8 years old or older; Table 9.2). This is a high rate of mortality for capuchins in this population and may reflect the reality of risks associated with solitary living. By comparison, females from ages of 6–8 experience a total of 15.8% mortality, which itself is markedly higher than just before or after this period (Fig. 9.3). If the estimated dispersal phase male mortality of 38% is accurate, then the life expectancy of a male at birth is only about 7 years, somewhat less than that of a female; nearly 2/3 of males born will die without ever reaching dominant status, whereas only about 40% of females born will die before giving birth at least once.

After a male migrates into a social group, attaining the alpha status may entail both mortality risks and large reproductive rewards. Although a slim majority (11 of 20) of males that attain alpha status do so by challenging the group's dominant male in one or more bouts of contact aggression, nearly half of all males attain the alpha position indirectly (Table 9.2). It is not easy to calculate the real frequency of alpha challenges and their chances of success because if a challenge does not succeed, no change in the social structure occurs and both the dominant and the challenger may sustain only minor wounds. If a successful challenge occurs, the challenger is always wounded, sometimes (2 of 11 cases) severely enough that he cannot immediately assume the alpha status even if he mortally wounds the dominant. In one case (MAC, 2009), the putative challenger (HOM) appears to be have been so severely wounded that he died a few weeks later – none of the other males in the group was severely wounded and none quickly took on the alpha role after the dominant was deposed.

After achieving the dominant status, a male can expect to retain it for 5 years; once he is deposed, it appears that he usually dies (although we can absolutely confirm the death in only two cases). In two cases, the former dominant survived the challenge; in one of these (DE), he recovered from his wounds and won back the dominant position after about 6 weeks, and in the other (ST), he left with one female as a fission product of the group after his deposition, but this daughter group was last seen 3 months after the dominant was defeated. The net result of the these challenges and their attendant mortality is that male life expectancy after immigrating to a social group at age 8 is estimated to be only an additional 10.9 years (Janson et al. unpublished data), roughly half the remaining life expectancy of 19 years of an adult female that reaches 8 years of age. Based on these differences in survival rates, life history theory predicts that male capuchins from this population should show more rapid senescence than should females, even under conditions of low environmental and social stress (e.g., captive breeding colonies).

9.6.2 Female Philopatry, Matrilines, and Group Membership

Capuchin monkeys in this population conform to some but not all of the generalities about matrilineal group structure derived from Old World primates. Like baboons and macaques, capuchin females and their kin in this population tend to support each other in aggressive interactions, daughters come to rank close to (although sometimes above) their mother, and group fissions tend to follow matrilineal divisions (but one or more members of a matriline may stay behind in the natal group). However, despite clear feeding priority and central positioning of the alpha female F2 (e.g., Di Bitetti and Janson 2001a), recruitment success into her matriline was notably lower than that of lower-ranked females (Fig. 9.3); likewise, when DO's matriline became dominant in 2002, it began actually to decline in size, whereas previously, it had been increasing. In both cases, low recruitment into the matriline was seemingly due to lower survival of offspring, and in F2's case, this was compounded with a male-biased sex ratio among her offspring. The low recruitment success of the dominant matriline does not appear to occur in recently founded daughter groups, perhaps because agonistic relationships among these related females are relatively infrequent and of low intensity. Whatever the reason for the low recruitment success (and given the small sample size of matrilines for which we have long-term data, it could be coincidence), F2's small matriline was vulnerable to marked rank change when the matriarch died; a year after F2's death, the lone surviving female of that matriline (UR, Table 9.1) was the lowest-ranked adult female.

The ability of low-ranking adult females to grow their matrilines in the face of contest competition for food may be an unusual feature of this population. Low predation risk on adult capuchins because of a lack of large monkey-eating eagles means that low-ranking subordinates may have the ability to avoid contest competition by foraging at the periphery of the group on alternative resources that might yield one or a few monkeys as much food *per capita* as do the larger and more productive food sources favored (and contested) by the dominants (Di Bitetti and Janson 2001b). Although not easily amenable to an experimental test within this population, this hypothesis makes clear predictions: the relative reproductive success of subordinate females compared to dominants should be negatively correlated with the predation rate on adult females if the major predators preferentially attack animals at the periphery of a prey group.

If the lability of matriline dominance observed in MAC is a more general feature of other groups and populations of *Cebus apella*-like monkeys, this could also explain some general features of their social behavior. For instance, even though they squabble frequently at food sources, *apella*-like capuchins have been shown to be relatively socially tolerant compared to rhesus macaques (de Waal 1986; Brosnan and de Waal 2003). Second, females have not been observed to fight over rank (nearly all agonism is clearly in the context of access to food, or defense of kin or offspring against males) or to engage in the kinds of aggressive tactics expected of animals using aggression to maintain rank position (e.g., Silk 2002).

Both these features could be expected outcomes of a social system in which matrilineal support in contest competition is real, but high matrilineal rank is not easily preserved and thus not worth risking escalated contests to acquire. High matriline rank might still be worth acquiring if the breeding success of males from high-ranking matrilines is markedly higher than that of sons from low-ranking matrilines, but we have no data to test this question. A further outcome of increased mobility of matriline rank and its lack of correlation with fecundity is that females, while preferring to be "philomatric" (staying with their matriline), can take a chance to establish or join a new social group, as illustrated by the fusion and subsequent fission of the RUB group in 1993–1995 and the frequent association of group fission with male replacement.

9.6.3 Group Life Cycles

Despite some variation in details, there is an overriding pattern in group changes that profoundly affects their composition, demography, and behavior. The resulting cyclical sequence of events can be described as a group "life cycle" (cf. colony life cycles of social insects, e.g., Oster and Wilson 1978). This is summarized in a "stage" diagram (Fig. 9.7) in which each distinct configuration of group size and recency of dominant male is represented as a "stage" in the cycle, with stages connected by probabilities of moving from one stage to another. The probabilities are not shown in the diagram but could be entered into a matrix of transition probabilities based on our data. Not all the connections or possible outcomes are encompassed in this diagram, but it is a way to visualize the regularities of change in group structure that emerge from predictable individual-level behaviors. As such, it provides a way to connect natural selection on individual behaviors with their consequences at the level of group structure (cf. Hemelrijk 2002). Note that Fig. 9.7 is based on a small sample and may represent only the life cycle of MAC, which contributed most of the data; nonetheless, it provides a hypothesis against which to compare long-term demographic data from other capuchin and monkey populations.

In our study, cyclical change is driven by replacement of the group's alpha male and the attendant almost inevitable infanticide that follows (Fig. 9.7; see also Fedigan 2003). If a male attains the alpha role (by whatever means) and survives the first 6 months in that role, he is extremely likely to remain as alpha for about 5 years, although a minority can last considerably longer, even over 11 years. The lack of opportunity to replace the alpha during this period is the apparent reason that other non-natal males in the group at the time nearly always migrate away within days to months of the replacement; they are not driven out by invading males, as occurs in white-faced capuchins (Gros-Louis et al. 2003; Fedigan and Jack 2004). The lack of adult males in the group appears to attract immigration from non-natal males, especially into groups with large numbers of females. In the meantime, the alpha male will have eliminated most or all of the infants less than 8 months of age

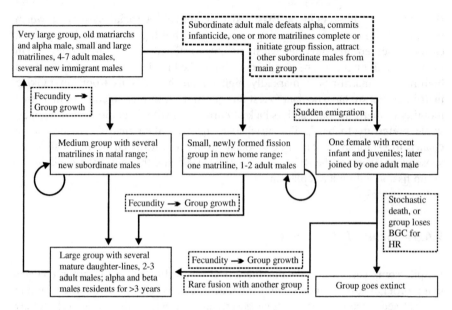

Fig. 9.7 Schematic diagram of group life cycle. Each box with a solid border is a distinct group size or composition, connected by *arrows* representing possible future outcomes of processes that are described in boxes with *dashed outlines*. These processes include net fecundity (exceeding mortality), male replacement, group fission, and risk of infanticide. *Circular arrows* pointing back to the same box represent the likelihood that a group stays in the same state from 1 year to the next. In our population, we did not observe the decline of well-established groups toward smaller sizes and extinction, but this possibility would be easy to include using *arrows* leading from larger groups to smaller ones

within 1–2 months of his having established himself as the undisputed alpha male (a process which may take up to 6 months, depending on his injuries, the presence of challengers, etc.).

Some females use the male replacement to initiate permanent group splits, either by leaving the core area of the natal group as individuals with their offspring ("splinter" groups: three cases) or by forming first temporary, then increasingly durable, foraging subgroups attended by one or two adult males from the natal group. These "daughter" groups, typically (but not always) composed of females from a single matriline, eventually move to areas that are adjacent to or slightly overlapping the natal group's home range. Splinter groups typically contain a female with her infant of the year and appear to be one way for a female to avoid infanticide by the new alpha male in her natal group. Two of the three known such splinter groups were never found again, so it may be that these females eventually join some other group far from their natal group or die as isolated individuals. The new matriarchs of fission groups experience higher birth rates in the period following the group's establishment, although the overall growth rates of fission groups are not distinguishable from that of the natal group. In typical groups where the alpha male lasts at least 5 years, group size can increase quickly, doubling within a decade, thus providing a more attractive target for new male immigration and sowing the eventual seeds of the next male replacement. If the group is still relatively small at the time of a male replacement, the group may stay together for another "round" until the following replacement. Thus, there is an overall cycle to group size and alpha male stability that is coordinated by the events that attend the replacement of a group's alpha male (Fig. 9.7).

Understanding the group life cycle not only illuminates the importance of rare sources of selection, such as infanticide, but also is necessary to place any short-term study of a particular group into a proper perspective. Given the dynamic and cyclical nature of group size and composition, observers cannot hope to study a "typical" group or even a typical set of groups. For instance, much of the variation that we observed between groups at the start of our study in 1991 was likely an effect of observing them in different stages of their life cycles. Acknowledging and documenting group life cycle should allow researchers to detect group-specific traits that reveal potentially important relationships between extrinsic traits (e.g., habitat quality) and group characteristics. Failure to understand group life cycles may lead to the false rejection of hypotheses relating within-population variation of group size to food availability, predation risk, patchiness, or other variables thought to affect sociality.

9.7 Conclusions

Long-term studies of animal populations have been important in revealing rare sources of strong natural selection, long-term demographic cycles or epidemics, and effects of slow habitat change. For slowly maturing, long-lived animals like most primates, long-term studies are also essential to document important events or phases of individual life histories that are critical to explaining variation in individual reproductive success. In our study, the rarest and most dramatic event in individual life histories is the assumption of the alpha position by males; only about 1/3 of all males survive long enough to challenge for the alpha position; the outcome of such challenges is usually fatal for one of the participants, and the consequences of a successful takeover ramify through the group via infanticide, high turnover in non-natal males, and likely group fission (Fig. 9.7). The result of these social upheavals is that male takeovers essentially "reset the clock," often producing daughter groups with all new immigrant males, females and offspring that are of a single matriline, an immediate loss of most infants followed by a year in which nearly all females have babies, and a relatively young adult alpha male that is likely to remain as alpha for 5 or more years, long enough for incest avoidance to become an important part of mating strategies. Subsequent growth of groups is associated with lower average kinship among females, greater aggressive competition between the descendants of the original adult females, larger group size, and increased attraction of immigrant males to the group, which in turn sets the stage for new male challenges to the alpha male and possible group fission. Recognizing this group life cycle places a different interpretation on differences in group size. Rather than interpreting these differences as stochastic variation around a stationary mean (e.g., Cohen 1971), or reflecting differences in habitat quality (e.g., Gillespie and Chapman 2001) or group competitiveness (Robinson 1988), various groups may simply be at distinct points in relatively predictable life cycles. The coordinated changes that accompany transition through the stages of a group's life cycle should predictably affect social behaviors (female–female agonism, existence of juvenile cohorts, female mating preferences). If predictable group life cycles emerge as a feature of other primate populations, documenting and understanding these life cycles should help to clarify the differences in social organization and structure between species, how these differences relate to ecological variables and fundamental life-history traits, and help to integrate male and female social strategies in a broader understanding of primate social systems.

Acknowledgments This work could not have been completed without the kind permission of the Delegacion Regional Nordeste Argentina of the Argentine National Parks Administration, especially Paula Cichero. Over 20 researchers contributed observations to the long-term database; the most important of these were Paula Tujague, Fermino Silva, Brandon Wheeler, Barbara Tiddi, Clara Scarry, and Patricio Ramirez-Llorens. Major financial support for various portions of the long-term study was provided to CHJ by a Fulbright scholarship, a grant from the Committee on Research and Exploration of the National Geographic Society, NSF grants BNS-9009023, IBN-9511642, BNS-9870909, and BCS-0515007, as well as CONICET predoctoral fellowships to MDB and MCB.

References

Agostini I, Visalberghi E (2005) Social influences on the acquisition of sex-typical foraging patterns by juveniles in a group of wild tufted capuchin monkeys (*Cebus nigritus*). Am J Primatol 65:335–351

Baldovino MC (2010) Desarrollo de los infantes de mono caí (*Cebus apella nigritus*): ontogenia de la habilidad motora y comportamientos alomaternales. PhD thesis, Universidad Nacional de Córdoba, Córdoba, Argentina

Baldovino MC, Di Bitetti MS (2008) Allonursing in tufted capuchin monkeys (*Cebus nigritus*): milk or pacifier? Folia Primatol 79:79–92

Barton RA, Capellini I (2011) Maternal investment, life histories, and the costs of brain growth in mammals. Proc Natl Acad Sci USA 108:6169–6174

Brosnan SF, de Waal FBM (2003) Monkeys reject unequal pay. Nature 425:297–299

Brown AD, Zunino GE (1990) Dietary variability in *Cebus apella* in extreme habitats: evidence for adaptability. Folia Primatol 54:187–195

Calder WA III (1996) Size, function, and life history. Dover Press, Mineola, NY

Charnov EL, Berrigan D (1993) Why do female primates have such long lifespans and so few babies? or life in the slow lane. Evol Anthropol 1:191–194

Cohen J (1971) Casual groups of monkeys and men. Harvard University Press, Cambridge, MA

de Waal FBM (1986) Class-structure in a rhesus-monkey group - the interplay between dominance and tolerance. Anim Behav 34:1033–1040

9 The Group Life Cycle and Demography of Brown Capuchin Monkeys 211

Di Bitetti MS (1997) Evidence for an important social role of allogrooming in a platyrrhine primate. Anim Behav 54:199–211

Di Bitetti MS (2005) Food-associated calls and audience effects in tufted capuchin monkeys, *Cebus apella nigritus*. Anim Behav 69:911–919

Di Bitetti MS (2009) Estacionalidad en la abundancia de artrópodos del sotobosque en el Parque Nacional Iguazú, Argentina. In: Carpinetti B, Garciarena M (eds) Contribuciones para la conservación y Manejo en el Parque Nacional Iguazú. Buenos Aires, Administración de Parques Nacionales, pp 191–204

Di Bitetti MS, Janson CH (2000) When will the stork arrive? Patterns of birth seasonality in neotropical primates. Am J Primatol 50:109–130

Di Bitetti MS, Janson CH (2001a) Reproductive socioecology of tufted capuchins (*Cebus apella nigritus*) in northeastern Argentina. Int J Primatol 22:127–142

Di Bitetti MS, Janson CH (2001b) Social foraging and the finder's share in capuchin monkeys, *Cebus apella*. Anim Behav 62:47–56

Di Bitetti MS, Paviolo A, De Angelo CD, Di Blanco YE (2008) Local and continental correlates of the abundance of a neotropical cat, the ocelot (*Leopardus pardalis*). J Trop Ecol 24:189–200

Fedigan LM (2003) Impact of male takeovers on infant deaths, births and conceptions in *Cebus capucinus* at Santa Rosa, Costa Rica. Int J Primatol 24:723–741

Fedigan LM, Jack KM (2004) The demographic and reproductive context of male replacements in *Cebus capucinus*. Behaviour 141:755–775

Fedigan LM, Rose LM (1995) Interbirth interval variation in three sympatric species of neotropical monkey. Am J Primatol 37:9–24

Gillespie TR, Chapman CA (2001) Determinants of group size in the red colobus monkey (*Procolobus badius*): an evaluation of the generality of the ecological-constraints model. Behav Ecol Sociobiol 50:329–338

Giraudo AR, Povedano H, Belgrano MJ, Krauczuk E, Pardiñas U, Miquelarena A, Ligier D, Baldo D, Castelino M (2003) Biodiversity status of the interior Atlantic forest of Argentina. In: Galindo-Leal C, Câmara IG (eds) The Atlantic forest of South America: biodiversity status, threats, and outlook. Island Press, Washington, DC, pp 160–180

Gros-Louis J, Perry S, Manson JH (2003) Violent coalitionary attacks and intraspecific killing in wild white-faced capuchin monkeys (*Cebus capucinus*). Primates 44:341–346

Hemelrijk CK (2002) Despotic societies, sexual attraction and the emergence of male 'tolerance': an agent-based model. Behaviour 139:729–747

Izar P, Ramos-da-Silva ED, de-Resende BD, Ottoni EB (2007) A case of infanticide in tufted capuchin monkeys (*Cebus nigritus*). Mastozool Neotrop 14:73–76

Izawa K (1980) Social behavior of the wild black-capped capuchin (*Cebus apella*). Primates 21:443–467

Jack KM, Fedigan LM (2004) Male dispersal patterns in white-faced capuchins, *Cebus capucinus*. Part 1: Patterns and causes of natal emigration. Anim Behav 67:761–769

Janson CH (1984) Female choice and mating system of the brown capuchin monkey *Cebus apella* (Primates: Cebidae). Z Tierpsychol 65:177–200

Janson CH (1985) Aggressive competition and individual food consumption in wild brown capuchin monkeys (*Cebus apella*). Behav Ecol Sociobiol 18:125–138

Janson CH (1990) Ecological consequences of individual spatial choice in foraging groups of brown capuchin monkeys, *Cebus apella*. Anim Behav 40:922–934

Janson CH (1996) Towards an experimental socioecology of primates: examples from Argentine brown capuchin monkeys (*Cebus apella nigritus*). In: Norconk MA, Rosenberger AL, Garber PA (eds) Adaptive radiations of neotropical primates. Plenum Press, New York, pp 309–325

Janson CH (1998) Experimental evidence for spatial memory in foraging wild capuchin monkeys, *Cebus apella*. Anim Behav 55:1229–1243

Janson CH (2007) Experimental evidence for route integration and strategic planning in wild capuchin monkeys. Anim Cogn 10:341–356

Janson CH, Di Bitetti MS (1997) Experimental analysis of food detection in capuchin monkeys: effects of distance, travel speed, and resource size. Behav Ecol Sociobiol 41:17–24

Kappeler PM, Fichtel C (2012) A 15-year perspective on the Social Organization and life history of Sifaka in Kirindy Forest. In: Kappeler PM (ed) Long-term field studies of primates. Springer, Heidelberg

Koenig A, Borries C, Doran-Sheehy DM, Janson CH (2006) How important are affiliation and cooperation? A reply to Sussman et al. Am J Phys Anthropol 131:522–523

Lynch Alfaro JW (2007) Subgrouping patterns in a group of wild *Cebus apella nigritus*. Int J Primatol 28:271–289

Lynch Alfaro JW, Boubli JP, Olson LE, Di Fiore A, Wilson B, Gutiérrez-Espeleta GA, Chiou KL, Schulte M, Neitzel S, Ross V, Schwochow D, Nguyen M, Farias I, Janson CH, Alfaro ME (in press) Explosive Pleistocene range expansion leads to widespread Amazonian sympatry between robust and gracile capuchin monkeys. J Biogeogr

Ministerio del Medio Ambiente (2005) Parque nacional Iguazú. In: Ministerio del Medio Ambiente (ed) Guía visual parques nacionales de la Argentina. Artgraf, Madrid, pp 70–79

Mitani JC, Watts DP, Pepper JW, Merriwether DA (2002) Demographic and social constraints on male chimpanzee behaviour. Anim Behav 64:727–737

Oster GF, Wilson EO (1978) Caste and ecology in the social insects. Princeton University Press, Princeton

Palombit RA, Cheney DL, Fischer J, Johnson S, Rendall D, Seyfarth RM, Silk JB (2000) Male infanticide and defense of infants in chacma baboons. In: van Schaik CP, Janson CH (eds) Infanticide by males and its implications. Cambridge University Press, Cambridge, pp 123–152

Paviolo A, Di Blanco YE, De Angelo CD, Di Bitetti MS (2009) Protection affects puma abundance and activity patterns in the Atlantic forest. J Mammal 90:926–934

Placci LG, Arditi SI, Ciotek LE (1994) Productividad de hojas, flores y frutos en el parque nacional iguazú. Yviraretá 5:49–56

Ramírez-Llorens P, Di Bitetti MS, Baldovino MC, Janson CH (2008) Infanticide in black capuchin monkeys (*Cebus apella nigritus*) in Iguazú National Park, Argentina. Am J Primatol 70:473–484

Robinson JG (1988) Group size in wedge-capped capuchin monkeys, *Cebus olivaceus*, and the reproductive success of males and females. Behav Ecol Sociobiol 23:187–197

Robinson JG, Janson CH (1987) Capuchins, squirrel monkeys, and atelines: socioecological convergence with Old World primates. In: Smuts BB, Cheney DL, Seyfarth RM, Wrangham RW, Struhsaker TT (eds) Primate societies. Chicago University Press, Chicago, pp 69–82

Silk JB (2002) Practice random acts of aggression and senseless acts of intimidation: the logic of status contests in social groups. Evol Anthropol 11:221–225

van Schaik CP (2000) Infanticide by male primates: the sexual selection hypothesis revisited. In: van Schaik CP, Janson CH (eds) Infanticide by males and its implications. Cambridge University Press, Cambridge, pp 27–60

Visalberghi E, Janson CH, Agostini I (2003) Response toward novel foods and novel objects in wild *Cebus apella*. Int J Primatol 24:653–675

Wheeler BC (2009) Monkeys crying wolf? Tufted capuchin monkeys use anti-predator calls to usurp resources from conspecifics. Proc R Soc Lond B 276:3013–3018

Part IV
Asia

Chapter 10
Social Organization and Male Residence Pattern in Phayre's Leaf Monkeys

Andreas Koenig and Carola Borries

Abstract The genus *Trachypithecus* (Colobinae, Presbytini) has previously been characterized by one-male groups and both male and female dispersal. Occasionally, males may mature in their natal groups, resulting in so-called age-graded multi-male groups. Our long-term observations of a population of Phayre's leaf monkeys (*Trachypithecus phayrei crepusculus*) in Thailand, while revealing values in group size and composition similar to other species, indicate a hitherto undescribed social organization, in which males mature and breed in their natal group or disperse and form new groups. Groups are not age-graded and multi-male groups are one phase of a dynamic social organization changing between multi-male and one-male constellations. The ways in which our views of the social organization of Phayre's leaf monkeys changed over a period of eight years underscore the importance of long-term studies for a full understanding of the behavioral ecology of long-lived species like primates.

10.1 Introduction

Until recently long-term studies on colobine monkeys, especially the tribe Presbytini (Asian colobines), were comparatively rare (see overview in Kirkpatrick 2007). Only two of the more than 50 species, in seven genera (Groves 2001), had been studied in detail, with multiple years of observation of identified individuals. These were Hanuman langurs (*Semnopithecus entellus*) at Jodhpur, India (e.g., Sommer and Rajpurohit 1989) and at Ramnagar, Nepal (e.g., Borries 2000), and Thomas langurs (*Presbytis thomasi*) at Ketambe, Indonesia (e.g., Steenbeek et al. 2000). This lack of long-term data is unfortunate, because the reliability of results from short-term studies is limited by default (Clutton-Brock and Sheldon 2010).

A. Koenig (✉) • C. Borries
Department of Anthropology, Stony Brook University, Stony Brook, NY, USA
e-mail: akoenig@notes.cc.sunysb.edu; cborries@notes.cc.sunysb.edu

P.M. Kappeler and D.P. Watts (eds.), *Long-Term Field Studies of Primates*,
DOI 10.1007/978-3-642-22514-7_10, © Springer-Verlag Berlin Heidelberg 2012

Infrequent events will rarely be documented and life history traits and demographic data may be biased, which may lead to spurious results, especially in comparative studies (Borries et al. 2011). In addition, the two Asian colobines studied on a long-term basis have very different social systems. Hence, it is unclear what the range of social systems is among the Presbytini and whether anyone pattern predominates (for a recent classification attempt, see Grueter and van Schaik 2010). Lastly, the lack of long-term studies of Presbytini is unfortunate, because available data on group size and composition, the ecology of female social relationships, and the residence patterns appear not to fit "classic" socio-ecological explanations.

10.1.1 Asian Colobines: (Relatively) Unexplored and Enigmatic

Most genera of the Presbytini, including *Presbytis, Simias, Trachypithecus*, form comparatively small groups of 20 or fewer individuals (cf. Table 4 in Kirkpatrick 2007). Given the expectation that folivorous primates such as colobines experience little to no feeding competition and hence should face only weak constraints on group size (review in Snaith and Chapman 2007), the small groups in these taxa posed the so-called folivore paradox (Janson and Goldsmith 1995; Steenbeek and van Schaik 2001). The resolution of the paradox might be that the upper limit of group size depends not just on the trade-off between ecological costs (i.e., within-group scramble competition) and benefits (i.e., predation avoidance), but also on social constraints, particularly the risk of male takeover and infanticide (Crockett and Janson 2000). Specifically, if the rate of male takeover increases with female group size, the risk of infanticide may increase accordingly and limit maximum group size. Some evidence indeed supports this idea (Crockett and Janson 2000), but other explanations are plausible, either in addition or alternatively (Janson and Goldsmith 1995; Snaith and Chapman 2005, 2007). Thus, studies that unravel the constraints on group size in colobines, and in folivorous primates more generally, could also improve explanations of the folivore paradox.

Ecological models of female social relationships (e.g., van Schaik 1989) predict competitive regimes and female social structure reasonably well, but seem to be particularly weak in predicting dispersal patterns (Koenig 2002; Koenig and Borries 2009). Specifically, the models suggest that in female dispersal species contest competition and linear dominance hierarchies should be absent. However, in several such species females form linear dominance hierarchies and they may contest for food (e.g., mountain gorillas: Robbins et al. 2005; overview in Koenig et al. 2004). Hence, disclosing additional evidence for the links between feeding competition, social relationships, and female dispersal could improve the explanatory power of socio-ecological models.

The Presbytini also pose challenges for a comparative understanding of primate mating systems. The mating system of many nonhuman primates involves female defense polygyny, in which the number of adult males in bisexual groups is positively associated with the number of adult females and their overlap in receptivity (Emlen and Oring 1977; Nunn 1999; Kappeler 2000). In these species, males usually leave their natal groups upon reaching maturity and aggressively take over or immigrate into other bisexual groups. Among the Presbytini, specifically the genera *Presbytis* and *Trachypithecus*, the situation is slightly more complicated. The mating system has been described as female defense polygyny (van Schaik et al. 1992) and groups usually contain relatively few females, which should make them easy for single males to defend (van Schaik and Hörstermann 1994). However, surprisingly often groups contain multiple males. At least for some populations, groups can be described as age-graded (*sensu* Eisenberg et al. 1972) with male tolerance allowing maturing males to stay (Sterck and van Hooff 2000). For unknown reasons the percentage of "true" multi-male groups and age-graded multi-male groups as well as male residence vary across Presbytini, and the costs and benefits of multi-male stages are poorly understood (Sterck and van Hooff 2000).

Exploring these three aspects of primate social systems requires investigating individual social strategies, reproductive decisions, and reproductive success of known individuals over multiple groups and years; thus, it requires a long-term approach.

10.1.2 Research Goals and Expectations

These questions on the constraints of group size, female feeding competition and dispersal, and the social organization and residence pattern of Asian colobines led us to search for an appropriate study species and site in 1998. The few published reports that were available at that time indicated that the genus *Trachypithecus* in general and the species *T. phayrei* in particular might be a good fit. This and other Southeast Asian species seemed to display group sizes (13–27 on average) intermediate between those of *Semnopithecus* (ca. 26 on average) and *Presbytis* (ca. 8 on average; based on Bennett and Davies 1994); the *Trachypithecus* values are close to a proposed switch point between strong and weak risk of infanticide and, conversely, between weak and strong feeding competition (see details in Crockett and Janson 2000). At the same time it seemed likely that females would disperse (Sterck 1998) and occasional multi-male groups of *T. phayrei* had been reported (Bennett and Davies 1994).

Following explorations in Northeast India and mainland Southeast Asia in 1998 and 1999, we settled on the Phu Khieo Wildlife Sanctuary (PKWS), Thailand, as study area and began to habituate the first group of *Trachypithecus phayrei crepusculus* in October 2000 (note that there is variation in nomenclature (*T. phayrei* vs *T. holotephreus*; Groves (2001) *contra* Brandon-Jones et al. (2004)) and uncertainty in subspecies/species assignment (*T. phayrei crepusculus* vs *T. crepusculus*; Groves (2001) *contra* Roos (2004); i.e., here we follow the nomenclature provided in Groves (2001)).

In the following, we will summarize main results of our study focusing on social organization and male residence pattern reporting how, over a period of 8 years,[1] our views changed. As in other Asian colobines, we expected male Phayre's leaf monkeys to exhibit female defense polygyny (van Schaik et al. 1992) with occasional age-graded multi-male groups (Sterck and van Hooff 2000). Because groups contain relatively few females and become multi-male due to tolerance and not due to changes in monopolization potential, we expected at most a weak positive relationship between the number of males and the number of females. In a strictly age-graded system, one would further predict that males remain in their natal groups for some time following maturation and that they can be ranked according to age (Eisenberg et al. 1972). We expected both sexes to disperse, but changes in male membership to occur primarily via male immigration and takeover (Sterck and van Hooff 2000). Alternatively, one might predict a true multi-male pattern with male immigration and takeover as in Hanuman langurs (Borries 2000), or a pattern in which groups form and disband through female dispersal as in Thomas langurs (Sterck 1997). However, even in Thomas langurs male takeovers have been observed occasionally (Steenbeek et al. 2000).

10.2 Field Site: History and Methods

10.2.1 Study Area and Site

The study area, PKWS, is located in Northeast Thailand (16°5'-35' N, 101°20'-55' E, Chaiyaphum Province, elevation: 300–1,300 m asl) and comprises an area of 157,300 ha as part of the Western Isaan Forest Complex, a conservation area of 598,400 ha in total (Kumsuk et al. 1999). The area became a wildlife sanctuary (the highest protection status in Thailand) in 1979 and is effectively protected via ranger patrols and helicopter surveys, although illegal logging, collection of aloewood (*Aquilaria crassna*), and poaching still occur occasionally (Grassman et al. 2005). The vegetation has been classified primarily as hill and dry evergreen forest (75%) in addition to some other plant communities (Grassman et al. 2005). PKWS harbors a diverse animal community that includes Asian elephant, Asiatic black bear, Malayan sun bear, Asian forest bison (gaur), and four deer species (Kumsuk et al. 1999). With eight out of the nine felids found in Thailand (e.g., tiger, leopard, clouded leopard, golden cat), two canids (jackal and dhole), ten viverrids, five larger raptor species, and two python species, the predator community is plentiful (Grassman et al. 2005).

---

[1] Systematic data collection ran from January 2001 to January 2009 when it was discontinued due to a lack of funding.

We selected a part of the dry evergreen forest at Huai Mai Sot Yai (16°27' N, 101°38' E; 600–800 m asl) as our study site for two reasons. First, the area is slightly hilly, but not too steep to follow arboreal primates. More importantly, cursory surveys indicated a relatively high diversity and density of primates. The primate community at Huai Mai Sot Yai consists of *T. phayrei* and *Hylobates lar*, three macaque species (*Macaca assamensis, M. leonina, M. mulatta*) and northern slow loris (*Nycticebus bengalensis*; Borries et al. 2002; Hassel-Finnegan et al. 2008).

The study site is accessible through a network of trails encompassing more than 100 km. Most of these trails were made by elephants and gaurs, with occasional connections cut between them. The trails were measured, marked, and GPS-mapped every 100 m. To put these data points on a map, we digitized the topographic maps of the area turning it into a digital elevation model.

10.2.2 Facilities

The presence of elephants did not allow maintaining a field camp close to the field site and facilities at the headquarters of the sanctuary could only be used for short periods of time. Hence, with support from the National Science Foundation we established a field station at the sanctuary's headquarters that consisted of a kitchen and lab, an office, and two residential buildings (four rooms each).

The disadvantage of this arrangement was a daily "commute" of 11 km (one way) on a small paved road to and from the headquarters located in the center of the sanctuary. However, the advantage of this arrangement was to have electricity (4 h a day via generator), which allowed running a freezer and other electrical gear (e.g., drying oven, mechanical food tester, computer, battery charging equipment, etc.). In addition, the size of the field station allowed the permanent presence of several assistants and students as well as smaller laboratory procedures and storage. Lastly, the location of the field station in the headquarters allowed researchers immediate contact with the sanctuary authorities, an important aspect for a smooth coordination of research activities. In addition, the headquarters has a helipad for emergency evacuation in case of accidents with elephants, gaurs, bears, or venomous snakes.

10.2.3 Data Collection

From the start, our project was designed as a long-term study with a multidisciplinary approach to primate behavioral ecology, particularly questions to group size constraints, the ecology of female social relationships, and female and male reproductive strategies. Accordingly, we collected data on primate community ecology along with data on the ecology, demography, life history, behavior, hormones, and genetics of Phayre's leaf monkeys. Our approaches and procedures rested on published descriptions (see below) and our past field experience in India and Nepal. In addition, we profited from material and descriptions kindly made available by colleagues (e.g., unpublished monitoring guide by J. Altmann, S. Altmann, and G. Hausfater).

10.2.3.1 Ecological Data

Weather data that included temperature, humidity, and rainfall were recorded via data loggers. Temperature and humidity were recorded directly in the forest via two loggers (2 h intervals; one logger as a back-up). To measure rainfall, a flow-through rain gauge was set up ca. 6 km away from the study site at a ranger station. A rain gauge initially installed at the field site itself was destroyed by elephants after only a few months. A second rain gauge at the headquarters served as backup.

To estimate the forest composition and to measure plant distribution, we used a stratified random approach to establish 33 botanical plots, each 50×50 m; this represented ca. 3% of the home ranges of our study groups (Struhsaker 1975). Within each plot we measured all trees of ≥ 10 cm in diameter and all climbers of ≥ 5 cm in diameter (total of 4,538 stems). Botanical work was done primarily by the staff of the sanctuary, because in Thailand only forest personnel are permitted to collect plant parts (we requested and received a special permit for botanical work).

From the botanical plots we selected a subsample of trees and climbers for phenology data collection. We included as many plant species as possible, because in the beginning we did not know exactly, which species were langur food. The sample consisted of 546 trees and climbers from 121 species. If possible, we included 5–10 mature individuals per common species. Rare species ($N < 5$ in botanical plots) were included if the leaf monkeys were known to use them. Data for different phytophases were collected once a month in the middle of the month using a point scale (from 0 to 3) and a semiquantitative index based on $\log_{10}$ (i.e., 0 for 1–9, 1 for 10–99 etc.; Janson and Chapman 1999). To circumvent problems with interobserver reliability, data were collected by two researchers.

10.2.3.2 Primate Community

To describe the primate community of the site, we conducted line transect sampling for four consecutive days each month on a 4-km transect. We discontinued the data collection after 480 km had been walked, when cumulative density analysis indicated no further improvement in data quality for the most common species (Borries et al. 2002; Hassel-Finnegan et al. 2008).

10.2.3.3 Habituation and Identification

We habituated four groups of Phayre's leaf monkeys. The area has experienced some hunting in the past, and all monkeys initially fled from observers. It took several months to reliably count and identify all individuals, and habituation to the point when observers were ignored took 7–12 months per group.

The federal laws and regulations for research in forested areas in Thailand make it very hard to receive permission for capturing wild animals. Thus, to identify individuals we relied on traditional methods based on physical characteristics (National Research Council 1981). All group members were identified via the shape of their crest, eye rings, white muzzle, and the shape of depigmented skin below the belly button. We established an identification sheet for each individual and a library of digital images of the markers. These tools facilitated learning the identity of the monkeys within 1–2 months. Importantly, changes in physical characteristics required an annual update of the ID charts.

10.2.3.4 Demography and Life History

In general, we followed groups from dawn to dusk for a minimum of 4 days per month (mean: 8.7 days). During every follow, we completed at least one full group count and identified all members. We also recorded births, immigrations, disappearances, emigrations, injuries, and nipple contact. Once a month we assessed immature individuals to demarcate landmarks in growth by comparing their sitting height or head-body length to adult group members (National Research Council 1981). Individuals were considered juvenile if they were smaller than adult females. Subadult males were as tall as adult females, but smaller than adult males. Both males and females became adult, when they had reached the height/length of adult males or females, respectively. Once adult, individuals were assessed for several more months to assure that they had ceased growing. The demographic data allowed for compilations of weaning ages, interbirth intervals, and rates of maturation and dispersal. Altogether the study included 277 group-contact months and 23,677 contact hours (Borries et al. 2011).

Behavioral Data

Individual behavioral data were collected via 20 min focal animal sampling (Martin and Bateson 2007), in which we combined instantaneous sampling at 1-min intervals with continuous recording. Behavioral data emphasized feeding and social behavior of adults and, sometimes, juveniles. The length of a focal sample was determined based on the median time an observer could follow an individual monkey without interruption and the median duration of certain behavioral states such as grooming (E. Larney unpublished data). Agonistic and sexual behavior was also collected *ad libitum*. Depending on the research questions, these data were supplemented with data on grooming bouts, allomaternal care, feeding rates, nutritional data, food physical properties (Lucas et al. 2003), and other variables.

At the group level, we used scan sampling (Martin and Bateson 2007) at 30-min intervals to collect data on mutually exclusive activities (feeding, traveling, resting, social) and on the height of individuals above the ground. We noted the behavior of all identified individuals except infants within 10 min. At the start and end of a group follow and on the hour and the half hour, we collected ranging data at the approximate geometric center of the group via a handheld GPS (UTM coordinates and error reading). We also collected group-level data on food patch depletion (focal tree samples; Snaith and Chapman 2005) and intergroup encounters.

10.2.3.5 Hormones and Genetics

We and our students collected fecal samples to investigate reproductive hormones (Lu 2009; Lu et al. 2010) and relatedness and paternity (Larney unpublished). Sample collection was noninvasive and followed standard procedures that either involved freezing (Lu et al. 2010) or a two-step ethanol-silica method (Nsubuga et al. 2004). As with botanical work, fecal sample collection required a special permit as well as CITES clearance for export.

10.2.3.6 Data Consistency

To ensure standardized data collection and interobserver reliability, we first established an ethogram for the species based on the behavioral repertoire for Hanuman langurs (Dolhinow 1978). The behavioral categories, including standard abbreviations and descriptions, were listed in a field manual that also explained all observational, sample collection, and data processing procedures and definitions. Such a manual is an essential tool in training and re-training of observers to ensure consistency in data collection over time. Consistency can be improved if training is conducted by the same individuals (in our study, ourselves and long-term rangers). In addition, we encouraged our assistants to specialize in certain tasks so that not everyone had to become an expert in all methods. Lastly, we conducted interobserver reliability tests (Martin and Bateson 2007) during training.

10.3 General Characteristics, Life History, and Social Organization

10.3.1 General Characteristics of Phayre's Leaf Monkeys

Phayre's leaf monkeys are midsized nonhuman primates. Adult individuals weigh about 6–8 kg with a moderate sexual size dimorphism, i.e., males weigh about 8 kg and females ca. 6–7 kg (Smith and Jungers 1997). Measurements, which we could take for one adult female (7.0 kg), confirmed the value for females.

Like many other Asian colobines, Phayre's leaf monkeys are primarily arboreal spending most of their time at heights between 5 and 50 m. During parts of the winter and spring (January–March), all groups come to the ground to eat soil and to drink. In October, they sometimes come to the ground to feed on bamboo shoots.

As members of the subfamily of Colobinae, Phayre's leaf monkeys are characterized by foregut fermentation (Bauchop and Martucci 1968) and with 46% of time feeding on leaves (data for adult individuals for three groups over 1 year; Borries et al. 2011) their diet fits the criterion for a folivorous primate (at least 40–45%; Leigh 1994). However, the amount of leaves is relatively small compared to other Asian colobines, which commonly have over 50% leaves in their diet (Kirkpatrick 2007). Conversely, Phayre's leaf monkeys devote a relatively high proportion of feeding time (35%) to fruits and seeds.

10.3.2 Life History

A recent compilation highlights similarities of life history traits of Phayre's leaf monkey with other wild Asian colobines (Lu et al. 2010; Borries et al. 2011). Female Phayre's leaf monkeys have their first infants at an average age of 5.3 years compared to 5.4 to 6.7 years for other Presbytini. The average duration of gestation is 205 days, in the middle of the range for other wild Asian colobines (198–212 days).

As in most other species of *Trachypithecus*, infants are born with a flamboyant natal coat (bright orange), which gradually changes to the adult gray coat over a period of 26 weeks (Borries et al. 2008; Larney and Koenig unpublished). Weaning (defined here as cessation of nipple contact) takes place at 19–21 months, and weaning age increases with group size (Borries et al. 2008). As in other colobines, Phayre's leaf monkey females nurse their infants almost until the next parturition (Borries et al. 2001, 2011). Thus, with an average of 22 months the interbirth interval following a surviving infant is only slightly longer than the mean weaning age (Borries et al. 2008). Interbirth intervals are significantly longer in larger groups. Because infant survival is independent of group size, these differences in reproductive rates may lead to differences across groups in mean female fitness.

These group size effects on reproductive rates stand in contrast to the absence of group size effects in folivorous mountain gorillas (Robbins et al. 2007) and are instead similar to those reported for frugivorous or omnivorous primates (van Noordwijk and van Schaik 1999; Altmann and Alberts 2003). In contrast to other folivores, increases in group size might have negative effects on reproduction in Phayre's leaf monkeys because much of their food comes from depletable patches (cf. Snaith and Chapman 2005, 2007). Alternatively or in addition, these group size effects may reflect co-variation of group size and habitat quality (Dunbar 1987; Harris and Chapman 2007).

10.3.3 Social Organization

In this population, Phayre's leaf monkeys formed bisexual groups averaging 19 individuals (Table 10.1). The mean sizes of our focal groups ranged from 12.1 to 25.7 individuals (range: 6–33), including means of 1.2–2.7 adult males (range: 1–5; Table 10.1). During one month the group PB had no adult male, because the single adult male was absent due to an injury. One-male and multi-male social organizations were about equally likely: in 48.4% ($N = 134$) of group-months, groups contained single adult males. The second (24.9%, $N = 69$) and third (15.9%, $N = 44$) most common constellations were two-male and three-male groups. Groups contained ca. 7 adult females on average; group means ranged from 4.3 to 10.4 (range: 3–12; Table 10.1). In most months (86.3%), the number of adult females ranged between four and ten. Groups typically contained close to 5 subadult or juvenile individuals and ca. 5.5 infants.

The size and composition of our focal groups varied considerably over the course of the study period (Table 10.1). Group size varied by a factor of 1.7 (PB) to a factor of 3.1 (PS) and female group size by a factor of 1.3 (PS) to a factor of 3.3 (PA). However, rather than a consistent direction of change, like the general increase in group size documented for muriquis (Strier and Mendes 2012), the changes in group size or female group size followed U, inverted U, J, or S shapes (results not shown). Only in group PA did total group size and the number of females increase overall during the study, although the increase was not steady. In addition to births and maturation, much of the variation was due to female dispersal (Borries et al. 2004). It took 16 months from the start of the study before the first female immigration could be documented; this apparently low rate was probably a habituation effect, given that we now know that female immigrations occurred at a rate of 2–3 per group-year.

These general characteristics of group size and female dispersal more or less matched our expectations: group size was indeed intermediate between *Semnopithecus* and *Presbytis* (cf. Kirkpatrick 2007), and, as in many other *Trachypithecus*

Table 10.1 Composition of the study groups until January 2009 (inclusively) arranged by group size

Group	Start	Adult males	Adult females	Subadults and juveniles	Infants	Group size
PS	Mar 2002	1.2 (1–2)	4.3 (3–7)	3.1 (0–7)	3.4 (1–5)	12.1 (6–19)
PA	Jan 2001	2.7 (1–4)	6.1 (3–10)	6.1 (3–9)	4.7 (1–9)	19.6 (14–27)
PO	Aug 2005	2.7 (1–5)	7.9 (7–9)	3.9 (1–6)	6.0 (3–8)	20.5 (15–26)
PB	Aug 2003	1.2 (0–3)[a]	10.4 (9–12)	5.6 (0–12)	8.6 (2–11)	25.7 (20–33)
Unweighted / weighted average		1.95 / 1.92	7.18 / 6.84	4.68 / 4.77	5.68 / 5.43	19.48 / 18.95

"Start" indicates the month of the first reliable demographic record. Mean values are given with ranges in parentheses. Results are based on 277 group-months totaling 2,405 contact days and 23,677 contact hours. For details on contact times see Borries et al. (2011)
[a]One month without an adult male; the only adult male had disappeared temporarily.

and *Presbytis* species (Sterck 1998), females regularly dispersed. However, compared to previous reports (Sterck and van Hooff 2000), the high frequency of multi-male groups (over 50%) was unexpected.

10.3.4 Female Group Size and the Number of Males

One-male groups were slightly smaller (17.57 ± 7.44 SD) than multi-male groups (20.22 ± 3.77 SD), but a mixed model ANOVA of group size per study year with "group identity" as random factor, hierarchically nested in social organization (fixed effect), showed no effect of the one-male *versus* the multi-male condition ($F = 0.19, p = 0.675$; Fig. 10.1a). Instead, "group identity" was the driving factor ($F = 136.58, p < 0.001$). The number of females in one-male groups was marginally higher (6.95 ± 2.98 SD) than the number in multi-male groups (6.73 ± 2.15 SD). Again, a mixed model ANOVA showed no effect of the one-male *versus* the multi-male condition ($F = 0.01, p = 0.937$), while the random factor "group identity" had a significant effect ($F = 150.93, p < 0.001$). Only for group PS did the number of females differ significantly between the one-male and the multi-male stage, with more females present when the group had multiple males (Fig. 10.1a).

The socio-ecological model (Emlen and Oring 1977) predicts that the number of males per group is positively related to the number of females. However, monthly data from 276 group-months (excluding one month with no adult male in PB) gave only a weak positive association between the number of females and the number of males (Pearson's $r = 0.117, p = 0.052$; Fig. 10.1b) that explained less than 2% of the variance. The number of males was much better predicted by a quadratic fit in the form of an inverted U-shape: it initially increased with the number of females,

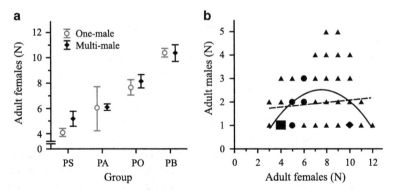

Fig. 10.1 Relationship between the number of adult females and males. (**a**) Number of adult females per group in relation to social organization (one-male *vs* multi-male) of the four focal groups. Group size increasing from left to right. Mean values and 95% confidence limits are given. (**b**) Number of males in relation to the number of females. Size and style of markers indicate the number of group-months (*triangle*: 1–10, *circle*: 11–20, diamond: 21–30, *square*: >40). Linear fit (*hatched line*): $y = 1.58 + 0.05x$; quadratic fit (*solid line*): $y = 0.41 + 0.11x + 0.01x^2$

but then decreased (similar shapes were found for fitted functions such as Lowess or Weighted Least Squares).

Overall, these results indicate that the number of males was not strongly affected by the monopolization potential of females, contrary to the socio-ecological model (Emlen and Oring 1977) and to results from Nunn's (1999) comparative analysis of data from many primate species. Instead the results were similar to other *Trachypithecus* species (Sterck and van Hooff 2000), in which multi-male groups are primarily age-graded.

10.4 Male Residence Pattern and Group Dynamics: Benefits of a Long-Term Approach

While data on group composition can answer questions on social organization and interrelationships between the number of males and females, they cannot answer questions about stability and age-gradedness of multi-male constellations. How do groups form and how are multi-male groups (or one-male groups) maintained? Do males immigrate and/or take over groups? Do they form new groups? Are males tolerated beyond maturation? Particularly questions that relate to rare events as well as to stability and maturation can only be answered with long-term data. In the following, we will therefore describe male residence patterns and group dynamics in our study population.

10.4.1 Male Residence Pattern: The First 5 Years

We started habituating the first study group (PA) in October 2000. PA's home range (bold outline; Fig. 10.2) was surrounded by those of five other bisexual groups (gray outline; Fig. 10.2). At least one of these groups (to the northeast) was a multi-male group, although its exact composition was unknown. Also, in the southeast of PA we occasionally encountered adult or nonadult males, who might have belonged to an all-male band. As of January 2001, PA itself had seven adult females with offspring. It also contained one adult and two subadult males – one bigger, one smaller – and thus had the potential to turn into an age-graded group. However, we did not know whether the oldest male was the father of the younger males and whether the males were natal. In 2002, the all-male band in the southeast of PA became a bisexual group (called R), but it was not clear, how this group had formed. Also, the group to the northeast of PA seemed to have fissioned into a small easily recognizable group with one adult male (group L) and a second group, farther to the northeast, with several males and females. During this year, male membership in PA remained stable and we habituated the second study group, PS. In the following 3 years (2003–2005) we habituated two more groups, PB and PO.

10 Social Organization and Male Residence Pattern in Phayre's Leaf Monkeys

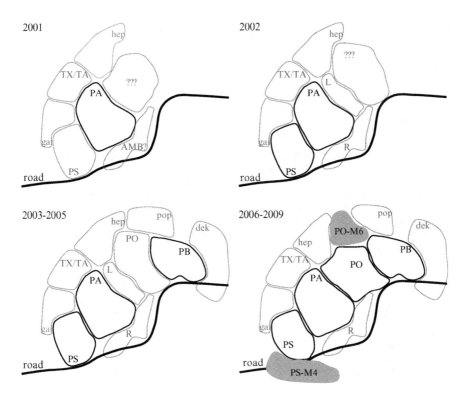

Fig. 10.2 Approximate location of home ranges of the study groups (*bold outline*) and their neighbors (*gray outline*) during the course of the study. Minor shifts of home ranges and their generally small overlap are not shown for simplicity. *Gray* areas indicate two new groups formed in 2008 by males from PO and PS (details see text)

In our four focal groups, male membership remained rather stable over the first 5 years of study. Four natal juvenile males (two in PA, two in PB) and one adult male from PA, who had matured in that group, disappeared (Table 10.2). In addition, one subadult natal male in group PA was twice temporarily absent. The absences of the subadult male were triggered by fights among the males. In the end, however, he returned to his natal group and, when adult, became the alpha male.

Thus, over the first 5 years of our study (133 group-months) not much happened with regard to male dispersal. We neither witnessed male takeover nor male immigration except as a return of a natal male. Because we had regularly witnessed female immigrations, this absence of male immigrations was presumably not caused by a lack of habituation. The question of age-gradedness was hard to answer, because only two males (one natal and one potentially natal) had matured and stayed. However, their continued residence was not a result of male tolerance as suggested for an age-graded structure. Maturing males had frequent, occasionally severe fights (Fig. 10.3) with each other or with older males (one older male lost an

Table 10.2 Disappearances and dispersal of males between January 2001 and January 2009

Years	Event	Age classes		
		Juvenile	Subadult	Adult
2001–2005	Disappearance	4		1
(133 group-months)	Temporary absence		2	
2006–2009	Disappearance	5	1	2
(144 group-months)	Temporary absence		2	4
	Emigration[a]	2	1	5
	Group fusion[b]			1

For each male all events were included, i.e., some males contributed more than once to the dataset. Emigration indicates that males had been relocated after they had left their group. Temporary absence indicates that males had been seen outside their group, but that they returned after a mean absence of 34 days (range: 2–90 days)

[a] All eight events refer to the formation of two new groups (see text for details)

[b] The small nonstudy group L (1 adult male, 1 adult female, 1 juvenile female) fused with group PO, from which it had likely fissioned several years earlier

Fig. 10.3 A subadult male from PS after a severe fight with the only adult male in the group, in which his right shoulder and neck were severely wounded. When adult, he challenged the male again and became the alpha male. Several months later the former alpha male left with three immature males and formed a new group (cf. Fig. 10.2). Photo © Andreas Koenig

eye). The timing of male rank ascendance always coincided with males reaching adulthood and was likely related to a power shift between maturing and aging males. The presence of peers seemed not to have influenced rank ascendance. If anything, fights among maturing males might have prevented or delayed ascendance. In general, contrary to our expectations and instead similar to Ugandan red colobus monkeys (*Piliocolobus tephrosceles*; Struhsaker 2010), males seemed to be philopatric. As in red colobus monkey, males occasionally emigrated from their natal groups, sometimes only temporarily, but unlike in red colobus we had not seen immigrations and group dissolution.

10.4.2 Male Residence Pattern: The Next 3 Years

In the following years (2006 to January 2009; 144 group-months), most male membership changes fitted the pattern described earlier: several juvenile, subadult, and adult males disappeared or were temporarily absent (Table 10.2). Temporary absences often occurred in connection with fights among group males. In addition, group L (1 adult male, 1 adult female, 1 juvenile female), which was sandwiched between groups PA and PO, fused with PO and its male became a member of PO (Fig. 10.2). While this event technically represented an "immigration" by an adult male, it resulted from a group fusion, during which the male returned to his original group. Given our previous observation we presume that the groups split in 2002 (or the male left and formed a new group; see below) and 4 years later the groups fused again. It seems noteworthy that none of the infants born in the small group L survived.

Events observed in the following years would change our perception of the male residence pattern further. In 2007, four of the five adult males of PO left (Table 10.2). Again this happened after severe fights. In contrast to other cases, in which males disappeared from the area, we encountered these four males occasionally at the periphery of the home range of PO. In 2008, the males had been joined by females, forming a new group north of PO (called PO-M6; Fig. 10.2). Because we did not follow this group regularly (all females were unknown to us and unhabituated), it is not entirely clear whether and how much area PO or other groups "lost" in the process of group formation. Importantly, some of the females in this new group had relatively old infants with adult coats, indicating that these infants had been born prior to group formation. Thus, females with infants must have joined the males. Some months later, an adult male (M4), plus a subadult and two juvenile males left group PS and moved south, leaving PS with a single natal adult male (Fig. 10.2; Table 10.2). These four males were joined by females with older infants, forming a new group (PS–M4). As in the case of PO, none of the females was from the males' former group.

These events, which happened after more than 7 years of study, showed clearly that some males manage to form new groups, while others emigrate but eventually return to their natal group. Thus, males have more reproductive options than breeding in their natal group. Interestingly, both mass emigrations of males seemed unrelated to male rank ascendance, but took place after the mating season and the new groups were established before the beginning of the next mating season. Since most females in their old group were pregnant and would not conceive in the next mating season, emigration and group formation might have been the result of poor reproductive prospects. Why infanticide did not occur in the context of group formation (Sterck et al. 2005) remains an open question.

Importantly, based on our initial observations one would have concluded that the adult males within a group were more or less closely related to each other (depending on group size, reproductive skew, and extra-group paternity; Lukas et al. 2005). However, during group formation, males were joined by females with infants, indicating that males co-residing in these groups may not be related at all. How extensively relatedness among group males varies is a question we hope to answer with the analysis of DNA samples. In any case, the residence pattern emerging from these observations also helped us to better understand the variable social organization of groups.

10.4.3 Group Dynamics and Group Life Cycle

One aspect of social organization, i.e., the number of adult males in our study groups, varied considerably through time (Fig. 10.4). Some of the groups had extended periods with only a single adult male, while others contained multiple males for extended periods. In all groups, this variation arose solely through male emigration or disappearance and maturation.

In PS, one adult male was present most of the time until one of the natal males matured in 2007. This natal male eventually became the alpha male, and later the only male, when in 2008 the former alpha male and 3 younger males left to form a new group (see above). Thus, the group switched from one-male to multi-male, then back to one-male. In PA, the number of adult males varied between one and four; it gradually increased during the study, and the group was multi-male 98% of the time. Changes in the number of adult males occurred through occasional

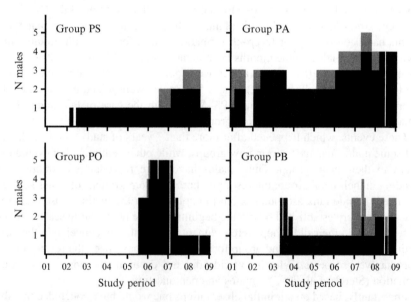

Fig. 10.4 Number of adult and subadult males of the four study groups over the study period 2001–2009. With one exception (see text) changes did not occur through immigrations but through maturation of natal males and emigrations/disappearances. *Black bars*: adult males; *gray bars*: subadult males; no column = no data available (except for group PB in September 2008 when the only adult male disappeared temporarily)

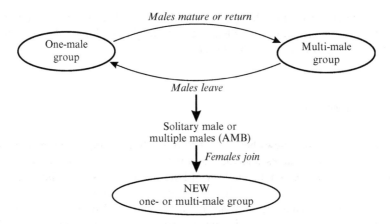

Fig. 10.5 Life cycle of Phayre's leaf monkey groups

disappearances and through maturation of natal males. When our observation of PO started, the group had multiple males. Four of these males left and formed a new group, and PO became one-male. Lastly, PB was initially a multi-male group. However, after adult males disappeared, it was one-male until natal males matured and it turned multi-male. Unfortunately, in this group the adult male died (likely due to predation by a clouded leopard or a leopard) right at the time when the first natal males matured. Thus, we do not know whether one of the maturing males would have overthrown the old male to become the new alpha male.

In essence, it seems that groups follow a rather simple "life cycle" (Fig. 10.5). Groups may change between a one-male and a multi-male stage either when males mature and breed in their natal group or when maturing natal males emigrate, either permanently or temporarily. In the latter process a group may or may not become one-male, depending on the number of males leaving a multi-male group. Dispersing males may either form new one-male or new multi-male groups. In contrast to other Asian colobines (Sterck and van Hooff 2000), we have never witnessed a takeover by a strange male or immigrations of males that were not natal (or likely natal). Similarly, in contrast to other long-term studies we have not seen groups dissolve through female dispersal (Sterck 1997). While we cannot be entirely sure that takeover, immigrations, or group dissolution will never happen, after 277 group-months we can be reasonably sure that they will be rare.

The emerging picture of the social organization of Phayre's leaf monkeys and the "life cycle" of a group is reminiscent of group dynamics in mountain gorillas (Watts 2000; Robbins 2007). Mountain gorillas are similarly characterized by natal and secondary dispersal of females and conditional male dispersal. While some groups have been found to be age-graded, in other cases natal males may become dominant over (presumed) fathers and brothers. However, unlike the nonterritorial female defense polygyny system of mountain gorillas, male Phayre's leaf monkeys actively defend areas with little overlap between groups, i.e., they defend territories (Gibson and Koenig unpublished). This pattern is more consistent with the resource defense polygyny system of chimpanzees (Williams et al. 2004). In the end, male reproductive strategies in Phayre's leaf monkeys might possibly best be described as a mixture of gorilla and chimpanzee strategies.

10.5 Conclusions

The ways in which our views have changed through time emphasize the tentative nature of short-term studies and the importance of long-term studies for a full understanding of the behavioral ecology of long-lived species like primates.

Overall, group size and composition in Phayre's leaf monkeys were similar to other species of the genus *Trachypithecus* (Sterck and van Hooff 2000). However, only our long-term results revealed a social organization differing from other Asian colobines in several aspects: unlike the "true" multi-male groups described for Hanuman langurs, variation in the number of males was not affected by the number of females and one-male groups did not turn into multi-male groups *via* immigrations (Borries 2000). But Phayre's leaf monkeys also did not resemble the one-male structure with occasional age-graded groups of Thomas langurs, in which the multi-male phase may be a transitional stage after sons have matured and before a group dissolves or is taken over (Steenbeek et al. 2000). In Phayre's leaf monkeys, males may breed in their natal groups or they may disperse and form new groups. Even adult (breeding) males may disperse to form new groups. Multi-male groups were a regular part of a dynamic system that changed back and forth between a multi-male and a one-male stage with the occasional formation of new groups. Unlike species such as Thomas langurs (Steenbeek et al. 2000), however, multi-male stages were not age-graded (Eisenberg et al. 1972). Rather, dominance rank in relation to age followed an inverted U-shape, indicating that dominance rank depends on male resource holding potential as in baboons (Packer 1979).

Grueter and van Schaik (2010) recently proposed three main categories of social organization for Presbytini: (1) one-male groups with male immigrations (with occasional age-graded multi-male groups; most *Presbytis* spp., *Trachypithecus* spp.), (2) "true" multi-male groups with male immigrations (with a variable proportion of one-male groups; *Semnopithecus* spp.), and (3) multi-level societies (most snub-nosed monkeys). This scheme must now be expanded to include a fourth form of social organization: one- and multi-male groups with male philopatry and new group formation (*T. phayrei crepusculus*). Given the fragmentary and short-term nature of data for most Asian colobines, it is possible that the social organization described here is not unique. For example, multi-male groups have consistently been found in red-shanked Douc langurs (Lippold and Vu 2008). Given that male dispersal might be conditional, depending on the costs of dispersal and breeding opportunities in multi-male groups (Watts 2000), the possibility exists that these groups resemble the pattern of Phayre's leaf monkeys. Only increased efforts to conduct more long-term studies with Asian colobines could offer an answer.

Acknowledgments We would like to thank Peter Kappeler and his team for organizing an inspiring conference, Peter Kappeler and David Watts for putting this volume together, and Charles Janson, Peter Kappeler, and David Watts for their helpful comments on the manuscript. Our field project would not have been possible without the support of many institutions and people. We wish to thank the National Research Council of Thailand, the Department of National Parks, Wildlife and Plant Conservation, and the Phu Khieo Wildlife Sanctuary (K. Kreetiyutanont, M. Kumsuk, T. Naknakced, K. Nitaya, and J. Prabnasuk) for support and permission. We gratefully acknowledge support and cooperation by J. Beehner (University of Michigan), N. Bhumpakphan and W. Eiadthong (Kasetsart University), W. Brockelman (Mahidol University), N. Czekala (Papoose Foundation), T. Disotell and A. DiFiore (New York University), N. Dominy (UC at Santa Cruz), G. Gale and T. Savini (King Mongkut University at Thonburi), P. Lukas and E. Vogel (George Washington University). For help with the data collection we thank our sanctuary rangers, volunteer research assistants, predoc, and postdoc researchers: A. Bprasapmu, R. Bunting, J. Burns, K. Carl, A. Derby-Lewis, S. Dtubpraserit, W. Engelke, V. Ethier, J. Garten, L. Gibson, C. Gilbert, K. Jenks, J. Kamilar, S. Kropidlowski, E. Larney, E. Lloyd, A. Lu, E. McCullough, J. Mizen, S. Myers, W. Nathongbo (†), K. Ossi-Lupo, G. Pages, G. Preece, Z. Primeau, L. Sarringhaus, S. Suarez, A. Suyanang, P. Terranova, B. Wheeler, T. Whitty, A. Yamee, and R. Zulueta. For help with plant identification we are grateful to N. Bhumpakphan, W. Brockelman, A. Chunchaen, W. Eiadthong, P. Ketdee, and M. Kumsuk. For technical support in Thailand we thank E. Akarachaiyasak, C. Kanwutanakip-Savini, N. Naksathit, R. Phoonjampa, K. Roongadulpisan, U. Suwanvecho, and M. Umponjan. The study was financially supported by the National Science Foundation (BCS-0215542 & BCS-0542035), the National Science Foundation DDIG (BCS-0452635 to C. Borries & A. Lu; BCS-0647837 to A. Koenig & K. Ossi-Lupo), the Leakey Foundation (to A. Koenig & C. Borries, A. Lu, K. Ossi-Lupo), the American Society of Primatologists (to E. Larney, A. Lu, K. Ossi-Lupo), the Wenner-Gren Foundation (to E. Larney, K. Ossi-Lupo), and Stony Brook University (Department of Anthropology & Office of the Vice President for Research to AK). The research was approved by the Institutional Animal Care and Use Committee (IACUC) Stony Brook University (IDs: 20001120 to 20081120) and complied with the current laws of Thailand and the USA.

References

Altmann J, Alberts SC (2003) Intraspecific variability in fertility and offspring survival in a nonhuman primate: behavioral control of ecological and social sources. In: Wachter KW, Bulatao RA (eds) Offspring: human fertility behavior in biodemographic perspective. National Academies Press, Washington, DC, pp 140–169

Bauchop T, Martucci RW (1968) Ruminant-like digestion of the langur monkey. Science 161:698–700

Bennett EL, Davies AG (1994) The ecology of Asian colobines. In: Davies AG, Oates JF (eds) Colobine monkeys: their ecology, behaviour and evolution. Cambridge University Press, Cambridge, pp 129–171

Borries C (2000) Male dispersal and mating season influxes in Hanuman langurs living in multimale groups. In: Kappeler PM (ed) Primate males: causes and consequences of variation in group composition. Cambridge University Press, Cambridge, pp 146–158

Borries C, Koenig A, Winkler P (2001) Variation of life history traits and mating patterns in female langur monkeys (*Semnopithecus entellus*). Behav Ecol Sociobiol 50:391–402

Borries C, Larney E, Kreetiyutanont K, Koenig A (2002) The diurnal primate community in a dry evergreen forest in Phu Khieo Wildlife Sanctuary, Northeast Thailand. Nat Hist Bull Siam Soc 50:75–88

Borries C, Larney E, Derby AM, Koenig A (2004) Temporary absence and dispersal in Phayre's leaf monkeys (*Trachypithecus phayrei*). Folia Primatol 75:27–30

Borries C, Larney E, Lu A, Ossi K, Koenig A (2008) Costs of group size: lower developmental and reproductive rates in larger groups of leaf monkeys. Behav Ecol 19:1186–1191

Borries C, Lu A, Ossi-Lupo K, Larney E, Koenig A (2011) Primate life histories and dietary adaptations: a comparison of Asian colobines and macaques. Am J Phys Anthropol 144:286–299

Brandon-Jones D, Eudey AA, Geissmann T, Groves CP, Melnick DJ, Morales JC, Shekelle M, Stewart C-B (2004) Asian primate classification. Int J Primatol 25:97–164

Clutton-Brock TH, Sheldon BC (2010) The seven ages of *Pan*. Science 327:1207–1208

Crockett CM, Janson CH (2000) Infanticide in red howlers: female group size, male membership, and a possible link to folivory. In: van Schaik CP, Janson CH (eds) Infanticide by males and its implications. Cambridge University Press, Cambridge, pp 75–98

Dolhinow P (1978) A behavior repertoire for the Indian langur monkey (*Presbytis entellus*). Primates 19:449–472

Dunbar RIM (1987) Habitat quality, population dynamics, and group composition in colobus monkeys (*Colobus guereza*). Int J Primatol 8:299–329

Eisenberg JF, Muckenhirn NA, Rudran R (1972) The relation between ecology and social structure in primates. Science 176:863–874

Emlen ST, Oring LW (1977) Ecology, sexual selection, and the evolution of mating systems. Science 197:215–223

Grassman LI Jr, Tewes ME, Silvy NJ, Kreetiyutanont K (2005) Ecology of three sympatric felids in a mixed evergreen forest in north-central Thailand. J Mammal 86:29–38

Groves CP (2001) Primate taxonomy. Smithsonian Institution Press, Washington, DC

Grueter CC, van Schaik CP (2010) Evolutionary determinants of modular societies in colobines. Behav Ecol 21:63–71

Harris TR, Chapman CA (2007) Variation in diet and ranging of black and white colobus monkeys in Kibale National Park, Uganda. Primates 48:208–221

Hassel-Finnegan HM, Borries C, Larney E, Umponjan M, Koenig A (2008) How reliable are density estimates for diurnal primates? Int J Primatol 29:1175–1187

Janson CH, Chapman CA (1999) Resources and primate community structure. In: Fleagle JG, Janson CH, Reed KE (eds) Primate communities. Cambridge University Press, Cambridge, pp 237–267

Janson CH, Goldsmith ML (1995) Predicting group size in primates: foraging costs and predation risks. Behav Ecol 6:326–336

Kappeler PM (2000) Primate males: history and theory. In: Kappeler PM (ed) Primate males: causes and consequences of variation in group composition. Cambridge University Press, Cambridge, pp 3–7

Kirkpatrick RC (2007) The Asian colobines: diversity among leaf-eating monkeys. In: Campbell CJ, Fuentes A, MacKinnon KC, Panger M, Bearder SK (eds) Primates in perspective. Oxford University Press, New York, pp 186–200

Koenig A (2002) Competition for resources and its behavioral consequences among female primates. Int J Primatol 23:759–783

Koenig A, Borries C (2009) The lost dream of ecological determinism: time to say goodbye? ... or a white queen's proposal? Evol Anthropol 18:166–174

Koenig A, Larney E, Lu A, Borries C (2004) Agonistic behavior and dominance relationships in female Phayre's leaf monkeys - preliminary results. Am J Primatol 64:351–357

Kumsuk M, Kreetiyutanont K, Suvannakorn V, Sanguanyat N (1999) Diversity of wildlife vertebrates in Phu Khieo Wildlife Sanctuary, Chaiyaphum Province. Wildlife Conservation Division, Royal Forest Department, Bangkok, Thailand

Leigh SR (1994) Ontogenetic correlates of diet in anthropoid primates. Am J Phys Anthropol 94:499–522

10 Social Organization and Male Residence Pattern in Phayre's Leaf Monkeys

Lippold LK, Vu NT (2008) The time is now: survival of the Douc langurs of Son Tra, Vietnam. Primate Conserv 23:75–79

Lu A (2009) Mating and reproductive patterns in Phayre's leaf monkeys. PhD thesis, Stony Brook University, Stony Brook

Lu A, Borries C, Czekala NM, Beehner JC (2010) Reproductive characteristics of wild female Phayre's leaf monkeys. Am J Primatol 72:1073–1081

Lucas PW, Osorio D, Yamashita N, Prince JF, Dominy NJ, Darvell BW (2003) Dietary analysis. I. Food physics. In: Setchell JM, Curtis DJ (eds) Field and laboratory methods in primatology: a practical guide. Cambridge University Press, Cambridge, pp 184–198

Lukas D, Reynolds V, Boesch C, Vigilant L (2005) To what extent does living in a group mean living with kin? Mol Ecol 14:2181–2196

Martin P, Bateson P (2007) Measuring behaviour. An introductory guide, 3rd edn. Cambridge University Press, Cambridge

National Research Council (1981) Techniques for the study of primate population ecology. National Academy Press, Washington DC

Nsubuga AM, Robbins MM, Roeder AD, Morin PA, Boesch C, Vigilant L (2004) Factors affecting the amount of genomic DNA extracted from ape faeces and the identification of an improved sample storage method. Mol Ecol 13:2089–2094

Nunn CL (1999) The number of males in primate social groups: a comparative test of the socioecological model. Behav Ecol Sociobiol 46:1–13

Packer C (1979) Inter-troop transfer and inbreeding avoidance in *Papio anubis*. Anim Behav 27:1–36

Robbins MM (2007) Gorillas. Diversity in ecology and behavior. In: Campbell CJ, Fuentes A, MacKinnon KC, Panger M, Bearder SK (eds) Primates in perspective. Oxford University Press, New York, pp 305–321

Robbins MM, Robbins AM, Gerald-Steklis N, Steklis HD (2005) Long-term dominance relationships in female mountain gorillas: strength, stability and determinants of rank. Behaviour 142:779–809

Robbins MM, Robbins AM, Gerald-Steklis N, Steklis HD (2007) Socioecological influences on the reproductive success of female mountain gorillas (*Gorilla beringei beringei*). Behav Ecol Sociobiol 61:919–931

Roos C (2004) Molecular evolution and systematics of Vietnamese primates. In: Nadler T, Streicher U, Ha TL (eds) Conservation of primates in Vietnam. Haki Publishing, Hanoi, pp 23–28

Smith RJ, Jungers WL (1997) Body mass in comparative primatology. J Hum Evol 32:523–559

Snaith TV, Chapman CA (2005) Towards an ecological solution to the folivore paradox: patch depletion as an indicator of within-group scramble competition in red colobus monkeys (*Piliocolobus tephrosceles*). Behav Ecol Sociobiol 59:185–190

Snaith TV, Chapman CA (2007) Primate group size and interpreting socioecological models: do folivores really play by different rules? Evol Anthropol 16:94–106

Sommer V, Rajpurohit LS (1989) Male reproductive success in harem troops of Hanuman langurs (*Presbytis entellus*). Int J Primatol 10:293–317

Steenbeek R, van Schaik CP (2001) Competition and group size in Thomas's langurs (*Presbytis thomasi*): the folivore paradox revisited. Behav Ecol Sociobiol 49:100–110

Steenbeek R, Sterck EHM, de Vries H, van Hooff JARAM (2000) Costs and benefits of the one-male, age-graded, and all-male phases in wild Thomas' langur groups. In: Kappeler PM (ed) Primate males: causes and consequences of variation in group composition. Cambridge University Press, Cambridge, pp 130–145

Sterck EHM (1997) Determinants of female dispersal in Thomas langurs. Am J Primatol 42:179–198

Sterck EHM (1998) Female dispersal, social organization, and infanticide in langurs: are they linked to human disturbance? Am J Primatol 44:235–254

Sterck EHM, van Hooff JARAM (2000) The number of males in langur groups: monopolizability of females or demographic processes? In: Kappeler PM (ed) Primate males: causes and consequences of variation in group composition. Cambridge University Press, Cambridge, pp 120–129

Sterck EHM, Willems EP, van Hooff JARAM, Wich SA (2005) Female dispersal, inbreeding avoidance and mate choice in Thomas langurs (*Presbytis thomasi*). Behaviour 142:845–868

Strier KB, Mendes SL (2012) The Northern Muriqui (*Brachyteles hypoxanthus*): lessons on behavioral plasticity and population dynamics from a critically endangered species). In: Kappeler PM, Watts DP (eds) Long-term field studies of primates. Springer, Berlin

Struhsaker TT (1975) The red colobus monkey. University of Chicago Press, Chicago, IL

Struhsaker TT (2010) The red colobus monkeys: variation in demography, behavior, and ecology of endangered species. Oxford University Press, Oxford

van Noordwijk MA, van Schaik CP (1999) The effects of dominance rank and group size on female lifetime reproductive success in wild long-tailed macaques, *Macaca fascicularis*. Primates 40:105–130

van Schaik CP (1989) The ecology of social relationships amongst female primates. In: Standen V, Foley RA (eds) Comparative socioecology: the behavioral ecology of humans and other mammals. Blackwell, Oxford, pp 195–218

van Schaik CP, Hörstermann M (1994) Predation risk and the number of adult males in a primate group: a comparative test. Behav Ecol Sociobiol 35:261–272

van Schaik CP, Assink PR, Salafsky N (1992) Territorial behavior in southeast Asian langurs: resource defense or mate defense? Am J Primatol 26:233–242

Watts DP (2000) Causes and consequences of variation in male mountain gorilla life histories and group membership. In: Kappeler PM (ed) Primate males: causes and consequences of variation in group composition. Cambridge University Press, Cambridge, pp 169–179

Williams JM, Oehlert GW, Carlis JV, Pusey AE (2004) Why do male chimpanzees defend a group range? Anim Behav 68:523–532

Chapter 11
White-Handed Gibbons of Khao Yai: Social Flexibility, Complex Reproductive Strategies, and a Slow Life History

Ulrich H. Reichard, Manoch Ganpanakngan, and Claudia Barelli

Abstract Long-term field research on wild animals is essential for understanding life history and social systems of long-lived organisms like primates. Gibbons (family Hylobatidae) live surprisingly slow lives, given their relatively small body mass. Following an approximately 7-year-long juvenile period, one of the longest among all primates, Khao Yai white-handed gibbon females begin reproducing at an average age of 10.5 ± 1.2 years. This is much later than in monkeys of at least the same body mass and, remarkably, at about the same age as in mountain gorillas. Our long-term research also revealed remarkable social flexibility analogous to that seen in other apes. At Khao Yai, white-handed gibbons form pairs or small two-male/one-female reproductive units, although individuals may temporarily also live in single-male/multi-female groups, and here we report a novel, semi-solitary life stage of two older males for the first time. Mating patterns also turned out to be flexible, with males and females mating polygamously, including extra-pair copulations and regular polyandrous mating of females living in multi-male groups. We have also found that in accordance with this variability in male–female socio-sexual bonds, female gibbons at Khao Yai show cyclical sexual swellings that advertise the probability of ovulation without allowing males to exactly pinpoint the day of ovulation. After decades of research, we have come to recognize more clearly the importance of the gibbon community and feel confident that we understand the basic social and mating systems of the Khao Yai

---

U.H. Reichard (✉)
Department of Anthropology, Southern Illinois University, Carbondale, IL, USA
e-mail: ureich@siu.edu

M. Ganpanakngan
Khao Yai National Park, Bangkok, Thailand

C. Barelli
Department of Reproductive Biology, German Primate Center, Göttingen, Germany & Tropical Biodiversity Section, Trento Science Museum, Trento, Italy
e-mail: cbarelli@dpz.eu

P.M. Kappeler and D.P. Watts (eds.), *Long-Term Field Studies of Primates*,
DOI 10.1007/978-3-642-22514-7_11, © Springer-Verlag Berlin Heidelberg 2012

white-handed gibbon population, but we also continue to discover new details of the evolutionary forces that shape gibbons' complex social life.

11.1 Introduction

A unifying theme of early and current primate field studies is their "individual-centric" approach, which means that particular individuals and their lives become the focus of a researcher's attention and systematic data collections (e.g., Goodall 1986). Working with well-known individuals is a unique strength of long-term field studies and one that continuously draws students, volunteers, and periodically the media to our field. Hearing of the adventures of primate characters and following the fate of individuals through time often seems just as fascinating as vividly telling their stories and presenting data from the field (e.g., Perry and Manson 2008), which now sometimes even happens in "near-real-time" in the new format of primate field blogs. Beyond scientific curiosity and theoretically well-grounded questions, many primatologists, students, and professionals alike, feed off direct contact with well known, habituated individuals as their source of energy to write grant proposals, and involvement in the lives of their study subjects can bring researchers back to a field site year after year. Dedication and developing relationships with primate subjects and human communities living closest to them are emotional and intellectual reservoirs field workers use until a long-term study emerges, which is a necessary step to document life-history strategies of long-lived mammals.

11.2 History of the Khao Yai White-Handed Gibbon Study Site

Research on white-handed gibbons (*Hylobates lar*) of Khao Yai National Park (KY), Thailand began in 1977. Like other primate field projects, ours began small, but it gradually grew to become the longest ongoing gibbon study, and we have accumulated demography data on 14 habituated groups (Fig. 11.1). Like others, we believe that longitudinal research, although slow and difficult to maintain, is essential as it is often the only way to generate life history data, to decode strategies underlying complex behaviors in wild populations, like those that involve reciprocity, cooperation, conflict resolution, and to understand primate social dynamics more broadly (Wells 1991; Boesch and Boesch-Achermann 2000; Strier et al. 2002; Watts 2002). The complex social dynamics of KY white-handed gibbons would have been difficult to detect in a short-term study (see below), even if it covered several years. Lack of long-term documentation of gibbon demography, life-history strategies, and social dynamics until recently is the reason why the subtleties and complexity of their social organization remained unnoticed for a long time.

Over the years, many individuals have contributed to the ongoing demographic data collection (for a complete list, see Brockelman et al. 1998). Key people at the

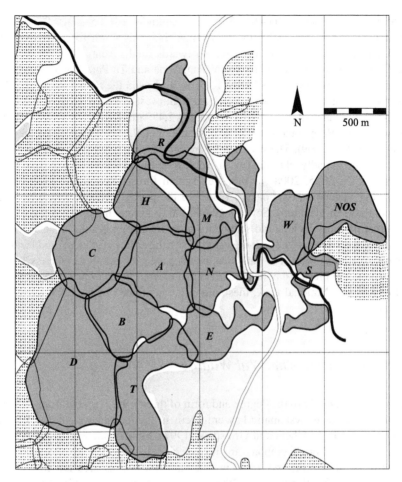

Fig. 11.1 Mo Singto study area with home range outlines of 13 habituated study groups (A-NOS). The home range of a fourteenth habituated study group is not shown on the map. *Thick solid line* = Lam Takhong river; *open line* = road; *fading areas* = tropical grasslands and low canopy regenerating forest

beginning were Treesucon (1984), Raemaekers and Raemaekers (1985) and W.Y. Brockelman, of whom only Brockelman continues to do research at the site. By the end of 1989, Reichard (1991) became involved, and in the mid 1990s, C. Barelli joined the research effort; since then, they have coordinated systematic recording of demography data. Quantitative data presented here were collected by C. Barelli and U.H. Reichard between 1989 and 2010.

Today, KY is a patch of 2,168 km^2 (Smitinand 1977) of forest surrounded by agricultural land on all sides except one. The park was established in 1962 and, in 2005, was included in Thailand's 6,199 km^2 large Dong Phayayen – Khao Yai Forest Complex (DPKY) as part of a new World Heritage site (UNESCO 2005) because it is a biodiversity hotspot in Asia (Lynam et al. 2006).

240 U.H. Reichard et al.

Located at latitudes 14°05′–14°15′ N and longitudes 101°05′–101°50′ E, KY is part of the Phanom Dongrak mountain range that runs north to south from the Thai–Laotian border before bending eastwards and eventually forming the Thai–Cambodian border in the region of Pang Sida and Ta Phraya National Parks. Elevation at KY ranges from ~250 to 1,351 m a.s.l., and the terrain is rugged. The climate is seasonally wet following the Asian southwest monsoon cycle (Singhrattna et al. 2005), with annual precipitation averaging 2,477 mm/year (range 2,038–3,111) (Tangtam 1992; Boonpragob et al. 1998; Kitamura et al. 2004, 2005, 2008; Bartlett 2009a; Gale et al. 2009). The wet months are March–October.

KY can be broadly classified as a tropical seasonal forest (Smitinand 1989; Kitamura et al. 2005, 2008) or moist evergreen forest (Round and Gale 2008), because this vegetation type occupies 64% of the park's land area found between 400 and 1,000 m elevation. Several gibbon study groups have established home ranges that partially include old secondary growth (i.e., groups A, H, and D). The gibbons have continuously and increasingly used these areas for travel and foraging since observations began. The park also includes areas of grassland where villagers living around the present day headquarters had cleared fields prior to the establishment of KY as a National Park; these are now maintained by annual burning and mowing.

11.2.1 Threats to Khao Yai Wildlife

Field sites vary greatly in the degree and form of threats they receive from humans. Due to its large size, systematic law enforcement is a constant challenge to park management at KY (Albers and Grinspoon 1997). Small-scale encroachment and hunting occur, although gibbons are not specifically targeted by poachers and, compared to other protected areas in Thailand, these pressures are low at KY (Lynam et al. 2006; Brodie et al. 2009). In our experience, the biggest threat to wildlife comes from selective, non-timber harvesting of *Mai hom* trees, *Aquilaria crassna* (Family Thymelaeaceae), by villagers and organized poacher groups. *Mai hom* trees produce agarwood, also known as aloewood or eaglewood, used by the perfume industry. The tree family occurs naturally in primary evergreen and semi-evergreen forests from ~600 to 1,400 m a.s.l. in many Southeast Asian countries and the commercially valuable resin is traditionally harvested by local people (Jensen and Meilby 2010).

At KY entire trees are sometimes felled, but more commonly mature trees are injured repeatedly to stimulate resin production (Zhang et al. 2008). Several months after a tree has been damaged, poachers return to chisel resin-soaked woodchips off until a tree eventually falls (Zhang et al. 2008). Large-scale harvest of *Mai hom* is obviously destructive because it involves bringing heavy machinery into the forest to fell and transport stems. But also small-scale poaching, i.e., poachers targeting specific trees and removing large quantities of woodchips, negatively affects wildlife because poachers stay in the forest for more days than they can carry food, and when their provisions are exhausted they hunt for food. Poachers often carry firearms, which makes encounters with them dangerous to park rangers, researchers, and our Thai field assistants alike. Poaching of *Aquilaria* trees also directly, although marginally affects gibbons, who feed on the tiny sprout of *Aquillaria* seeds during the trees' short fruiting period after biting off the thick husk with their long, sharp canines. Selective harvest of agarwood is not unique to KY; it also occurs at other protected sites in Thailand (Grassman et al. 2005). The market value of agarwood varies according to quality, and agarwood from KY consistently yields high market prices, which makes effective control of *Mai hom* harvest and trade difficult.

11.3 Highlights of Long-Term Gibbon Research

Identifying results and benefits of long-term research on KY white-handed gibbons is straightforward and well documented through numerous publications that span a wide variety of topics ranging from vocal communication (Raemaekers et al. 1984; Raemaekers and Raemaekers 1985) to ecology (Bartlett 2009a, b; Brockelman 2009), social behavior (Reichard 1995, 1998, 2003; Reichard and Sommer 1997; Brockelman et al. 1998; Sommer and Reichard 2000; Barelli et al. 2008a), reproduction (Barelli et al. 2007, 2008b; Barelli and Heistermann 2009), life history (Reichard and Barelli 2008), and cognition (Asensio et al. 2011).

In the following, we highlight advances in three areas of research on white-handed gibbons with which we have been particularly involved: (1) social organization, (2) reproductive strategies, and (3) life histories. Research on all of these topics substantially advances our knowledge about gibbons and helps shift understanding of gibbon social organization from a simplistic focus on monogamy to a more complex community model.

11.3.1 Flexible Social Organization

Our long-term research revealed a formerly unrecognized extent of social flexibility in white-handed gibbons. Although anecdotal reports of gibbon groups with more than one adult of one sex existed for some time (summarized in Fuentes (2000) and Reichard (2003)), systematic data allowing quantification of the frequency of social units not consisting of pairs became first available at KY (Barelli et al. 2007, 2008b; Reichard and Barelli 2008; Reichard 2009).

An important insight from our long-term observations is that white-handed gibbons are not *per se* committed to pair-living or other forms of social organization but instead respond in flexible ways to opportunities and actively pursue or passively accept changes in their social status. In the sample of 12 groups, 19 adult females and 22 adult males were residents at some point in time. Irrespective of the duration of these individuals' group membership, 42% of females and 68% of males experienced pair-living and at least one other type of group structure. Some individuals lived through multiple changes from pair-living to multi-male/single-female stages and back. For females who experienced non-pair-living periods, these times amounted to roughly 50% of the 12-year census period (range 33–100%); the corresponding value for males was ca. 60% (range 9–100%). These data illustrate that a wide spectrum from exclusive pair-living to exclusive multi-male/single-female grouping and various stages in between exist at KY and that non-pair-living is not a transitional stage, but for many adults represents a substantial portion of their prime reproductive years.

In summary, our long-term data indicate that, although a majority of gibbon groups are pair-living, breeding groups with more than two adults (excluding groups with adult offspring), particularly adult males, are no exception (Reichard 2009). In the sample of 12 well-known groups censused annually over 12 years (1999–2010, $N = 146$ units), we found an average of 25% of groups to be multi-male/single-female (Table 11.1). We believe these data are representative for the population as a whole because the values are similar to an earlier, larger census that included non-habituated groups (Reichard 2009). Importantly, some multi-male groups were always present, and in some years made up 33% or more of groups (Table 11.1). Based on long-term demographic records (Reichard 2009), most multi-male groups consisted of two adult males living with an unrelated female, i.e., a female neither one of the males had grown up with. Two groups were each composed of three adult males and one adult female and persisted for about 2 and 4 years, respectively. Group structures besides pair-living and multi-male/single-female units such as multi-female/multi-male, and multi-female/single-male have also been observed (Reichard 2009), but they are rare and, to our knowledge, have not resulted in stable breeding units.

Nevertheless, the occurrence of three multi-female/single-male groups is interesting as it illustrates the context-dependent social flexibility in this population. We twice discovered multi-female/single-male groups in which each of two females carried a nursing infant. Unfortunately in the first case, we did not know the group's social history and thus could not exclude the possibility that a daughter had conceived with the group's adult male, who had very likely replaced the female's presumed father. About 2 years and 4 months after the group had been discovered, one of the females disappeared with her offspring. In 2010, we witnessed a second group with two dependent infants. This time, we knew the social history of individuals and could confirm that a daughter gave birth a year after her mother. This was probably not the result of an incestuous mating, because the current male immigrated in 2007 and thus was unlikely to be the father of the female who had given birth recently. However, only a genetic study could confirm the kin relationships in this group. The third multi-female/single-male group formed after a young adult male and an adolescent female joined a young, unrelated adult female. The trio lived peacefully together for several years until the time of the younger female's sexual maturity, when the older female became increasingly aggressive. The younger female left before sexual behavior with the male was

Table 11.1 Group structure variation in Khao Yai white-handed gibbons, Thailand ($N = 146$ units)

		Census year[a]												Mean $\pm$ SD
		1999	2000	2001	2002	2003	2004	2005	2006	2007	2008	2009	2010	
P	(%)	83.3	75.0	83.3	75.0	70.0	63.6	66.7	66.7	75.0	75.0	71.4	53.3	71.2 $\pm$ 8.4
MM	(%)	8.3	16.7	16.7	25.0	30.0	36.4	33.3	33.3	25.0	25.0	21.4	33.3	25.3 $\pm$ 8.5
MF	(%)	8.3	8.3	0	0	0	0	0	0	0	0	0	0	1.4 $\pm$ 3.2
SS	(%)	0	0	0	0	0	0	0	0	0	0	7.1	13.3	2.1 $\pm$ 4.2
Total units (N)		12	12	12	12	10	11	12	12	12	12	14	15	

[a]Data collection period October–December

P Pairs, *MM* Multi-male/single-female groups, *MF* Multi-female/single-male groups, *SS* Semi-solitary individuals spending time alone or associated with an established group witnessed. Thus, so far we can only confirm pairs and multi-male/single-female groups as reproductive units in the KY population.

Our knowledge of the complexity of social flexibility still continues to grow. Most recently, for example, we began to recognize yet another formerly unknown status of individuals. The surprising observation is that two males sometimes associate with a group and at other times spend long periods by themselves; we have termed this "semi-solitariness". The situation is radically different from "floating" commonly used to describe a period when an unmated individual seeks a mate following natal dispersal. In contrast, the two semi-solitary males are older and come from established multi-male groups. Whether they are searching for mates is unclear. For example, Frodo is a nearly 30 year-old male who was thought to have left his multi-male group permanently in 2007, after he was absent from the group for several months. In 2008, however, he re-appeared and occasionally traveled again with his former group. At first, we speculated that he was perhaps visiting while transitioning into another group; a phenomenon we have repeatedly witnessed with young adult males during the process of natal dispersal. Over the past 2 years, however, we realized that he sometimes foraged alone in the familiar home range. His periods alone lasted from a few hours to several days. He did not attempt to immigrate into or even contact a group other than his previous group. Interestingly, he could re-join this social group peacefully and was tolerated without signs of agonism by the resident male and female. At present, Frodo lives partly with a group and partly alone and thus is semi-solitary.

The second case of semi-solitariness concerns Cassius II, who is also at least 30 years old. In early 2010, his putative son secondarily dispersed into a neighboring group and shortly thereafter Cassius II also appeared in this group. Unlike Frodo, he either spends time with his former group or the neighboring group or is by himself. He commutes between the two groups primarily during intergroup encounters and presently shows a preference for staying in the overlap area between the two adjacent home ranges. From our observations, it seems that he travels temporarily with whichever group is in the overlap area and he rarely follows either group deeper into its home range. Like Frodo, his integration into both groups seems unproblematic, with his arrival often preceded by soft hoots and loud vocalizations, but without aggression. However, both of these semi-solitary males seem subordinate to the resident males in the groups they join because they do not call during duets and also otherwise behave like secondary males in multi-male groups (Barelli et al. 2008b).

Overall, semi-solitariness seems to be rare, although we believe previous cases might have passed unnoticed because we never expected individuals of the 12-year census period to live alone almost secretively, and our data collection has always focused on individuals in identifiable groups. In the past, phases of semi-solitariness, if they occurred, were categorized as "transitional" and thus did not make it into publication, even when we were not sure about the whereabouts of "transient individuals" until they reappeared in other groups. The reasons for semi-solitariness are unclear. Perhaps it is an alternative strategy to the subordinate, secondary status in multi-male groups, because both semi-solitary males are affiliated with multi-male/

single-female groups. The presence of mature sons in the neighborhood may also importantly influence flexible group membership in this population, but further speculation must await knowledge of kin relations.

The recent observations highlight the great importance of time depth in understanding social dynamics and evolutionary forces of male–male competition and female mate choice that shape reproductive strategies in primates, perhaps particularly in apes, who express an impressive range of behavioral flexibility (van Schaik et al. 2004). Interpretations of group dynamics would have been very different had our study ceased after 5 or 10 years. We illustrate this point with an example of known transitions in and out of study group "A" (Fig. 11.2), although the argument applies to the entire study population. At each 5-year interval, the group composition of several groups involved would have looked different and consequently would have been interpreted differently with regard to the social and mating system of the population (Table 11.2).

Finally, we can ask why this flexibility (particularly in forming small multi-male units) was not recognized in earlier studies of wild gibbons. Perhaps when social histories of individuals were not known well, all too often additional adult males were considered adult sons of a breeding pair. At KY, however, longitudinal records of many groups allowed us to detect the presence of multi-male/single-female groups.

11.3.2 Female Reproductive Strategies

Our understanding of reproductive strategies of white-handed gibbon females has undergone dramatic changes during the past two decades. Although they were initially thought to be passive and monogamous recipients of males' socio-sexual strategies, it is now clear that gibbon females actively pursue their own reproductive interests, just like other mammalian females who are pair-living or form small polyandrous groups (Griffith et al. 2002; Wolff and Macdonald 2004; Munshi-South 2007). Following a plethora of molecular studies of female reproductive strategies in pair-living birds, the classic concept of female sexual monogamy has been shattered in most pair-living species. Recent molecular genetics and endocrinology studies have changed the perception of female reproductive interests, to which white-handed gibbon females are no exception.

Primate females may generally gain from multiple mating. Polyandrous mating during fertile periods might increase the probability of conception (van Noordwijk and van Schaik 2000) or of having their offspring sired by males who produce the most competitive sperm (Small 1989; Dixson 1998). Copulating with many males may also function to confuse paternity, which is advantageous in species with a high risk of infanticide (Hrdy 1979; Nunn 1999; van Schaik et al. 2000). Moreover, if a female preferentially copulates with her social partner compared to other males, as we found for KY gibbons, she might additionally benefit from her mate's raised paternity probability because her mate will be the most likely protector should her

246　U.H. Reichard et al.

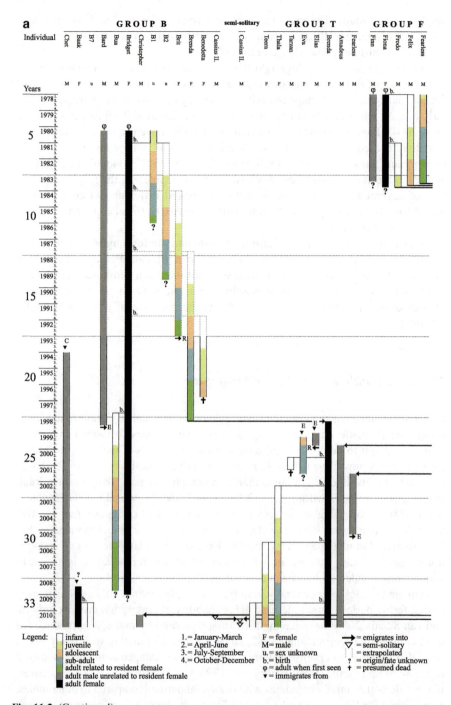

Fig. 11.2 (Continued)

11 White-Handed Gibbons of Khao Yai

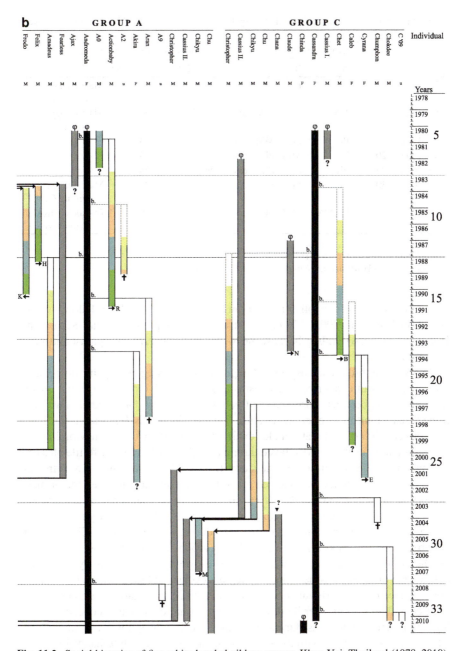

Fig. 11.2 Social histories of five white-handed gibbon groups, Khao Yai, Thailand (1978–2010)

Table 11.2 Social organization in five neighboring white-handed gibbon groups in 5-year intervals (1978–2010)

Study period	Years[a]	Group composition					Social organization		Event
		A	B	C	F	T	PL (%)	MM (%)	
1978–1982	5	PL	PL	PL	PL		100	0	Male change (group C)
1983–1987	10	MM	PL	MM	dis.		33	67	Male change (group A)
1988–1992	15	PL	PL	MM	dis.		67	33	
1993–1997	20	PL	MM	PL	dis.		67	33	
1998–2002	25	PL	PL	PL	dis.	MM	75	25	
2003–2007	30	MM	PL	PL	dis.	PL	75	25	Male change (group C)
2008–2010	33	PL/MM[b]	MM	PL	dis.	PL	50	50	Female change (groups B & C)

[a]Cumulative

[b]Occasionally joined by a semi-solitary male

PL Pair-living, *MM* Multi-male/single-female, *dis.* Dissolved

next infant be attacked by other males (van Schaik et al. 1999, 2004; Palombit et al. 2000; Buchan et al. 2003; Moscovice et al. 2009) or predators (van Schaik and Hörstermann 1994) and/or because he will defend a territory against intruders (Goldizen 2003).

Analyses of proximate aspects of reproductive strategies depend on reliable information about endocrine mechanisms and reproductive physiology that underlie interactions between hormonal and behavioral factors. Our studies have confirmed that monitoring ovarian function in wild gibbons is feasible (Barelli and Heistermann 2009), and that females exhibit behavioral and non-behavioral reproductive status cues that are displayed during both the fertile and non-fertile phase of the ovarian cycle. During a recent study (2003–2005), we found that although females' mating activity was skewed toward one preferred male (i.e., the primary male), half of the studied females ($N = 10$) lived in multi-male groups and each one also copulated with the second, subordinate male (i.e., the secondary male) in her group. Mating with a primary male increased during the fertile phase (Barelli et al. 2008b). Primary males in multi-male gibbon groups performed most copulations and had priority of access to fertile females. However, copulations by secondary males were distributed widely through female cycles, and these males had mating opportunities during periovulatory periods (Fig. 11.3). Copulating with both males even continued into non-fertile days of the menstrual cycle when conception was highly unlikely (as well as during pregnancy when conception was impossible), which contrasts strongly with the still widespread view that white-handed gibbons are socio-sexually monandrous and focused on single partners; instead, KY white-handed gibbon females are often sexually polyandrous (Barelli et al. 2008b, Reichard 2009).

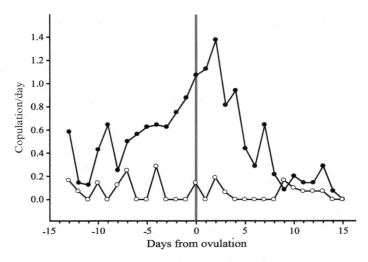

Fig. 11.3 Frequency of female copulation (number of copulations/day in which copulations occurred) with primary males (*black dots*) and secondary males (*white dots*) related to the day of ovulation (day '0'). Frequency of female copulation is calculated by first averaging the frequency for each female separately, and secondly across these individual values to yield a representative composite frequency that equally balanced individual contributions (see Barelli et al. 2008a, b)

Mating during pregnancy has also been suggested as a mechanism to confuse paternity and reduce the risk of infanticide in case of a male change (van Noordwijk and van Schaik 2000; van Schaik et al. 2000). It is noteworthy here that the only lactating females we have so far witnessed to become sexually active were two females who had just experienced male changes. While still carrying nursing infants, the females developed sexual swellings. In one case, the relationship with the new male was tense despite copulations. A few weeks after the male take-over, the female increasingly refused copulation attempts and stayed out of close proximity, and in the days prior to the disappearance of the female's infant some grappling and screaming was noticed. This female's sexual activity might have been a tactic to protect her suckling infant from harm by deceptively signaling receptivity to the new male. The anecdote is consistent with predictions of the sexual selection hypothesis for male infanticide (van Schaik et al. 2000): (1) The new male was unknown in the area and therefore can be assumed to have had a zero probability of having fathered the female's current offspring. (2) The new male had an increased chance of fathering the female's subsequent offspring because he remained with the female (and still is paired with her) and she mated with him during her subsequent cycle. (3) The female gave birth faster again than she would have had the infant survived. For this female, two of the three previous interbirth intervals between surviving infants were 3 years and one was 3 years and 8 months long, but the interbirth interval following the disappearance of the infant was only 2 years and 3 months.

11.3.2.1 Gibbon Sexual Swellings and Their Functional Significance

In strictly pair-living, monandrous females, sexual swellings are not expected to evolve (Nunn 1999) because male–male competition is low or absent and no selection pressure exists for females to advertise their fertile periods to a pair mate. However, gibbon females often mate polyandrously, and groups at KY frequently have two adult males that both maintain sexual relationships with the group female. It is thus not surprising that white-handed gibbons have sexual swellings (Barelli et al. 2007). These cyclical sexual swellings are admittedly small compared to the well-known, exaggerated sexual swellings of chimpanzees, baboons, and some macaques, but despite their modest size they follow the same physiological principles. Based on faecal progestogen profiles of 8 females over 15 menstrual cycles, we found that in 80% of cycles, ovulation overlapped tightly with the maximum swelling phase (duration: Ø 9.3 days; 42.8% of cycle length). In fact, the probability of ovulation peaked on average on day three of the maximum swelling period, although the timing between maximum swelling and probability of ovulation varied between days − 1 to day 13 of the swelling period and three times an ovulation fell outside the maximum swelling phase (Barelli et al. 2007). Thus, in analogy to sexual swelling patterns in primates living in multi-male social systems (Deschner et al. 2004; Engelhardt et al. 2005; Fürtbauer et al. 2010), KY gibbons also exhibit cyclical sexual swellings during their menstrual cycle that do not precisely indicate the day of ovulation.

To understand sexual swellings occurring outside the menstrual cycle better, we also tested five pregnant and six lactating females. Surprisingly, different swellings phases were noticeable also in pregnant females (and in similar proportions compared to cycling females), but not in lactating females, who were rarely swollen. We conclude that despite their smaller size, gibbons' sexual swellings probably serve functions similar to those suggested for primates with exaggerated swellings. In support of such an interpretation, female sexual activity corresponds with the size of the sexual swelling. Primary males, but not secondary males, copulate more frequently with cycling and pregnant females who are maximally swollen than with those females who are not or are only partially swollen (Barelli et al. 2008b).

Over the last 30 years, several hypotheses have been proposed to explain the evolution of conspicuous sexual swellings in species in which females mate with multiple males (reviewed in Zinner et al. 2004), whereas moderate or small sexual swellings have rarely been considered. Exaggerated swellings are hypothesized to increase paternity certainty, reliably advertise changes in female reproductive status ("obvious-ovulation hypothesis": Hamilton 1984), or provide information on female quality ("reliable-quality indicator hypothesis": Pagel 1994). They may also function to confuse paternity if ovulation does not precisely occur at peak swelling and thereby allow females to mate with multiple males when potentially fertile ("best-male hypothesis": Clutton-Brock and Harvey 1976; "many-male hypothesis": Hrdy 1981; Hrdy and Whitten 1987). Lastly, the "graded signal hypothesis" (Nunn 1999) posits that exaggerated swellings indicate the probability of ovulation, without allowing a male to precisely pinpoint the day of ovulation, thus giving a female more freedom to manipulate males' mating interests, particularly in species where males are larger than and dominant to females. Following this last hypothesis, the highest probability of ovulation should occur close to peak swelling size, but because of the prolonged duration of receptivity associated with a prolonged display of the sexual signal, females might mate with other males when ovulation is less likely but still possible (Nunn 1999). Our sexual swelling data on gibbons are in line with the graded signal hypothesis suggesting that it can also be applied to less conspicuous swellings (Barelli et al. 2008b; Reichard 2009). The occurrence of sexual swellings in gibbons (Nadler et al. 1993; Cheyne and Chivers 2006) may be related to the widespread flexibility in social organization revealed by recent research (Fuentes 2000; Lappan 2007a; Malone and White 2008; Reichard 2009).

Although "the graded signal hypothesis" offers the most comprehensive explanation for the patterns of sexual swellings, it does not explain the presence of sexual swellings during pregnancy and lactation. Developing a swelling during pregnancy may help maintain the male's sexual interest and mating activity, which can create paternity confusion and reduce the risk of infanticide (van Schaik et al. 1999; Engelhardt et al. 2005) and perhaps decrease a male's interest in EPCs and thereby allow the female to benefit from his permanent presence. Thus, flexible mating behavior and imprecise sexual swelling signals in wild gibbons are consistent with the theory of paternity confusion. Moreover, the clear association between sexual swelling size and copulation frequency supports our interpretation that the small swellings in gibbons attracts male sexual interest and are analogous to the exaggerated swellings of Old World monkeys and great apes.

11.4 White-Handed Gibbon Life History

Understanding primate life history strategies depends critically on data from wild, unprovisioned, and naturally reproducing populations (Brockman and van Schaik 2005; Cords and Chowdhury 2010; Fürtbauer et al. 2010). Our gibbon project has now reached a time-depth that allows us to begin to assess some life history variables. Perhaps knowledge of basic life history parameters should generally guide our perception of the time-depth of field studies because the number of generations contributing to a data set may be biologically more meaningful than the number of field seasons, a common proxy often used in relation to the labels "long-term" and "short-term" study. For example, a short-term study of a few years on a mouse lemur population represents a greater biological time-depth and perhaps sample size than a decade-long study of a few individuals of an orangutan population.

From a life-history perspective, our study is still in its infancy. Gibbons' adult group sizes are small and their life history is extremely slow for a primate of such small size (Ross 2004; Reichard and Barelli 2008). We still lack, for example, data on the maximum or even average lifespan. A few old individuals whom we have known for a long time have disappeared and probably died, but others who appear to be of similar age are still alive, and reliable, systematic birth records only exist for the population since the early 1990s (Reichard and Barelli 2008). To estimate a minimum age of the oldest adult females with unknown birth dates, we used long-term records of date of first appearance in the population and added to this the average years until first reproduction (see below). The data indicate that females may live to age 40 or older, although they tend to begin to "disappear" by this age, probably because they die (Table 11.3). The oldest female in our sample is alive at age 43 and some females continued to reproduce between 30 and 40 years of age, although the sample of females alive past 30 years of age is small. It is clear that wild gibbon females enjoy a long life span compared to other primates of similar mass (i.e., 5–6 kg). Unfortunately, females with known birth dates will still be nowhere near the end of their potential reproductive careers or lives at ages of 15–25 years, so our knowledge of female life histories is still incomplete (Table 11.3).

Data on age at first reproduction are available for five females, who gave birth for the first time on average at age 10.5 ± 1.2 years (range 8.4–12.8 years). We don't know the exact onset of menarche yet, but for two sub-adult females who displayed their first elongated vulva with a conspicuous mass of pink tissue at approximately 8 years of age (Hima: 8 years and 109 days; Rung: 8 years and 49 days; Barelli et al. 2007), no distinct cyclic pattern in progestogen levels (follicular and luteal components of the menstrual cycle) was detected by that

Table 11.3 Minimum age estimates of wild white-handed gibbon females at Khao Yai National Park, Thailand

Females with unknown birth date				Current status
Group	Female	Minimum age estimates (years)	Estimated age at last *or* most recent birth (years)	
A	Andromeda	43	40	Present
C	Cassandra	41	40	Disappeared
B	Bridget	39	28	Disappeared
N	Natasha	38	32	Disappeared
S	Sofi	36	35	Present
H	Hannah	32	29	Present
D	Daow	27	24	Present
R	Brit	27	22	Present
W	Wolga	25	25	Present
J	Jenna	23	18	Disappeared
NOS	Nasima	23	21	Present
Females with known birth date				
T	Brenda	25	24	Present
N	Hima	15	14	Present
M	Rung	14	12	Present

age. Two precisely known gestation periods of two females lasted 184 and 195 days respectively (Barelli et al. 2007), which is shorter than the commonly assumed 210 days gestation period for white-handed gibbons (van Tienhoven 1983). If we subtract gestation length from age at first birth we can conservatively measure sexual maturity to occur at the latest at the age of first conception, which occurred on average at the age of 10.0 ± 1.5 years (range 7.8–12.2 years, $N = 5$) in these females. An interesting difference existed among the five females because the female with the earliest onset of reproduction (8.4 years) was the only female still residing in her natal group. This group had two simultaneously breeding females for several months, because her mother had given birth a year earlier (see above), until the daughter's infant disappeared for unknown reasons. At the time of writing, the young female still resides with her natal group at age 9 years and 3 months. Pre-dispersal reproduction is exceptional in our population and most females disperse at the age of 7–8 years (Brockelman et al. 1998). Overall, white-handed gibbons at KY begin reproducing much later than monkeys of similar body mass and, remarkably, at about the same age as female mountain gorillas (Okamoto et al. 2000; Nakagawa et al. 2003; Wich et al. 2004; Hsu et al. 2006; Fürtbauer et al. 2010; Di Fiore et al. 2011).

Detailed data are also available for female interbirth intervals (IBI). The average population IBI between surviving offspring of habituated KY females ($N = 11$, 1990–2009) was 3.4 ± 0.7 years (range 34–71 months, $N = 22$ IBI). Adding one exceptional IBI of 14.4 years (173 months) increases the average IBI to 3.9 ± 0.4 years ($N = 23$ IBI). The one exceptionally long IBI was surprising because copulations were observed across most years. Prior to her most recent birth, the female was considered post-reproductive for 9.8 years according to Caro et al. (1995). The anecdote illustrates the danger of assessing female reproductive status behaviorally, which might be particularly misleading in long-lived apes. Death of a suckling infant significantly shortens birth intervals to an average 2.2 ± 0.7 years (range 11–36 months, $N = 9$ IBI; t-test: $t_{(29)} = 4.64$, $p < 0.001$), although great variation naturally exists in this measure because it depends on variables such as infant age at death or a mother's age or body condition. The shortest IBI recorded after an infant's death was 11 months, which meant that a female conceived only 3 months after she had lost an infant, and the longest was 3.1 years, which closely resembles the mean IBI in the population.

We can also calculate infant and juvenile mortality and the length of the juvenile period in our population. Infant mortality during the first year was 11.1% ($N = 54$ infants born) and until weaning it was 25.6% ($N = 43$ infants surviving from birth to weaning), which is moderate to low, compared to many other primates (Wright 1995; Boesch and Boesch-Achermann 2000; Robbins et al. 2004; Strier et al. 2006; Carter et al. 2008; Cords and Chowdhury 2010). Juvenile mortality between weaning and 5 years of age remained low at 8.8% ($N = 34$ weaned infants) but rose to 13.6% ($N = 22$ juveniles older than 5 years) if the period between weaning and dispersal is considered. Overall, the juvenile period in gibbons is very long. Considering that weaning occurs between 24 and 30 months (Treesucon 1984) and ends the latest with first conception (see above), female gibbons spend about 7 years as non-reproductive immatures.

11.5 Conclusions

The most dramatic change in our understanding of gibbons, as we see it, has been the shift from a socio-sexual monogamy model toward a dynamic community based-model in which individuals, although living in small social units and on small, group-specific home ranges, are connected to a much larger social sphere that involves permanent exchanges and interactions across core social unit's socio-spatial boundaries. Individuals call to each other in loud songs, they frequently meet in overlapping areas between group home ranges, and females visibly signal fertile periods to males in their vicinity with modest sexual swellings. Males seem to be more socially mobile than females, as predicted by sexual selection theory (Altmann 1990), given that they move more frequently between groups than females do. Females are more often the long-term occupants of home ranges and female take-overs of breeding groups usually involve younger females taking over the home range of old females whom they oust from the groups. Interestingly, so far we have not encountered an ousted female again, whereas ousted males frequently reappear in other groups and our long-term records show that some males transfer three and four times.

The dynamic community model is well suited to incorporate the recent wealth of unexpected findings that have emerged across gibbon taxa (Palombit 1994a, b; Malone and Oktavinalis 2006; Lappan 2007a, b; Lappan and Whittaker 2009). The social dynamics of gibbon communities are also in line with new findings of female reproductive strategies. Females mate polyandrously (Barelli et al. 2008b; Reichard 2009) and their moderate sexual swellings (Barelli et al. 2007) probably allow them to increase male–male competition to achieve EPCs, and more broadly to manipulate male sexual activities, all of which may benefit their own reproductive interests. However, reproductive strategies will not be fully understood until we have molecular paternity data.

References

Albers HJ, Grinspoon E (1997) A comparison of the enforcement of access restrictions between Xishuangbanna Nature Reserve (China) and Khao Yai National Park (Thailand). Environ Conserv 24:351–362

Altmann J (1990) Primate males go where the females are. Anim Behav 39:193–195

Asensio N, Brockelman WY, Malaivijitnond S, Reichard UH (2011) Gibbon travel paths are goal oriented. Anim Cogn 14:395–405

Barelli C, Heistermann M (2009) Monitoring female reproductive status in white-handed gibbons (Hylobates lar) using fecal hormone analysis and patterns of genital skin swellings. In: Lappan

S, Whittaker DJ (eds) The gibbons: new perspectives on small ape socioecology and population biology. Springer, Berlin, pp 313–325

Barelli C, Heistermann M, Boesch C, Reichard UH (2007) Sexual swellings in wild white-handed gibbon females (*Hylobates lar*) indicate the probability of ovulation. Horm Behav 51:221–230

Barelli C, Boesch C, Heistermann M, Reichard UH (2008a) Female white-handed gibbons (*Hylobates lar*) lead group movements and have priority of access to food resources. Behaviour 145:965–981

Barelli C, Heistermann M, Boesch C, Reichard UH (2008b) Mating patterns and sexual swellings in pair-living and multimale groups of wild white-handed gibbons, *Hylobates lar*. Anim Behav 75:991–1001

Bartlett TQ (2009a) The gibbons of Khao Yai: seasonal variation in behavior and ecology. Pearson Prentice Hall, Upper Saddle River, NJ

Bartlett TQ (2009b) Seasonal home range use and defendability in white-handed gibbons (*Hylobates lar*) in Khao Yia National Park, Thailand. In: Lappan S, Whittaker DJ (eds) The gibbons: new perspectives on small ape socioecology and population biology. Springer, Berlin, pp 265–275

Boesch C, Boesch-Achermann H (2000) The chimpanzees of the Taï forest: behavioural ecology and evolution. Oxford University Press, Oxford

Boonpragob K, Homchantara N, Coppins BJ, McCarthy PM, Wolseley PA (1998) An introduction to the lichen flora of Khao Yai National Park, Thailand. Bot J Scotland 50:209–219

Brockelman WY (2009) Ecology and the social system of gibbons. In: Lappan S, Whittaker DJ (eds) The gibbons: new perspectives on small ape socioecology and population biology. Springer, Berlin, pp 211–239

Brockelman WY, Reichard U, Treesucon U, Raemaekers JJ (1998) Dispersal, pair formation and social structure in gibbons (*Hylobates lar*). Behav Ecol Sociobiol 42:329–339

Brockman DK, van Schaik CP (2005) Seasonality and reproductive function. In: Brockman DK, van Schaik CP (eds) Seasonality in primates: studies of living and extinct human and non-human primates. Cambridge University Press, Cambridge, pp 269–306

Brodie JF, Helmy OE, Brockelman WY, Maron JL (2009) Bushmeat poaching reduces the seed dispersal and population growth rate of a mammal-dispersed tree. Ecol Appl 19:854–863

Buchan JC, Alberts SC, Silk JB, Altmann J (2003) True paternal care in a multi-male primate society. Nature 425:179–181

Caro TM, Sellen DW, Parish A, Frank R, Brown DM, Voland E, Borgerhoff Mulder M (1995) Termination of reproduction in nonhuman and human female primates. Int J Primatol 16:205–220

Carter ML, Pontzer H, Wrangham RW, Peterhans JK (2008) Skeletal pathology in *Pan troglodytes schweinfurthii* in Kibale National Park, Uganda. Am J Phys Anthropol 135:389–403

Cheyne SM, Chivers DJ (2006) Sexual swellings of female gibbons. Folia Primatol 77:345–352

Clutton-Brock TH, Harvey PH (1976) Evolutionary rules and primates societies. In: Bateson PPG, Hinde RA (eds) Growing points in ethology. Cambridge University Press, Cambridge, pp 195–237

Cords M, Chowdhury S (2010) Life history of *Cercopithecus mitis stuhlmanni* in the Kakamega Forest, Kenya. Int J Primatol 31:433–455

Deschner T, Heistermann M, Hodges K, Boesch C (2004) Female sexual swelling size, timing of ovulation, and male behavior in wild West African chimpanzees. Horm Behav 46:204–215

Di Fiore A, Link A, Campbell CJ (2011) The Atelines: behavioral and socioecological diversity in a New World radiation. In: Campbell CJ, Fuentes A, MacKinnon KC, Bearder SK, Stumpf RM (eds) Primates in perspective, 2nd edn. Oxford University Press, Oxford, pp 155–188

Dixson AF (1998) Primate sexuality: comparative studies of the prosimians, monkeys, apes and human beings. Oxford University Press, Oxford

Engelhardt A, Hodges JK, Niemitz C, Heistermann M (2005) Female sexual behavior, but not sex skin swelling, reliably indicates the timing of the fertile phase in wild long-tailed macaques (*Macaca fascicularis*). Horm Behav 47:195–204

Fuentes A (2000) Hylobatid communities: changing views on pair bonding and social organization in hominoids. Yrbk Phys Anthropol 43:33–60

Fürtbauer I, Schülke O, Heistermann M, Ostner J (2010) Reproductive and life history parameters of wild female *Macaca assamensis*. Int J Primatol 31:501–517

Gale GA, Round PD, Pierce AJ, Nimnuan S, Pattanavibool A, Brockelman WY (2009) A field test of distance sampling methods for a tropical forest bird community. Auk 126:439–448

Goldizen AW (2003) Social monogamy and its variations in callitrichids: do these relate to the costs of infant care? In: Reichard UH, Boesch C (eds) Monogamy: mating strategies and partnerships in birds, humans and other mammals. Cambridge University Press, Cambridge, pp 232–247

Goodall J (1986) The chimpanzees of Gombe: patterns of behavior. Belknap, Cambridge, MA

Grassman LI Jr, Tewes ME, Silvy NJ, Kreetiyutanont K (2005) Ecology of three sympatric felids in a mixed evergreen forest in north-central Thailand. J Mammal 86:29–38

Griffith SC, Owens IPF, Thuman KA (2002) Extra-pair paternity in birds: a review of interspecific variation and adaptive function. Mol Ecol 11:2195–2212

Hamilton WJ III (1984) Significance of paternal investment by primates to the evolution of adult male-female associations. In: Taub DM (ed) Primate paternalism. Van Nostrand Reinhold, New York, pp 309–335

Hrdy SB (1979) Infanticide among animals: a review, classification, and examination of the implications for the reproductive strategies of females. Ethol Sociobiol 1:13–40

Hrdy SB (1981) The women that never evolved. Harvard University Press, Cambridge, MA

Hrdy SB, Whitten PL (1987) Patterning of sexual activity. In: Smuts BB, Cheney DL, Seyfarth RM, Wrangham RW, Struhsaker TT (eds) Primate societies. University of Chicago Press, Chicago, pp 370–384

Hsu MY, Lin JF, Agoramoorthy G (2006) Effects of group size on birth rate, infant mortality and social interactions in Formosan macaques at Mt Longevity, Taiwan. Ethol Ecol Evol 18:3–17

Jensen A, Meilby H (2010) Returns from harvesting a commercial non-timer forest product and particular characteristics of harvesters and their strategies: *Aquilaria crassna* and Agarwood in Lao PDR. Econ Bot 64:34–45

Kitamura S, Yumoto T, Poonswad P, Noma N, Chuailua P, Plongmai K, Maruhashi T, Suckasam C (2004) Pattern and impact of hornbill seed dispersal at nest trees in a moist evergreen forest in Thailand. J Trop Ecol 20:545–553

Kitamura S, Suzuki S, Yumoto T, Chuailua P, Plongmai K, Poonswad P, Noma N, Maruhashi T, Suckasam C (2005) A botanical inventory of a tropical seasonal forest in Khao Yai National Park, Thailand: implications for fruit-frugivore interactions. Biodiv Conserv 14:1241–1262

Kitamura S, Yumoto T, Noma N, Chuailua P, Maruhashi T, Wohandee P, Poonswad P (2008) Aggregated seed dispersal by wreathed hornbills at a roost site in a moist evergreen forest of Thailand. Ecol Res 23:943–952

Lappan S (2007a) Social relationships among males in multimale siamang groups. Int J Primatol 28:369–387

Lappan S (2007b) Patterns of dispersal in Sumatran siamangs (*Symphalangus syndactylus*): preliminary mtDNA evidence suggests more frequent male than female dispersal to adjacent groups. Am J Primatol 69:692–698

Lappan S, Whittaker DJ (2009) The gibbons: new perspectives on small ape socioecology and population biology. Springer, Berlin

Lynam AJ, Round PD, Brockelman WY (2006) Status of birds and large mammals in Thailand's Dong Phayayen - Khao Yai Forest Complex. Biodiversity Research and Training (BRT) Program and Wildlife Conservation Society, Bangkok

Malone NM, Oktavinalis H (2006) The socioecology of the silvery gibbon (*Hylobates moloch*) in the Cagar Alam Leuweung Sanchang (CALS), West Java, Indonesia. Am J Phys Anthropol 129 (suppl 42):124

Malone NM, White FJ (2008) The socioecology of Javan gibbons *(Hylobates moloch)*: tests of competing hypotheses. Am J Phys Anthropol 135(suppl 46):148

Moscovice LR, Heesen M, Di Fiore A, Seyfarth RM, Cheney DL (2009) Paternity alone does not predict long-term investment in juveniles by male baboons. Behav Ecol Sociobiol 63:1471–1482

Munshi-South J (2007) Extra-pair paternity and the evolution of testis size in a behaviorally monogamous tropical mammal, the large treeshrew (*Tupaia tana*). Behav Ecol Sociobiol 62:201–212

Nadler RD, Dahl JF, Collins DC (1993) Serum and urinary concentrations of sex hormones and genital swelling during the menstrual cycle of the gibbon. J Endocrinol 136:447–455

Nakagawa N, Ohsawa H, Muroyama Y (2003) Life-history parameters of a wild group of West African patas monkeys (*Erythrocebus patas patas*). Primates 44:281–290

Nunn CL (1999) The evolution of exaggerated sexual swellings in primates and the graded-signal hypothesis. Anim Behav 58:229–246

Okamoto K, Matsumura S, Watanabe K (2000) Life history and demography of wild moor macaques (*Macaca maurus*): summary of 10 years of observations. Am J Primatol 52:1–11

Pagel M (1994) The evolution of conspicuous oestrous advertisement in Old World monkeys. Anim Behav 27:1333–1341

Palombit RA (1994a) Dynamic pair bonds in hylobatids: implications regarding monogamous social systems. Behaviour 128:65–101

Palombit RA (1994b) Extra-pair copulations in a monogamous ape. Anim Behav 47:721–723

Palombit RA, Cheney DL, Fischer J, Johnson S, Rendall D, Seyfarth RM, Silk JB (2000) Male infanticide and defense of infants in chacma baboons. In: van Schaik CP, Janson CH (eds) Infanticide by males and its implications. Cambridge University Press, Cambridge, pp 123–152

Perry S, Manson JH (2008) Manipulative monkeys: the capuchins of Lomas Barbudal. Harvard University Press, Cambridge, MA

Raemaekers JJ, Raemaekers PM (1985) Field playback of loud calls to gibbons (*Hylobates lar*): territorial, sex-specific and species-specific responses. Anim Behav 33:481–493

Raemaekers JJ, Raemaekers PM, Haimoff EH (1984) Loud calls of the gibbon (*Hylobates lar*): repertoire, organisation and context. Behaviour 91:146–189

Reichard UH (1991) Zum Sozialverhalten einer Gruppe freilebender Weißhandgibbons (*Hylobates lar*). MSc thesis, Göttingen University, Göttingen

Reichard UH (1995) Extra-pair copulation in a monogamous gibbon (*Hylobates lar*). Ethology 100:99–112

Reichard UH (1998) Sleeping sites, sleeping places, and presleep behavior of gibbons (*Hylobates lar*). Am J Primatol 46:35–62

Reichard UH (2003) Social monogamy in gibbons: the male perspective. In: Reichard UH, Boesch C (eds) Monogamy: mating strategies and partnerships in birds, humans and other mammals. Cambridge University Press, Cambridge, pp 190–213

Reichard UH (2009) Social organization and mating system of Khao Yai white-handed gibbons, 1992–2006. In: Lappan S, Whittaker DJ (eds) The gibbons: new perspectives on small ape socioecology and population biology. Springer, Berlin, pp 347–384

Reichard UH, Barelli C (2008) Life history and reproductive strategies of Khao Yai *Hylobates lar*: implications for social evolution in apes. Int J Primatol 29:823–844

Reichard UH, Sommer V (1997) Group encounters in wild gibbons (*Hylobates lar*): agonism, affiliation, and the concept of infanticide. Behaviour 134:1135–1174

Robbins MM, Bermejo M, Cipolletta C, Magliocca F, Parnell RJ, Stokes E (2004) Social structure and life-history patterns in western gorillas (*Gorilla gorilla gorilla*). Am J Primatol 64:145–159

Ross C (2004) Life histories and the evolution of large brain size in great apes. In: Russon A, Begun DR (eds) The evolution of thought: evolutionary origins of great ape intelligence. Cambridge University Press, Cambridge, pp 122–139

Round PD, Gale GA (2008) Changes in the status of *Lophura* pheasants in Khao Yai National Park, Thailand: a response to warming climate? Biotropica 40:225–230

Singhrattna N, Rajagopalan B, Kumar KK, Clark M (2005) Internannual and interdecadal variability of Thailand summer monsoon season. J Climate 18:1697–1708

Small MF (1989) Female choice in nonhuman primates. Yrbk Phys Anthropol 32:103–127

Smitinand T (1977) Plants of Khao Yai National Park. New Thammada Press Ltd., Bangkok

Smitinand T (1989) Thailand. In: Campbell DG, Hammond HD (eds) Floristic inventory of tropical countries: the status of plant systematic, collections, and vegetation, plus recommendations for the future. The New York Botanical Gardens, New York, pp 63–82

Sommer V, Reichard U (2000) Rethinking monogamy: the gibbon case. In: Kappeler PM (ed) Primate males: causes and consequences of variation in group composition. Cambridge University Press, Cambridge, pp 159–168

Strier KB, Dib LT, Figueira JEC (2002) Social dynamics of male muriquis (*Brachyteles arachnoides hypoxanthus*). Behaviour 139:315–342

Strier KB, Boubli JP, Possamai CB, Mendes SL (2006) Population demography of northern muriquis (*Brachyteles hypoxanthus*) at the Estação Biológica de Caratinga/Reserva particular do Patrimônio Natural-Felìciano Miguel Abdala, Minas Gerais, Brazil. Am J Phys Anthropol 130:227–237

Tangtam N (1992) Khao Yai ecosystem: the hydrological role of Khao Yai National Park. In: Proceedings of the International Workshop on Conservation and Sustainable Development, AIT/Bangkok and Khao Yai National Park, Bangkok, Thailand, 22–26 April 1991. Asian Institute of Technology, Bangkok, pp 345–363

Treesucon U (1984) Social development of young gibbons (*Hylobates lar*) in Khao Yai National Park, Thailand. MSc thesis, Mahidol University, Bangkok

van Noordwijk MA, van Schaik CP (2000) Reproductive patterns in eutherian mammals: adaptations against infanticide? In: van Schaik CP, Janson CH (eds) Infanticide by males and its implications. Cambridge University Press, Cambridge, pp 322–360

van Schaik CP, Hörstermann M (1994) Predation risk and the number of adult males in a primate group: a comparative test. Behav Ecol Sociobiol 35:261–272

van Schaik CP, van Noordwijk MA, Nunn CL (1999) Sex and social evolution in primates. In: Lee PC (ed) Comparative primate socioecology. Cambridge University Press, Cambridge, pp 204–240

van Schaik CP, Hodges JK, Nunn CL (2000) Paternity confusion and the ovarian cycles of female primates. In: van Schaik CP, Janson CH (eds) Infanticide by males and its implications. Cambridge University Press, Cambridge, pp 361–387

van Schaik CP, Preuschoft S, Watts DP (2004) Great ape social systems. In: Russon AE, Begun DR (eds) The evolution of thought: evolutionary origins of great ape intelligence. Cambridge University Press, Cambridge, pp 190–209

van Tienhoven A (1983) Reproductive physiology of vertebrates. Cornell University Press, Ithaca, NY

Watts DP (2002) Reciprocity and interchange in the social relationships of wild male chimpanzees. Behaviour 139:343–370

Wells RS (1991) The role of long-term study in understanding the social structure of a bottle nose dolphin community. In: Pryor K, Norris KS (eds) Dolphin societies: discoveries and puzzles. University of Berkley Press, Berkley, pp 199–226

Wich SA, Utami-Atmoko SS, Mitra Setia T, Rijksen HD, Schürmann C, van Hooff JARAM, van Schaik CP (2004) Life history of wild Sumatran orangutans (*Pongo abelii*). J Hum Evol 47:385–398

Wolff JO, MacDonald DW (2004) Promiscuous females protect their offspring. Trends Ecol Evol 19:127–134

Wright PC (1995) Demography and life history of free-ranging *Propithecus diadema edwardsi* in Ranomafana National Park, Madagascar. Int J Primatol 16:835–854

Zhang L, Brockelman WY, Allen MA (2008) Matrix analysis to evaluate sustainability: the tropical tree *Aquilaria crassna*, a heavily poached source of agarwood. Biol Conserv 141:1676–1686

Zinner DP, Nunn CL, van Schaik CP, Kappeler PM (2004) Sexual selection and exaggerated sexual swellings of female primates. In: Kappeler PM, van Schaik CP (eds) Sexual selection in primates: new and comparative perspectives. Cambridge University Press, New York, pp 71–89

Part V
Africa

Chapter 12
The Amboseli Baboon Research Project: 40 Years of Continuity and Change

Susan C. Alberts and Jeanne Altmann

Abstract In 1963, Jeanne and Stuart Altmann traveled through Kenya and Tanzania searching for a baboon study site. They settled on the Maasai-Amboseli Game Reserve (later Amboseli National Park) and conducted a 13-month study that laid the groundwork for much future research. They returned for a short visit in 1969, and came again in July 1971 to establish a research project that has persisted for four decades. In July 1984 Susan Alberts joined the field team, later becoming a graduate student and eventually a director. Over the years, we have tackled research questions ranging from feeding ecology to behavioral endocrinology, from kin recognition to sexual selection, and from aging research to functional genetics. A number of our results have explicitly depended upon the longitudinal nature of the research. Without decades worth of individual-based data we would not have known, for instance, that the presence of fathers influenced the maturation rates of their offspring, that maternal dominance rank had pervasive effects on the physiology of sons, or that the social behavior of a female influenced her infants' survival. Here we summarize the major research themes that have characterized each of the past four decades, and our directions for the future, emphasizing the scientific insights that the longitudinal nature of the study has made possible.

S.C. Alberts (✉)
Department of Biology, Duke University, Durham, NC, USA
e-mail: alberts@duke.edu

J. Altmann
Department of Ecology and Evolutionary Biology, Princeton University, Princeton, NJ, USA
e-mail: altj@princeton.edu

P.M. Kappeler and D.P. Watts (eds.), *Long-Term Field Studies of Primates*,
DOI 10.1007/978-3-642-22514-7_12, © Springer-Verlag Berlin Heidelberg 2012

12.1 Introduction

In 1963, Jeanne and Stuart Altmann took a reconnaissance trip through Kenya and Tanzania, searching for the best site to study baboons, genus *Papio* (Altmann and Altmann 1970). They traveled in a Land Rover with their infant son, Michael, in the back, and they visited six national parks and reserves. One stood out to them as ideal: the Maasai-Amboseli Game Reserve, later Amboseli National Park, in southern Kenya immediately north and west of Mt. Kilimanjaro. The reserve had thousands of baboons, was dominated by an acacia tree woodland interspersed with areas of open grassland, and had a full complement of herbivores and predators. The fever tree woodlands were full of wildlife, visibility was very good (Fig. 12.1), and it was possible to drive through most of the terrain (often a problem in other parks and reserves because of thick vegetation or rocks). In addition, the yellow baboons, *Papio cynocephalus*, that inhabited the reserve neither approached nor fled from humans, as in other areas with high baboon populations. The Altmanns carried out a 13-month study, describing in detail the baboons' demographic structure, social system, ranging patterns, and feeding ecology (Altmann and Altmann 1970). With other studies of wild baboons that were being carried out in this decade, the groundwork was laid for baboons to become one of the best-studied nonhuman primates in the wild (e.g., Hall and DeVore 1965; Kummer 1968; Stoltz and Saayman 1970).

The Altmanns left Kenya in 1964, and stayed away for 5 years. When they returned for a visit in 1969, they encountered a vastly changed landscape in Amboseli, most strikingly characterized by a dramatic decline of the fever tree

Fig. 12.1 Baboons in Amboseli, Kenya. Visibility is very good in the ecosystem, and after four decades of research the animals are well habituated to the presence of neutral human observers. Photo © Susan Alberts

woodlands, a pattern that has continued since then. Several factors probably contributed to this woodland decline, possibly including natural aging of the woodlands, pastoralist grazing patterns and associated burning, rising water table with a concomitant rise in a salt layer in the soil and, increasingly, the impact of an growing elephant population that both kills trees and prevents woodland regeneration (Western and Sindiyo 1972; Struhsaker 1973; Western and van Praet 1973; Western and Maitumo 2004; Western 2007). The Altmanns described "walking around in shock" for the first several days of their return, confirming that the baboon population had decreased dramatically since 1964. They repeatedly re-censused the baboon population beginning in 1969, eventually confirming a drop in population size of more than 90% from the early 1960s (Altmann et al. 1985).

In spite of the changes, Amboseli remained a wonderful place to watch baboons. The baboon population was small, but was stabilizing in spite of the population decline (Altmann et al. 1985), the terrain was manageable, and the visibility was still excellent. In July 1971, the Altmanns and Ph.D. student Glenn Hausfater began observing Alto's Group, which consisted of 35 members, thus establishing a research project that has, against the odds, persisted for 4 decades.

Another study group (Hook's Group) was added to intensive study in 1980 with Michael Pereira's Ph.D. research, roughly doubling the size of the population under study. In 1984, in work led by Amy Samuels and by Phillip Muruthi, then a B.Sc. student at the University of Nairobi, we also began monitoring a social group living near a tourist lodge that subsisted partially on the lodge's refuse. Observations of Lodge Group lasted only 12 years (through 1996), but greatly enriched our understanding of the flexibility of baboon behavior and life history by demonstrating that baboons can substantially accelerate their life histories, growth rates, and reproductive rates when nutritional conditions are good, and can adapt behaviorally and ecologically to a wide range of environmental conditions (e.g., Altmann and Muruthi 1988; Muruthi et al. 1991; Altmann and Alberts 2005).

The subset of the Amboseli baboon population that is under intensive study (hereafter the "study population") currently consists of ~350 extant animals (with life history and behavioral data on an additional ~1,000 that have died) in six social groups. All six current study groups are either fission products of the two original groups (Alto's, which fissioned in 1990–1991, and Hook's, which fissioned in 1995) or are fission products of daughter groups of Alto's (Dotty's Group, which fissioned in 1999, or Nyayo's Group, where a fission is in process in 2011 at this writing; Fig. 12.2 and see also Altmann and Alberts 2003; Van Horn et al. 2007). All members of the study population subsist entirely on wild foods and experience no human management (Lodge Group is no longer a study group; instead, we monitor its demography opportunistically, as we do ~8 other non-study groups in the Amboseli basin).

The Amboseli study population is part of an extensive, continuous baboon population that occupies a large part of eastern Africa. From the perspective of baboons, the Amboseli basin is locally bounded on the west and the north by inhospitable stretches of land, but males move to and from social groups living to the south and to the east of the basin. Thus, the study population experiences gene

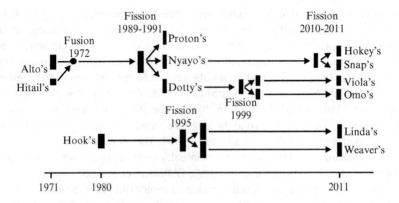

Fig. 12.2 Timeline of research on wild-feeding study groups in Amboseli. Research began on Alto's Group in 1971 and on Hook's Group in 1980. Four subsequent natural group fissions produced the 2011 study groups. Proton's Group (one product of Alto's fission) was dropped from the study in 1993 for logistical reasons

flow with the surrounding local population and has a large estimated effective population size (Storz et al. 2002). The baboon population in the Amboseli basin appears to be demographically stable or moderately growing, although it still persists at a level far below that of the early 1960s (Altmann et al. 1985; Samuels and Altmann 1991).

Today the Amboseli Baboon Research Project (ABRP) is a longitudinal, coordinated series of studies of baboons in the Amboseli region with a broad range of research interests. It has supported 23 Ph.D. or M.Sc. theses, including six from Kenyan students, with another six in progress at this writing. We still rely heavily on observational techniques that have changed little since 1971, although we have also employed key technological innovations (some of which are described below) that have enriched the basic behavioral and ecological data enormously. In addition, we have invested heavily in data preservation and accessibility by developing our database, BABASE (see Sect. 12.4). This has allowed us to readily engage in multi-decade and lifespan analyses as well as shorter term cross-sectional studies.

Since the mid 1990s, the longitudinal data collection (demographic, behavioral, and ecological) has depended heavily on our dedicated and highly skilled Kenyan field team. Raphael Mututua began working for the project in 1981 as a field assistant; he developed into an outstanding observer and researcher and now fills the essential role of the project's field manager. His leadership in the field is of the highest quality. Serah Sayialel (beginning in 1989) and Kinyua Warutere (beginning in 1995) fill out our senior Kenyan research team. The project would be unimaginable without them (Fig. 12.3). Several other staff members (Gideon Marinka, Benard Ochieng Oyath, and Longida Siodi) play important roles as field assistants to Mututua, Sayialel, Warutere, and visiting researchers.

Here we summarize some of the major research themes that the Amboseli Baboon Research Project has tackled since its inception. This review is by no means exhaustive; it is not possible to include all of the more than 200 publications

Fig.12.3 Raphael Mututua (*left*), field manager of the Amboseli Baboon Research Project, observing baboons with senior Kenyan team members Serah Sayialel (*center*) and Kinyua Warutere (*right*). Photo © Jeanne Altmann

on the Amboseli baboons here (see a full list of publications on the Amboseli Baboon Research Project website; http://www.princeton.edu/~baboon/). Instead, our goal is to highlight the major themes that have informed, and been informed by, the long-term and longitudinal nature of the study.

12.2 The 1970s: Foundations

Several research projects in the 1970s focused on the question of how the baboons made a living in a semi-arid environment that received only an average of ~350 mm of rain per year. The answer was: "with difficulty". These early studies showed that the baboons spent a large fraction of their time foraging – that is, moving and feeding (e.g., Altmann 1980; Post 1981). Two studies in particular suggested that the nutrition obtained by the baboons during natural foraging was minimally sufficient for growth and reproduction at a rate that would support population replacement. Specifically, Jeanne Altmann showed that mothers spent as much time feeding as they possibly could by the time their infants were 5–6 months of age, and that even when maternal feeding time was at a maximum, maintaining their own body weight would be difficult (Altmann 1980). She inferred that this constraint essentially forced the infants to begin the transition to nutritional independence at this age (Altmann 1980; see also Altmann and Samuels 1992). Further, her calculations indicated that wild baboons were subsisting on a diet that, in captivity, was considered so dangerously inadequate that experimental diets that approximated the Amboseli baboons' diet were discontinued because of health concerns for the animals (Altmann 1980). In a second, concurrent study, Stuart Altmann showed that most infants, during the transition to nutritional independence, fell well below the optimal diet that he predicted they could obtain (Altmann 1991, 1998). Decades after the infant data were collected, he was able to use the life history data that had accumulated on the female infants in his study to show that their nutritional shortfall during infancy predicted, with astonishing accuracy, their future reproductive success; such a result was only possible because of the longitudinal and long-term nature of the study (Altmann 1991, 1998). These studies of feeding ecology were important in setting the stage for future work on this population, because they framed a major component of the underlying ecological problem the baboons had to solve (the other major ecological problem, predation, provides a second, but harder to study, theme that we have not yet investigated thoroughly in Amboseli; see Alexander 1974; van Schaik 1983). Ecology provided the context in which all other aspects of behavior and life history are played out.

A second important foundation laid in the 1970s was behavioral. One of the most obvious features of baboon society, evident to any observer who spends much time watching baboons, is the agonistic interactions that regularly occur between individuals. While many less obvious features of baboon society are equally important, there is no question that the dominance relations that arise out of these agonistic interactions have a profound and pervasive influence on many aspects of baboon life. One of the first studies that the Altmanns and Hausfater carried out was a careful analysis of aggressive and submissive interactions that occurred between individuals, and the subsequent construction of dominance hierarchies based on the direction of "wins" in these interactions.

During his yearlong study, Hausfater and the Altmanns developed the system of assigning dominance ranks in Amboseli that is still used by the research project today (Hausfater 1975). With respect to male mating success, Hausfater found that higher-ranking males enjoyed greater mating success than lower ranking males, but the relationship was not perfect and the highest-ranking male in his study did not experience the highest mating success. Consequently, Hausfater's study did not provide a definitive answer to the question "what is the importance of high rank for male primates?" This question preoccupied primate researchers for several decades to come (e.g., Strum 1982; Cowlishaw and Dunbar 1991; Kutsukake and Nunn 2006; Port and Kappeler 2010; Alberts in press). However, Hausfater's study brought to the fore a model of dominance-based priority of access that Stuart Altmann had developed years before during research on rhesus monkeys (Altmann 1962). The priority-of-access model – which posits that dominance rank functions as a queue in which males wait for reproductive opportunities – still provides an important "null model" for how dominance rank works to enhance male reproductive success in multi-male social groups. It has provided measures of the relative importance of dominance rank in male reproductive success in numerous social systems (extending beyond primates), and departures from the predictions of the model have provided important insights about how factors other than dominance rank affect male reproductive success (e.g., Boesch et al. 2006; Kappeler and Port 2008; Ostner et al. 2011; Surbeck et al. 2011).

The 1970s also saw the publication of Jeanne's landmark 1974 paper on behavioral sampling methods (Altmann 1974). The paper provided methodological guidance for behavioral studies of all kinds, and was motivated by a longstanding recognition of the need for data collection methods suited to answering questions through quantitative, model-testing analyses. It was also, to a considerable extent, grounded in Jeanne's experience studying the Amboseli baboons.

By the end of the 1970s, a longitudinal, individual-based perspective had emerged among the Amboseli researchers (especially Jeanne, whose Ph.D. research on mothers and infants had impressed upon her the importance of a life history perspective on behavior). During this time period, the Altmanns and Hausfater formalized many aspects of data collection and wrote the first version of the standardized protocols used by the project, the Monitoring Guide for the Amboseli Baboon Research Project (Altmann and Alberts 2004). The intention was that participating researchers would contribute to the longitudinal data as well as collecting data for their own short-term projects and benefitting from the monitoring data gathered by others in previous years. The current version of the Monitoring Guide (Altmann and Alberts 2004) has this to say about the longitudinal data collection and the manner in which visiting researchers contribute to it:

> Of the data sets described in this guide, some (i.e., demographic data) have been ongoing since 1971, while others extend back for somewhat shorter periods. Almost all data types that we currently collect extend back to at least 1980. Still other types of data were collected for shorter periods of time and are no longer a focus of our research efforts. Whatever the data set, the value of the data collected at the Project lies in its consistency and in its consistently high quality across time. This guidebook ... is meant as a guide for the permanent staff in Amboseli, for short-term visitors to the Project, and for visiting researchers pursuing their own projects (Ph.D. students, post-docs and other collaborators that stay long enough to learn the baboon ID's and contribute to the long-term data). It is absolutely essential that everyone who contributes to the Project's data set collect the data in accordance with the guidelines laid out here. Visiting researchers will collect additional data for their own specific research questions, which will extend beyond the monitoring data described here; these visiting researchers will still contribute to the monitoring data collection that is described in these procedures.
>
> When you contribute to the data of the Amboseli Baboon Research Project, you are contributing to a data set that we believe is unique in its time depth, breadth and detail. It is important to us that you take this responsibility very seriously. Never be satisfied with your data collection; always strive for more data of higher quality.

12.3 The 1980s: Females

When Jeanne Altmann began her landmark study on the ecology of motherhood and infancy in the mid-1970s, the topics of motherhood, and of female behavior in general, were not seen as a particularly important focus for the primate studies that were emerging at the time. Although infant development and maternal care were intensively studied among developmental psychologists in laboratory settings and a few initial descriptions were made in field studies, the questions posed were primarily about social and cognitive growth, with a view to understanding human development. However, change was in the air, and a number of scientists studying wild primates were beginning to pose important questions about females and their infants (e.g., Jay 1963; Hrdy 1977; Seyfarth 1977; Pusey 1978; Fossey 1979).

Jeanne's study of baboon mothers and infants in Amboseli (Altmann 1980) emphasized an evolutionary perspective on behavior and highlighted how important the evolutionary ecology of motherhood and infancy are for understanding a species' behavior and ecology. It also highlighted the importance of ecological constraints on social behavior, a topic of growing interest throughout the 1970s (e.g., Crook and Gartlan 1966; Clutton-Brock and Harvey 1977; Emlen and Oring 1977). In addition, the study was grounded in the notion, unusual at the time but now widespread in studies of evolutionary ecology, that viewing behavior in the context of life history is important, in particular for understanding how the things an animal does today are both shaped by and shape its reproductive and social trajectory.

Female dominance rank also came under scrutiny in Amboseli in the 1980s. Although the importance of female dominance rank for reproductive success was not yet fully understood, evidence of its importance for females of many primate species had begun to accumulate (reviewed in Silk 1987). Baboon females, like many cercopithecine primate females, typically attain a social dominance rank just below that of their mothers, with the help of both kin and unrelated females (Walters 1987). This phenomenon had been known for some time in cercopithecines (reviewed in Melnick and Pearl 1987); the Amboseli study revealed that its consequence was cross-generational, long-term consistency of dominance rank relationships in wild populations. Further, the Amboseli work identified this long-term consistency of female dominance ranks as the most important source of stability in baboon social structure, a phenomenon generalizable to other species with maternal rank inheritance (Hausfater et al. 1982).

Some years later, however, the accumulating longitudinal data provided an even richer picture of female dominance relationships, revealing that long periods of stability could be punctuated by short periods of rapid change when some matrilines permanently fell and other permanently rose in rank (Samuels et al. 1987). Indeed, some females, by targeting higher-ranking females, were able to raise the ranks of all their female family members, while in other cases entire matrilines fell in rank, and occasionally matrilines split, with one female maintaining a higher rank than the rest of her family (Samuels et al. 1987).

Research on female social relationships and maternal behavior has remained a major theme in Amboseli through the years (Silk et al. 2003, 2004, 2006a, b; Van Horn et al. 2007; Gesquiere et al. 2008; Nguyen et al. 2008, 2009). A major step forward in understanding the strong and complex social relationships of female baboons came in 2003, when we discovered, with collaborator Joan Silk, that strong affiliative relationships have direct adaptive value for females: infants of more socially integrated females experienced higher survival than infants of more socially isolated females (Fig. 12.4; Silk et al. 2003). The effect of social integration on infant survival was later replicated in the long-term study of a chacma baboon

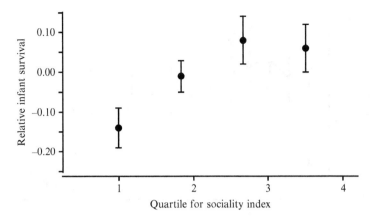

Fig.12.4 Females that were more socially integrated (females in the upper quartiles of a composite sociality index) experienced higher survival of their infants. Redrawn from Silk et al. (2003)

group in the Okavango Delta in Botswana (Silk et al. 2009), and those authors went on to demonstrate that social bonds enhanced the survival of the females themselves, not just of their infants (Silk et al. 2010). Again, without longitudinal data, these discoveries would have remained beyond our reach.

12.4 The 1990s: Males

After the first study on male dominance rank and mating behavior in the 1970s (Hausfater 1975) male social behavior took a decided back seat to female social behavior and ecological questions in the research priorities of the Amboseli team. Michael Pereira's work on the social behavior of juvenile males and females was an exception; he examined juvenile behavior in the context of the life history, and found sex differences in behavior that could be explained by the different demands made on males and females as adults (Pereira 1988a, b, 1989). Nonetheless, male social behavior was not intensively studied in Amboseli during this period, partly because of the rich vein of research on female behavior that had been tapped, and partly because some of the most interesting behavioral questions on adult males required more life history information and longitudinal data than were available in the 1970s and early 1980s.

Ronald Noë and Betty Sluijter were the first to study social behavior of adult males in Amboseli after Hausfater (Noë 1986, 1992; Noë and Sluijter 1990, 1995). They also brought a life history perspective to the study of males, in the now-growing tradition of Amboseli research, and showed that over the course of adulthood, males changed their levels of investment from an emphasis on direct male–male competition in young adulthood to a greater investment in relationships with adult females and juveniles in middle age (Noë and Sluijter 1990). These changes were associated with changes in male dominance rank, and as we amassed

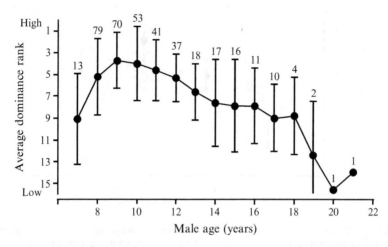

Fig.12.5 Male baboons reach their highest dominance rank in early adulthood when they are in their physical prime, and then steadily fall in rank as they age. Points show the mean (±SD) dominance rank for males in each age class, considering only males born into Amboseli study groups (i.e., males of known age); numbers above the error bars represent the number of males that contributed to the value for that age class. See also Alberts et al. (2003)

longitudinal data on known-aged individuals, it became clear how strongly age-based they were, with dominance rank peaking in young adulthood and declining steadily throughout the rest of life in a pattern very different from the relative stability of female dominance ranks (Fig. 12.5; see also Packer et al. 2000; Alberts et al. 2003).

By the late 1980s, male life history data had accumulated sufficiently for Susan Alberts to take male maturation and dispersal as the theme of her Ph.D. research. Dispersal represents an intense physical challenge at the onset of male adulthood, requiring that males leave the groups into which they were born and move into groups in which they have no prior relationships; in those groups, they must establish social relationships and obtain reproductive opportunities. Susan found that during maturation, males were subject to constraints of their mothers' dominance ranks, just as their sisters were (Alberts and Altmann 1995a). Male baboons, like most male primates, are independent of their mothers well before puberty (for an interesting primate exception to this rule see Surbeck et al. 2011); also, subadult male baboons are much larger than females and do not receive assistance from females in attaining or maintaining their dominance ranks. Indeed, analyses from another baboon population indicate that maternal dominance rank has no impact on the eventual dominance rank that sons attain (Packer et al. 2000). For these reasons, the dependence of male maturation on maternal dominance rank was surprising. However, this finding was supported by a later analysis of growth rates, which demonstrated an impact of maternal rank on growth rates for both sons and daughters (Altmann and Alberts 2005) and echoed similar findings in a few other primate species (e.g., Paul et al. 1992; Bercovitch et al. 2000; van Noordwijk and van Schaik 2001). In Amboseli, it presaged an even more profound effect of maternal dominance rank on male physiology at the onset of adulthood (Onyango et al. 2008; see Sect. 12.5).

Susan's analyses of dispersal revealed that males often spent time alone during dispersal while searching for other groups to join, and consequently both experienced elevated mortality risks (from 3 to 10 times higher than the mortality risk for males living in groups) and missed mating opportunities. However, males also appeared to experience reproductive costs if they remained in their natal group: although the sample size was small and based on observational rather than genetic estimates of paternity, the available data suggested high mortality of offspring for whom the natal males were likely fathers (Alberts and Altmann 1995b). This occurred in spite of strong evidence for close inbreeding avoidance between mothers and sons, and between maternal siblings. Finally, Susan's analysis indicated that natal males undergoing their first dispersal, and older males undergoing secondary dispersal, dispersed in response to the availability of females and to their own mating success in a given group (Alberts and Altmann 1995b; see also Altmann 2000).

By the end of the 1990s, substantial life history and behavioral data had accumulated for both males and females, and the groundwork had been laid for integrative studies of both sexes. In addition, we had begun to employ digital data loggers for collecting focal animal samples and hand-held GPS devices for collecting locational data. Among the most important advances of the 1990s was the development of BABASE, our longitudinal, individual-based database (Pinc et al. 2009). Field data are returned to the US either weekly (for digital data collected with hand-held data loggers and GPS machines) or monthly (for paper data) and are incorporated into BABASE in twice-yearly updates. BABASE is a web-based, PostgreSQL database that now houses most of our field data, including demographic, reproductive, behavioral, locational, ecological, and meteorological data. It is continually growing as we incorporate additional data sets into the original database design. These developments set the stage for the next important phase of the research, which involved getting "under the skin" with genetic and endocrine research.

12.5 The 2000s: Under the skin

In the late 1980s and early 1990s, two methodological developments occurred in genetics and physiology that had an enormous impact on primate field research, including the Amboseli baboon project. These were the development of a technique to extract DNA from feces (e.g., Höss et al. 1992; Gerloff et al. 1995) and the development of a technique to extract metabolites of steroid hormones from feces (e.g., Wasser et al. 1988). For the first time, we could study physiological responses and patterns of genetic relatedness – especially paternity – without invasive methods. This enabled us to pursue questions that were simply impossible to approach using naturalistic behavior observations alone. It would be 5–10 years after the first development of these techniques before they were perfected and applied to the Amboseli study (e.g., Khan et al. 2002; Buchan et al. 2003, 2005; Lynch et al. 2003), but once these techniques were established (with genetics analysis taking place in Susan's lab and hormonal analysis in Jeanne's), they transformed our understanding of baboon behavior and ecology (similar advances were being made in other primate field studies; e.g., Ziegler et al. 1997; Borries et al. 1999; Launhardt et al. 2001).

12.5.1 Hormones

Our hormone data have revealed several interesting surprises. In an analysis that combined endocrine data with data on male mating behavior and female reproductive states, we found that alpha males, but not males of other ranks, differentiated conceptive from non-conceptive cycles in their mate guarding (consortship) behavior, and that both the size of the sexual swelling and levels of circulating estrogen in the females provided potential cues about female fecundability (Gesquiere et al. 2007). This endocrine work nicely complemented the fecal DNA-based paternity analysis, which also revealed evidence of male mate choice (see Sect. 12.5.2 and Alberts et al. 2006).

In a study of pregnant females, we found evidence of a hormonal signature of impending fetal losses among pregnant females. This signature was evident for up to 2 months before the loss itself occurred (losses occurred in 91 of 656 pregnancies), ~14% (Beehner et al. 2006). In an additional instance of longitudinal data use, we discovered that maturing males whose mothers had been low-ranking when the males were born had higher fecal glucocorticoid levels than males whose mothers had been high-ranking, even when the glucocorticoid levels were measured 6–8 years after the males' births, years after males were independent of their mothers and in many cases years after the mothers had died (Fig. 12.6; Onyango

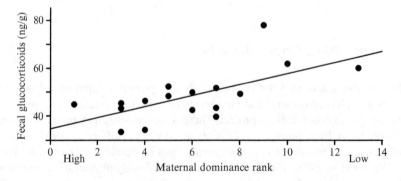

Fig.12.6 Subadult males born to higher ranking mothers had lower levels of fecal glucocorticoids than males born to lower ranking mothers, even when the fecal glucocorticoids were measured years after the period of offspring dependence on the mother. $R^2 = 0.421$, $P = 0.005$. Redrawn from Onyango et al. (2008)

et al. 2008). Glucocorticoids are products of the hypothalamic-pituitary axis (HPA) and are important in regulating the stress response and in mobilizing metabolic activity (Nelson 2005); our results indicated a long-term effect of maternal dominance rank on offspring physiology. Fecal glucocorticoid levels also predicted the peri-parturitional behavior of new mothers (specifically their responsiveness to their new infants), confirming findings that had previously only been documented in captive animals (Nguyen et al. 2008). Finally, we identified physiological effects of Amboseli's harsh dry seasons on both adult males and adult females (Gesquiere et al. 2008, 2011).

Most recently, we identified a surprising relationship between male dominance rank and endocrine profiles (Fig. 12.7; Gesquiere et al. 2011). While fecal testosterone (fT) levels declined as a function of dominance rank (with high-ranking males having the highest fT and low-ranking males the lowest), fecal glucocorticoids (fGC) presented a different and unexpected relationship with male dominance rank. Alpha males exhibited the highest fGC levels, and beta

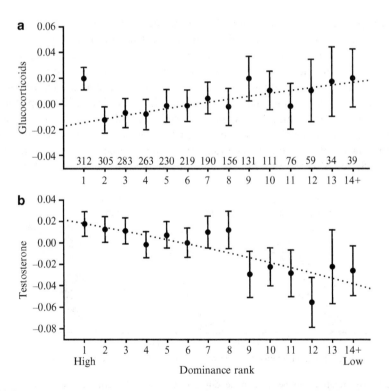

Fig. 12.7 Highest ranking (alpha) males had very high glucocorticoids, much higher than second-ranking males and a striking departure from the pattern for glucocorticoids to increase (**a**) and testosterone to decrease (**b**) as dominance rank declined. Values (mean ± SE across male monthly averages) represent the residuals obtained from a statistical model of log-transformed hormone concentrations that accounted for age, environmental factors, and hierarchy stability as fixed factors, male identity as a random factor. After Gesquiere et al. (2011)

(second-ranking) males exhibited the lowest fGC levels, with a monotonic increase in fGC below rank 2. This striking difference between alpha and beta males has not been described before, possibly because researchers often group these males together as "high ranking males." Moreover, despite predictions in the literature that alpha males should experience elevated glucocorticoids only when the dominance hierarchy is unstable, we found no effect of hierarchy stability on the relative endocrine levels of alpha and beta males although overall levels of stress hormones were higher during instability. Thus, regardless of how stable male rank relationships are, being at the very top of a social hierarchy may be more stressful than being immediately below, probably because of energetic costs associated with mating and with male–male competitive interactions (Gesquiere et al. 2011).

12.5.2 Paternity

The first problem we tackled using DNA extracted from feces was the problem of paternity and its impact on social relationships. The question of whether male dominance rank mattered for male reproductive success had been a vexing one for decades; different primate studies produced different answers to this question and a resolution to the contradictions was slow to emerge (see reviews in Kutsukake and Nunn 2006; Port and Kappeler 2010; Alberts in press). Using 32 group-years of behavioral data on mating success in Amboseli, we had shown that the apparent contradictions across different studies were probably explained by real variance in the importance of male dominance rank over time for many if not most primates (Fig. 12.8; see also Strum 1982 for an early prediction of this sort). Both male

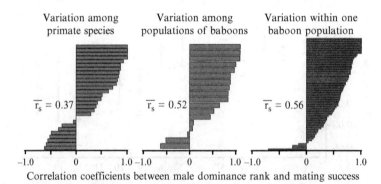

Fig.12.8 The relationship between male dominance rank and mating success shows similar variance at all levels of analysis; among primate species (*left panel*, data from Cowlishaw and Dunbar 1991), among populations of a single species, baboons (*center panel*, data from Bulger 1993), and within a single population of baboons, Amboseli (*right panel*, data from Alberts et al. 2003). Each graph shows the rank-ordered distribution of Spearman correlation coefficients for the relationship between male rank and mating success across studies (*left and center panels*) or across time periods in Amboseli (*right panel*). Redrawn from Altmann and Alberts (2003)

density and the distribution of male competitive abilities within groups affected the relationship between male dominance rank and mating success in Amboseli (Alberts et al. 2003; see also Cowlishaw and Dunbar 1991, Kutsukake and Nunn 2006; Port and Kappeler 2010).

However, an important question remained: what would the data look like if we were able to measure reproductive success using actual paternity data instead of just mating success? Our first paternity analysis, focusing on Lodge Group (the food-supplemented group that we studied from 1984 to 1996) and using DNA extracted from blood obtained during a darting project (see Sect. 12.6.4), had indicated that genetic paternity results were nicely predicted from behavioral data (Altmann et al. 1996). Would the same hold true in the wild-feeding groups? Using fecal DNA, we assigned paternity to 208 offspring conceived across a range of demographic conditions in wild-feeding groups. We found that male reproductive success, like mating success, depended on dominance rank but in a density-dependent manner and also depended on relative male competitive ability. In addition, we found little evidence for successful surreptitious mating (although subadult males occasionally produced offspring using this strategy), and no evidence for differential sperm success or sperm depletion. Most strikingly, we found clear evidence that alpha males fathered more offspring than expected based on their mating behavior, and that this excess of offspring resulted from male mate choice for female experiencing conceptive rather than non-conceptive cycles (Alberts et al. 2006). These results supported behavioral evidence of male mate choice in other baboons (Bulger 1993; Weingrill et al. 2003) and complemented our work on male mate choice based on female endocrine profiles and sexual swellings (Gesquiere et al. 2007).

The most surprising result from our paternity analyses was the discovery that male baboons differentiated their own offspring from the offspring of other males, and supported them disproportionately during agonistic interactions (Buchan et al. 2003). A number of previous researchers had hypothesized that paternal care occurred in baboons, and some had provided strong circumstantial evidence for its occurrence (Ransom and Ransom 1971; Altmann 1980; Stein 1984; Palombit et al. 1997). In addition to our confirmation of paternal care in baboons, we later discovered that offspring who resided with their fathers longer during their juvenile periods reached maturity earlier than offspring whose fathers dispersed or died earlier in their juvenile periods (Charpentier et al. 2008a). Neither of these results on paternal effects would have been discovered without longitudinal data; the implication was that paternal presence had an impact not only on the juvenile's daily interactions, but on an important life history component (see Altmann et al. 1988; Altmann and Alberts 2003 for data on the importance of age at maturity in this population). However, much remains to be done before we fully understand male–juvenile interactions in multi-male primate species. Some care by male baboons cannot be explained as paternal investment (Smuts 1985; Buchan et al. 2003; Moscovice et al. 2009; Nguyen et al. 2009), raising the possibility that males are engaged in mating effort when they care for young (incentivizing future mating with the mother; e.g., Smuts 1985; van Schaik and Paul 1996). Furthermore, the extent of paternal care in most other multi-male primate species remains an open question, with few data either refuting or confirming its occurrence (but see Paul et al. 1996; Ménard et al. 2001; Lehmann et al. 2006; Wroblewski 2010).

The paternal care that male baboons provide had an important corollary in our research; female baboons differentiated paternal sisters from non-kin, and formed preferential relationships with them (Smith et al. 2003; Silk et al. 2006a), supporting Jeanne's early predictions about paternal kin selection in primate groups (Altmann 1979). These relationships between paternal sisters were typically not as strong as relationships between maternal sisters, but were measurably stronger than relationships among non-kin, and sometimes played a role in patterns of permanent group fission (Van Horn et al. 2007). Significant relationships between paternal sisters were first documented in rhesus macaques (Widdig et al. 2001), suggesting that paternal kin networks may be a robust phenomenon in many primates (see, e.g., Watts 1997 for data on relationships among paternal siblings in gorillas, but see Langergraber et al. 2007 for different results in chimpanzees). The genesis of relationships between paternal sisters in baboons remains obscure, and fathers may well play a role in establishing them when they provide care to their offspring.

12.6   2010 Forward: New directions

At the start of this decade our energy and interest will be focused on four areas involving major new collaborations with Elizabeth Archie and Jenny Tung. These areas are aging, disease transmission, hybridization, and functional genetics and genomics. These investigations expand the multidisciplinary nature of our program in ways that explicitly allow us to capitalize on the longitudinal, long-term nature of our study.

12.6.1   Aging

With 40 years of demographic, behavioral, and ecological data accumulated, we are in a position to examine not only mortality patterns, but also behavioral and health correlates of aging that contribute to mortality; this represents a long-time goal that is finally within sight. To our knowledge, the work we have recently initiated is the first systematic study of how health and behavior change with age in a natural primate population, and of whether genetic and social predictors of these changes can explain individual differences in survival and longevity. Sex differences in health and survival during aging are major topics of interest in medicine, epidemiology, demography, and evolutionary biology. Despite this pervasive interest, and despite a wealth of data on aging in humans and a few well-studied model organisms, patterns of aging in wild animals remain largely undescribed. Not only are there large gaps in our knowledge of age-related changes in survival in wild animals, but virtually nothing is known about age-related changes in physiology, behavior, or other aspects of health and functioning for animals in the wild (Brunet-Rossinni and Austad 2006). We argue that studies of aging in wild animal populations, especially in our primate relatives, can provide a comparative perspective on human aging (Bronikowski et al. 2011), generate new questions, produce insights into the answers to old ones, and identify opportunities for alleviating the adverse consequences of aging. The longitudinal nature of our data will allow us to analyze not only declines in survival with age (demographic senescence; Bronikowski et al. 2002, 2011) but to produce a systematic description of how multiple indicators of health and functioning ability change with age in a wild primate (e.g., Altmann et al. 2010; Galbany et al. 2010, 2011).

12.6.2 Disease Transmission

Biologists currently have a poor empirical understanding of how infectious agents spread within and between social groups of wild animals, because of the logistical challenges of directly tracking the movements of infectious agents in the wild. Our parasite research, led by Elizabeth Archie, is designed to map the movements of common infectious agents onto the social landscape of the Amboseli baboons, using tools from social network analysis and population genetics.

One of our key hypotheses involves whether the transmission of nematodes or bacteria is socially structured at the level of social groups (i.e., whether baboons are most likely to be infected by group members for some or all parasites), a rarely tested but critical assumption of most research on the disease-related costs of group living (Altizer et al. 2003; Nunn and Altizer 2006). We will also test whether social networks predict the movements of infectious organisms within social group, and whether and how infection risk varies among individuals. Finally, we will examine the ecological correlates of between-group parasite transmission. This research will employ fecal samples that were collected during the past 15 years, as well as new samples, to provide a unique longitudinal perspective on parasite transmission in this population.

The result will be a picture of how different infectious agents, with a variety of transmission modes and fitness effects, move between and within social groups. This in turn will enhance our ability to understand individual differences in the risk of infection in more detail than has been possible before in natural primate populations. Socially structured disease transmission has important implications for understanding the evolution of sociality – a basic feature of our primate lineage – because exposure to disease is thought to be a major evolutionary cost of group-living (Alexander 1974; Altizer et al. 2003). Also, current models of disease transmission tend to be highly sensitive to variation in transmission patterns; hence, to predict the dynamics of an epidemic in social wildlife, biologists need accurate information on how disease spreads in natural populations of social animals.

12.6.3 Hybridization

The Amboseli baboon population comprises primarily yellow baboons (*P. cynocephalus*), specifically the "ibean" morphotype of yellow baboons. This morphotype shares some morphological similarities with anubis (olive) baboons, possibly because of anubis admixture in the ibean lineage over the course of evolutionary history (Jolly 1993). Further, Amboseli is situated on the boundary between the ranges of yellow and anubis baboons, with yellow baboons to the south and east and anubis baboons to the north and west (Jolly 1993; Kingdon 1997; Newman et al. 2004). Six anubis males have immigrated into Amboseli study groups over the course of the study, and one small (ca. 18) mixed-sex group of anubis baboons also entered the basin in the early 1980s (Samuels and Altmann 1986; Tung et al. 2008). Hybrids now occur both in study and non-study groups in Amboseli, resulting both from the anubis immigrations we have detected (and probably other, undetected anubis immigrants into non-study groups) and from the movement of hybrid males between study and non-study groups and successful reproduction by these males. In Amboseli, as in all baboon hybrid zones studied thus far, mating occurs freely and results in viable and fertile offspring; little or no evidence of hybrid dysgenesis has been found (reviewed in Tung et al. 2008).

Strikingly, in Amboseli, both males and females with a higher proportion of genetic admixture from anubis baboons (measured via a genetic "hybrid score"; see (Tung et al. 2008)) reached physical maturity earlier than animals with a yellow baboon genetic background (another result that depended on longitudinal data). Males showed a particularly pronounced effect of admixture, especially for age at natal dispersal (Fig. 12.9; Charpentier et al. 2008b). We are now poised to examine

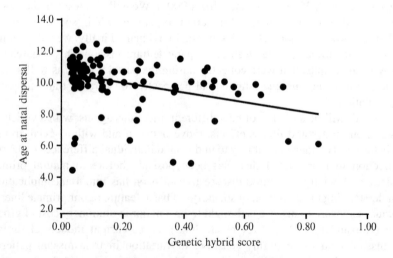

Fig.12.9 Male baboons in Amboseli with higher genetic hybrid scores (more anubis ancestry) are likely to disperse at a younger age than males with lower genetic hybrid scores (less anubis ancestry). $R^2 = 0.216, P < 0.0001$. After Charpentier et al. (2008a, b)

the consequences of hybridization further, considering both life history traits and behavioral traits that are associated with life history (specifically dominance, mating, and dispersal) as a function of hybridity.

12.6.4 *Functional Genetics and Genomics*

By combining the detailed behavioral and ecological data available for the Amboseli baboons with genetic and genomic research, we have the opportunity to place genetic inferences in the context of a well-understood natural primate system. Indeed, we have argued that as genomic resources for primate species accumulate, natural populations of primates should increasingly be targets of research on evolutionary and functional genetics and genomics because they offer the prospect of identifying functional genetic variation that influences traits of known ecological and adaptive significance (Tung et al. 2010). While work on captive populations can identify genetic systems of suspected significance, only field studies can confirm whether and how the systems in question are variable or relevant in nature (e.g., Keele et al. 2009; Tung et al. 2009); this in turn can have important implications for understanding both primate and human evolution.

In 1989, we learned from Robert Sapolsky how to temporarily anesthetize and immobilize baboons for drawing blood and taking morphometric measurements. We employed the technique at the time for both endocrine and genetic analyses, darting ~150 animals in a 3-year period (e.g., Sapolsky and Altmann 1991; Altmann et al. 1996; Sapolsky et al. 1997). We have continued to employ this technique occasionally over the years to affix radio collars to single females in each group. Beginning in 2006, we resumed a more extensive darting effort in order to obtain high quality DNA and RNA samples for as many adults as we could in the population. Because our research depends heavily on not affecting the behavior, health, survival, or reproductive success of the study animals, we have invested considerable effort in our time- and labor-intensive approach, which is focused on safety and on maintaining habituation in our study animals.

We dart individual baboons, one at a time, with an anesthetic-bearing dart, using a hand-held blowpipe. We dart no more than two animals per day, and no study group is darted more than once per week. On the day of a planned darting, the target animal (an adult male or an adult female that is not carrying a dependent infant and not past early pregnancy) is followed, often for many hours, until an appropriate opportunity arises. This occurs when the air is still, when neither the target animal nor other baboons will witness the darting, and when the animal is seated, standing still, or walking slowly. After the animal is darted and falls asleep (within 3–10 min), we wait for the other group members to move on (only rarely do other group members take any interest in the sleeping individual) and then quickly pick up the sleeping animal and move it to a shady processing site. We weigh the animal, carry out various other body measurements, and collect blood and skin samples in DNA- and RNA-preserving buffers. After the protocol is complete, the animal is allowed to recover in a holding cage in the back of our pickup truck until fully awake (after 3–4 h), when it is released at some distance from its social group but within sight. We have darted over 100 animals since 2006 in this manner, with no injuries and no untoward incidents. Blood and tissue samples remain in Amboseli for up to 48 h in an evaporatively-cooled hut before we send them to Nairobi (via air) for temporary freezing, and then transport them to the US (Tung et al. 2011).

Our functional genetics and genomics research, which uses the blood and skin samples that we collect during darting, has already begun to bear fruit even though it is in its early stages. For instance, in research led by Jenny Tung we identified variation in the *cis*-regulatory region of the baboon *FY* (Duffy) gene that was associated with variation in susceptibility to *Hepatocystis*, a malaria-like pathogen common in baboons (Tung et al. 2009). We also found evidence, in several genes involved in immunity, of allele-specific gene expression patterns (i.e., cases in which segregating genetic variants cause differences in the levels of expression of particular genes), and in one of these genes (*CCL5*) we identified an influence of maternal dominance rank on the extent to which a given *cis*-regulatory variant affected gene expression (Tung et al. 2011). Much of our effort in functional genetic analyses going forward is focused on early life effects on adult gene expression patterns.

12.7 Conclusions

The Amboseli Baboon Research Project, like the other long-term primate studies described in this volume, represents a labor of love. The researchers who have worked on the Amboseli project, particularly its directors, have made enormous personal sacrifices to keep the project going over the long term. Funding for the project has invariably come in grants of short duration, typically 2–3 years but occasionally, when we have been very lucky, 5 years. We have no NGO or endowment, and the intellectual growth of the project depends on us being able to convince grant reviewers that we really do have something new to study every few years (we have fielded questions of the sort "haven't you studied this species enough yet? Couldn't this be done more cheaply on captive rodents?"; other authors in this volume have surely had similar experiences). Yet, as Clutton-Brock and Sheldon (2010) have noted, long-term studies of primates and other animals offer our best opportunity for novel and innovative research on behavioral ecology. Our study provides ample evidence of this. The novel maternal and paternal effects we have described, our evidence for the functional importance of female social relationships, the patterns of continuity and change in dominance ranks that we have documented for both sexes and their implications for lifetime fitness, and a range of other results described here, have helped to shape current research in evolutionary ecology in ways that simply would not have been possible for a comparably long series of independent short-term studies. The same is true for the other longitudinal studies described in this volume. And, as we have argued elsewhere, longitudinal studies also offer our best opportunity for understanding the evolution of important traits by combining genetic, phenotypic, and environmental data on the same individuals (Tung et al. 2010). Surely, a challenge for the twenty-first century involves raising the profile of long-term primate studies and ushering in an era in which the value of such studies, as evidenced by the chapters in this volume, are more widely acknowledged.

Acknowledgments We gratefully acknowledge the National Science Foundation for generous support over the years, most recently through DEB 0846286 and DEB 0919200 to S.C. Alberts, DEB 0846532 to J. Altmann, and IOS 1053461 to E. Archie. We are also very grateful for support from the National Institute of Aging (R01AG034513-01 and P01AG031719). We also thank the Princeton Center for the Demography of Aging (funded through P30AG024361), the Chicago Zoological Society, the Max Planck Institute for Demographic Research, the L.S.B. Leakey Foundation and the National Geographic Society for support at various times over the years.
We thank the Kenya Wildlife Services, Institute of Primate Research, National Museums of Kenya, and members of the Amboseli-Longido pastoralist communities for their cooperation and assistance in Kenya. A number of people, too numerous to list here, have contributed to the long-term data collection over the years, and we are grateful to all of them for their dedication and contributions. Particular thanks go to Stuart Altmann, the late Amy Samuels, and the Amboseli Baboon Project long-term field team (Raphael S. Mututua, Serah N. Sayialel, and J. Kinyua Warutere), as well as to Vera Somen and Tim Wango for their untiring assistance in Nairobi.
Karl Pinc has provided expertise in database design and management for many years and we are grateful for his seminal contributions to the development of BABASE. We also thank the database technicians who have provided assistance with BABASE over the years, most recently D. Onderdonk, C. Markham, T. Fenn, N. Learn, and L. Maryott. Our Amboseli research is approved by the IACUC at Princeton University and at Duke University and has adhered to all the laws and guidelines of Kenya.

References

Alberts SC (in press) Magnitude and sources of variation in male reproductive performance. In: Mitani J, Call J, Kappeler PM, Palombit R, Silk JB (eds) The evolution of primate societies. University of Chicago Press, Chicago

Alberts SC, Altmann J (1995a) Preparation and activation: determinants of age at reproductive maturity in male baboons. Behav Ecol Sociobiol 36:397–406

Alberts SC, Altmann J (1995b) Balancing costs and opportunities: dispersal in male baboons. Am Nat 145:279–306

Alberts SC, Watts HE, Altmann J (2003) Queuing and queue-jumping: long-term patterns of reproductive skew in male savannah baboons, *Papio cynocephalus*. Anim Behav 65:821–840

Alberts SC, Buchan JC, Altmann J (2006) Sexual selection in wild baboons: from mating opportunities to paternity success. Anim Behav 72:1177–1196

Alexander RD (1974) The evolution of social behavior. Annu Rev Ecol Syst 5:325–383

Altizer S, Nunn CL, Thrall PH, Gittleman JL, Antonovics J, Cunningham AA, Dobson AP, Ezenwa V, Jones KE, Pedersen AB, Poss M, Pulliam JRC (2003) Social organization and parasite risk in mammals: integrating theory and empirical studies. Annu Rev Ecol Syst 34:517–547

Altmann SA (1962) A field study of the sociobiology of rhesus monkeys, *Macaca mulatta*. Ann NY Acad Sci 102:338–435

Altmann J (1974) Observational study of behavior: sampling methods. Behaviour 49:227–266

Altmann J (1979) Age cohorts as paternal sibships. Behav Ecol Sociobiol 6:161–164

Altmann J (1980) Baboon mothers and infants. Harvard University Press, Cambridge, MA

Altmann SA (1991) Diets of yearling female primates (*Papio cynocephalus*) predict lifetime fitness. Proc Natl Acad Sci USA 88:420–423

Altmann SA (1998) Foraging for survival: yearling baboons in Africa. University of Chicago Press, Chicago

Altmann J (2000) Models of outcome and process: predicting the number of males in primate groups. In: Kappeler PM (ed) Primate males: causes and consequences of variation in group composition. Cambridge University Press, Cambridge, pp 236–247

Altmann J, Alberts SC (2003) Intraspecific variability in fertility and offspring survival in a nonhuman primate: behavioral control of ecological and social sources. In: Wachter KW, Bulatao RA (eds) Offspring: human fertility behavior in biodemographic perspective. National Academies Press, Washington, DC, pp 140–169

Altmann J, Alberts SC (2004) Monitoring guide for the Amboseli baboon research project: protocols for long-term monitoring and data collection. http://www.princeton.edu/~baboon/monitoring_guide.htm

Altmann J, Alberts SC (2005) Growth rates in a wild primate population: ecological influences and maternal effects. Behav Ecol Sociobiol 57:490–501

Altmann SA, Altmann J (1970) Baboon ecology: African field research. University of Chicago Press, Chicago

Altmann J, Muruthi P (1988) Differences in daily life between semiprovisioned and wild-feeding baboons. Am J Primatol 15:213–221

Altmann J, Samuels A (1992) Costs of maternal care: infant carrying in baboons. Behav Ecol Sociobiol 29:391–398

Altmann J, Hausfater G, Altmann SA (1985) Demography of Amboseli baboons, 1963-1983. Am J Primatol 8:113–125

Altmann J, Altmann SA, Hausfater G (1988) Determinants of reproductive success in savannah baboons, *Papio cynocephalus*. In: Clutton-Brock TH (ed) Reproductive success: studies of individual variation in contrasting breeding systems. University of Chicago Press, Chicago, pp 403–418

Altmann J, Alberts SC, Haines SA, Dubach J, Muruthi P, Coote T, Geffen E, Cheesman DJ, Mututua RS, Saiyalel SN, Wayne RK, Lacy RC, Bruford MW (1996) Behavior predicts genetic structure in a wild primate group. Proc Natl Acad Sci USA 93:5797–5801

Altmann J, Gesquiere L, Galbany J, Onyango PO, Alberts SC (2010) Life history context of reproductive aging in a wild primate model. Ann NY Acad Sci 1204:27–138

Beehner JC, Nguyen N, Wango EO, Alberts SC, Altmann J (2006) The endocrinology of pregnancy and fetal loss in wild baboons. Horm Behav 49:688–699

Bercovitch FB, Widdig A, Nürnberg P (2000) Maternal investment in rhesus macaques (*Macaca mulatta*): reproductive costs and consequences of raising sons. Behav Ecol Sociobiol 48:1–11

Boesch C, Kohou G, Néné H, Vigilant L (2006) Male competition and paternity in wild chimpanzees of the Taï forest. Am J Phys Anthropol 130:103–115

Borries C, Launhardt K, Epplen C, Epplen JT, Winkler P (1999) DNA analyses support the hypothesis that infanticide is adaptive in langur monkeys. Proc R Soc Lond B 266:901–904

Bronikowski AM, Alberts SC, Altmann J, Packer C, Carey KD, Tatar M (2002) The aging baboon: comparative demography in a non-human primate. Proc Natl Acad Sci USA 99:9591–9595

Bronikowski AM, Altmann J, Brockman DK, Cords M, Fedigan LM, Pusey AE, Stoinski T, Morris WF, Strier KB, Alberts SC (2011) Aging in the natural world: comparative data reveal similar mortality patterns across primates. Science 331:1325–1328

Brunet-Rossinni AK, Austad SN (2006) Senescence in wild populations of mammals and birds. In: Masoro EJ, Austad SN (eds) Handbook of the biology of aging. Elsevier, Amsterdam, pp 243–266

Buchan JC, Alberts SC, Silk JB, Altmann J (2003) True paternal care in a multi-male primate society. Nature 425:179–181

Buchan JC, Archie EA, Van Horn RC, Moss CJ, Alberts SC (2005) Locus effects and sources of error in noninvasive genotyping. Mol Ecol Notes 5:680–683

Bulger JB (1993) Dominance rank and access to estrous females in male savanna baboons. Behaviour 127:67–103

Charpentier MJE, Van Horn RC, Altmann J, Alberts SC (2008a) Paternal effects on offspring fitness in a multimale primate society. Proc Natl Acad Sci USA 105:1988–1992

Charpentier MJE, Tung J, Altmann J, Alberts SC (2008b) Age at maturity in wild baboons: genetic, environmental and demographic influences. Mol Ecol 17:2026–2040

Clutton-Brock TH, Harvey PH (1977) Primate ecology and social organization. J Zool Lond 183:1–39

Clutton-Brock TH, Sheldon BC (2010) The seven ages of *Pan*. Science 327:1207–1208

Cowlishaw G, Dunbar RIM (1991) Dominance rank and mating success in male primates. Anim Behav 41:1045–1056

Crook JH, Gartlan JS (1966) Evolution of primate societies. Nature 210:1200–1203

Emlen ST, Oring LW (1977) Ecology, sexual selection, and the evolution of mating systems. Science 197:215–223

Fossey D (1979) Development of the mountain gorilla (*Gorilla gorilla beringei*): the first thirty-six months. In: Hamburg DA, McCown E (eds) Perspectives on human evolution, vol 5, The great apes. Benjamin, Cummings, Menlo Park, CA, pp 139–184

Galbany J, Dotras L, Alberts SC, Pérez-Pérez A (2010) Tooth size variation related to age in Amboseli baboons. Folia Primatol 81:348–359

Galbany J, Altmann J, Pérez-Pérez A, Alberts SC (2011) Age and individual foraging behavior predict tooth wear in Amboseli baboons. Am J Phys Anthropol 144:51–59

Gerloff U, Schlötterer C, Rassmann K, Rambold I, Hohmann G, Fruth B, Tautz D (1995) Amplification of hypervariable simple sequence repeats (microsatellites) from excremental DNA of wild living bonobos (*Pan paniscus*). Mol Ecol 4:515–518

Gesquiere LR, Wango EO, Alberts SC, Altmann J (2007) Mechanisms of sexual selection: sexual swellings and estrogen concentrations as fertility indicators and cues for male consort decisions in wild baboons. Horm Behav 51:114–125

Gesquiere LR, Khan M, Shek L, Wango TL, Wango EO, Alberts SC, Altmann J (2008) Coping with a challenging environment: effects of seasonal variability and reproductive status on glucocorticoid concentrations of female baboons (*Papio cynocephalus*). Horm Behav 54:410–416

Gesquiere LR, Onyango PO, Alberts SC, Altmann J (2011) Endocrinology of year-round reproduction in a highly seasonal habitat: environmental variability in testosterone and glucocorticoids in baboon males. Am J Phys Anthropol 144:169–176

Gesquiere LR, Learn NH, Simao MCM, Onyango PO, Alberts SC, Altmann J (2011) Life at the top: Rank and stress in wild male baboons. Science 333:357–360

Hall KRL, DeVore I (1965) Baboon social behavior. In: DeVore I (ed) Primate behavior: field studies of monkeys and apes. Holt, Rinehardt, and Winston, New York, pp 53–110

Hausfater G (1975) Dominance and reproduction in baboons (*Papio cynocephalus*): a quantitative analysis. Karger, Basel

Hausfater G, Altmann J, Altmann SA (1982) Long-term consistency of dominance relations among female baboons (*Papio cynocephalus*). Science 217:752–755

Höss M, Kohn M, Pääbo S, Knauer F, Schröder W (1992) Excrement analysis by PCR. Nature 359:199

Hrdy SB (1977) The langurs of Abu: female and male strategies of reproduction. Harvard University Press, Cambridge, MA

Jay P (1963) Mother-infant relations in langurs. In: Rheingold HL (ed) Maternal behavior in mammals. Wiley, New York, pp 282–304

Jolly CJ (1993) Species, subspecies, and baboon systematics. In: Kimbel WH, Martin LB (eds) Species, species concepts, and primate evolution. Plenum Press, New York, pp 67–107

Kappeler PM, Port M (2008) Mutual tolerance or reproductive competition? Patterns of reproductive skew among male redfronted lemurs (*Eulemur fulvus rufus*). Behav Ecol Sociobiol 62:1477–1488

Keele BF, Jones JH, Terio KA, Estes JD, Rudicell RS, Wilson ML, Li Y, Learn GH, Beasley TM, Schumacher-Stankey JC, Wroblewski EE, Mosser A, Raphael J, Kamenya S, Lonsdorf EV, Travis DA, Mlengeya T, Kinsel MJ, Else JG, Silvestri G, Goodall J, Sharp PM, Shaw GM, Pusey AE, Hahn BH (2009) Increased mortality and AIDS-like immunopathology in wild chimpanzees infected with SIVcpz. Nature 460:515–519

Khan MZ, Altmann J, Isani SS, Yu J (2002) A matter of time: evaluating the storage of fecal samples fo steroid analysis. Gen Comp Endocrinol 128:57–64

Kingdon J (1997) The Kingdon field guide to African mammals. Academic, London

Kummer H (1968) Social organization of hamadryas baboons: a field study. University of Chicago Press, Chicago

Kutsukake N, Nunn CL (2006) Comparative tests of reproductive skew in male primates: the roles of demographic factors and incomplete control. Behav Ecol Sociobiol 60:695–706

Langergraber KE, Mitani JC, Vigilant L (2007) The limited impact of kinship on cooperation in wild chimpanzees. Proc Natl Acad Sci USA 104:7786–7790

Launhardt K, Borries C, Hardt C, Epplen JT, Winkler P (2001) Paternity analysis of alternative male reproductive routes among the langurs (*Semnopithecus entellus*) of Ramnagar. Anim Behav 61:53–64

Lehmann J, Fickenscher G, Boesch C (2006) Kin biased investment in wild chimpanzees. Behaviour 143:931–955

Lynch JW, Khan MZ, Altmann J, Njahira MN, Rubenstein N (2003) Concentrations of four fecal steroids in wild baboons: short-term storage conditions and consequences for data interpretation. Gen Comp Endocrinol 132:264–271

Melnick DJ, Pearl MC (1987) Cercopithecines in multimale groups: genetic diversity and population structure. In: Smuts BB, Cheney DL, Seyfarth RM, Wrangham RW, Struhsaker TT (eds) Primate societies. University of Chicago Press, Chicago, pp 121–134

Ménard N, von Segesser F, Scheffrahn W, Pastorini J, Vallet D, Gaci B, Martin RD, Gautier-Hion A (2001) Is male-infant caretaking related to paternity and/or mating activities in wild barbary macaques (*Macaca sylvanus*)? Neurosciences 324:601–610

Moscovice LR, Heesen M, Di Fiore A, Seyfarth RM, Cheney DL (2009) Paternity alone does not predict long-term investment in juveniles by male baboons. Behav Ecol Sociobiol 63:1471–1482

Muruthi P, Altmann J, Altmann SA (1991) Resource base, parity, and reproductive condition affect females' feeding time and nutrient intake within and between groups of a baboon population. Oecologia 87:467–472

Nelson RJ (2005) An introduction to behavioral endocrinology. Sinauer Associates, New York

Newman TK, Jolly CJ, Rogers J (2004) Mitochondrial phylogeny and systematics of baboons (*Papio*). Am J Phys Anthropol 124:17–27

Nguyen N, Gesquiere LR, Wango EO, Alberts SC, Altmann J (2008) Late pregnancy glucocorticoid levels predict responsiveness in wild baboon mothers (*Papio cynocephalus*). Anim Behav 75:1747–1756

Nguyen N, Van Horn RC, Alberts SC, Altmann J (2009) "Friendships" between new mothers and adult males: adaptive benefits and determinants in wild baboons (*Papio cynocephalus*). Behav Ecol Sociobiol 63:1331–1344

Noë R (1986) Lasting alliances among adult male savannah baboons. In: Else JG, Lee PC (eds) Primate ontogeny, cognition and social behaviour. Cambridge University Press, Cambridge, pp 381–392

Noë R (1992) Alliance formation among male baboons: shopping for profitable partners. In: Harcourt AH, de Waal FBM (eds) Coalitions and alliances in humans and other animals. Oxford University Press, Oxford, pp 285–321

Noë R, Sluijter AA (1990) Reproductive tactics of male savanna baboons. Behaviour 113:117–170

Noë R, Sluijter AA (1995) Which adult male savanna baboons form coalitions? Int J Primatol 16:77–105

Nunn CL, Altizer S (2006) Infectious disease and primate social systems. In: Nunn CL, Altizer S (eds) Infectious diseases in primates: behavior, ecology, and evolution. Oxford University Press, New York, pp 176–213

Onyango PO, Gesquiere LR, Wango EO, Alberts SC, Altmann J (2008) Persistence of maternal effects in baboons: mother's dominance rank at son's conception predicts stress hormone levels in subadult males. Horm Behav 54:319–324

Ostner J, Heistermann M, Schülke O (2011) Male competition and its hormonal correlates in Assamese macaques (*Macaca assamensis*). Horm Behav 59:105–113

Packer C, Collins DA, Eberly LE (2000) Problems with primate sex ratios. Philos Trans R Soc Lond B 355:1627–1635

Palombit RA, Seyfarth RM, Cheney DL (1997) The adaptive value of 'friendships' to female baboons: experimental and observational evidence. Anim Behav 54:599–614

Paul A, Kuester J, Arnemann J (1992) Maternal rank affects reproductive success of male Barbary macaques (*Macaca sylvanus*): evidence from DNA fingerprinting. Behav Ecol Sociobiol 30:337–341

Paul A, Kuester J, Arnemann J (1996) The sociobiology of male-infant interactions in Barbary macaques, *Macaca sylvanus*. Anim Behav 51:155–170

Pereira ME (1988a) Agonistic interactions of juvenile savanna baboons. I. Fundamental features. Ethology 79:195–217

Pereira ME (1988b) Effects of age and sex on intra-group spacing behaviour in juvenile savannah baboons, *Papio cynocephalus cynocephalus*. Anim Behav 36:184–204

Pereira ME (1989) Agonistic interactions of juvenile savanna baboons. II. Agonistic support and rank acquisition. Ethology 80:152–171

Pinc KO, Altmann J, Alberts SC (2009) BABASE: Technical specifications for the Amboseli Baboon Project Data Management System. http://papio.biology.duke.edu/babase_system.html

Port M, Kappeler PM (2010) The utility of reproductive skew models in the study of male primates, a critical evaluation. Evol Anthropol 19:46–56

Post DG (1981) Activity patterns of yellow baboons (*Papio cynocephalus*) in the Amboseli National Park, Kenya. Anim Behav 29:357–374

Pusey AE (1978) Age-changes in the mother-offspring association of wild chimpanzees. In: Chivers DJ, Herbert J (eds) Recent advances in primatology. Academic, London, pp 119–123

Ransom TW, Ransom BS (1971) Adult male-infant relations among baboons (*Papio anubis*). Folia Primatol 16:179–195

Samuels A, Altmann J (1986) Immigration of a *Papio anubis* male into a group of *Papio cynocephalus* baboons and evidence for an *anubis-cynocephalus* hybrid zone in Amboseli, Kenya. Int J Primatol 7:131–138

Samuels A, Altmann J (1991) Baboons of the Amboseli basin: demographic stability and change. Int J Primatol 12:1–19

Samuels A, Silk JB, Altmann J (1987) Continuity and change in dominance relations among female baboons. Anim Behav 35:785–793

Sapolsky RM, Altmann J (1991) Incidence of hypercortisolism and dexamethasone resistance increases with age among wild baboons. Biol Psychiatr 30:1008–1016

Sapolsky RM, Alberts SC, Altmann J (1997) Hypercortisolism associated with social subordinance or social isolation among wild baboons. Arch Gen Psychiatr 54:1137–1143

Seyfarth RM (1977) A model of social grooming among adult female monkeys. J Theor Biol 65:671–698

Silk JB (1987) Social behavior in evolutionary perspective. In: Smuts BB, Cheney DL, Seyfarth RM, Wrangham RW, Struhsaker TT (eds) Primate societies. University of Chicago Press, Chicago, pp 318–329

Silk JB, Alberts SC, Altmann J (2003) Social bonds of female baboons enhance infant survival. Science 302:1231–1234

Silk JB, Alberts SC, Altmann J (2004) Patterns of coalition formation by adult female baboons in Amboseli, Kenya. Anim Behav 67:573–582

Silk JB, Altmann J, Alberts SC (2006a) Social relationships among adult female baboons (*Papio cynocephalus*). I. Variation in the strength of social bonds. Behav Ecol Sociobiol 61:183–195

Silk JB, Altmann J, Alberts SC (2006b) Social relationships among adult female baboons (*Papio cynocephalus*). II. Variation in the quality and stability of social bonds. Behav Ecol Sociobiol 61:197–204

Silk JB, Beehner JC, Bergman TJ, Crockford C, Engh AL, Moscovice LR, Wittig RM, Seyfarth RM, Cheney DL (2009) The benefits of social capital: close social bonds among female baboons enhance offspring survival. Proc R Soc Lond B 276:3099–3104

Silk JB, Beehner JC, Bergman TJ, Crockford C, Engh AL, Moscovice LR, Wittig RM, Seyfarth RM, Cheney DL (2010) Strong and consistent social bonds enhance the longevity of female baboons. Curr Biol 20:1359–1361

Smith K, Alberts SC, Altmann J (2003) Wild female baboons bias their social behaviour towards paternal half-sisters. Proc R Soc Lond B 270:503–510

Smuts BB (1985) Sex and friendship in baboons. Aldine, Hawthorne, NY

Stein DM (1984) The sociobiology of infant and adult male baboons. Ablex, Norwood, NJ

Stoltz LP, Saayman GS (1970) Ecology and behaviour of baboons in the northern Transvaal. Ann Transvaal Mus 26:99–143

Storz JF, Beaumont MA, Alberts SC (2002) Genetic evidence for long-term population decline in a savannah-dwelling primate: inferences from a hierarchical Bayesian model. Mol Biol Evol 19:1981–1990

Struhsaker TT (1973) A recensus of vervet monkeys in the Masai-Amboseli Game Reserve, Kenya. Ecology 54:930–932

Strum SC (1982) Agonistic dominance in male baboons: an alternative view. Int J Primatol 3:175–202

Surbeck M, Mundry R, Hohmann G (2011) Mothers matter! Maternal support, dominance status and mating success in male bonobos (*Pan paniscus*). Proc R Soc Lond B 278:590–598

Tung J, Charpentier MJE, Garfield DA, Altmann J, Alberts SC (2008) Genetic evidence reveals temporal change in hybridization patterns in a wild baboon population. Mol Ecol 17:1998–2011

Tung J, Primus A, Bouley AJ, Severson TF, Alberts SC, Wray GA (2009) Evolution of a malaria resistance gene in wild primates. Nature 460:388–391

Tung J, Alberts SC, Wray GA (2010) Evolutionary genetics in wild primates: combining genetic approaches with field studies of natural populations. Trends Genet 26:353–362

Tung J, Akinyi MY, Mutura S, Altmann J, Wray GA, Alberts SC (2011) Allele-specific gene expression in a wild nonhuman primate population. Mol Ecol 20:725–739

Van Horn RC, Buchan JC, Altmann J, Alberts SC (2007) Divided destinies: group choice by female savannah baboons during social group fission. Behav Ecol Sociobiol 61:1823–1837

van Noordwijk MA, van Schaik CP (2001) Career moves: transfer and rank challenge decisions by male long-tailed macaques. Behaviour 138:359–395

van Schaik CP (1983) Why are diurnal primates living in groups? Behaviour 87:120–144

van Schaik CP, Paul A (1996) Male care in primates: does it ever reflect paternity? Evol Anthropol 5:152–156

Walters JR (1987) Transition to adulthood. In: Smuts BB, Cheney DL, Seyfarth RM, Wrangham RW, Struhsaker TT (eds) Primate societies. University of Chicago Press, Chicago, pp 358–369

Wasser SK, Risler L, Steiner RA (1988) Excreted steroids in primate feces over the menstrual cycle and pregnancy. Biol Reprod 39:862–872

Watts DP (1997) Agonistic interventions in wild mountain gorilla groups. Behaviour 134:23–57

Weingrill T, Lycett JE, Barrett L, Hill RA, Henzi SP (2003) Male consortship behaviour in chacma baboons: the role of demographic factors and female conceptive probabilities. Behaviour 140:405–427

Western D (2007) A half a century of habitat change in Amboseli National Park, Kenya. Afr J Ecol 45:302–310

Western D, Maitumo D (2004) Woodland loss and restoration in a savanna park: a 20-year experiment. Afr J Ecol 42:111–121

Western D, Sindiyo DM (1972) The status of the Amboseli rhino population. East Afr Wildl J 10:43–57

Western D, van Praet C (1973) Cyclical changes in the habitat and climate of an East African ecosystem. Nature 241:104–106

Widdig A, Nürnberg P, Krawczak M, Streich WJ, Bercovitch FB (2001) Paternal relatedness and age proximity regulate social relationships among adult female rhesus macaques. Proc Natl Acad Sci USA 98:13769–13773

Wroblewski EE (2010) Paternity and father-offspring relationships in wild chimpanzees, *Pan troglodytes schweinfurthii*. PhD thesis, University of Minnesota, Minneapolis

Ziegler TE, Santos CV, Pissinatti A, Strier KB (1997) Steroid excretion during the ovarian cycle in captive and wild muriquis, *Brachyteles arachnoides*. Am J Primatol 42:311–321

Chapter 13
The 30-Year Blues: What We Know and Don't Know About Life History, Group Size, and Group Fission of Blue Monkeys in the Kakamega Forest, Kenya

Marina Cords

Please note the erratum to this chapter at the end of the book.

Abstract Long-term studies uniquely allow researchers to investigate phenomena that play out over long periods, as well as rare events that accumulate slowly into a respectable sample. This chapter takes both approaches in reporting on a 30-year study of blue monkeys (*Cercopithecus mitis stuhlmanni*), reviewing life-history data mainly for females, which can live up to 33 years, and presenting data related to group fission, a rare event. Compared to close relatives, blue monkeys appear to have an exceptionally slow life history, related to low levels of mortality in forest environments. Group fissions show variable patterns, occurring at variable group sizes, and usually involving the splitting of a few family units, including mothers and young daughters. Ecological factors such as feeding competition do not appear to explain why fission occurs, and females do not seem to increase reproductive rates, improve infant survival or reduce the likelihood of male takeovers after fission.

13.1 Introduction

Long-term studies of animal populations uniquely allow two kinds of investigation. First, one can document phenomena that occur over long periods; second, one can examine patterns in rare events, which accumulate slowly. This chapter takes both perspectives in reporting on a 30-year study of blue monkeys (*Cercopithecus mitis stuhlmanni*), an African forest-dwelling guenon.

Most African guenons have not been well studied even on a short-term basis. Most likely, this reflects the practical difficulties of observation in the forested habitat that most species inhabit: dense vegetation, rather small body size, hairy faces, and the fact

---

M. Cords (✉)
Department of Ecology, Evolution and Environmental Biology, Columbia University,
New York, NY, USA
e-mail: mc51@columbia.edu

P.M. Kappeler and D.P. Watts (eds.), *Long-Term Field Studies of Primates*,
DOI 10.1007/978-3-642-22514-7_13, © Springer-Verlag Berlin Heidelberg 2012

that many are hunted makes field study, and especially individual identification, a significant challenge. Long-term study offers a potential solution because it permits habituation, at least in unhunted populations. Habituation allowed me to address questions and use field methods that were unthinkable at the beginning of our study.

After describing briefly some of the conditions of the study, I first summarize what we know about the life history of this species, based on records from individually identified animals monitored over three decades. This period approaches a natural lifetime. The data allow a robust understanding of basic life-history parameters, at least for females, and contribute to a comparative view of life-history variation in the primate order. Second, I address the related topics of group fission and group size, by examining aspects of the circumstances and consequences of rare fission events. Even after 30 years, the sample is small. The data suggest some common patterns, but also present puzzles that even longer-term study may help to resolve.

13.2 Study Site, Population, and Research History

13.2.1 Study Site

The study population inhabits the Kakamega Forest, western Kenya, a rainforest at 1,765 m with a gazetted area of 238 km^2 (Mitchell et al. 2009). Annual rainfall averaged 1,973 $\pm$ 310 mm over the 1979–2009 study period (unpublished KFS and BIOTA records). Combining elements of central African lowland and Afromontane forests, Kakamega is a relatively young forest (~10–12,000 years old), more isolated than many others in the East African region from similar forest patches. This history, together with decades of human influence (Mitchell 2004), has left the forest as an island amidst densely populated farmland, and diverse in the vegetative assemblages represented. Of the total gazetted area, less than half is natural forest, with the remaining portion plantation, bush- and grassland, and even farmland.

Our ~2 km^2 study site, located around the Isecheno forest station (0°14′11″ N, 34°52′02″ E), comprises mainly near-natural and old secondary forest (Fig. 13.1). This area was selectively logged in the 1930s, with enrichment planting of some indigenous species and some exotic species (which largely failed) in the 1940s (Mitchell 2004; see Watts (2012), for a similar history in another East African forest). Trees with the highest importance values (≥ 1.4, where IV $=$ unitless sum of relative density, relative basal area and frequency, with 3 as a maximum value; Grieg-Smith 1983) in each monkey home range include *Antiaris toxicaria* and *Trilepisium madagascariensis* (Moraceae), *Croton megalocarpus* (Euphorbiaceae), *Funtumia africana* (Apocynaceae), *Olea capensis* (*Oleaceae*), and *Polyscias fulva* (Araliaceae; Card 2010). In the last 5 years, after several group fissions and home range shifts in the study population, we have added other forest types to the areas occupied at least sometimes by certain study groups, including mature (70 year old)

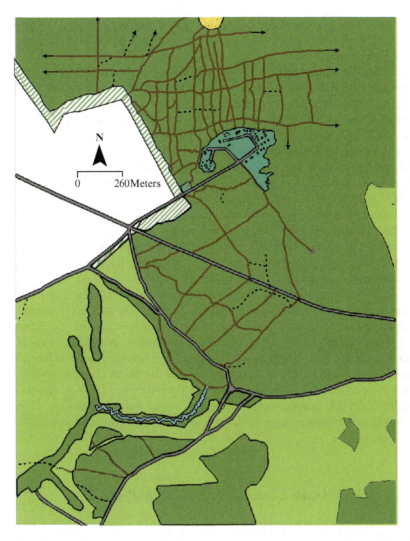

Fig. 13.1 The Isecheno study area. *Dark green* area shows forest cover (habitat used by *blue monkeys*), with footpaths used by researchers indicated in *brown*, dirt roads in *grey*. *Pale green* shows scrubland with some scattered trees, which the monkeys rarely cross; these areas of government land have varied histories, some having been plantations and/or areas of shifting cultivation. *Green striped area* represents a tea plantation ("Nyayo tea zone"), added in 1986, and white areas adjacent to tea show areas of human habitation with small-scale agriculture (private land). The forest station is indicated in turquoise. *Small yellow* area at top of the figure shows a natural grassland

"mixed plantations" (Mitchell et al. 2009) where *Prunus africana* and *Zanthoxylum gilletti* are the two most important species, and exotic plantations of *Bischofia javanica*, *Cupressus lusitanica*, *Grevillea robusta* and *Pinus patula*, all of which offer food to blue monkeys.

The Kakamega Forest as a whole has undergone significant habitat change even over the 30 years of this study, with anthropogenic increases in fragmentation and forest loss in many areas, and successional recovery in others that are better protected (Lung and Schaab 2006). The actual study area used for primate research appears to have been somewhat buffered from these forest-wide trends, probably because of its proximity to the forest station and perhaps because of the long-term presence of researchers (Fashing et al. 2004). Locally, habitat change has mainly taken the form of occasional (and illegal) tree- and liana-cutting. Two more major changes during the 30-year study include the 1986 razing of 60 m of edge forest to create the Nyayo tea zone, intended as a buffer area, and the 2008 razing of 20 m of forest along two roads passing through the study area for installation of power lines. Establishment of the tea zone decimated what was likely a considerable portion of the home range of what became one of our study groups 6 years later, but details of its effect on the monkeys are unknown as this group was not being closely monitored at the time. The 2008 cutting of forest for power lines led the three affected groups to change their ranging behavior, with each expanding into areas that were seldom or never used previously.

Natural predators of blue monkeys still occur in the forest. Alarm responses to raptors are common, typically a near daily occurrence. The African crowned eagle (*Stephanoaetus coronatus*), a confirmed predator elsewhere (Lawes et al. in press), is regularly if rarely seen, most often in aerial displays. In the first years of the study, I witnessed actual eagle attacks, but these have become much less frequent; it is possible that eagles prefer to hunt monkeys that are not as close to humans. The only witnessed predation involved a Gaboon viper (Förster 2008). We have occasionally observed humans, accompanied by dogs, hunting monkeys illegally. Circumstantial evidence suggests that a few of our study animals were killed by such hunters.

13.2.2 The Primate Community and Study Population

Common members of the Kakamega primate community include blue and redtail (*Cercopithecus ascanius*) guenons, as well as guerezas (*Colobus guereza*), which are the most commonly sighted diurnal species (Fashing and Cords 2000) and pottos (*Perodicticus potto*), which are regularly seen at night (K. Davey, W. Okeka, and E. Pimley personal communication). Rare or spatially restricted species include de Brazza's monkey (*Cercopithecus neglectus*) along particular river courses (Muriuki and Tsingalia 1990; Chism and Cords 1997/1998), olive baboons (*Papio hamadryas anubis*), and the occasional vervet (*Chlorocebus aethiops*; personal observation). Of these, deBrazza's monkeys are the only species never observed in our study area, which does not contain the riverine habitat they favor. While the more common species have been the subject of detailed study (Cords 1987; Wahome et al. 1993; Fashing 2001a, b, c, 2002; Chapman et al. 2002; Fashing et al. 2007), long-term individual-based records are available only for blue monkeys (Cords and Chowdhury 2010). The study population, at 198–242 individuals per km^2, is relatively dense (Fashing and Cords 2000), and it has been holding steady, or possibly increasing slightly, over the study period (Fashing et al. in press).

13.2.3 History of the Study

Blue monkeys at Kakamega live in groups of 7–65 members, with a single adult male most of the time (although other adult males may join during the breeding season, Cords 2002a). When I began research in July 1979, I studied just one group (T) of ~45 individuals (as well as one group of redtail monkeys). The T group fissioned in 1984 (Cords and Rowell 1986), but one daughter group moved into inaccessible habitat by 1989, forcing us to truncate records for these animals. In 1992 we began working with a neighboring group (G), which fissioned in 1999. Three subsequent fissions (2005–2009), described in more detail below (Table 13.1), left us with six groups in the study population in 2009.

Several aspects of the study conditions changed gradually over the three decades. First, the animals became more habituated to human observers, with particularly noticeable changes during the first 15–20 years. Although blue monkeys are primarily arboreal, I sometimes found them in low vegetation and even on the ground during my first 12 months, but they were skittish and retreated into the canopy when I came close. After 20 years, by contrast, the presence of observers seldom had any noticeable effect on their movements or location. Second, increasing habituation facilitated individual recognition, as it allowed the close-up inspections needed to distinguish individuals based on minutiae of their physical appearance. Our study never included any kind of capture or marking for this purpose; instead, we used features such as the shape of the tail end, nose and ears, hairiness of ears, and subtle differences in skin or hair color. For adult females,

Table 13.1 Group fissions in the Kakamega blue monkeys from July 1979 to July 2010

Size of parent group (name)[a]	Date of fission	Size of larger daughter group (name)[a]	Size of smaller daughter group (name)[a]	Fraction of matrilineal family units that broke apart[b]	Ranks of matrilineal units in the smaller group[c]
46 (T)	1984	33 (Tw)	13 (Te)	–	–
49 (G)	Oct 1999	28 (Gs)	21 (Gn)	–	–
61 (Tw)	Apr 2005	44 (Tws)	17 (Twn)	2/12	1, 2*, 3*
37 (Gs)	Oct 2008	30 (Gsa)	7 (Gsb)	2/8	4*, 5*, 6, 8
31 (Gsa)	Nov 2009	22 (Gsaa)	9 (Gsc)	2/5	2*, 4, 5*

[a]Group sizes exclude the resident adult male
[b]See Fig. 13.2 for details
[c]Highest-ranking matriline = 1; asterisks indicate matrilines whose members ended up in different daughter groups

nipple length and coloration, and for males, scars and stiff fingers, were also useful features, documented in recent years with digital photographs. Third, habituation allowed close following of individual animals and more continuous monitoring of their movement and activity patterns. Focal samples would have been unthinkable early on, but were possible by the late 1990s (Pazol and Cords 2005). Some of the results presented below derive from focal animal samples taken in recent years. Fourth, the observation schedule changed: after 2 years (1979–1981) of observations averaging 12 days per month, there followed 16 years in which observer presence was more intermittent, typically 2–5 months of near daily records per year. Beginning in mid-1997, however, observations occurred continually on a near daily basis (Cords and Chowdhury 2010 present further details). Fifth, the research focus shifted, reflecting what was possible logistically. Initially I studied interspecific sociality, examining associations of blue monkeys with redtails (Cords 1987, 1990a, b), and aspects of the mating behavior of adults (Cords 1988, 2002a; Cords et al. 1986). Only after 20–25 years was it possible to conduct detailed studies of social behavior of younger animals (Förster and Cords 2002, 2005; Ekernas and Cords 2007; Cords et al. 2010) and to amass rare events into an informative sample (Cords 2007; Cords and Fuller 2010). Lastly, the number of monitored groups increased throughout the study, from one in 1979 to six in 2009. This increase reflects the facts that greater habituation facilitated observational study generally and that we were motivated to continue monitoring long-studied individuals after natural group fission events.

13.3 Long-Term Data: Life History

13.3.1 Summary of Life-History Patterns

Beginning in late-1979, we collected data on births, disappearances and deaths, immigrations and emigrations of individually recognized animals. These data allowed derivation of basic life-history variables for this population (Cords and Chowdhury 2010), especially for females, the philopatric sex. Males emigrate from their natal groups at puberty (7 years; Ekernas and Cords 2007), and thereafter live for several years either alone or in loose associations with other males away from groups with females. After emigration, they therefore become difficult to monitor longitudinally. Here, I summarize the main findings of our study (details in Cords and Chowdhury 2010).

Blue monkeys have an extremely slow life history in the context of other cercopithecines (Cords and Chowdhury 2010). The mean, median, and modal age of first birth for females is 7 years. Most interbirth intervals lasted 24–36 months, averaging 25 months during the later years when observations were not intermittent; in these years, observers were not likely to miss births followed by neonatal death. This figure masks the usual strong effect of the first infant's fate on the interbirth interval, however. The mean interval increased from 18 months ($N = 53$) when the first infant died within 12 months to 31 months ($N = 193$) when it survived the first year (Cords and Chowdhury 2010). Given moderately strong seasonality in births (Cords and Chowdhury 2010), this means that most females produce a new infant after 2–3 years when the first one survives its first year.

Age-specific survivorship showed patterns common to many primates (Fig. 13.2), with higher mortality for infants (23% died, 5% were right-censored, i.e. had unknown fates) and a fairly constant rate of mortality from young adulthood onward. Annual mortality rates for juveniles (aged 2–4 years, all prereproductive) averaged $5 \pm 8\%$ ($N = 29$ years), an identical value for annual mortality of adults (aged 5 years or older, $5 \pm 5\%$, $N = 29$ years). The oldest female of known age lived to at least 33.5 years; several others that apparently died of natural causes (as opposed to a few cases in which human intervention was suspected) lived into their late 20s, and several females of this age are alive at the time of writing.

For animals with such long lifetimes, even 30 years is insufficient to document maximal lifespan with a large sample, and an estimate of average lifespan would be biased toward animals that died young. Despite sparse information on old females, our data suggest that female blue monkeys that live into their 20s and beyond may have postreproductive lifespans of several years. The female with the longest known lifespan did not give birth during the last 11 years of her life. Observations were not intermittent, so the long nonreproductive period cannot be attributed to undetected births followed by neonatal death. Other females also had long periods (8.0, 6.4, and 3.7 years) between the birth of their last offspring and their own deaths but intermittent observations, at least in the first two cases, make it impossible to ascertain that intervening births (followed by neonatal death) did not occur. Some other individuals, however, showed no evidence of reduced reproductive

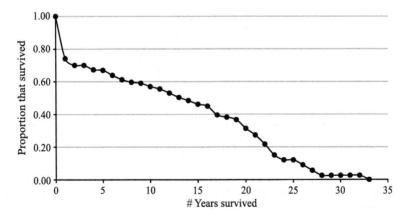

Fig. 13.2 Age-specific survivorship curve, based on minimum survival data from 418 individuals. Age class 0 refers to newborns, 1–0–1 year olds, and so forth. For further details, see Cords and Chowdhury (2010). Reprinted with kind permission from Springer Science + Business Media, Cords and Chowdhury (2010; Fig. 2)

rates even though they lived into their late 20s. Two females lived to ≥27 years and each gave birth for the last time within 2 years of her death.

We evaluated reproductive termination using Caro et al.'s (1995) criterion, by which a female is considered to have terminated reproduction if the difference between age at death and age at last birth exceeds the mean of her interbirth intervals by ≥2 standard deviations. Only 14 females with known or estimated ages died after giving birth at least three times, allowing calculation of mean birth intervals and standard deviation. Two of the 14 terminated reproduction, according to Caro et al.'s criterion. One of these females had the longest recorded lifespan at Kakamega and the other died at about 21 years. However, included in this analysis were four females who we suspected were victims of poaching; if their deaths were thus untimely, they may inflate the number of females that did not show slowdowns at the end of their lives.

Age-specific fecundity rose fairly steeply during years 5–8, and appeared to decrease gradually from about age 13 years (Fig. 13.3). The most advanced age at which a female gave birth was 26 years when we included only females whose ages were known to the nearest year. The fecundity curve shows nonzero values at ages 27 and 28 because there was some probability of a female being 27–28 years old at the birth of an offspring in cases where maternal age was estimated with greater uncertainty (see Cords and Chowdhury 2010 for further details on age estimation). Even though our data spanned a 29-year period, sample sizes for these old females were small.

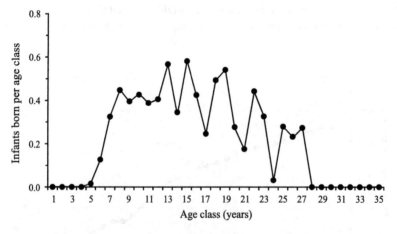

Fig. 13.3 Age-specific fecundity. Data include male and female infants born to 65 females. Greater fluctuations at later ages likely reflect reduced sample sizes. Reprinted with kind permission from Springer Science + Business Media, Cords and Chowdhury (2010; Fig. 3)

13.3.2 Blue Monkey Life History in a Comparative Context

The extreme slowness of the blue monkey life history is apparent when one compares these data with reports from other taxa. Ross (1992a) compiled data from multiple primate species on age at first reproduction, birth rate (interbirth interval), and maximal longevity to estimate r_{max}, the maximum potential rate of population increase. Data came from both captive and wild animals, but it is noteworthy that she identified blue monkeys as the second slowest breeding cercopithecid, after the closely related *C. ascanius*. Only apes had r_{max} values lower than these two forest-dwelling guenons.

Isbell et al. (2009, 2011) examined Ross' values in the context of body size variation to emphasize the extremely fast life history of patas monkeys relative to their body size. I repeated their analysis, using my data to calculate r_{max} for blue monkeys ($r_{max} = 0.11$). As Fig. 13.4 shows, among guenons some species breed considerably faster than expected for their body sizes, and others breed slower. Isbell et al.'s (2009) conclusion about the extraordinarily fast patas monkey is robust: among haplorrhines, only callitrichines have even higher r_{max} values relative to body size. Blue monkeys join *C. ascanius* and *Miopithecus talapoin* in having r_{max} values that deviate most negatively from the haplorrhine regression line, i.e. that are the lowest relative to expectations (Fig. 13.4). Thus the guenon tribe of cercopithecines includes both the fastest- and slowest-breeding Old World monkeys in this sample.

Cords and Chowdhury (2010) examined comparative data as well, but limited their consideration to *wild* cercopithecines. Even uncorrected for body mass (and *C. mitis* are among the smaller animals in this clade), blue monkeys had the latest age at first reproduction and were among three species with the longest mean

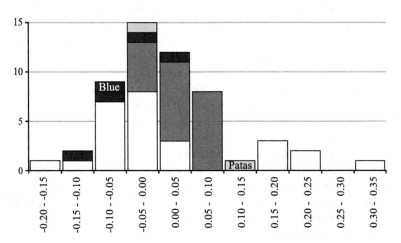

Fig. 13.4 Histogram of residuals of regression between ln body mass and ln r_{max} for haplorhine primates, using data from Ross (1992a). Forest guenons shown in *black*, open-country guenons in light *grey*. *Dark grey* represents other cercopithecoids, *white* shows noncercopithecoids, i.e., platyrrhines and apes

interbirth interval. Data on mortality after infancy require long-term study and are therefore scarce for wild populations, preventing a thorough analysis. However, a smaller-scale comparison within the guenon (Cercopithecini) tribe, which is thus more phylogenetically controlled, supports theoretical predictions (Promislow and Harvey 1990; Reznick et al. 2002) in relating life-history "speed" to levels of juvenile and adult mortality (Cords and Chowdhury 2010). When adult mortality is relatively high, selection should favor individuals breeding relatively early and often. Savanna-woodland guenons (vervets and patas monkeys) have annual juvenile and adult mortality rates three to six times higher than those of forest-dwelling blue monkeys: for example, Isbell et al. (2009) reported average annual adult mortality of 15% for vervets and 22–33% for patas, while the equivalent figure for Kakamega blue monkeys is only 5%. Similarly, maximal lifespans are very different, with vervets and patas living only into their teens (Isbell et al. 2009), while blue monkeys live into their 20s and sometimes into their 30s. Vervet and patas females first give birth much younger than blue monkeys (3 years patas, 4.5–5.7 years vervets), and interbirth intervals are considerably shorter (13–14 months). Finally, it is worth noting that these differences in life-history characters appear to be evolved ones: Rowell and Richards (1979) compared the same three guenons held in a single captive location and found the same relative patterns of age at first reproduction, interbirth interval and longevity as data from the wild provide.

Environmental variables are expected to affect life-history speed through their effects on mortality rates, but analyses of primate life histories in the context of environmental variation have been somewhat contradictory, perhaps reflecting limitations on both the data (both on the animals and their habitats) and the analytical methods (Ross and Jones 1999). The guenons are an excellent primate group in which to consider how habitat-related variation in mortality drives life-history speed, as this group includes both obligate forest dwellers, such as blue monkeys, and species that inhabit more open savanna-woodland environments. Mortality levels appear to be very different in these environments, even though very little information is available on causes of mortality in forests (Cords and Chowdhury 2010). It would be valuable to corroborate the preliminary cross-species comparisons with data from additional forest-dwelling guenons, but to my knowledge, there are no ongoing long-term studies that include individual-based life-history records. Replicating such analyses with data from wild populations of other primates may occur sooner: the macaques, a similarly species-rich group whose members inhabit a wide variety of habitats, would be good candidates. Indeed, Ross (1992b, but see Ross and Jones 1999), using data largely from captive macaques, found interspecific variation in life-history speed consistent with expectations from variation in the habitats characterizing wild populations. Long-term data from multiple wild populations are not yet available, but should provide an informative comparison.

13.4 Rare Events: Group Fission and Group Size

For social animals that spend their entire lifetimes in single groups, studying the adaptive benefits of group-living is difficult: one cannot experimentally manipulate the variable of interest, and even natural variation may be nonexistent. In blue monkeys, for example, females remain in their natal groups for life, never spending time alone or transferring to new groups. A female's group identity changes only during group fissions and fusions. These events offer a potentially revealing window through which to consider the costs and benefits of group-living (Dittus 1988; Van Horn et al. 2007), as both the circumstances surrounding the fission or fusion, and the way in which individuals realign themselves in new groups, may indicate what makes a group successful or not for its members. Such events are typically uncommon, however, and gathering even a modest sample of occurrences from which to generalize patterns can be a challenge. In 30 years, only five fissions occurred in our study groups, and the first example of a fusion occurred in the 30th year. Here I report on the fission events, and how they may inform an understanding of blue monkey society.

13.4.1 Basic Features of Blue Monkey Fissions

Group fissions occurred episodically over the 30-year period (Table 13.1). The rate of fission would be entirely different if one compared the first vs. last few years of the study. In addition, the critical group size at which fission occurs appeared to be 45–50 animals after the first two fissions (agreeing with reports from another wild guenon, *C. ascanius*; Struhsaker and Leland 1988; Windfelder and Lwanga 2002), but then group Tw grew to 61 animals before it split in 2005. By contrast, the two most recent fissions occurred in groups that were considerably smaller than in previous cases (Table 13.1), suggesting that the processes driving fission are not simple consequences of group size.

A common feature of all group fissions was that the parent group split unequally (see also Perry et al. 2012). The larger daughter group averaged 70% of original group members (Table 13.1). In all cases, the larger daughter group claimed the larger portion of the original group's territory. Another common characteristic of fissions was that one of the daughter groups expanded its territory within a year after the fission, engaging in a series of aggressive intergroup encounters with neighboring groups that had previously occupied the new area. In four of five cases, it was the smaller daughter group, relegated to the smaller portion of the original range, that did this, suggesting that an insufficient supplying area for food drove the territorial expansion. Territory boundaries remained remarkably stable except after fissions: indeed, some boundaries between groups have involved the very same trees over 30 years (Cords 2007). Similar range changes have been reported for redtail monkeys, with range expansion sometimes involving the larger and sometimes the smaller fission product (Struhsaker and Leland 1988; Windfelder and Lwanga 2002).

Unfortunately we had little background knowledge of neighboring nonstudy groups, which would assist a more detailed understanding of how these range expansions come about. A new group that is too small may face particular challenges. For example, 6 months after the most recent fission (2009), the ten-member Gsc group had managed to secure only a very small area of forest (approximately 3 ha) where it had priority over its neighbors. In contrast, other groups typically have areas of exclusive use that are five to ten times larger. The addition of nine new group members to Gsc (two adult females, seven mixed-sex juveniles), in the first fusion ever witnessed in blue monkeys, had not changed this situation 14 months later. Gsc seemed to move through the forest at the mercy of other groups, usually retreating from any they encountered. Struhsaker and Leland (1988) reported a similar consequence for an unusually small (~15 member) group formed after a redtail monkey group fission.

Females appeared to engineer the process of group fission in every case: over a period of days to months, they formed temporary subgroups with unstable membership, with no noticeable increase in aggression (see also Perry et al. 2012). Eventually, they settled into parties that remained apart from one another and aggressively defended their portion of the original group's territory against the new sister group. We have taken the date of the first aggressive territorial encounter between new sister groups as the date of fission in our study. Females also sometimes appeared to take an active role in deciding how the original groups divide. This was most obvious when they directed aggression at one another, seemingly trying to deter a former group-mate from joining their new sub-party. In contrast, no aspect of male behavior suggested that males attempted to influence the process of fission. The adult male resident in the original group joined the larger daughter group twice, and the smaller group three times.

Female cercopithecine monkeys typically remain with their female kin for life. One would thus expect fission to occur along kinship lines, as reported for baboons and several macaques (Dittus 1988; Oi 1988; Ménard and Vallet 1993; Okamoto and Matsumura 2001; Widdig et al. 2006; Van Horn et al. 2007). The three most recent fissions in our study population occurred when we had sufficiently deep pedigree records to evaluate whether close maternal kin (grandmother–grandoffspring, mother–offspring and sibling pairs) remained together after fission. While matrilines generally did stay together, each fission included 2–3 matrilineal units that broke apart (Table 13.1; Fig. 13.5). In six of these seven cases, mothers ended up in different groups than some or all of their daughters, and all six involved the separation of a mother and at least one *juvenile* daughter, even though juveniles are socially close to their mothers as indicated by frequent grooming (Cords 2000a; Cords et al. 2010). The seventh case involved three sisters whose mother was no longer alive at the time of group fission: the youngest (aged 4) did not join her older sisters after the split. A particularly puzzling observation was that one adult female (indicated by an asterisk in Fig. 13.5), abandoned her mother, sisters and 2.4 year old daughter (who was occasionally seen to suckle just weeks

13 The 30-Year Blues: What We Know and Don't Know About Life History

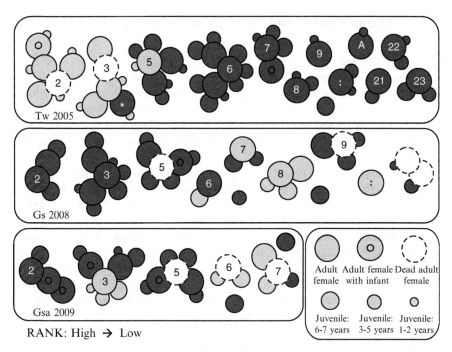

Fig. 13.5 Maternal kinship, rank, and group fission. For each of three fissions, individuals are represented as *circles*, with *shading* (*light vs. dark grey*) indicating group membership after fission, and size proportional to age (see legend). To indicate maternal relatedness, a large circle (mother) overlaps the circles representing her offspring. Matriline rank decreases from left to right, indicated by integer values (1 = highest). Matriline rank was derived from dyadic agonistic interactions among adult females, 9–12 months before fission; matrilines represented only by juveniles therefore have no rank and are randomly placed. Individual marked by *asterisk* is discussed in text

before the split) to become the sole member of her matriline in her new group. Although her family was second-ranking in the original group, and retained this position in their new daughter group, this female sunk to the bottom of the hierarchy in her new group, where she was regularly harassed. She seemed highly motivated to join this group, despite social obstacles and disadvantages. Most oddly, as the two daughter groups engaged in aggressive territorial fights, thus dividing up their former joint territory, this female crossed back on six occasions to fight with her family against the group in which she now lived.

Exceptions to the rule of kin remaining together after fission have been noted in rhesus and Japanese macaques (also capuchins, Perry et al. 2012), and attributed to paternal kinship and social bonds (Van Horn et al. 2007). Analyses now in progress may allow us to evaluate these factors as explanations for the severing of bonds with close kin in our study population.

13.4.2 Group Fission and the Costs of Feeding Competition

Primatologists often interpret group fission as an ecological necessity, inevitable when a group is too large for its individual members to move and forage efficiently because competition for food is too high (Koenig 2002; Sussman and Garber 2007). To see whether data from Kakamega support this interpretation of group fission, I compared movement and time budgets of adult females in groups of different size, and rates of aggression and reproductive output for the same females before and after fission.

As more animals in one place are likely to exhaust the food supply more quickly, larger groups should move farther and faster than smaller ones, or cover larger areas to meet their resource demands, assuming equal food density and abundance (Snaith and Chapman 2007). We examined these predictions using data on group travel, in which an hourly "center of mass" was plotted on a map of the study area, and all 50×50 m^2 quadrats used on a given day were noted. We found no difference in the area occupied per day among groups whose sizes differed up to twofold (Table 13.2a). Furthermore, while the group's daily travel distance varied significantly among three groups, the pattern was opposite the expectation, with the smallest group moving farthest. This pattern was not caused by variation in habitat quality, estimated as the basal area of food trees (m^2/ha), which would lead one to predict even longer routes in larger groups (Table 13.2a).

I also examined travel distances of individual females, as opposed to a center of mass of the entire group, to provide a higher-resolution analysis of foraging effort. Contrary to expectation, however, females in larger groups did not cover longer distances (Table 13.2b). Feeding activity – along with associated movement – has a diurnal rhythm, with peaks in the morning and afternoon hours (Cords 1987). When I analyzed the travel data separately for the morning, midday and afternoon hours, weak group effects emerged for morning and afternoon, but these were opposite those expected, with smaller groups moving farther than larger ones (Table 13.2b). Differences in habitat quality (measured as basal area of major food trees, Table 13.2b) could not explain these results: for example, females in larger groups might in principle reduce travel if they occupied better-quality habitat, but this was not the case. A possible explanation for greater travel in the two smaller groups (TWN, GN) relates to the attractiveness of particular resources at the edges of their home ranges: GN ate soil at a particular spot while TWN used exotic fruiting trees in the forest station.

To examine changes in time budgets, I compared females present both before and after two fissions (Gs group fission in October 2008, $N = 13$ females and Gsa group fission in October 2009, $N = 9$ females). We conducted focal samples (averaging 32 h per female in the 4 months before and in the 4 months after the 2008 fission, and 34 h per female in the 4 months before, 24 h per female in the 4 months after the 2009 fission) in which activity (moving, feeding, resting, social) was noted at 1-min intervals. I conducted before-vs.-after comparisons and present results separately for females in the two daughter groups, since fission might benefit

Table 13.2 Ranging variables and habitat quality for groups of different size

(a)	Gn: 23	Gs: 33	Tw: 46	Comparison
Area used by group (ha/day)	4.12 (2.75–5.73, 35)	4.25 (3.61–5.02, 29)	3.47 (2.62–4.88, 18)	Kruskal–Wallis, $p = 0.302$
Daily travel path of group (m)	733 (590–853, 39)	639 (536–761, 31)	583 (495–646, 17)	Kruskal–Wallis $H = 8.15$, $p = 0.017$
Habitat quality: basal area of food trees[a] (m^2/ha)	49.0	48.4	43.2	–

(b)	Twn: 16.5	Gn: 30.5	Tws: 43	Gs: 47	Comparison
Individual travel: estimated marginal means ± SEM	20.0 ± 1.2	18.8 ± 1.1	17.3 ± 2.5	19.1 ± 2.3	GLMM[b] Group: $F = 0.36$, $p = 0.75$; Period: $F = 6.28$, $p = 0.002$; Month: $F = 14.35$, $p < 0.001$
Individual travel AM (m per 10 min)	20.4 ± 2.2 (98)	23.8 ± 3.9 (130)	11.4 ± 1.7 (55)	16.8 ± 1.9 (23)	GLMM[c] Group: $F = 2.26$, $p = 0.093$
Individual travel MD (m per 10 min)	18.5 ± 2.3 (86)	16.6 ± 1.2 (119)	14.9 ± 3.6 (21)	14.0 ± 3.1 (29)	GLMM[c] Group: $F = 1.02$, $p = 0.383$
Individual travel PM (m per 10 min)	23.3 ± 1.9 (112)	23.1 ± 2.8 (117)	16.3 ± 5.2 (19)	10.7 ± 1.8 (13)	GLMM[c] Group: $F = 2.20$, $p = 0.090$
Habitat quality: basal area of food trees[d] (m^2/ha)	44.0	57.0	54.5	34.3	–

Each group indicated by name and total membership at time of data collection. (a) Area used and travel distance by group on days with ≥ 10.5 h observations, March–August 2001. Values reported are medians, with inter-quartile range and N of days in brackets. (b) Travel distance of individual females in 10 min samples (means ± SEM for each of N (in brackets) females). Row 1 shows estimated marginal means for each group as derived from full dataset; rows 2–4 show observed values from each of three periods, morning (AM: 0700–1030), midday (MD: 1030–1430) and afternoon (PM: 1430–1800)

[a] Trees included were the top 30, according to plant feeding records made over an annual period (Cords 1987), and including *Maesopsis eminii* which was identified later as a major feeding tree. These species accounted for 89% of all feeding records

[b] Data from 822 10-min focal samples collected from 54 females in four groups over 8 months (July 2005–May 2006). GLMM included group ($N = 4$) and period of day ($N = 3$) as fixed factors, month as covariate, and individual as random effect

[c] GLMM with auto-regressive covariance structure included group as fixed effect, date as repeated effect, and individual as random effect. Pairwise comparisons distinguished TWS from TWN and GN groups in the morning ($p \leq 0.024$), and GS from TWN and GN groups in the afternoon ($p \leq 0.02$), although these differences became insignificant with Bonferroni correction

[d] Trees included each accounted for $\geq 0.1\%$ of annual plant feeding records, as per Cords (1987). Together, they constituted 92% of all feeding records

one new group but not the other. In 2008, the five females that ended up in the smaller group showed no significant change in time budgets; for the eight that joined the larger group, moving time increased after fission from $18 \pm 6\%$ to $22 \pm 8\%$ of point samples (Wilcoxon matched-pairs signed-ranks test, $W = -34$, two-tailed $p < 0.02$), while other activities showed no significant changes. To the extent that moving time varies positively with energetic costs, an increase in smaller groups is just opposite to expectations. In 2009, the smaller group contained only two females, whose shifts in activity did not coincide. The seven females in the larger group increased feeding time (from $44 \pm 2\%$ to $48 \pm 2\%$, $W = -24$, two-tailed $p < 0.05$) and decreased resting time (from $26 \pm 2\%$ to $23 \pm 3\%$, $W = 28$, two-tailed $p < 0.02$) after fission. If food is harder to find in larger groups, and thus requires greater foraging effort, these changes are also opposite to expectations.

The relationship of contest competition to group fission can be evaluated by mapping dominance ranks onto fission dynamics and by comparing rates of aggression before and after fission. Neither approach suggests a clear relationship between fission and direct competition. We could evaluate dominance relationships for adult females for the four most recent fissions. In two cases (2005, 2008), fission separated higher-ranking from lower-ranking females (Fig. 13.5), although high-ranking matrilines formed the smaller daughter group in 2005 and the larger one in 2008. In the 1999 and 2009 fissions, high- and low-ranking females in the original group did not separate cleanly from one another: daughter groups included both high- and low-rankers, who generally maintained their relative positions in the new groups (Klass 2010; Fig. 13.5; n.b. the 1999 fission is not shown because kinship was not well known).

The lack of a consistent rank-related fission pattern agrees with the observation that the behavioral process of fission did not seem to involve one subgroup (presumably of low-ranking individuals) *seceding* from the main group, as has been reported in other cercopithecines where dominance rank is generally a more important predictor of social behavior (Malik et al. 1985; Dittus 1988). In blue monkeys, high- and low-ranking individuals seem to differ little socially, ecologically and reproductively (Cords 2000b; Pazol and Cords 2005, but see Förster et al. 2011); the lack of clear rank-stratification in group fissions is consistent with this general pattern, and agrees with the report of a single fission in relatively tolerant moor macaques (Okamoto and Matsumura 2001).

Blue monkeys generally exhibit low rates of aggression, but most aggressive acts – more than expected by chance – occur in a feeding context (Cords 2000b; Pazol and Cords 2005), suggesting that females compete directly for food. Therefore we also checked whether rates of aggression received decreased after fission for individuals in our study groups; such a pattern would suggest that fission reduces contest competition. Again we used focal animal data for 4 months before and after two group fissions, in which all aggressive interactions of focal subjects were noted. In one case (Gs fission in 2008), rates of aggression received did not change significantly (Wilcoxon matched-pairs signed-ranks test, two-tailed $p > 0.05$). For the eight females in the larger group, the average rate ($\pm$SEM) actually increased from 0.084 ($\pm$0.033) to 0.165 ($\pm$0.037) acts per hour, but this difference was not significant. Females that ended up in the smaller group received aggression at higher rates than those in the larger group, but the rate did not decrease significantly for them after fission either (before: 0.334 ($\pm$0.079), after: 0.309 ($\pm$0.092) acts per hour, $N = 5$). Altogether, 9 of 13 females received aggression at higher rates after fission. The results did not change if we considered all aggression, both received and given (data not shown).

In the second case (Gsa fission in 2009), the females that ended up in the larger group received aggression at lower rates after fission (before: 0.100 ($\pm$0.032) vs. after: 0.039 ($\pm$0.021) acts per hour; $W = 21$, $N = 7$, $p \leq 0.031$), suggesting that contest competition was reduced. Six of seven individuals received less aggression. The two females in the smaller group showed an inconsistent pattern, with one receiving aggression at a higher rate, and one at a lower rate, after fission. For all individuals, results were very similar if we considered all aggression, both received and given (data not shown).

The decrease in rates of received aggression that occurred in one new group after the second fission likely reflected changes in the monkeys' feeding behavior from the pre- to postfission periods. In 2009 only, the proportion of time spent feeding on fruits was lower after fission (median: 7% of observation time) than before fission (11% of observation time; Wilcoxon matched-pairs signed ranks test, $N = 9$, $W = 41$, $p \leq 0.027$); no such decrease occurred in 2008. Aggression occurs disproportionately when blue monkeys feed on fruit (Cords 2000b), so this change in diet might have driven changes in aggression rates. Indeed, rates of aggression relative to time spent feeding on fruit were not different before vs. after fission in either 2008 ($W = 44$, $N = 13$, $p \leq 0.946$) or 2009 ($W = 21$, $N = 9$, $p \leq 0.910$). Overall, then, analyses of aggressive rates did not support the hypothesis that rates of contest competition decreased after fission.

13.4.3 Group Fission and Female Reproduction

Ultimately, behavioral costs and benefits should be reflected in measures of reproductive output. In some cercopithecine monkeys (e.g., redtails: Windfelder and Lwanga 2002; baboons: Altmann and Alberts 2003), but not all, females have shown higher reproductive rates after fission. For three of the later fissions we observed, life-history data allowed us to examine whether reproductive success improved for individual adult females after fission. Figure 13.6 shows the birth rate for females that were monitored 4–6 years before and after three group fissions (1984, 1999, 2005). We deliberately chose a fairly long multi-year window before and after the fission to dampen effects of random variation in a species with interbirth intervals of 2–3 years. None of these comparisons revealed a significant difference (Wilcoxon test, all $p > 0.05$). We also compared rates of infant survival before vs. after fission (2005, 2008), but there were also no differences (Fig. 13.7). Overall, the results from these measures of reproductive success agreed with our behavioral measures in showing no advantage to females living in smaller groups after fission.

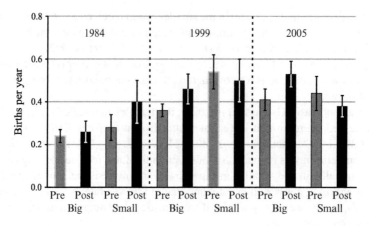

Fig. 13.6 Birth rate of females in big and small daughter groups before and after three group fissions. Females were included only if present both before and after fission for 4–6 years. No differences were statistically significant (see text)

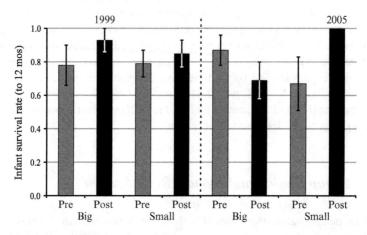

Fig. 13.7 Infant survival (to 12 months) for females in big and small daughter groups before and after two group fissions. Females were included only if they gave birth to at least one offspring within 4–6 years both before and after fission. No differences were statistically significant (see text)

13.4.4 Group Fission and Male Group Membership Changes

However, group size may affect reproductive success in other ways. In some primates, for example, infanticide risk is greater in larger groups (Crockett and Janson 2000). Blue monkey males in this population and others sometimes kill infants when they take over groups (Butynski 1982; Fairgrieve 1995; Cords and Fuller 2010). If group fission reduces the risk of infanticide, male takeovers should occur less often after fission than before. To date, however, there is no evidence that

13 The 30-Year Blues: What We Know and Don't Know About Life History

Table 13.3 Dynamics of adult male group membership before and after group fission

Group and fission date	Parent group	Larger daughter group	Smaller daughter group	Years of observation
(a) Rate of male replacement (# singular resident males per year)[a]				
T 1984	0.2	0.2	0.25	5
G 1999	0.6	0.6	0.6	5
Tw 2005	0.2	0.2	0.4	4.5
(b) Number of breeding season male influxes per breeding season				
T 1984	0.4	0	0	5
G 1999	0.6	0	0	5
Tw 2005	0.8	0.2	0.6	5

[a]Singular male residents lived in the group for some period with no other adult males present, in contrast to some male influx participants who were never the *only* male present

the rate of male takeovers or the occurrence of male influxes during the breeding season (which also introduce unfamiliar males to the group) declines after fission (Table 13.3). Reports of fission in other (redtailed) guenons actually showed increased rates of male membership changes after fission (Struhsaker and Leland 1988; Windfelder and Lwanga 2002).

13.5 Conclusions

Clearly 30 years is not enough to understand fully the group dynamics in a study population like ours. Five fissions is still a small sample. Furthermore, making inferences about the effects of group size based on comparisons of groups that vary in size can be difficult when groups are large and animals are hard to see and habituate. In such cases, the number of groups that researchers can monitor at once is limited. When our study began, we were also unaware that group size varied extensively, given that occasional counts of neighboring groups suggested numbers like those in our single study group. The process of fission, along with background demographic processes influencing age and sex composition, has actually expanded the range of variation in a small cluster of neighboring groups, and raised questions that were not even apparent early on. It took 30 years for us to detect a group with only ten members.

Further research should resolve some of the unanswered questions. While current evidence does not seem to support within-group competition for food as a factor stimulating fission or regulating group size, the advantages of smaller group size might be apparent only episodically or when evaluated over a longer period. The immediate stimulus for group fission may not be ecological disadvantage, but the disorganized and uncoordinated movement that typifies the largest groups (also Perry et al. 2012), or less connected social networks within groups that fission. Also, we have not thoroughly documented the extent of between-group competition, and its relationship to group size. Frequent aggressive intergroup encounters over feeding trees or feeding areas suggest an important role for between-group contest competition. We know already, however, that success in individual inter-group encounters depends primarily on where the encounter occurs rather than on relative group size, with even a single animal from the "home" group able to evict a larger party of intruders from its territory (Cords 2002b). This pattern raises a different question: how are territorial boundaries established, and when do they change? For blue monkeys, group fission is apparently an important part of the answer. Relative group sizes are set at the moment of fission, and group sizes in turn seem to determine territory sizes of the new groups; thus fission provides an opportunity for individuals to reconfigure their distribution – as groups – on the landscape.

Our ongoing study will address such possibilities, as well as the ways in which life-history variables respond to variation in group size. Ultimately, some signal in life history is expected to reflect the benefits that females gain from living in relatively small groups, hence to explain the propensity of groups to fission when they become large. Only long-term data offer the possibility to detect group size effects on fitness in a species with a life history as slow as that of blue monkeys.

Acknowledgments Funding for this research has come from the National Science Foundation (NSF-GF, SBR 95–23623, BCS 98–08273 and 05–54747), L.S.B. Leakey Foundation, Wenner Gren Foundation, Columbia University, the University of California, AAAS and the Ford Foundation. I am very grateful to the government of Kenya for research clearance, and the University of Nairobi Zoology Department, Institute of Primate Research, National Museums of Kenya, Moi University Department of Wildlife Management, and Masinde Muliro University of Science and Technology Biological Sciences Department for local sponsorship. I warmly acknowledge the local forestry officials and personnel for their on-the-ground help of many sorts. This long-term study would have been impossible without the fieldwork contributions of very many people, who added in large and small ways to the database. For their help over relatively long periods, I particularly thank P. Akelo, M. Atamba, S. Brace, B. Brogan, C. Brogan, S. Chowdhury, N. Cohen, S. Förster, M. Gathua, A. Fulghum, J. M. Gathua, J. Glick, S.F. Ihwagi, Kalabata, J. Kirika, K. McFadden, K. MacLean, S. Maisonneuve, C. Makalasia, S. Mbugua, C. Mitchell, N. Mitchell, S. Mugatha, C. Oduor, C. Okoyo, J. Omondi, B. Pav, K. Pazol, A. Piel, S. Roberts, T. Rowell, E. Shikanga, M. Tsingalia, and E. Widava. Many others have contributed over shorter time frames in the field, or to data management in the USA. I particularly thank N. Cohen, S. Mason, and K. Ross for contributing data on movement and activity patterns in groups of different size, K. Klass for her work on dominance hierarchies, and S. Chowdhury for her work on life-history data. J. Fuller kindly prepared Fig. 13.1. I thank T. Sleator for formatting all figures, and tolerating years of my field time.

References

Altmann J, Alberts SC (2003) Intraspecific variability in fertility and offspring survival in a nonhuman primate: behavioral control of ecological and social sources. In: Wachter KW, Bulatao RA (eds) Offspring: human fertility behavior in biodemographic perspective. National Academies Press, Washington, DC, pp 140–169

Butynski TM (1982) Harem-male replacement and infanticide in the blue monkey (*Cercopithecus mitis stuhlmanni*) in the Kibale Forest, Uganda. Am J Primatol 3:1–22

Card L (2010) Assessing fruit availability for a generalist frugivore, the blue monkey (*Cercopithecus mitis*), in the Kakamega Forest, Kenya. MA thesis, Columbia University

Caro TM, Sellen DW, Parish A, Frank R, Brown DM, Voland E, Borgerhoff Mulder M (1995) Termination of reproduction in nonhuman and human female primates. Int J Primatol 16:205–220

Chapman CA, Chapman LJ, Cords M, Gathua JM, Gautier-Hion A, Lambert JE, Rode K, Tutin CEG, White LJT (2002) Variation in the diets of *Cercopithecus* species: differences within forests, among forests, and across species. In: Glenn ME, Cords M (eds) The guenons: diversity and adaptation in African monkeys. Kluwer Academic, Plenum Publishers, New York, pp 325–350

Chism J, Cords M (1997/1998) De Brazza's monkeys *Cercopithecus neglectus* in the Kisere National Reserve, Kenya. Afr Primates 3:18–22

Cords M (1987) Mixed-species association of *Cercopithecus* monkeys in the Kakamega Forest, Kenya. Univ Calif Pub Zool 117:1–109

Cords M (1988) Mating systems of forest guenons: a preliminary review. In: Gautier-Hion A, Bourlière F, Gautier J-P, Kingdon J (eds) A primate radiation: the evolutionary biology of the African guenons. Cambridge University Press, Cambridge, pp 323–339

Cords M (1990a) Mixed-species association of East African guenons: general patterns or specific examples? Am J Primatol 21:101–114

Cords M (1990b) Vigilance and mixed-species association of some East African forest monkeys. Behav Ecol Sociobiol 26:297–300

Cords M (2000a) Grooming partners of immature blue monkeys (*Cercopithecus mitis*) in the Kakamega Forest, Kenya. Int J Primatol 21:239–254

Cords M (2000b) Agonistic and affiliative relationships of adult females in a blue monkey group. In: Whitehead PF, Jolly CJ (eds) Old World monkeys. Cambridge University Press, Cambridge, pp 453–479

Cords M (2002a) When are there influxes in blue monkey groups? In: Glenn ME, Cords M (eds) The guenons: diversity and adaptation in African monkeys. Kluwer Academic, Plenum Publishers, New York, pp 189–201

Cords M (2002b) Friendship among adult female blue monkeys (*Cercopithecus mitis*). Behaviour 139:291–314

Cords M (2007) Variable participation in the defense of communal feeding territories by blue monkeys in the Kakamega Forest, Kenya. Behaviour 144:1537–1550

Cords M, Chowdhury S (2010) Life history of *Cercopithecus mitis stuhlmanni* in the Kakamega Forest, Kenya. Int J Primatol 31:433–455

Cords M, Fuller JL (2010) Infanticide in *Cercopithecus mitis stuhlmanni* in the Kakamega Forest, Kenya: variation in the occurrence of an adaptive behavior. Int J Primatol 31:409–431

Cords M, Rowell TE (1986) Group fission in blue monkeys of the Kakamega Forest, Kenya. Folia Primatol 46:70–82

Cords M, Mitchell BJ, Tsingalia HM, Rowell TE (1986) Promiscuous mating among blue monkeys in the Kakamega Forest, Kenya. Ethology 72:214–226

Cords M, Sheehan MJ, Ekernas LS (2010) Sex and age differences in juvenile social priorities in female philopatric, nondespotic blue monkeys. Am J Primatol 72:193–205

Crockett CM, Janson CH (2000) Infanticide in red howlers: female group size, male membership, and a possible link to folivory. In: van Schaik CP, Janson CH (eds) Infanticide by males and its implications. Cambridge University Press, Cambridge, pp 75–98

Dittus WPJ (1988) Group fission among wild toque macaques as a consequence of female resource competition and environmental stress. Anim Behav 36:1626–1645

Ekernas LS, Cords M (2007) Social and environmental factors influencing natal dispersal in blue monkeys, *Cercopithecus mitis stuhlmanni*. Anim Behav 73:1009–1020

Fairgrieve C (1995) Infanticide and infant eating in the blue monkey (*Cercopithecus mitis stuhlmanni*) in the Budongo Forest Reserve, Uganda. Folia Primatol 64:69–72

Fashing PJ (2001a) Activity and ranging patterns of guerezas in the Kakamega Forest: intergroup variation and implications for intragroup feeding competition. Int J Primatol 22:549–577

Fashing PJ (2001b) Feeding ecology of guerezas in the Kakamega Forest, Kenya: the importance of Moraceae fruit in their diet. Int J Primatol 22:579–609

Fashing PJ (2001c) Male and female strategies during intergroup encounters in guerezas (*Colobus guereza*): evidence for resource defense mediated through males and a comparison with other primates. Behav Ecol Sociobiol 50:219–230

Fashing PJ (2002) Population status of black and white colobus monkeys (*Colobus guereza*) in Kakamega Forest, Kenya: are they really on the decline? Afr Zool 37:119–126

Fashing PJ, Cords M (2000) Diurnal primate densities and biomass in the Kakamega Forest: an evaluation of census methods and a comparison with other forests. Am J Primatol 50:139–152

Fashing PJ, Forrestel A, Scully C, Cords M (2004) Long-term tree population dynamics and their implications for the conservation of the Kakamega Forest, Kenya. Biodivers Conserv 13:753–771

Fashing PJ, Dierenfeld ES, Mowry CB (2007) Influence of plant and soil chemistry on food selection, ranging patterns, and biomass of *Colobus guereza* in Kakamega Forest, Kenya. Int J Primatol 28:673–703

Fashing, P, Nguyen, N, Luteshi, P, Opondo, W, Cash, J & Cords, M (in press) Evaluating the suitability of planted forests for African forest monkeys: A case study from Kakamega Forest, Kenya. Am J Primatol

Foerster S (2008) Two incidents of venomous snakebite on juvenile blue and Sykes monkeys (*Cercopithecus mitis stuhlmanni* and *C. m. albogularis*). Primates 49:300–303

Foerster S, Cords M, Monfort S (2011) Social behavior, foraging strategies and fecal glucocorticoids in female blue monkeys (*Cercopithecus mitis*): potential fitness benefits of high rank in a forest guenon. Am J Primatol 73:1–13

Förster S, Cords M (2002) Development of mother-infant relationships and infant behavior in wild blue monkeys (*Cercopithecus mitis stuhlmanni*). In: Glenn ME, Cords M (eds) The guenons: diversity and adaptation in African monkeys. Kluwer Academic, Plenum Publishers, New York, pp 245–272

Förster S, Cords M (2005) Socialization of infant blue monkeys (*Cercopithecus mitis stuhlmanni*): allomaternal interactions and sex differences. Behaviour 142:869–896

Grieg-Smith P (1983) Quantitative plant ecology, 3rd edn. University of California Press, Berkeley

Isbell LA, Young TP, Jaffe KE, Carlson AA, Chancellor RL (2009) Demography and life histories of sympatric patas monkeys, *Erythrocebus patas*, and vervets, *Cercopithecus aethiops*, in Laikipia, Kenya. Int J Primatol 30:103–124

Isbell LA, Young TP, Jaffe KE, Carlson AA, Chancellor RL (2011) Erratum to: Demography and life histories of sympatric patas monkeys, *Erythrocebus patas*, and vervets, *Cercopithecus aethiops*, in Laikipia, Kenya. Int J Primatol 32:268–269

Klass K (2010) Dominance and agonism among wild blue monkeys (*Cercopithecus mitis stuhlmanni*) in the Kakamega Forest, Kenya. MA thesis, Columbia University, New York

Koenig A (2002) Competition for resources and its behavioral consequences among female primates. Int J Primatol 23:759–783

Lawes MJ, Cords M, Lehn C (in press) *Cercopithecus mitis* profile. In: Butynski TM, Kingdon J & Kalina J (eds) Mammals of Africa, Volume 2: Primates. Bloomsbury Publishing, London

Lung T, Schaab G (2006) Assessing fragmentation and disturbance of west Kenyan rainforests by means of remotely sensed time series data and landscape metrics. Afr J Ecol 44:491–506

Malik I, Seth PK, Southwick CH (1985) Group fission in free-ranging rhesus monkeys of Tughlaqabad, northern India. Int J Primatol 6:411–422

Ménard N, Vallet D (1993) Dynamics of fission in a wild Barbary macaque group (*Macaca sylvanus*). Int J Primatol 14:479–500

Mitchell N (2004) The exploitation and disturbance history of Kakamega Forest, western Kenya. BIOTA East Report No. 1. Bielefelder Ökologische Beiträge 20:1–77

Mitchell N, Schaab G, Wägele JW (2009) Kakamega Forest ecosystem: an introduction to the natural history and the human context. BIOTA East Africa Report No. 5. Karlsruher Geowissenschaftlicher Schriften A, 17:1–58

Muriuki JW, Tsingalia HM (1990) A new population of de Brazza's monkey in Kenya. Oryx 24:157–162

Oi T (1988) Sociological study on the troop fission of wild Japanese monkeys (*Macaca fuscata yakui*) on Yakushima Island. Primates 29:1–19

Okamoto K, Matsumura S (2001) Group fission in moor macaques (*Macaca maurus*). Int J Primatol 22:481–493

Pazol K, Cords M (2005) Seasonal variation in feeding behavior, competition and female social relationships in a forest dwelling guenon, the blue monkey (*Cercopithecus mitis stuhlmanni*), in the Kakamega Forest, Kenya. Behav Ecol Sociobiol 58:566–577

Perry S, Godoy I, Lammers W (2012) The Lomas Barbudal Monkey Project: two decades of research on *Cebus capucinus*. In: Kappeler PM (ed) Long-term field studies of primates. Springer, Heidelberg

Promislow DEL, Harvey PH (1990) Living fast and dying young: a comparative analysis of life history variation among mammals. J Zool Lond 220:417–437

Reznick D, Bryant MJ, Bashey F (2002) *r*- and *K*-selection revisited: the role of population regulation in life-history evolution. Ecology 83:1509–1520

Ross C (1992a) Environmental correlates of the intrinsic rate of natural increase in primates. Oecologia 90:383–390

Ross C (1992b) Life history patterns and ecology of macaque species. Primates 33:207–215

Ross C, Jones KE (1999) Socioecology and the evolution of primate reproductive rates. In: Lee PC (ed) Comparative primate socioecology. Cambridge University Press, Cambridge, pp 73–110

Rowell TE, Richards SM (1979) Reproductive strategies of some African monkeys. J Mammal 60:58–69

Snaith TV, Chapman CA (2007) Primate group size and interpreting socioecological models: do folivores really play by different rules? Evol Anthropol 16:94–106

Struhsaker TT, Leland L (1988) Group fission in redtail monkeys (*Cercopithecus ascanius*) in the Kibale Forest, Uganda. In: Gautier-Hion A, Bourlière F, Gautier J-P, Kingdon J (eds) A primate radiation: the evolutionary biology of the African guenons. Cambridge University Press, Cambridge, pp 364–388

Sussman RW, Garber PA (2007) Cooperation and competition in primate social interactions. In: Campbell CJ, Fuentes A, MacKinnon KC, Panger M, Bearder SK (eds) Primates in perspective. Oxford University Press, New York, pp 636–651

Van Horn RC, Buchan JC, Altmann J, Alberts SC (2007) Divided destinies: group choice by female savannah baboons during social group fission. Behav Ecol Sociobiol 61:1823–1837

Wahome JM, Rowell TE, Tsingalia HM (1993) The natural history of de Brazza's monkey in Kenya. Int J Primatol 14:445–466

Watts DP (2012) Long-Term Research on Chimpanzee Behavioral Ecology in Kibale National Park, Uganda. In: Kappeler PM (ed) Long-term field studies of primates. Springer, Heidelberg

Widdig A, Nürnberg P, Bercovitch FB, Trefilov A, Berard JB, Kessler MJ, Schmidtke J, Streich WJ, Krawczak M (2006) Consequences of group fission for the patterns of relatedness among rhesus macaques. Mol Ecol 15:3825–3832

Windfelder TL, Lwanga JS (2002) Group fission in red-tailed monkeys (*Cercopithecus ascanius*) in Kibale National Park, Uganda. In: Glenn ME, Cords M (eds) The guenons: diversity and adaptation in African monkeys. Kluwer Academic, Plenum Publishers, New York, pp 147–159

Chapter 14
Long-Term Research on Chimpanzee Behavioral Ecology in Kibale National Park, Uganda

David P. Watts

Abstract Long-term data are crucial for addressing questions about the behavior and ecology of chimpanzees because of their slow life histories. Despite the long history of field research on chimpanzees, the number of sites that have provided long-term data on multiple communities in the same population is still small. Long-term data on two habituated chimpanzee communities in Kibale National Park, Uganda, have provided important insights into variation in chimpanzee behavioral ecology and life-history strategies. Long-term data on diet, phenology, and forest composition indicate that Ngogo is better habitat for chimpanzees; this helps explain why chimpanzee population density is three times higher there than at Kanyawara, why the Ngogo community is three times as large as that at Kanyawara, and why female gregariousness is higher at Ngogo. Both sites have provided important data on sex differences in gregariousness and in space use, on long-term social bonds, on hunting, on intergroup aggression, and on other important topics in behavioral ecology. The large size of the Ngogo community offers valuable insights into demographic influences on behavior and on male reproductive success and into chimpanzee-red colobus predator–prey dynamics. In this chapter, I summarize some of the major findings of this research and compare Kibale data to those from other long-term chimpanzee research sites.

14.1 Introduction

Chimpanzees (*Pan troglodytes*) are well-studied compared to most other primates, as evidenced by other chapters in this volume (Gombe: Wilson 2011; Mahale: Nakamura and Nishida 2011) and by ongoing long-term studies at Taï (Boesch and Boesch-Achermann 2000; Boesch 2009), Budongo (Reynolds 2005), and Bossou

---

D.P. Watts (✉)
Department of Anthropology, Yale University, New Haven, CT, USA
e-mail: david.watts@yale.edu

P.M. Kappeler and D.P. Watts (eds.), *Long-Term Field Studies of Primates*,
DOI 10.1007/978-3-642-22514-7_14, © Springer-Verlag Berlin Heidelberg 2012

(Matsuzawa 2002), younger ongoing projects elsewhere (e.g., Kalinzu: Hashimoto et al. 2001; Furuichi et al. 2011; Goualougo: Morgan and Sanz 2006), and multiple shorter studies (reviewed in Stumpf 2007). Phylogenetic closeness to humans makes chimpanzees a favored "referential model" (Moore 1996) for early hominins and partly explains the relatively large number of field studies. Detailed demographic, behavioral and ecological data exist for several chimpanzee populations, sometimes supplemented by noninvasively collected hormonal and genetic data used to address questions about reproductive ecology (e.g., Emery Thompson et al. 2007a), physiological costs of male–male competition (e.g., Muller and Wrangham 2005; Muehlenbein et al. 2004), male reproductive skew (e.g., Wroblewski et al. 2009), and other topics.

Paradoxically, chimpanzees are still poorly understood in many important respects. They are difficult to habituate, partly because fission–fusion sociality makes contacting individuals (especially females) repeatedly and reliably difficult. Like most other primates, chimpanzees have slow life histories; interbirth intervals and maximum lifespan are particularly long (Bronikowski et al. 2011). Thus many years are required to collect adequate data on survival, reproductive success, and other variables for direct tests of evolutionary hypotheses. For example, males compete and cooperate in complex ways within communities, and questions about the fitness consequences of alliance formation, grooming tactics, maneuvering for rank acquisition, and other aspects of male social strategies remain open despite years of study (Muller and Mitani 2005; Mitani 2009a; Fig. 14.1).

Ecological complexity compounds the challenges that social complexity poses for understanding chimpanzee behavior. Chimpanzees eat varied diets, and ecological flexibility allows them to occupy a wide range of habitats (Caldecott and Miles 2005). Variation in food abundance and distribution, predation risk, and socially transmitted extractive foraging techniques contribute to differences in socioecology among and within populations (e.g., greater female gregariousness at Taï than at Gombe or Mahale; Lehmann and Boesch 2009; Boesch 2009). But the number of

Fig. 14.1 Adult male chimpanzees at Ngogo, Kibale National Park, Uganda (Photo © David Watts)

sites with habituated communities is small and most long-term data come from single communities at each, which limits knowledge of within-population variation and hampers assessment and explanation of between-population variation. Researchers at Taï (Herbinger et al. 2001) habituated three adjacent communities, but subsequent population declines due to disease and poaching unfortunately limited comparative study. Mahale researchers habituated two communities, but have followed only one since the demise of K Group early in the study (Nakamura and Nishida 2012). All three communities at Gombe are now under study, but only two are subject to direct observations (Wilson 2012).

In this chapter, I review some highlights of two long-term chimpanzee research projects in Kibale National Park, Uganda. Kibale is the second largest remaining forest in Uganda and has the largest Ugandan population of chimpanzees. It has a long and distinguished history of research on primate behavioral ecology and African rainforest ecology (Struhsaker 1997; Chapman et al. 2005). Chimpanzee research there is now in its third decade at Kanyawara and well into its second at Ngogo. These two study areas are separated by only 10–12 km, but the resident chimpanzee communities differ vastly in size (see Sect. 14.3.1). Long-term comparative data on demography, diet, and the phenology of important food species at both sites thus provide important insights into socioecological variation. Kanyawara–Ngogo comparisons also raise intriguing questions about variation in chimpanzee life-history strategies. Moreover, results of Kibale chimpanzee research are relevant to many issues in behavioral ecology that can only be addressed with long-term data. These include the question of whether members of primate groups maintain consistent, long-term social relationships. Other questions concern hunting and meat eating: chimpanzees engage in group hunts of a variety of prey species, but determining whether hunting and meat transfers are cooperative and investigating predator–prey dynamics requires long-term data. A third set of issues concerns intergroup aggression, which involves within-group cooperation in chimpanzees: inter-group hostility seems universal in chimpanzees and can involve cooperative "patrolling" of territory boundaries and incursions into neighboring territories that sometimes lead to lethal attacks on members of other groups (Wrangham 1999). Patrols and incursions are infrequent events and observations of lethal attacks are rare (Wilson and Wrangham 2003; Mitani et al. 2010; below); thus long-term observations are necessary to assess their functions. Finally, as for most primates, measuring and explaining variation in reproductive success requires long-term data.

14.2 The Study Site: Kibale National Park

Kibale has had protected status since 1932, first as a Forest Reserve and eventually as a National Park. Research is conducted under the auspices of the Makerere University Biological Field Station (MUBFS), which maintains two principle research camps. In this section, I briefly describe research infrastructure, describe the habitat, and summarize the history of research, focusing on chimpanzees.

14.2.1 Research Infrastructure

Researchers require permission from the Makerere University Institute for the Environment and Natural Resources, which operates MUBFS; the Uganda Wildlife Authority, which manages the park; and the Ugandan National Council on Science and Technology. MUBFS maintains research facilities (including extensive trail grid systems) at Kanyawara and Ngogo. Facilities at Kanyawara include concrete houses for researchers, offices, a small library, dormitories used to host students enrolled in Tropical Biology Association field courses and participants in occasional conferences, and power from the main electricity grid. Individual researchers make their own arrangements for lab facilities, sample storage, computer use, etc. Ngogo camp is smaller; researchers stay in wooden houses or tents, have additional open-air work and eating space, and rely on solar panels for electricity and on propane for refrigeration and for freezing biological samples (e.g., chimpanzee urine).

14.2.2 Ecology of Kibale Forest

Kibale is in southwestern Uganda just north of the equator, between $0°13'-0°41'$ N and $30°19'-30°32'$ E (Fig. 14.2). The habitat follows decreasing north–south gradients in altitude (from 1,590 m to 1,110 m) and rainfall. Data on rainfall and minimum/maximum temperatures are collected daily at both sites by chimpanzee research project personnel and those of other projects; data from earlier research and historic records from nearby Fort Portal are also available (Struhsaker 1975, 1997; Chapman et al. 2010). A typical year includes two rainy seasons (September–November, March–May) and two dry seasons (December–February, June–August), although the amount and timing of rainfall varies considerably among years (Struhsaker 1997; Chapman et al. 2010). Mean annual rainfall at Kanyawara was 1,475 mm between 1970 and 1991 (Struhsaker 1997), but 1,749 mm between 1990 and 2001 (Chapman et al. 2010). Annual rainfall at Ngogo was $1,393 \pm 200$ mm from 1998 to 2009 (Mitani and Watts, unpublished data).

Kibale is a mid-altitude, moist semi-deciduous forest, with plant species composition and diversity intermediate between lowland and montane African moist evergreen forests (Struhsaker 1975, 1997; Butynski 1990; Chapman and Lambert 2000). The vegetation comprises a mix of old growth forest (about 57% of total surface area; Chapman et al. 2010); anthropogenic grassland; woodland; swamp forest; *Papyrus* swamp; and young forest at various stages of regeneration from human disturbance (Struhsaker 1975, 1997; Butynski 1990; Lwanga 2003; Chapman et al. 2010). Commercial logging occurred in some of the Kanyawara study area from the 1930s until the early 1990s (below); the Ngogo study area was never subjected to commercial logging (Struhsaker 1997) and consequently contains proportionately more old growth forest. However, even old growth forest at Ngogo shows signs of human disturbance (e.g., grindstones on the forest floor),

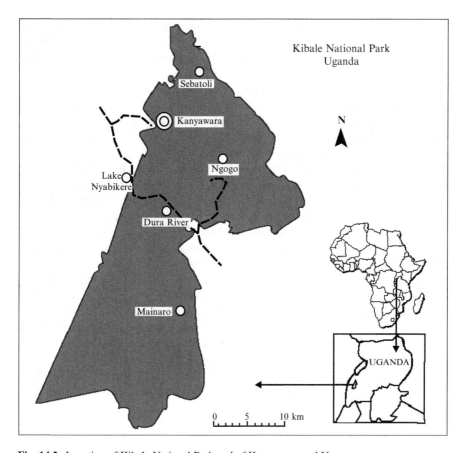

Fig. 14.2 Location of Kibale National Park and of Kanyawara and Ngogo

and Kibale provides a case study of potential forest recovery from anthropogenic disturbance that belies the mythology of "pristine" tropical forests (Struhsaker 1997; Lwanga 2003; Chapman et al. 2010).

Field assistants at both sites collect monthly data on the phenology of important chimpanzee food species. Comparative data on other common tree species, including those important in the diets of monkeys, are available from previous research (e.g., Struhsaker 1997) and other ongoing projects (e.g., Chapman et al. 2005). At Ngogo, the phenology sample includes 20 stems each of 20 tree species from which the chimpanzees eat fruit. These data serve to calculate a monthly ripe fruit score (RFS), given by (Mitani et al. 2002a):

$$\text{RFS} = \sum_{i=1}^{20} p_i \cdot d_i \cdot s_i,$$

where p_i = percentage of the ith tree species possessing ripe fruit, d_i = density of the ith tree species (stems per ha), and s_i = mean DBH (cm) of the ith tree species. The sample includes six fig species, and thus also provides phenology scores for figs and for nonfig-fruit separately.

Phenology data show complex patterns of variation in fruit abundance. For example, on average, 1.17–6.67% of stems of the 20 most common tree species at four Kibale sites fruited per month in each of the 12 years from 1990 to 2001 (Chapman et al. 2005), with no significant difference in the proportion of trees fruiting in wet- versus dry-season months. Inter-monthly variation in the availability of ripe fruit from species important in the Ngogo chimpanzee diet is bimodally distributed, with a moderate peak in May to July and a higher peak in September through December, but inter-annual variation in monthly and total annual fruit availability is pronounced (Mitani et al. 2002a; Watts et al. in press a, b; Fig. 14.3). Fruit production varies more at Kanyawara than Ngogo (Chapman et al. 1999) and many important chimpanzee food species are more abundant at Ngogo (Potts et al. 2009), which helps explain differences in chimpanzee population density between the sites (see Sect. 14.3.1). Some important chimpanzee food species fruit predictably in certain months, although not necessarily annually; for example, *Uvaropsis congensis* has a major fruiting peak around April–June in most years at Ngogo and a secondary peak around November–December in some years (Watts et al. in press b). Fruiting seasonality is less evident or absent in other cases (Struhsaker 1997; Chapman et al. 2005). *Chrysophyllum albidum*, an important chimpanzee food species at Ngogo, is a mast fruiting species; masting events have occurred in 2000, 2005, 2010, whereas in other years since 1995 this species has produced little

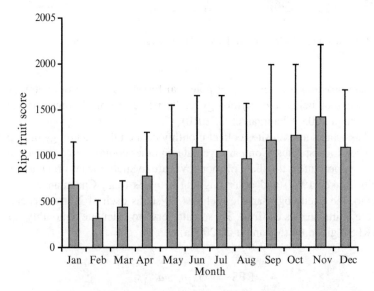

Fig. 14.3 Inter-monthly and inter-annual variation in ripe fruit availability at Ngogo, 1998–2009. Values are monthly mean ripe fruit index scores; error bars show one standard deviation

fruit. Neither the remarkable inter-annual variation in fruit availability nor the gradual changes in tree communities associated with regeneration from human disturbance, the aging of mature forest, and climate change would be evident without long-term phenology data and the rich history of work on vegetation dynamics (e.g., Chapman et al. 1999, 2005, 2010; Struhsaker 1997; Lwanga et al. 2000; Lwanga 2003; Omeja et al. 2009).

Seven diurnal primate species besides chimpanzees occur in Kibale: three guenons (redtailed monkeys, *Cercopithecus ascanius schmidti*; blue monkeys, *Cercopithecus mitis stuhlmanii*; L'hoest's monkey, *Cercopithecus l'hoesti*), two colobines (red colobus, *Procolobus rufomitratus tephrosceles*; black and white colobus, *Colobus guereza*), baboons (*Papio anubis*), and grey-cheeked mangabeys (*Lophocebus albigena*). Vervets (*Chlorocebus aethiops*) occur along the forest margin and in grasslands in some areas. Nocturnal primates include pottos (*Perodicticus potto*), Demidov's dwarf galago (*Galagoides demidovii*), and the eastern needle-clawed bushbaby (*Galago inustus*). Leopards might have preyed on chimpanzees historically, but are now absent. Kibale is home to a variety of forest ungulates, including several on which chimpanzee prey (blue duiker, *Cephalophus monticola*; red duiker, *Cephalophus callipygus*; bushbuck, *Tragelaphus scriptus*; bushpig: *Potomochoerus porcus*; Watts and Mitani 2001), and to diverse rodent, avian, and invertebrate faunas (reviewed in Struhsaker 1997). Elephants are abundant; their feeding on saplings and pole-sized stems and trampling of seedlings influences forest structure and, combined with rodent and insect predation on seeds and seedlings, can prevent forest regeneration in large gaps (Struhsaker et al. 1996).

14.2.3 A Brief History of Kibale

The Government of the Uganda Protectorate gazetted Kibale as a Crown Forest, to be managed for timber production by the Uganda Forest Department, in 1932 (Struhsaker 1997). Management included various intensities of selective commercial logging in some forest compartments and clear-cutting and replacement of native species with plantations of exotic pine and cypress in some others, while other areas were left intact or subjected to only small-scale pit sawing for local use. The Kanyawara area experienced a mix of all these options. In contrast, the Ngogo area was not subject to commercial logging, although it shows other effects of human disturbance (e.g., anthropogenic grasslands). Struhsaker (*ibid.*; cf. Omeja et al. 2009) provides a detailed history of how logging affected vegetation composition and populations of arboreal cercopithecoid primates, ungulates, and rodents, and summarizes the conflicts between wildlife conservation and management for timber extraction.

Conflict heightened in 1964 with the establishment of a wildlife corridor between Kibale and Queen Elizabeth National Park to the southwest, but was resolved in 1993, when Kibale was declared a National Park. This brought it under the administrative control of the Uganda Wildlife Authority, ended all legal timber extraction once remaining plantation forest had been clear-cut, and shifted management priority to wildlife and habitat protection and ecotourism development. Recent research by Omeja et al. (2009) has shown that the forest is regenerating naturally in previously clear-cut areas at Kanyawara, and that enrichment planting of indigenous trees does not improve regeneration success. Likewise, Jeremiah Lwanga's (2003) work at Ngogo shows that forest can regenerate from grassland naturally so long as animal seed dispersers are still present and the young forest/grassland mosaic is protected from fire.

In 1970, Thomas Struhsaker initiated the Kibale Forest Project at Kanyawara; in 1972, he established a second research area at Ngogo, in the forest interior (Struhsaker 1975, 1997; Fig. 14.2). His primary focus was forest ecology and the behavioral ecology of red colobus monkeys (Struhsaker 1975, 2010), but he also studied the behavior ecology of redtailed monkeys and grey-cheeked mangabeys and the effects of logging. Since 1971, many expatriate and Ugandan students, often mentored by Struhsaker, and many independent researchers have studied primate behavioral ecology, forest dynamics, rodent population biology, elephant ecology, and other topics (Struhsaker et al. 1996; Chapman and Chapman 2000).

Notable among the numerous long-term research efforts besides Struhsaker's is work on primate ecology, complemented by regular primate censuses, by Colin Chapman and his students. This has mainly focused on Kanyawara, but covers multiple sites, including forest fragments outside the park (e.g., Chapman et al. 2002). Among its important results has been demonstration of unexpectedly high within-habitat variation in red colobus diet composition and food nutritional quality (Chapman et al. 2002, 2010). To summarize this rich body of research and the many other long-term Kibale projects is beyond the scope of this chapter, but it is clearly important for understanding chimpanzee behavioral ecology and for conservation.

14.2.4 Chimpanzee Research

Michael Ghiglieri conducted the first Kibale chimpanzee research, at Ngogo, in 1976–1978 (Ghiglieri 1984, 1986). He collected valuable information on diet opportunistically and by watching large trees that had attractive fruit crops, especially figs. He confirmed that chimpanzees at Ngogo had a fission–fusion social system and that males were relatively highly gregarious, and he noted that males associated and groomed more with each other than with females (Ghiglieri 1984). However, limited habituation prevented him from following the chimpanzees consistently or from obtaining accurate information on community size and composition. He did not see chimpanzee hunts, nor did he see boundary patrols or interactions with neighboring communities (*ibid.*). This was ironic given later documentation of extremely high predation on red colobus and of high rates of patrolling and lethal inter-community aggression at Ngogo (see Sects. 14.3.4 and 14.3.5), especially because Power (1990), citing Ghiglieri, argued that the supposed absence of hunting and lethal intergroup aggression at Ngogo showed that such behaviors at Gombe and Mahale were artifacts of disturbance by humans.

A hiatus on chimpanzee research at Ngogo followed Ghiglieri's fieldwork. Meanwhile, Gilbert Isibirye-Basuta initiated research on chimpanzee behavioral ecology at Kanyawara in 1985. He habituated the chimpanzees sufficiently to gather quantitative data on diet, on the relationship of party size to food patch size, and on how gregariousness varied in relation to variation in fruit abundance (Isibirye-Basuta 1989). Subsequently, Richard Wrangham started the Kibale Chimpanzee Project at Kanaywara in 1987. The Kanyawara chimpanzee community has been observed continuously since then and its members have been well habituated for most of that time. Under Wrangham's direction and, since 2008, the co-direction of Martin Muller, researchers affiliated with the project have studied many aspects of behavior and ecology (e.g., male behavioral endocrinology: Muller and Wrangham 2004a, b; competition between resident and immigrant females: Kahlenberg et al. 2008a, b) and compiled a long-term demographic and life-history data base that substantially adds to comparative data from other long-term chimpanzee research sites (Emery Thompson et al. 2007b; Hawkes 2010).

Chimpanzee research resumed at Ngogo in 1991 when Bettina Grieser-Johns, helped by expatriate research assistants and a Ugandan field assistant, started habituation efforts (Grieser Johns 1996). Simultaneously, R. Wrangham and C. Chapman employed four other Ugandan field assistants to work on habituation and to conduct comparative research on chimpanzee and baboon feeding ecology that combined direct observational with collection of fecal samples (Wrangham et al. 1991). David Watts spent 2 months at Ngogo in 1993 to investigate prospects for a long-term study; he returned with John Mitani in 1995 to initiate the Ngogo Chimpanzee Project. In collaboration with Jeremiah Lwanga (since 1997) and with the help of doctoral and postdoctoral researchers and Ugandan field assistants, they have since maintained a continuous research presence. Steady progress with habituation led to the identification of all community members; to extensive databases on hunting, meat eating, and inter-community aggression; and to detailed studies of feeding ecology, mating strategies, and many aspects of social relationships. At 16 years, the Ngogo demographic and life-history database is still shallow, but it allows preliminary comparisons with data from other chimpanzee research sites (see Sect. 14.3.2).

Both chimpanzee research projects have provided extensive research and training opportunities for Ugandan scientists. For example, Emily Otali received her doctorate from Makerere University for research on chimpanzee social behavior at Kanaywara and is currently (2010) Project Manager for the Kibale Chimpanzee Project. Jeremiah Lwanga, who had earlier done Master's research (Makerere University) on blue monkey behavioral ecology and doctoral research (University of Florida) on forest ecology in Kibale, became Project Manager for the Ngogo Chimpanzee Project in 1997; he is currently (2010) Interim Director of MUBFS. Both projects have sponsored Master's research by Makerere students. Researchers and Project Managers at both sites work with UWA officials and park rangers to help coordinate anti-poaching efforts, and the research projects have contributed formally and informally to financing these efforts and to maintenance of park infrastructure. Researchers provided information and advice that facilitated successful habituation of chimpanzees for tourism at Kanyanchu, a Kibale site reserved for that purpose, and they participate in planning of park management strategy and of chimpanzee and forest conservation efforts in Uganda. As at other African research sites where great apes are habituated, research and tourism bring risks of disease transmission from humans, but also help to protect the animals and their habitats (Köndgen et al. 2008).

Some individual research projects have focused on feeding ecology (e.g., Wrangham et al. 1998; Potts 2008), and Ugandan field assistants at both sites routinely collect data on diet via scan sampling; these data plus phenology records allow long-term assessment of dietary variation. Field assistants systematically collect data on party size and composition, home range use, and aspects of social behavior on a near-daily basis and collect data specific to particular research questions on an as-needed basis. All researchers and field assistants contribute to long-term demography and life-history data bases and provide data on uncommon, but important events such as hunts and meat eating, boundary patrols, and inter-community interactions. Kanyawara researchers have established a relational database that covers many aspects of behavior and ecology (e.g., diet; Gilby et al. 2010b) and a website with information about research and conservation (http://www.fas.harvard.edu/~kibale/index.html). Ngogo researchers have not yet developed such resources, nor have all aspects of data collection been standardized between Kanyawara and Ngogo; collaboration in data collection and management should be a priority for the future.

14.3 Important Results of Long-Term Research

14.3.1 Feeding Ecology, Community Size, and Population Density

Gaining accurate information on community size and composition and on the size of the respective home ranges required years of effort. Since the late 1980s, the Kanyawara chimpanzee community has numbered about 40–55, including 8–12 adult males. In contrast, the Ngogo community has had from about 142–165 members, with between 22 and 32 adult males; these numbers are unprecedented in chimpanzee research. The home range of the Ngogo community was about 29 km^2 between 1998 and 2009 (Mitani et al. 2010). The Kanyawara community has a home range of about 35 km^2 (Emery Thompson and Wrangham 2008); thus, chimpanzee population density is three times higher at Ngogo. Such large differences in community size and population density are startling given the proximity of the sites and caution against assuming that data on the behavioral ecology of a single community represent its population. They would not have been evident had research lasted only a few years, nor would the large extent of the contrasts in forest ecology and chimpanzee feeding behavior over this small geographic scale that help to explain them.

Studies of primate feeding ecology typically span a single annual cycle, but even several years may be inadequate to document the full range of dietary flexibility in species that have diverse diets and in whose habitats food availability does not follow predictable annual cycles. Long-term Kibale data are consistent with those from other sites indicating that chimpanzees are highly frugivorous, but also highlight the absence of universal constraints on their foraging strategies. Fruit, including figs, accounts for most chimpanzee feeding time at both sites (Fig. 14.4; Conklin-Brittain et al. 1998; Wrangham et al. 1996, 1998; Potts et al. 2009; Watts et al. in press a).

At Kanyawara, data from one annual cycle showed that feeding time devoted to nonfig fruit was positively related to its abundance, consistent with the argument that chimpanzees are ripe fruit specialists (Wrangham et al. 1998). Long-term data from Ngogo substantiate this finding: the percentage of feeding time devoted to nonfig fruit was positively related to its abundance over a 125-month period (Watts et al. in press a). Figs also have great importance and account for the second highest percentage of chimpanzee feeding time at both sites (Wrangham et al. 1993; Watts et al. in press a). More detailed consideration of fig use shows a crucial difference between the sites, however: figs from *Ficus mucuso* are quantitatively the most important food at Ngogo, whereas this species is absent at Kanyawara. Availability

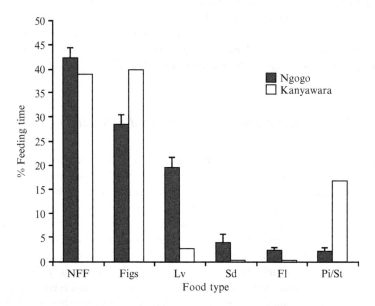

Fig. 14.4 Percent of annual feeding time devoted to major plant food categories at Ngogo and Kanyawara. Ngogo data cover eight consecutive years; data from Kanyawara cover four consecutive years (Wrangham et al. 1996). Column height shows annual mean; error bars give 1 sd for Ngogo data. *NFF* non-fig fruit; *Figs* all fig species; *Lv* leaves; *Sd* seeds; *Fl* flowers; *Pi/St* pith and stems. Minor items (e.g., rotting wood) and meat not included

of figs from *F. mucuso*, which fruits nonseasonally, is independent of the availability of nonfig fruit, and the proportion of feeding time devoted to *F. mucuso* varies inversely with time feeding on nonfig fruit. It is thus a "filler" fallback food, *sensu* Marshall et al. (2009). Kanyawara chimpanzees also use figs as fallbacks (Wrangham et al. 1993), but no species there provide the enormous fruit crops of *F. mucuso*, which has a relatively high stem density in the Ngogo study area and produces some fruit during all months of an average year. Terrestrial piths and stems are also fallbacks at Kanyawara (Wrangham et al. 1993), but not at Ngogo, where leaves supplement figs as fallbacks (Watts et al. in press b). Leaves of *Pterygota mildbraedii* saplings, which are highly abundant at Ngogo, are among the most important foods there, and seeds from the fruit of mature *P. mildbraedii* are also important; this species is virtually absent at Kanyawara. In all, chimpanzees at Ngogo devote most of their feeding time to foods that, for reasons that are not well understood, are common there but rare or absent at Kanyawara (Potts et al. 2011; Watts et al. in press a).

Observational data on diet composition are also available from studies at six other chimpanzee research sites, most covering much shorter periods than studies of feeding ecology in Kibale (summarized in Morgan and Sanz 2006). Considerable cross-site variation exists, but inter-annual variation in Kibale nearly encompasses the total range. For example, the proportion of annual feeding time devoted to fruit (including figs) varies from 59% (Gombe: Wrangham 1977) to 87% (Ngogo, females only: Wakefield 2010); data collected independently of Wakefield's (*ibid.*) at Ngogo gave a community-wide range of 63.4–76.2% over eight consecutive years (Watts et al. in press a), and Kanyawara values over four consecutive years varied from 74.5% to 84.5% (Wrangham et al. 1996). Such extensive variation in one population highlights chimpanzee ecological flexibility.

14.3.2 Female Reproductive Ecology and Life-History Strategies

Emery Thompson and Wrangham (2008) used 12 years of data on ovarian steroid hormones, female reproductive parameters, fruit availability, and diet to show that ovarian functioning of "central" females at Kanyawara varied with their intake of drupaceous fruit and that they were more likely to cycle and to conceive when drupe intake, and thus relative energy intake, was high. Long-term Kanywara data also show inter-individual variation in ovarian function and reproductive rates associated with variation in female core area quality (Emery Thompson et al. 2007a) and dominance ranks (Kahlenberg et al. 2008b), presumably due to variation in net energy intake mediated by differential access to good feeding areas and to individual feeding sites. Given lower variation in fruit availability at Ngogo (Chapman et al. 1999), questions arise concerning whether Ngogo female reproductive rates are higher than those of Kanyawara central females, mirroring the faster central vs. slower "northern" Kanyawara female contrast (Emery Thompson et al. 2007a), and whether survivorship is higher at Ngogo. These are important

14 Long-Term Research on Chimpanzee Behavioral Ecology

with regard to reconstructions of human life-history evolution that compare pooled chimpanzee data from Gombe, Mahale, Tai, Bossou, and Kibale-Kanyawara to data on recent hunter–gatherers (e.g., Kaplan et al. 2000; Hawkes 2010). The chimpanzee data show a population in decline, a condition that cannot have held over evolutionary time (Hill et al. 2001).

Preliminary Ngogo data indicate that reproductive rates may indeed be high. Ngogo records now include 20 completed intervals between the birth of a surviving infant and the birth of the female's subsequent infant. Mean interbirth interval length is 62.9 months (sd $= 8.31$, median $= 63$ months). This is significantly shorter than the mean of 69.1 months (sd $= 8.58$, median $= 68$ months) for 31 intervals at Mahale (Nishida et al. 1985; unpaired t-test, $t = 2.52$, df $= 49$, $p = 0.015$), and shorter, although not significantly so, than the mean of 69.1 months (sd $= 17.4$, median $= 65$ months) for 33 intervals at Taï (Boesch and Boesch-Achermann 2000; unpaired t-test, $t = 1.75$, $p = 0.087$). The Ngogo mean is substantially lower than the mean of 73 months (sd $= 13.5$) for central females at Kanyawara (Emery Thompson et al. 2007a), although the Kanyawara values include incomplete intervals and are thus not directly comparable. The mean length of completed intervals following surviving births at Gombe is around 6 years (Fig. 1 in Jones et al. 2010). Excluding censored intervals tends to decrease estimates of interval length (*ibid.*). Still, a Kaplan–Meier survival analysis of Ngogo data that includes incomplete intervals of at least 24 months ($N = 48$) yields a median interval of 62.5 months, considerably shorter than the means for Kanyawara and Gombe. Restricting the analysis to incomplete intervals of at least 48 months, as Emery Thompson et al. (2007a) did for Kanyawara, also gives a median of 62.5 months.

Infant survival to 1 year is not unusually high at Ngogo (77.6% of 98 births, including three within-community infanticides), nor is survival to 5 years (71.8% of 75 births) high compared to Kanyawara (c. 85% survival to 5 years; Emery Thompson et al. 2007b) or to Taï before predation and disease epidemics greatly increased mortality (88% survival to 5 years; Boesch and Boesch-Achermann 2000). However, mortality at Ngogo has been concentrated among known first offspring (9/15 births, including one infanticide) and has been lower for infants of multiparous females. Perhaps females respond to favorable ecological conditions by starting their reproductive careers before they are fully grown, then face tradeoffs between investing in offspring growth or in their own growth and survival, as Altmann (1980) described for baboons. In contrast, adult female survival seems relatively high at Ngogo. Assigning females to 5-year age categories based on known or estimated ages and calculating deaths per female per year gives an approximate value of 69.9% survival to age 30, considerably higher than the value of 46% that Emery Thompson et al. (2007b) calculated using data from five other sites. However, this tentative result awaits confirmation with longer-term data and proper comparative analysis.

14.3.3 Social Relationships

Long-term research confirms Ghiglieri's (1984) inference that on average, males are more gregarious than females in Kibale (Mitani et al. 2000; Gilby et al. 2008; Langergraber et al. 2009). However, habituation led to the realization that Ngogo females are generally more gregarious than those at Mahale and Gombe and as gregarious as those at Taï, which shows that no east/west dichotomy of female chimpanzee gregariousness exists. Also, variance in "pairwise dyadic affinity" indices is higher for males than females, with both the lowest and the highest values belonging to female dyads (Langergraber et al. 2009).

Males invest considerable effort in competing for status and form dominance hierarchies (Watts 1998; Watts and Mitani 2001; Muller and Wrangham 2004a, b, 2005; Mitani 2009a). Male rank is positively related to reproductive success at Ngogo, but reproductive skew is low, as expected given the large number of males (Langergraber et al. 2010; cf. Wroblewski et al. 2009). Male–male social relationships are differentiated, with some dyads associating in parties, remaining in close proximity, grooming, and forming coalitions more than others (Watts 2000a, b; Mitani et al. 2002b; Langergraber et al. 2007; Gilby et al. 2008; Mitani 2009a, b). Ngogo has provided some of the most extensive data on social exchange among male chimpanzees: males show reciprocity in grooming, coalition formation, and meat sharing and interchange between grooming and coalitionary support, grooming and meat sharing, and coalitionary support and meat sharing (Watts 2000b, 2002; Mitani and Watts 2001; Mitani 2009a). Males also engage in boundary patrols most often with others who are their main grooming and coalition partners and with whom they most often participate in hunts (Watts and Mitani 2001).

Langergraber et al. (2007) used noninvasive genetics sampling at Ngogo to help resolve long-standing questions about the influence of kinship on male–male social relationship. Compared to chance expectations, maternal brothers associated in the same parties more, spent more time in close proximity, groomed more, and formed coalitions, shared meat, and jointly participated in boundary patrols more often. However, most dyads with high scores on these measures were nonrelatives. No evidence of paternal kinship effects exists yet, although this is a subject of ongoing research.

Long-term data on male social relationships support the argument that non-human primates establish and maintain social bonds – i.e., invest differentially in social relationships with particular individuals (Silk 2007; Silk et al. 2010) – rather than merely interacting in biological markets in which individuals seek to maximize net gains from competitive and cooperative interactions, and differentiation of social networks reflects variation in partner availability and value (Henzi and Barrett 2007). Social exchange and alliance formation between male chimpanzees is consistent with biological markets theory, but males at Ngogo also show long-term consistency in association and grooming and considerable consistency in coalition formation (Mitani 2009b), although alliances form and dissolve tactically (Watts and Mitani unpublished data). Male dyads also show long-term consistency in party association and time spent in close proximity at Kanyawara, although association patterns underwent considerable realignment following a change of alpha males (Gilby and Wrangham 2008). Data on long-term consistency in grooming and coalition formation at Kanyawara are not yet available.

Female dyads at Kanyawara also show long-term consistency in association (*ibid.*), a finding at odds with the standard idea that female–female social bonds are unimportant in eastern chimpanzees. Frequent and consistent association between some female dyads may help them to maintain relatively high quality core areas and to resist attempts by immigrants to establish core areas: residents are often aggressive to new immigrants and males often intervene in such female–female conflicts to protect recent immigrants (Kahlenberg et al. 2008b). Data on fertility and infant survival at Kanyawara indicate that female lifetime reproductive success should vary positively with core area quality (Emery Thompson et al. 2007a). Joint harassment of immigrants may represent mutualism: females who often associate for other reasons are acting in self-interest, but their interests are similar.

Females at both Kibale sites form spatial "neighborhoods" (Emery Thompson et al. 2007a; Wakefield 2008; Langergraber et al. 2009). At least at Ngogo, social cliques form; these comprise females who associate more often than expected by chance, independently of core area overlap, and who often groom each other (Wakefield 2008). Thus females have differentiated social relationships similar to those at Taï (Boesch and Boesch-Achermann 2000; Lehmann and Boesch 2008, 2009). We do not have long-term quantitative data on Ngogo females to rival those on males, but some female cliques apparently can persist for over a decade (although not necessarily with constant membership) and some preferred partnerships last for many years.

Long-term data on habitat use and demography at Ngogo, combined with paternity sampling, have revealed unexpected complexity in male socio-spatial and reproductive strategies. As at other eastern chimpanzee research sites (e.g., Mahale), males range more widely than parous females and use all of the territory. However, individuals tend to use the territory unevenly and to form spatial subgroups, and grooming and coalition formation is more common within than between subgroups (Mitani and Amsler 2003). Such socio-spatial substructuring has not been reported from other sites and may result from a tendency of adult males to concentrate their activities near where their mothers had core areas when the males were young (as documented at Gombe; Murray et al. 2008) combined with the unusual demography of Ngogo. Substructuring enters into male mating strategies: the probability that a male sires a female's offspring varies positively with the extent to which he associates with her when she is not in estrus, independently of the positive effects of dominance rank on paternity probabilities (Langergraber et al. 2010).

14.3.4 Hunting and Meat Sharing

Both Kibale sites have produced extensive data on hunting and meat sharing. Ngogo researchers have documented predation on 12 species, including all seven other diurnal primates, but over 80% of hunts have been of red colobus. The chimpanzees succeed in over 80% of red colobus hunts and average about four kills per successful hunt (maximum = 13); thus red colobus have accounted for close to 90% of all kills (Mitani and Watts 1999, 2001; Watts and Mitani 2002a, b; unpublished Ngogo data). At Kanyawara, 49% of 152 red colobus hunts between 1990 and 2003 were successful; most led to single kills, with a maximum of seven (Gilby and Wrangham 2007). Kanyawara values are similar to those for Mahale, Gombe, and Taï; the high Ngogo values reflect the large size of typical hunting parties there (Mitani and Watts 1999; Watts and Mitani 2002a, b).

Long-term data from both sites have provided new insights into hunting decisions and the function of meat transfers. At Kanyawara, hunts are more likely to follow red colobus encounters when drupaceous fruit is abundant than when it is scarce, independently of the number of male chimpanzees present per encounter. This supports the hypothesis that chimpanzees engage in this risky and energetically costly behavior more when they can easily gain energy from other food sources (Gilby and Wrangham 2007). At Ngogo, the frequency of red colobus hunts also varies positively with ripe fruit availability (Watts and Mitani 2002a). This relationship seems to be driven by a positive relationship between fruit availability and the number of males per party; both hunting success and meat offtake increase with the number of males present (Mitani and Watts 1999; Watts and Mitani 2002b), which is the best predictor of whether chimpanzees hunt red colobus on encounter (Mitani and Watts 2001). In contrast, the presence of particular males who were most highly motivated to pursue monkeys was the best predictor of whether Kanyawara chimpanzees hunted on encounter; the absolute number of males present had no significant effect (Gilby et al. 2008).

At both Ngogo (Watts and Mitani 2002b) and Kanyawara (Gilby et al. 2008), the number of males who receive meat increases with the number of males present after successful hunts, but per capita meat availability does not increase with male party size at Ngogo (Watts and Mitani 2002a, b). Individuals not present at hunts and who thus did not contribute to prey captures nevertheless sometimes receive meat. Also, successful beggars often receive only small amounts, a finding consistent with the hypothesis that the main nutritional importance of meat eating is to provide micronutrients (the "meat scrap hypothesis"; Tennie et al. 2009). Data on hunting decisions and meat transfers provide evidence against the hypothesis that males exchange meat for mating opportunities: males are less likely to hunt on encounter when sexually receptive females are present (Mitani and Watts 2001; Watts and Mitani 2002a; Gilby et al. 2010a); they do not preferentially share meat with sexually swollen females, and copulations associated with meat transfers are uncommon (Mitani and Watts 2001; Gilby et al. 2010a). Moreover, meat transfers to swollen females at Ngogo do not influence the probability of mating in subsequent cycles (Mitani and Watts 2001). Meat transfers at Ngogo occur preferentially between maternal brothers and between males who are important grooming and/or coalition partners, consistent with the "male politics" hypothesis (Mitani and Watts 2001). Meat transfers sometimes involve aggression or persistent begging that interrupts consumption; over 60% of bouts at which multiple individuals consume meat include active (voluntary) meat transfers (Watts and Mitani unpublished data). Given that transfers mostly occur between closely bonded social partners, they probably do not function simply to reduce harassment, contrary to Gilby's (2006) "harassment" hypothesis.

The availability of long-term primate census data at Ngogo has allowed assessment of the impact of predation by chimpanzees on the local red colobus population. Red colobus population density declined steeply between 1975 and 2007, whereas the densities of redtails and mangabeys remained stable (Teelen 2007; Lwanga et al. 2011; Fig. 14.5). Teelen used data on red colobus diet, tree species stem densities, and long-term data on tree mortality (Lwanga et al. 2000) to argue that changes in forest composition could not explain the red colobus decline and that predation by chimpanzees was probably the explanation (cf. Lwanga et al. 2011). She substantiated this with simulation models, based on red colobus group composition and life-history data, that predicted the likelihood of population persistence for 100 years given predation that varied from zero to the extreme levels documented in 2002 (Teelen 2008). The model with zero predation showed a healthy population with extremely low probability of extinction. Low levels of predation were sustainable, especially if few adult males were killed. However, actual predation rates – even moderate ones – led to extremely low probabilities of persistence, and continued predation at the extreme 2002 rate led rapidly to extinction.

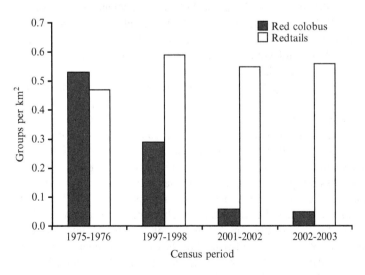

Fig. 14.5 Density of red colobus and redtail groups in the Ngogo study area, 1975–2003. Data from Teelen (2007)

Chimpanzees are not obligate carnivores, and hunting rates should decline as prey encounter rates decline and the chimpanzees must expend increasingly more time and energy to find situations where success is highly likely, especially given that most hunts occur during hunting "patrols" that can last up to several hours (Watts and Mitani 2002b). Indeed, both encounters between chimpanzees and red colobus and the frequency of red colobus hunts have decreased in recent years; such hunts account for a declining proportion of all hunts, and predation intensity has correspondingly declined. This may eventually allow the red colobus population to recover, but ongoing censuses show no evidence of recovery yet (J. Lwanga personal communication 2010).

14.3.5 Territoriality and Intergroup Aggression

Relations between neighboring chimpanzee communities are hostile, and chimpanzees are among the few mammals that regularly engage in lethal coalitionary intergroup aggression (Manson and Wrangham 1991; Wrangham 1999; Wilson and Wrangham 2003; Muller and Mitani 2005). Males perpetrate most between-community aggression, and all the males of one community are allies in competition with outside males regardless of within-community competition. Long-term data on intergroup aggression from Kibale, especially from Ngogo, have been important in dispelling the argument that such aggression is a nonfunctional response to disturbance by humans. Neither chimpanzee community was provisioned, and the Ngogo chimpanzees never reach the forest edge, but chimpanzees at both sites regularly engage in boundary patrols and respond fearfully and/or aggressively to neighbors encountered during patrols or in other situations (Watts and Mitani 2001; Mitani and Watts 2005; Wilson et al. 2001; Wilson and Wrangham 2003). Lethal attacks during inter-community encounters have been documented at both sites (Wilson and Wrangham 2003; Watts et al. 2006; Muller and Mitani 2005; Mitani et al. 2010). Playback experiments at Kanyawara confirmed that males use auditory information to make numerical assessments in deciding how to respond to the proximity of neighbors (Wilson et al. 2001). This supports the "imbalance of power hypothesis" (Manson and Wrangham 1991), which holds that males are unlikely to attack neighbors without overwhelming numerical superiority that greatly reduces their own risks; the hypothesis is difficult to test otherwise unless observers follow habituated chimpanzees in neighboring communities simultaneously (Herbinger et al. 2001).

But the possibility of making low-risk attacks does not explain why they occur, nor is the "rival reduction hypothesis" (Wrangham 1999), which holds that lethal attacks on males reduce the strength of neighboring male "coalitions," an ultimate explanation for cooperative male aggression against neighbors. Coalitionary aggression is not risk free and patrolling has time and energy costs (Watts and Mitani 2001; Amsler 2010). Long-term Ngogo data have helped to clarify the function of such cooperation. At Gombe, reproductive rates of Kasakela females were higher when the community's territory was relatively large (because of expansion into areas previously used by the Kahama community) than when it was smaller; the expansion followed a series of lethal attacks on Kahama males (Williams et al. 2004; Wilson 2012). This finding supports the hypothesis that the main function of coalitionary intergroup aggression is to defend access to food that females need for successful reproduction and, if possible, to increase food availability.

Over a 10-year period at Ngogo, researchers documented 114 boundary patrols and recorded the location and outcome of all inter-community encounters during patrols and in other contexts; they also documented 21 lethal attacks by Ngogo males on outsiders (Mitani et al. 2010). Most patrols and encounters were concentrated in the northeast and the southwest of the Ngogo territory. Lethal attacks were concentrated in the northeast; if all targeted a single community, they constituted a major source of mortality. 2009 saw a major expansion of the Ngogo territory to the northeast, with many chimpanzees routinely traveling and feeding in a large area that they previously rarely entered except on boundary patrols or hunting patrols. This provides strong circumstantial support for the "food defense" hypothesis, although evaluating the reproductive payoffs of the territory expansion will require many more years of life-history data. Meanwhile, most encounters in the southwest have happened while chimpanzees from both communities were feeding near each other during major fruiting events by *C. albidum*, *U. congensis*, or *Aningeria altissima*, three of the most important food species at Ngogo (Potts et al. 2009; Watts et al. in press a). The high frequency of such intergroup feeding contests is also consistent with the food defense hypothesis. Feeding parties on both sides are typically large, so power imbalances are slight at most encounters. Despite frequent patrols and multiple lethal attacks in the southwest, the Ngogo territory has not obviously expanded in this high quality area (Mitani et al. 2010). Unfortunately, the size and composition of neighboring community is unknown, but it is clearly large and its males have killed at least three adult males from Ngogo (unpublished Ngogo data); its strength presumably makes expansion difficult.

14.4 Discussion

Chimpanzee research in Kibale is still young compared to projects at Gombe (Wilson 2012) and Mahale (Nakamura and Nishida 2012). Nevertheless, it has contributed important comparative data on many aspects of behavioral ecology, notably long-term data sets on feeding ecology, hunting and meat transfers, intergroup aggression, social relationships, and female reproductive competition and life-history strategies. Data on female reproductive success from Kanyawara are consistent with Gombe data (Pusey et al. 1997) in showing that competition for status can influence female reproductive success, presumably because high status helps females to establish high quality core areas in heterogeneous habitats and thereby gain nutritional benefits. Given that female chimpanzees can live more than 50 years and that intervals between surviving births are around 6 years (Emery

Thompson et al. 2007a, b; Bronikowski et al. 2011), measurement of female chimpanzee reproductive success clearly requires long-term research. Likewise, Ngogo data, obtainable only after a multi-year habituation effort, show that eastern chimpanzee females sometimes establish strong social bonds with each other and that no strict eastern vs. western chimpanzee dichotomy in female gregariousness and social bonding exists. Continued long-term research will show whether these strong bonds endure as long, or even longer, than those between certain male dyads (Langergraber et al. 2009; Mitani 2009b) and will also explore the fitness consequences of individual variation in sociality.

The extreme difference in community size between Ngogo and Kanyawara is one of the most remarkable findings of Kibale research. While similar size disparities have been documented elsewhere, this has been in the context of community declines due to human disturbance and/or disease (e.g., Tai: Boesch and Boesch-Achermann 2000) or to community dissolutions apparently associated with intergroup aggression (e.g., Mahale: Nishida et al. 1985). Kibale stands out because the communities at Ngogo and Kanyawara have been stable throughout the course of research. Findings from Ngogo extend the known range of viable community size for chimpanzees, and Ngogo-Kanyawara comparisons, combined with long-term Kibale data on forest composition and phenology, provide strong evidence that ecological variation both within and among habitats can influence community size and imply that this can translate into life-history variation.

These comparisons, and much of the research at Ngogo, repeatedly invoke the theme of demographic variation on chimpanzee behavior (Mitani and Watts 1999; Mitani 2006). High rates of boundary patrolling (Watts and Mitani 2001), high hunting success and prey offtake (Mitani and Watts 1999; Watts and Mitani 2002b), low male reproductive skew (Langergraber et al. 2010), substructuring among males in space use and social relationships (Mitani and Amsler 2003) and its effect on male mating strategies (Langergraber et al. 2010), and strong differentiation of female–female social relationships (Wakefield 2008, 2010; Langergraber et al. 2009) all point to ways in which variation in community size, thus in the number of social partners and competitors, can lead to variation in aspects of ecology and social behavior that can have important impacts on fitness. How great these impacts are is unclear, but is a crucial question for future research and a powerful justification for continuing research at both sites.

Kanyawara has already contributed importantly to understanding of female chimpanzee life-history strategies and to comparative understanding of human life-history evolution. Ngogo data offer the tantalizing possibility that this understanding requires substantial revision, especially because they come from a site that is ecologically highly favorable to chimpanzees and arguably the least disturbed of current long-term chimpanzee research sites (no history of provisioning or commercial logging, no crop raiding or direct contact with humans other than researchers and field assistants, no human-introduced disease epidemics). We can only hope that political and economic realities allow the research to continue long enough that we can determine how extensive any revision should be and allow the chimpanzees themselves to have a viable future.

Acknowledgments Research in Kibale is only possible with the consent of the Uganda Wildlife Authority, the Uganda National Council for Science and Technology, and Makerere University. Research on chimpanzees succeeds only with the consent and collaboration of the management and employees of the Makerere University Biology Field Station, notably Drs. John Kasenene and Gilbert Isibirye-Basuta, but not least including James Tibosiimwe and the others who maintain the trail grids and help to protect the chimpanzees and other forest denizens from poaching. Everyone who works in Kibale is indebted to Tom Struhsaker, who started it all and remains a font of knowledge about the history of research, the ecology of the forest, and the behavioral ecology of its primates. The late J. Barwogeza, C. Katangole, F. Mugurusi, the late D. Muhangyi, the late C. Muruuli, and P. T. Tuhaiwe deserve special thanks for data collection at Kanyawara, and research at Ngogo would not be possible without the dedicated efforts of G. Mbabazi, A. Mugoba, L. Ndagezi, and A. Tumusiime. Kanyawara research has benefited from the efforts of Field Managers Kim Duffy, Alain Houle, Carole Hooven, Katharin Pieta, and Michael Wilson; Ngogo chimpanzee researchers have benefited invaluably from the project management and research efforts of Dr. Jeremiah Lwanga and from his amazing knowledge of Kibale's ecology. Sylvia Amsler, Sholly Gunter, Kevin Langergraber, Kevin Potts, and Monica Wakefield deserve particular recognition for their contributions to long-term data sets on demography, ecology, life histories, habitat use, and territorial aggression at Ngogo. I particularly thank John Mitani for the best of all possible research collaborations. Research at Ngogo has been supported to U.S. National Science Foundation grants SBR-9253590, BCS-0215622 and IOB-0516644 and by the L.S.B. Leakey Foundation, the National Geographic Society, Primate Conservation Inc., the Wenner Gren Foundation for Anthropological Research, the University of Michigan, and Yale University. Research at Kanyawara has been supported by NSF, the H.F. Guggenheim Foundation, the L.S.B. Leakey Foundation, the National Geographic Society, and the Wenner Gren Foundation for Anthropological Research. Special thanks to Gary Aronsen for help with preparation of the figures.

References

Altmann J (1980) Baboon mothers and infants. Harvard University Press, Cambridge, MA

Amsler SJ (2010) Energetic costs of territorial boundary patrols by wild chimpanzees. Am J Primatol 72:93–103

Boesch C (2009) The real chimpanzee: sex strategies in the forest: behavioural ecology and evolution. Oxford University Press, Oxford

Boesch C, Boesch-Achermann H (2000) The chimpanzees of the Taï forest: behavioural ecology and evolution. Oxford University Press, Oxford

Bronikowski AM, Altmann J, Brockman DK, Cords M, Fedigan LM, Pusey AE, Stoinski T, Morris WF, Strier KB, Alberts SC (2011) Aging in the natural world: comparative data reveal similar mortality patterns across patterns. Science 331:1325–1328

Butynski TM (1990) Comparative ecology of blue monkeys (*Cercopithecus mitis*) in high- and low-density subpopulations. Ecol Monogr 60:1–26

Caldecott J, Miles L (2005) Atlas of great apes and their conservation. University of California Press, Berkeley

Chapman CA, Chapman LJ (2000) Constraints on group size in red colobus and red-tailed guenons: examining the generality of the ecological constraints model. Int J Primatol 21:565–585

Chapman CA, Lambert JE (2000) Habitat alteration and the conservation of African primates: case study of Kibale National Park, Uganda. Am J Primatol 50:169–185

Chapman CA, Wrangham RW, Chapman LJ, Kennard DK, Zanne AE (1999) Fruit and flower phenology at two sites in Kibale National Park, Uganda. J Trop Ecol 15:189–211

Chapman CA, Chapman LJ, Bjorndal KA, Onderdonck DA (2002) Application of protein-to-fiber ratios to predict colobine abundance on different spatial scales. Int J Primatol 23:283–310

Chapman CA, Chapman LJ, Struhsaker TT, Zanne AE, Clark CJ, Poulsen JR (2005) A long-term evaluation of fruiting phenology: importance of climate change. J Trop Ecol 21:31–45

Chapman CA, Chapman LJ, Jacob AL, Rothman JM, Omeja P, Reyna-Hurtado R, Hartter J, Lawes MJ (2010) Tropical tree community shifts: implications for wildlife conservation. Biol Conserv 143:366–374

Conklin-Brittain NL, Wrangham RW, Hunt KD (1998) Dietary response of chimpanzees and cercopithecines to seasonal variation in fruit abundance. II. Macronutrients. Int J Primatol 19:971–998

Emery Thompson M, Wrangham RW (2008) Diet and reproductive function in wild female chimpanzees (*Pan troglodytes schweinfurthii*) at Kibale National Park, Uganda. Am J Phys Anthropol 135:171–181

Emery Thompson M, Kahlenberg SM, Gilby IC, Wrangham RW (2007a) Core area quality is associated with variance in reproductive success among female chimpanzees at Kibale National Park. Anim Behav 73:501–512

Emery Thompson M, Jones JH, Pusey AE, Brewer-Marsden S, Goodall J, Marsden D, Matsuzawa T, Nishida T, Reynolds V, Sugiyama Y, Wrangham RW (2007b) Aging and fertility patterns in wild chimpanzees provide insights into the evolution of menopause. Curr Biol 17:2150–2156

Furuichi T, Idani G, Ihobe H, Hashimoto C, Tashiro Y, Sakamaki T, Mulavwa MN, Yangozene K, Kuroda S (2011) Long-Term Studies on Wild Bonobos at Wamba, Luo Scientific Reserve, D.R. Congo: Towards the Understanding of Female Life History in a Male-Philopatric Species. In: Kappeler PM (ed) Long-term field studies of primates. Springer, Heidelberg

Ghiglieri MP (1984) The chimpanzees of Kibale Forest: a field study of ecology and social structure. Columbia University Press, New York

Ghiglieri MP (1986) Feeding ecology and sociality of chimpanzees in Kibale Forest, Uganda. In: Rodman PS, Cant JGH (eds) Adaptations for foraging in nonhuman primates. Columbia University Press, New York, pp 161–194

Gilby IC (2006) Meat sharing among the Gombe chimpanzees: harassment and reciprocal exchange. Anim Behav 71:953–963

Gilby IC, Wrangham RW (2007) Risk-prone hunting by chimpanzees (*Pan troglodytes*) increases during periods of high diet quality. Behav Ecol Scoiobiol 61:1771–1779

Gilby IC, Wrangham RW (2008) Association patterns among wild chimpanzees (*Pan troglodytes schweinfurthii*) reflect sex differences in cooperation. Behav Ecol Sociobiol 62:1831–1842

Gilby IC, Eberly LE, Wrangham RW (2008) Economic profitability of social predation among wild chimpanzees: individual variation promotes cooperation. Anim Behav 75:351–360

Gilby IC, Pokempner AA, Wrangham RW (2010a) A direct comparison of scan and focal sampling methods for measuring wild chimpanzee feeding behaviour. Folia Primatol 81:254–264

Gilby IC, Emery Thomspson M, Ruane JD, Wrangham RW (2010b) No evidence of short-term exchange of meat for sex among chimpanzees. J Hum Evol 59:44–53

Grieser Johns B (1996) Responses of chimpanzees to habituation and tourism in the Kibale Forest, Uganda. Biol Conserv 78:257–262

Hashimoto C, Furuichi T, Tashiro Y (2001) What factors affect the size of chimpanzee parties in the Kalinzu Forest, Uganda? Examination of fruit abundance and number of estrous females. Int J Primatol 22:947–959

Hawkes K (2010) How grandmother effects plus individual variation in frailty shape fertility and mortality: guidance from human-chimpanzee comparisons. Proc Natl Acad Sci USA 107:8977–8984

Henzi SP, Barrett L (2007) Coexistence in female-bonded primate groups. Adv Stud Behav 37:43–81

Herbinger I, Boesch C, Rothe H (2001) Territory characteristics among three neighboring chimpanzee communities at Taï National Park, Côte d'Ivoire. Int J Primatol 22:143–167

Hill K, Boesch C, Goodall J, Pusey AE, Williams JM, Wrangham RW (2001) Mortality rates among wild-living chimpanzees. J Hum Evol 40:437–450

Isibirye-Basuta G (1989) Feeding ecology of chimpanzees in the Kibale Forest, Uganda. In: Heltne PG, Marquardt LA (eds) Understanding chimpanzees. Harvard University Press, Cambridge, MA, pp 116–127

Jones JH, Wilson ML, Murray CM, Pusey AE (2010) Phenotypic quality influences fertility in Gombe chimpanzees. J Anim Ecol 79:1262–1269

Kahlenberg SM, Emery Thompson M, Muller MN, Wrangham RW (2008a) Immigration costs for female chimpanzees and male protection as an immigrant counterstrategy to intrasexual aggression. Anim Behav 76:1497–1509

Kahlenberg SM, Emery Thompson M, Wrangham RW (2008b) Female competition over core areas in *Pan troglodytes schweinfurthii*, Kibale National Park, Uganda. Int J Primatol 29:931–947

Kaplan H, Hill K, Lancaster JB, Hurtado AM (2000) A theory of human life history evolution: diet, intelligence, and longevity. Evol Anthropol 9:156–185

Köndgen S, Kühl H, N'Goran PK, Walsh PD, Schenk S, Ernst N, Biek R, Formenty P, Mätz-Rensing K, Schweiger B, Junglen S, Ellerbrok H, Nitsche A, Briese T, Lipkin WI, Pauli G, Boesch C, Leendertz FH (2008) Pandemic human viruses cause decline of endangered great apes. Curr Biol 18:260–264

Langergraber KE, Mitani JC, Vigilant L (2007) The limited impact of kinship on cooperation in wild chimpanzees. Proc Natl Acad Sci USA 104:7786–7790

Langergraber KE, Mitani JC, Vigilant L (2009) Kinship and social bonds in female chimpanzees (*Pan troglodytes*). Am J Primatol 71:840–851

Langergraber KE, Mitani JC, Watts DP, Vigilant L (2010) Male-female social relationships and reproductive success in wild chimpanzees. Am J Phys Anthropol 50(suppl):151

Lehmann J, Boesch C (2008) Sex differences in chimpanzee sociability. Int J Primatol 29:65–81

Lehmann J, Boesch C (2009) Sociality of the dispersing sex: the nature of social bonds in West African female chimpanzees, *Pan troglodytes*. Anim Behav 77:377–387

Lwanga JS (2003) Forest succession in Kibale National Park, Uganda: implications for forest restoration and management. Afr J Ecol 41:9–22

Lwanga JS, Butynski TM, Struhsaker TT (2000) Tree population dynamics in Kibale National Park, Uganda 1975–1998. Afr J Ecol 38:238–247

Lwanga JS, Struhsaker TT, Struhsaker PJ, Nutynski TM, Mitani JC (2011) Primate population dynamics over 32.9 years at Ngogo, Kibale National Park, Uganda. Am J Primatol 73:1–15

Manson JH, Wrangham RW (1991) Intergroup aggression in chimpanzees and humans. Curr Anthropol 32:369–390

Marshall AJ, Boyko CM, Feilen KM, Boyko RH, Leighton M (2009) Defining fallback foods and assessing their importance in primate ecology and evolution. Am J Phys Anthropol 140:603–614

Matsuzawa T (2002) Chimpanzees of Bossou and Nimba 1976–2001: collection of articles in English. Primate Research Institute, Kyoto University, Kyoto

Mitani JC (2006) Demographic influences on the behavior of chimpanzees. Primates 47:6–13

Mitani JC (2009a) Cooperation and competition in chimpanzees: current understanding and future challenges. Evol Anthropol 18:215–227

Mitani JC (2009b) Male chimpanzees form enduring and equitable social bonds. Anim Behav 77:633–640

Mitani JC, Amsler SJ (2003) Social and spatial aspects of male subgrouping in a community of wild chimpanzees. Behaviour 140:869–884

Mitani JC, Watts DP (1999) Demographic influences on the hunting behavior of chimpanzees. Am J Phys Anthropol 109:439–454

Mitani JC, Watts DP (2001) Why do chimpanzees hunt and share meat? Anim Behav 61:915–924

Mitani JC, Watts DP (2005) Correlates of territorial boundary patrol behaviour in wild chimpanzees. Anim Behav 70:1079–1086

Mitani JC, Merriwether DA, Zhang C (2000) Male affiliation, cooperation and kinship in wild chimpanzees. Anim Behav 59:885–893

Mitani JC, Watts DP, Lwanga JS (2002a) Ecological and social correlates of chimpanzee party size and composition. In: Boesch C, Hohmann G, Marchant LF (eds) Behavioural diversity in chimpanzees and bonobos. Cambridge University Press, Cambridge, pp 102–111

Mitani JC, Watts DP, Pepper JW, Merriwether DA (2002b) Demographic and social constraints on male chimpanzee behaviour. Anim Behav 64:727–737

Mitani JC, Watts DP, Amsler SJ (2010) Lethal intergroup aggression leads to territorial expansion in wild chimpanzees. Curr Biol 20:R507–R508

Moore J (1996) Savanna chimpanzees, referential models and the last common ancestor. In: McGrew WC, Marchant LF, Nishida T (eds) Great ape societies. Cambridge University Press, Cambridge, pp 275–292

Morgan D, Sanz C (2006) Chimpanzee feeding ecology and comparisons with sympatric gorillas in the Goualougo Triangle, Republic of Congo. In: Hohmann G, Robbins MM, Boesch C (eds) Feeding ecology in apes and other primates: ecological, physiological and behavioural aspects. Cambridge University Press, Cambridge, pp 97–122

Muehlenbein MP, Watts DP, Whitten PL (2004) Dominance rank and fecal testosterone levels in adult male chimpanzees (*Pan troglodytes schweinfurthii*) at Ngogo, Kibale National Park, Uganda. Am J Primatol 64:71–82

Muller MN, Mitani JC (2005) Conflict and cooperation in wild chimpanzees. Adv Stud Behav 35:275–331

Muller MN, Wrangham RW (2004a) Dominance, aggression and testosterone in wild chimpanzees: a test of the 'challenge hypothesis'. Anim Behav 67:113–123

Muller MN, Wrangham RW (2004b) Dominance, cortisol and stress in wild chimpanzees (*Pan troglodytes schweinfurthii*). Behav Ecol Sociobiol 55:332–340

Muller MN, Wrangham RW (2005) Testosterone and energetics in wild chimpanzees (*Pan troglodytes schweinfurthii*). Am J Primatol 66:119–130

Murray CM, Gilby IC, Mane SV, Pusey AE (2008) Adult male chimpanzees inherit maternal ranging patterns. Curr Biol 18:20–24

Nakamura M, Nishida T (2012) Long-term field studies of Chimpanzees at Mahale Mountains National Park, Tanzania. In: Kappeler PM (ed) Long-term field studies of primates. Springer, Heidelberg

Nishida T, Hiraiwa-Hasegawa M, Hasegawa T, Takahata Y (1985) Group extinction and female transfer in wild chimpanzees in the Mahale National Park, Tanzania. Z Tierpsychol 67:284–301

Omeja PA, Chapman CA, Obua J (2009) Enrichment planting does not improve tree restoration when compared to natural regeneration in a former pine plantation in Kibale National Park, Uganda. Afr J Ecol 47:650–657

Potts KB (2008) Habitat heterogeneity on multiple spatial scales in Kibale National Park, Uganda: implications for chimpanzee population ecology and grouping patterns. PhD thesis, Yale University, New Haven, CT

Potts KB, Chapman CA, Lwanga JS (2009) Floristic heterogeneity between forested sites in Kibale National Park, Uganda: insights into the fine-scale determinants of density in a large-bodied frugivorous primate. J Anim Ecol 78:1269–1277

Potts KB, Watts DP, Wrangham RW (2011) Comparative feeding ecology of two communities of chimpanzees (*Pan troglodytes*) in Kibale National Park, Uganda. Int J Primatol 32:669–690

Power M (1990) The egalitarians – human and chimpanzee: an anthropological view of social organization. Cambridge University Press, Cambridge

Pusey AE, Williams JM, Goodall J (1997) The influence of dominance rank on the reproductive success of female chimpanzees. Science 277:828–831

Reynolds V (2005) The chimpanzees of the Budongo forest: ecology, behaviour, and conservation. Oxford University Press, Oxford

Silk JB (2007) The strategic dynamics of cooperation in primate groups. Adv Stud Behav 37:1–41

Silk JB, Beehner JC, Bergmann TJ, Crockford C, Engh AL, Muscovice LR, Wittig RM, Seyfarth RM, Cheney DL (2010) Female chacma baboons form strong, equitable, and enduring social bonds. Behav Ecol Sociobiol 64:1733–1747

Struhsaker TT (1975) The red colobus monkey. University of Chicago Press, Chicago

Struhsaker TT (1997) Ecology of an African rain forest: logging in Kibale and the conflict between conservation and exploitation. University Press of Florida, Gainesville

Struhsaker TT (2010) The red colobus monkeys: variation in demography, behavior, and ecology of endangered species. Oxford University Press, Oxford

Struhsaker TT, Lwanga JS, Kasenene JM (1996) Elephants, selective logging, and forest regeneration in Kibale Forest, Uganda. J Trop Ecol 12:45–64

Stumpf RM (2007) Chimpanzees and bonobos: diversity within and between species. In: Campbell CJ, Fuentes A, Mackinnon KC, Panger M, Bearder SK (eds) Primates in perspective. Oxford University Press, Oxford, pp 321–344

Teelen S (2007) Primate abundance along five transect lines at Ngogo, Kibale National Park, Uganda. Am J Primatol 69:1030–1044

Teelen S (2008) Influence of chimpanzee predation on the red colobus population at Ngogo, Kibale National Park, Uganda. Primates 49:41–49

Tennie C, Gilby IC, Mundry R (2009) The meat-scrap hypothesis: small quantities of meat may promote cooperative hunting in wild chimpanzees (*Pan troglodytes*). Behav Ecol Sociobiol 63:421–431

Wakefield ML (2008) Grouping patterns and competition among female *Pan troglodytes schweinfurthii* at Ngogo, Kibale National Park, Uganda. Int J Primatol 29:907–929

Wakefield ML (2010) Socioecology of female chimpanzees (*Pan troglodytes schweinfurthii*) in the Kibale National Park, Uganda: social relationships, association patterns, and costs and benefits of gregariousness in a fission-fusion society. PhD thesis, Yale University, New Haven, CT

Watts DP (1998) Coalitionary mate guarding by male chimpanzees at Ngogo, Kibale National Park, Uganda. Behav Ecol Sociobiol 44:43–55

Watts DP (2000a) Grooming between male chimpanzees at Ngogo, Kibale National Park, Uganda. I. Partner number and diversity and grooming reciprocity. Int J Primatol 21:189–210

Watts DP (2000b) Grooming between male chimpanzees at Ngogo, Kibale National Park, Uganda. II. Influence of male rank and possible competition for partners. Int J Primatol 21:211–238

Watts DP (2002) Reciprocity and interchange in the social relationships of wild male chimpanzees. Behaviour 139:343–370

Watts DP, Potts KB, Lwanga JS, Mitani JC (in press a) Diet of chimpanzees (*Pan troglodytes schweinfurthii*) at Ngogo, Kibale National Park, Uganda. 1. Diet composition and diversity. Am J Primatol

Watts DP, Potts KB, Lwanga JS, Mitani JC (in press b) Diet of chimpanzees (*Pan troglodytes schweinfurthii*) at Ngogo, Kibale National Park, Uganda. 2. Temporal variation and fallback foods. Am J Primatol

Watts DP, Mitani JC (2001) Boundary patrols and intergroup encounters in wild chimpanzees. Behaviour 138:299–327

Watts DP, Mitani JC (2002a) Hunting and meat sharing by chimpanzees at Ngogo, Kibale National Park, Uganda. In: Hohmann G, Boesch C, Marchant LF (eds) Behavioural diversity in chimpanzees and bonobos. Cambridge University Press, Cambridge, pp 244–255

Watts DP, Mitani JC (2002b) Hunting behavior of chimpanzees at Ngogo, Kibale National Park, Uganda. Int J Primatol 23:1–28

Watts DP, Muller MN, Amsler SJ, Mbabazi G, Mitani JC (2006) Lethal intergroup aggression by chimpanzees in the Kibale National Park, Uganda. Am J Primatol 68:161–180

Williams JM, Oehlert GW, Carlis JV, Pusey AE (2004) Why do male chimpanzees defend a group range? Anim Behav 68:523–532

Wilson ML (2012) Long-term studies of the Chimpanzees of Gombe National Park, Tanzania. In: Kappeler PM (ed) Long-term field studies of primates. Springer, Heidelberg

Wilson ML, Wrangham RW (2003) Intergroup relations in chimpanzees. Annu Rev Anthropol 32:363–392

Wilson ML, Hauser MD, Wrangham RW (2001) Does participation in intergroup conflict depend on numerical assessment, range location, or rank for wild chimpanzees? Anim Behav 61:1203–1216

Wrangham RW (1977) Feeding behaviour of chimpanzees in Gombe National Park, Tanzania. In: Clutton-Brock TH (ed) Primate ecology: studies of feeding and ranging behaviour in lemurs, monkeys and apes. Academic, London, pp 503–538

Wrangham RW (1999) Evolution of coalitionary killing. Yearbk Phys Anthropol 42:1–30

Wrangham RW, Conklin NL, Chapman CA, Hunt KD (1991) The significance of fibrous foods for Kibale Forest chimpanzees. Phil Trans R Soc Lond B 334:171–178

Wrangham RW, Conklin NL, Etot G, Obua J, Hunt KD, Hauser MD, Clark AP (1993) The value of figs to chimpanzees. Int J Primatol 14:243–256

Wrangham RW, Chapman CA, Clark-Arcadi AP, Isabirye-Basuta G (1996) Social ecology of Kanyawara chimpanzees: implications for understanding the costs of great ape groups. In: McGrew WC, Marchant LF, Nishida T (eds) Great ape societies. Cambridge University Press, Cambridge, pp 45–57

Wrangham RW, Conklin-Brittain NL, Hunt KD (1998) Dietary response of chimpanzees and cercopithecines to seasonal variation in fruit abundance. I. Antifeedants. Int J Primatol 19:949–970

Wroblewski EE, Murray CM, Keele BF, Schumacher-Stankey JC, Hahn BH, Pusey AE (2009) Male dominance rank and reproductive success in chimpanzees, *Pan troglodytes schweinfurthii*. Anim Behav 77:873–885

Chapter 15
Long-Term Field Studies of Chimpanzees at Mahale Mountains National Park, Tanzania

Michio Nakamura and Toshisada Nishida

Abstract Chimpanzee research in the Mahale Mountains, on the eastern shore of Lake Tanganyika, Tanzania, began in 1965. Although the Mahale Mountains did not initially have official protected status, researchers' conservation efforts and the financial support of the Japanese government led to the designation of Mahale as a national park in 1985. The Mahale project is the second-longest continuous field study of chimpanzees. Long-term demographic data show that the habituated chimpanzee group has decreased in size, largely due to disease outbreaks. Recent research has focused on variation in the behavioral repertoire of chimpanzees, producing a detailed audio-visual ethogram as well as evidence of social customs, some of which are candidates for cultural variation. Many primatologists are beginning to accept the notion that some behavioral elements of nonhuman animals are socially shaped. Our long-term studies of chimpanzee behavioral variation will hopefully contribute to a better understanding of the ways in which human and nonhuman primate behavior is shaped by the interaction between genes and culture.

15.1 History and Infrastructure of Mahale

15.1.1 Study Area

Mahale Mountains National Park is the largest protected area for wild chimpanzees (*Pan troglodytes*) in Tanzania and the world's second-oldest chimpanzee study site (Nishida 1990). It is located at 6°15′ S, 29°55′ E about 120 km south of Kigoma

---

M. Nakamura (✉)
Wildlife Research Center, Kyoto University, Kyoto, Japan
e-mail: nakamura@wrc.kyoto-u.ac.jp

T. Nishida
Japan Monkey Centre, Kanrin, Inuyama, Japan

P.M. Kappeler and D.P. Watts (eds.), *Long-Term Field Studies of Primates*,
DOI 10.1007/978-3-642-22514-7_15, © Springer-Verlag Berlin Heidelberg 2012

Town on the eastern shore of Lake Tanganyika. The national park protects 1,613 km^2 of semi-evergreen medium-altitude forest, montane forest, montane savanna, miombo woodland, and *Oxytenanthera* bamboo woodland (Nishida and Uehara 1981). The biological diversity of this area is characterized by the coexistence of fauna and flora of eastern and western Africa (Anonymous 1980). The terrain is mostly rugged and hilly, dominated by the Mahale mountain chain that runs from northwest to southeast across the western part of the park. The highest peak is Nkungwe (2,462 m a.s.l.).

The main area of long-term chimpanzee research (the Kasoje area) is relatively flat, and because the mountain chain blocks moisture from the lake, the area is dominated by continuous evergreen forest (Kasoje Forest). Chimpanzees and other animals also regularly frequent the well-developed riverine forests in steep areas at higher altitudes. *Isoberlinia* woodland dominates *Brachystegia* woodland in the eastern part of the park, and *Oxytenanthera* or *Oxytenanthera*/miombo mixed woodland is dominant in the central and eastern parts.

15.1.2 Access and Logistics

Mahale is more difficult to access than other chimpanzee research sites. It can be reached from Kigoma, the nearest town to the Gombe and Mahale sites and the capital of the Kigoma region. All research supplies must be brought in from Kigoma by boat via Lake Tanganyika, which takes at least 12 h. Most tourists fly directly into Mahale Mountains National Park on chartered flights from Arusha.

15.1.3 Brief History of Research at Mahale

Chimpanzee research by the Kyoto University team began in 1961, when Kinji Imanishi organized the Kyoto University Africa Primatological Expedition (KUAPE) and established a research camp at Kabogo Point on the shore of Lake Tanganyika (Azuma and Toyoshima 1961). Jun'ichiro Itani led the team from the fourth KUAPE expedition onward. These researchers carried out extensive surveys throughout western Tanzania, including in the Masito Escarpment and Ugalla areas (Suzuki 1969; Izawa 1970; Kano 1972; Itani 1979).

The Mahale Mountains Chimpanzee Research Project (MMCRP) was initiated in 1965, when Toshisada Nishida, then a graduate student at Kyoto University, first visited Mahale as a part of the fourth KUAPE expedition. Mahale was selected as a long-term research site after successful habituation of the chimpanzees by provisioning in 1966 (Nishida 1968, 1990). Figure 15.1 shows the number of researchers that have visited Mahale each year. In recent years, seven to ten researchers per year have visited Mahale; a total of 70 researchers had visited Mahale by 2009.

The researchers' activities have included efforts to conserve Mahale. Nishida had developed a proposal for a protected area as early as 1967. This vision gained

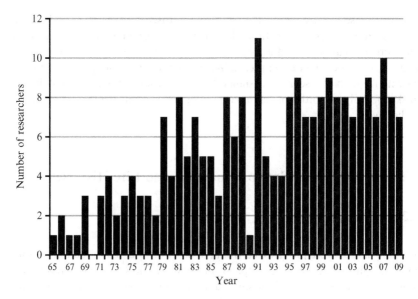

Fig. 15.1 The number of researchers who have visited Mahale each year between 1965 and 2009. Those who stayed over the new year were counted in both years

momentum in 1974, when the Japanese government provided financial support to the Japan International Cooperation Agency (JICA) for the surveys required to establish a protected area. Dr. Derek Bryceson, then Director of Tanzania National Parks (TANAPA), and Mr. Nobuyuki Nakashima, then Japanese ambassador to Tanzania, were earnest supporters of this project. During the JICA project (1975–1987), Mr. Raphael Jingu, then Director of the Game Division, established the Mahale Mountains Wildlife Research Centre (MMWRC) under the auspices of the Serengeti Wildlife Research Institute (now Tanzania Wildlife Research Institute). This center provided a base camp for patrolling the future park area and secured logistical support for the research project. When Mr. Eramus Tarimo became the Acting Director of MMWRC in 1979, practical preparations for the national park, including boundary demarcation, moved forward and Mahale was finally declared a national park in 1985 (Nishida and Nakamura 2008). In 1994, Tanzanian and Japanese researchers established a nongovernmental organization, the Mahale Wildlife Conservation Society (MWCS), with the goal of protecting wildlife at Mahale, mainly through local environmental education efforts.

15.1.4 Facilities and Basic Data Collection

The research camp at Mahale (Kansyana Camp) is situated in the Kasoje area, about 1 km from the lakeshore. It currently consists of two simple buildings with five bedrooms, a library, and storage rooms. Water is manually carried to a storage drum in the dry season and cooking is done using kerosene stoves and dead firewood collected from the forest. A limited solar-power system is available to charge computers and batteries, and a small weather station monitors daily temperatures and precipitation (Itoh et al. in press). We also collect monthly plant phenological data with the help of research assistants (Itoh and Nishida 2007; Itoh et al. in press) and keep daily records of chimpanzee sightings and demographic changes (see below).

15.2 Demography, Group Dynamics, and Life Histories

15.2.1 Study Groups

The chimpanzees were provisioned from the onset of research to facilitate habituation, until this practice was stopped in 1987. Following habituation of the chimpanzees in 1966, the initial study unit-group (or community) was called the Kajbara group (K group). A second group (Mimikire or M group) was habituated in 1968. Observation of these two adjacent habituated groups led to the identification of the natural fission–fusion social unit (Nishida 1968), description of antagonistic relationships between adjacent unit-groups, documentation of female transfer between groups (Nishida and Kawanaka 1972), and collection of detailed data on social behavior in K group (e.g., Nishida 1970, 1979). Following the extinction of K group in 1982 (Nishida et al. 1985), research efforts focused on M group. In the late 1990s, we noticed that an unidentified unit-group was utilizing the area made vacant by extinction of K group. In 2005, we began studying this new group (Miyako or Y group) living north of M group (Sakamaki and Nakamura 2007), but habituation had not yet been achieved by 2010.

15.2.2 Demographic Changes in M Group

One obvious benefit of long-term research is the collection of demographic data in combination with individual life histories. We have published demographic data through 1988 (Nishida et al. 1990) and 1999 (Nishida et al. 2003), respectively. In this section, we add data from 2000 to 2009 to provide updated information on changes in group size.

Figure 15.2 shows the demographic changes in M group between 1980 and 2009. Our study of M group began in 1965, but 50% of its members were not individually identified until 1980 (Hiraiwa-Hasegawa et al. 1984). The rapid increase in group size in the early 1980s was partially due to the immigration of K group females (with some immature offspring) in the course of that group's extinction. As a result, M group once contained more than 100 individuals. However, the size of M group

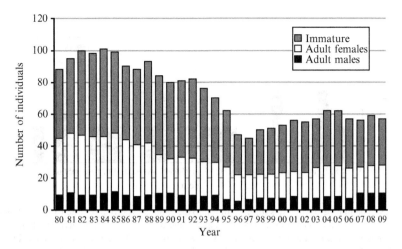

Fig. 15.2 The number of chimpanzees in the Mahale M group between 1980 and 2009. Numbers at the end of the year (31 December) represent those in the previous year. Adult males were defined as 16 years or older and adult females were defined as 13 years or older

decreased considerably between the late 1980s and 1997. Following this decline, M group gradually recovered until 2004 and remained nearly stable thereafter, except for a decrease in 2006. In 2009, the group had 57 members. Despite these fluctuations, the number of adult males (7–11) has remained relatively stable, except during 1995–1997, when it varied between 5 and 6. In contrast, the number of adult females has decreased by about two-thirds over the past 30 years. As a result, the socionomic sex ratio changed from 0.25 in 1980 to 0.48 in 2009. Such dynamics can only be documented by long-term studies spanning several decades.

Figure 15.3 shows the number of female immigrations into M group, excluding those of parous females from K group during that group's extinction. An average of 0.57 ± 0.73 females per year visited and eventually immigrated into M group. The number of immigrations has not declined in recent years; thus, the decrease in females in M group cannot be attributed to a reduction in female immigration. Although most females born into M group have emigrated to other groups during adolescence, four females born in the 1980s did not emigrate and subsequently gave birth in their natal group. These females were adolescents during the mid- to late-1990s, when M group was at its smallest. Thus, group size and composition may affect female transfer decisions. The ratio of females who emigrate from their natal group varies considerably among chimpanzee research sites (summarized by Nishida et al. 2003). Although costs for females are reportedly high in a new group (Nishida 1989; Williams et al. 2002; Kahlenberg et al. 2008; Pusey et al. 2008), they are only realized after immigration (or attempted immigration). The factors affecting female emigration or continuation in the natal group have not been fully investigated to date.

Figure 15.4 shows the number of male and female births in M group. Although the average number of male births (1.40 ± 1.25 per year) is smaller than that of

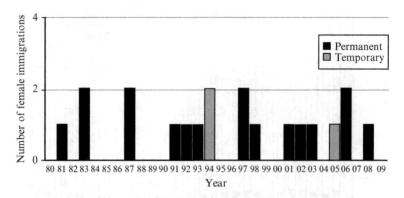

Fig. 15.3 The number of females immigrating into the Mahale M group between 1980 and 2009. Unknown adolescent females observed with M group members were identified as either temporary visitors or permanent immigrants. Temporary visitors were defined as those who stayed as long as several months, but disappeared before giving birth without apparent health problems or injury. Permanent immigrants were defined as those who stayed longer and usually gave birth within a few years

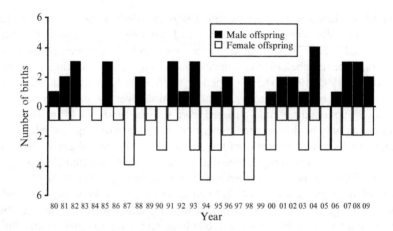

Fig. 15.4 The number of births into the Mahale M group between 1980 and 2009

female births (1.97 ± 1.33), this difference is not significant and has not changed over time. The overall trend of birth sex ratios that do not differ significantly from 1:1 but favor females slightly is similar to those reported for other chimpanzee study sites (Gombe: Goodall 1986; Taï: Boesch 1997).

The decrease in the size of M group thus cannot be explained by these factors. Most of the larger declines have been caused by disease outbreaks. In 1986, an epizootic that research assistants called an "AIDS-like disease" killed several chimpanzees in M group (Nishida 1990), and in 1993, a flu-like epizootic (severe coughing and sneezing) killed at least 11 chimpanzees (Hosaka 1995; Hosaka et al. 2000). From June to July 2006, a flu-like disease broke out in M group and likely

15 Long-Term Field Studies of Chimpanzees at Mahale Mountains

killed 12 individuals (Hanamura et al. 2008). Nishida et al. (2003) estimated that about 48% of all Mahale chimpanzee deaths to 1999 were caused by disease. An additional large decline in group size was caused by lion predation in 1989 (Tsukahara 1993).

Unfortunately, we do not know the cause for the largest decline in group size between 1995 and 1997, when as many as 20 individuals disappeared and only three bodies were recovered. Since there were no signs of disease outbreak and many mature males and females disappeared during this period, researchers hypothesized that M group had fissioned. Surveys around the periphery of M group's home range, however, located no "branch" or "splinter" group. The sources of this dramatic population loss remain unknown.

15.2.3 Life History of the Oldest Female

Studies of individual life histories can inform analyses of social and demographic data. Here, we present the life history of the current oldest female at Mahale, called Calliope. She was first identified in 1973 as a young adult. She was estimated to have been born in 1960, based on the age of her accompanying juvenile daughter. She gave birth to five infants (three females, two males) through the age of 37 years (Table 15.1). Her reproductive cycle resumed after her last offspring was weaned, but she did not give birth again. She stopped cycling in her 48th year. Although she has not given birth in the last 12 years, including 5 years during which she was nursing her last offspring, she remains alive and well. Another old female (Wakusi; ~49 years old in 2010) also stopped reproducing in 2005, indicating that Calliope's case is not exceptional. Although menopause may not be part of the typical life cycle of female chimpanzees (Nishida et al. 2003; Emery Thompson et al. 2007), some females experience several years of postreproductive life. Because it is often difficult to estimate the ages of older individuals and because older females are often shy toward human observers, insufficient data have been compiled on old females despite decades of research. We therefore need additional continuous

Table 15.1 Calliope's reproductive profile: an individual chimpanzee life history

Year	Event	Estimated age (years)
1976?	Gave birth to CB (f)	16
1981	Gave birth to CC (f)	21
1985	Gave birth to CT (m)	25
1991 Dec	Gave birth to CD (m)	31
1997 Sep	Gave birth to CR (f)	37
2001 Dec	Resumed reproductive cycle	41
2008 Mar	Final estrus	48
2010 Jul	Remains alive and well	50

observation to understand the entire chimpanzee life cycle and to document the extent and causes of inter-individual variation in life histories, including variation in female reproductive profiles.

15.3 Highlights of Studies Made Possible by Long-Term Research

15.3.1 Ethogram and Ethnography of Chimpanzees

One of the benefits of long-term research is the continuous accumulation of behavioral repertoire details through ongoing observation. The compilation of a detailed ethogram is a starting point for description of a species' behavioral diversity. The understanding of such behavioral diversity, in combination with socioecology, social relations, and the mind of the target species, may lead to the study of nonhuman "ethnography" (Wrangham et al. 1994; McGrew 2004). Although the use of the term "ethnographic" by researchers of animal behavior (e.g., Rendell and Whitehead 2001; Laland and Janik 2006) may be considered inappropriate by some cultural anthropologists (Ingold 2001), there is no reason to believe that researchers cannot conduct ethnography on nonhuman animals unless we define the term narrowly to exclude them.

Our 1999 ethogram of the Mahale chimpanzees (Nishida et al. 1999) listed 515 behavioral patterns. The new videographic version of the ethogram (Nishida et al. 2010) lists 891 behavioral patterns, constituting the largest ethogram of chimpanzees or any other primates. Such detailed data can form the basis for comparison of chimpanzee cultures (see Sect. 15.3.2) and for behavioral reconstructions of the common ancestor of *Pan* and *Homo*. Although not an originally intended application, the ethogram can also be used to assess behavioral enrichment in captivity by providing a reference for comparison of captive and wild behavioral repertories.

We have also published brief notes on rare behaviors by Mahale chimpanzees (see Nishida et al. 2009 for a detailed review). Single-case observations include colobus skin washing by an adult male (Nishida 1994), algae-feeding by a female (Sakamaki 1998), deception by an adult male to snatch a dead infant from its mother (Nishida 1998), knotted colobus skin "necklace" wearing (McGrew and Marchant 1998), self-medication in an attempt to remove sand fleas from a toe with a stick (Nishida 2002), playing with a squirrel (Zamma 2002a), exploratory/threatening behaviors toward a porcupine (Matsusaka 2007), use of wet hair to capture swarming termites (Kiyono-Fuse 2008), and collection and carrying of guinea-fowl feathers (Nakamura 2009). Although each of these observations may not be considered important on its own, the accumulation of such published cases allows us to grasp the full behavioral flexibility and variation of chimpanzees. After more than 40 years of research, we continue to observe new behavioral patterns.

Some initially anecdotal behaviors later became more common, suggesting behavioral innovation or the acquisition of a new habit (see also Perry et al. 2012). One example is chimpanzee predation on yellow baboons. Although chimpanzees at Gombe are known to occasionally hunt olive baboons (e.g., Teleki 1981), Mahale chimpanzees were first observed hunting and eating yellow baboons in 1996 (Nakamura 1997). Subsequent research documented several more incidents of predation on baboons, although this behavior was infrequent (Nishie 2004). The use of tools during hunting is another example of initially anecdotal behaviors that later became more commonly observed. Huffman and Kalunde (1993) reported tool-assisted squirrel hunting, but this was the only report of its kind until Pruetz and Bertolani (2007) documented 12 cases of hunting with tools by Fongoli chimpanzees. The Mahale case was assumed to be anecdotal and such habitual behavior was believed to be limited to Fongoli. However, several other cases at Mahale (Nakamura and Itoh 2008) suggested that the Mahale chimpanzees also habitually hunt with tools; this behavior is rarely observed because of difficult observation conditions. Similar explanations can be applied to a few observations of wild fruit sharing among adult males (Nakamura and Itoh 2001), genito-genital rubbing among females (Zamma and Fujita 2004), and nasal probing and nipple pressing (Marchant and McGrew 1999).

It is also important to record rare events, such as an unknown group's incursion into the center of M group's territory (Itoh et al. 1999) and the chimpanzees' attitudes toward a seriously weakened adolescent female. In the latter case, several adult females harassed the adolescent but other individuals intervened (Shimada and Matumula 2004). Finally, it is also important to accumulate observations of common but difficult to observe events, such as the behavior of a newly immigrated female. Although female immigrations are not rare, it is usually difficult to observe the behavior of such females because they are often very shy toward human observers. However, one immigrant female was continuously observed on the first and the second days of immigration (Nakamura and Itoh 2005). M group members, especially males, seemed to be excited to see the newcomer, which was groomed by three females and three males. Only one case of threatening by an adult female was observed during these 2 days. Although researchers have often emphasized the higher rates of aggression directed toward immigrant females by resident females (Nishida 1989; Kahlenberg et al. 2008), most resident females were not hostile to the newcomer in this case. Instead, resident males and females received the newcomer with "curiosity" and showed affiliative behaviors to her. Even a female who threatened the newcomer subsequently groomed her.

15.3.2 Cultural Behaviors

The formation of traditions and the genesis of fashions (seemingly new behaviors that may or may not become traditions) can only be identified through long-term study. The detailed ethogram allows us to detect cultural differences in behavior among chimpanzee populations. Although the use of various tools to obtain otherwise unreachable food resources have been a central topic of chimpanzee cultural studies (e.g., McGrew 1992), cultural variations also exist beyond the context of tool use. Comparison of our Mahale chimpanzee data with information from other chimpanzee study populations has revealed behavioral variation in several social domains, such as grooming and courtship displays (see Nishida et al. 2009 for a recent review).

15.3.2.1 Grooming

Hand-clasp grooming (McGrew and Tutin 1978; Nakamura 2002) is the classic example of cultural variation beyond the context of feeding technology. In this form of grooming, two chimpanzees sit face to face and clasp their right or left hands or wrists overhead to form an A frame; they then groom each other's lower arms. This behavioral pattern has been customarily performed in Mahale (M and K groups), Kibale (Kanyawara and Ngogo groups), Kalinzu, and Lopé, whereas only a few individuals have performed it in Taï and it has never been reported from Gombe, Budongo, or Bossou (Nakamura 2002). The pattern has also been observed in the captive colony of the Yerkes Primate Research Center (de Waal and Seres 1997; Bonnie and de Waal 2006).

McGrew et al. (2001) reported two grooming hand-clasp patterns: palm-to-palm and non-palm-to-palm. In the former pattern, the groomers clasp each other's palms, whereas in the latter pattern their wrists are often flexed and one hand rests on the other. The researchers argued that the palm-to-palm pattern was dominant in the Mahale K group but absent in M group. However, after reexamination with a larger data set, Nakamura and Uehara (2004) found that the palm-to-palm pattern was infrequently performed in M group.

Social scratching (Nakamura et al. 2000) is another common cultural behavior at Mahale. This simple behavior consists of one individual scratching another while grooming him/her. Adults most frequently scratch each others' backs, and mothers scratch various body parts while grooming their infants. Social scratching also occurs at Ngogo, but differs from that at Mahale (Nishida et al. 2004). A Mahale researcher who visited Gombe for a short time observed that three chimpanzees there scratched socially (Shimada 2002). However, there are no other reports of this simple behavior from other study sites.

Sounds produced during grooming are also known to differ among sites. When Nishida visited Ngogo, he noticed that chimpanzees there "sputtered" as though forcing air through their lips; Mahale chimpanzees do not (Nishida et al. 2004). Nakamura also observed this behavior among a few individuals at Bossou, although it was practiced primarily by a single juvenile male (Nakamura and Nishida 2006) and may not have reached the group level.

15.3.2.2 Courtship Display

Courtship displays (Nishida 1997) also vary among populations. Because the sexual context is usually obvious from the penile erection of the male and the sexual swelling of the female, any attention-getting courtship displays by males or females easily convey their intention to copulate. However, the frequent convergence of courtship displays within groups and variation among groups suggest underlying cultural processes. Leaf-clipping, in which a leaf (or leaves) is clipped to produce an audible sound, is a classic example of such a courtship display at Mahale (Nishida 1980). Bossou chimpanzees also exhibit this behavior, but Sugiyama (1981) argued that it is used there to express frustration. Taï chimpanzees also clip leaves, but only before they perform buttress drumming (Boesch 1996). The majority of leaf-clipping by Ngogo males is performed when soliciting estrous females, but it is also used before buttress drumming (Watts 2008). Thus, this pattern may not have been completely ritualized to a single context.

When Nakamura visited Bossou, he noticed that mature males often performed heel-tapping in the context of courtship (Nakamura and Nishida 2006). In this behavior, a tree bough, a rock, or the ground is rhythmically tapped with the heel to produce a conspicuous sound. Heel-tapping differs from stamping because the sole makes no contact with the substrate, but instead is held upright and facing forward; only the heel makes contact. Stamping is a common element of male courtship displays at Mahale (Nishida 1997), but heel-tapping has never been observed. Sugiyama (1989) observed heel-tapping among immature individuals at Bossou to invite play. Other courtship displays, such as shrub-bending (Nishida 1997) and knuckle-knocking (Boesch 1996), may be candidates for cultural variation.

15.3.2.3 Hygienic Behaviors

Grooming serves a social function in chimpanzees but also functions hygienically to remove ectoparasites and debris from the body. After removing an ectoparasite, chimpanzees often inspect and squash it. Behaviors associated with ectoparasite removal also vary among populations. Chimpanzees at Mahale perform leaf-grooming, in which they pick up ectoparasites with their lips during grooming, place them on leaves, and then squash them with their thumbs (Zamma 2002b). Chimpanzees in Gombe (Goodall 1965), Budongo (Assersohn et al. 2004), and Ngogo (Watts 2008) also leaf-groom. Chimpanzees at Bossou perform a similar behavior, but they place the parasites in the palm of one hand and smash them with the opposite index finger (Nakamura and Nishida 2006). Similarly, chimpanzees in Taï smash the parasites against a forearm with the index finger of the free hand (Boesch 1995).

These behaviors may have social functions, such as offering an opportunity for joint attention. Leaf-grooming at Mahale and index-to-palm squashing at Bossou sometimes attract the interest of other individuals, who gather around and watch. Moreover, leaf-grooming at Mahale sometimes occurs outside of the grooming context (Nakamura personal observation); this behavior may thus be performed without actual ectoparasite smashing.

15.3.2.4 Play

Variation in play is not well documented at most study sites, perhaps because the function of play is often difficult to define and young individuals are rarely the focus of study. However, leaf-pile pulling (Nishida and Wallauer 2003) at Mahale and Gombe may be an example of cultural variation in play behavior. In leaf-pile pulling, young chimpanzees walk backward while raking a pile of dry leaves with both hands down a slope, producing copious noise. Although this may be interpreted as solitary play, the performer often faces an individual immediately following him or her and attracts social attention.

Another local variant of play is the use of tools for drinking, whereby a chimpanzee uses leaves or sticks to obtain water. This tool use is usually not considered to be play at other study sites, but to function solely to obtain water (e.g., Tonooka 2001). This behavior had been only rarely observed at Mahale (Nishida 1990) until the early 1990s, but has since become more frequent among immature chimpanzees (Matsusaka et al. 2006). Most such behavior occurs during the rainy season, when running water is plentiful and tools are not necessary to obtain water. This tool use is often conducted in apparently playful contexts. The current form of drinking tool use at Mahale thus likely occurs as part of play, rather than out of necessity. It may lead in the future to more purposeful and efficient use of tools when obtaining water is difficult, in which case the playing would retro-spectively be called an exploratory behavior leading to innovation.

15.3.2.5 Feeding

Differences in feeding behaviors were studied at Gombe and Mahale in the early period of research (Nishida et al. 1983) but have not been systematically analyzed. Mahale chimpanzees never utilize oil palms, but began to utilize other species introduced by humans as food sources (e.g., mangoes, lemons, guavas; Takasaki 1983; Takahata et al. 1986; Nishida et al. 2009). If learning about new food items occurred at the individual level, some individuals would eat oil palms and some would not eat lemons. However, none of the Mahale chimpanzees eats oil palms and all eat lemons, suggesting that some form of social learning may play a role in the modification of food repertoires. Such factors are rarely considered in current studies of chimpanzee culture, perhaps because variation in feeding behavior is more parsimoniously explained by ecological differences (Byrne 2007). In contrast, Boesch et al. (2006) compared three groups at Taï and suggested that variation in feeding time relative to fruit availability may reflect cultural differences. Similarly,

Sakamaki et al. (2007) reported that two adjacent groups at Mahale used different food items, despite the availability and common occurrence of all items in both areas. Of course, unknown environmental factors may also explain such differences.

Behavioral variation outside tool-use contexts is infrequently described. For example, although hand-clasp grooming at Mahale was reported decades ago, it was not properly described using data from other sites where this behavioral pattern was known to exist. This omission forms a striking contrast to the detailed and repeated reports of the same type of tool use from different study sites. In addition, subtle behavioral patterns such as social scratching are seldom described separately. Such biases of behavioral description should be corrected to facilitate our understanding of the full range of behavioral and cultural variation in nonhuman primates.

15.4 Conclusions and Future Perspectives

Chimpanzee research at Mahale, on the eastern shore of Lake Tanganyika, has been ongoing since 1965. Long-term demographic monitoring has identified a decline in chimpanzee numbers, largely due to disease outbreaks. We must take the strongest precautions to prevent the introduction of human diseases into wild chimpanzee populations (Hanamura et al. 2008; Kaur et al. 2008). In 2006, we thus began requiring that all people visiting the chimpanzees wear masks (Hanamura et al. 2006), and we restricted the number of researchers and introduced quarantine periods for new researchers. With the assistance of the Frankfurt Zoological Society, TANAPA (2006) has finalized a general management plan that includes regulations for tourists intended to minimize the risk of disease introduction. Regrettably, some regulations are presently not well followed (Nakamura and Nishida 2009).

Long-term study allowed us to accumulate demographic, life history, and basic behavioral data on a long-lived species. These data are important not only within the context of academic knowledge, but also for the conservation of primate species, many of which face extinction. Long-term research is also personally rewarding to field workers; the term "nature smiles" appeared in an essay by the late Dr. Jun'ichiro Itani, the pioneer of Japanese primate field studies:

> "After half a century of field research, I often recall a few moments that I can never forget. They are the moments when almost all the important ideas came to me. . . . They are the jewels that we, field workers, dig out from the field with painstaking effort. We may say that they are the moments when nature smiles on the observer after bewilderingly long, long field work." (Itani 1993, our translation)

Our research projects have recently focused on the accumulation of comparative data on chimpanzee behavioral repertoires, especially on types of behavior that are candidates for cultural variation. Such work has produced a detailed chimpanzee ethogram and allowed us to gain insights into social customs. Although researchers have been studying chimpanzees for half a century, we still do not fully understand the variety of behaviors and the richness of chimpanzee societies. Encouraged by occasional smiles from nature, we continue to accumulate behavioral observations and to document the historical changes in chimpanzee society.

Acknowledgments We would like to thank Peter Kappeler for organizing the symposium "Long-term field studies of primates" held 8–11 December 2009, and for inviting MN to attend. We are also indebted to him and to David Watts for editing the volume and for providing useful comments on our manuscript. We thank the Tanzania Commission for Science and Technology, Tanzania National Parks, the Tanzania Wildlife Research Institute, Mahale Mountains National Park, and the Mahale Mountains Wildlife Research Centre for permission to conduct research in Mahale. It would not be possible to maintain the field site over decades without the collaboration of many researchers. We would like to thank the following researchers, who have participated in data collection and management of our research camp at Mahale: K Kawanaka, A Mori, S Uehara, J Itani, K Norikoshi, M Hiraiwa-Hasegawa, Y Takahata, T Hasegawa, H Hayaki, H Takasaki, MA Huffman, KD Hunt, T Tsukahara, K Masui, M Hamai, LA Turner, JC Mitani, K Hosaka, A Matsumoto-Oda, H Ihobe, N Itoh, T Sakamaki, N Corp, K Zamma, N Kutsukake, T Matsusaka, JV Wakibara, S Fujita, M Shimada, H Nishie, E Inoue, M Fujimoto, S Hanamura, M Kiyono-Fuse, T Kooriyama, and A Inaba. We also thank E Massawe, H Seki, H Baga, A Mutui, C Mwinuka, and T Clamsen for their logistical support and R Kitopeni, M Hamisi, M Bunengwa, K Athumani, H Bunengwa, H Katinkila, J Hassani, A Ramadhani, M Matumula, S Kabangula, M Mwami, R Hawazi, M Rashidi, B Hamisi, R Nyundo, M Seifu, M Masayuke, and B Athumani for assistance in the field. This study was financially supported by MEXT Grant-in-Aid (KAKENHI) (12375003, 16255007, 19255008 to TN, 19107007 to J Yamagiwa, and 21770262 to MN) and by the ITP-HOPE project of the Kyoto University Primate Research Institute, funded by JSPS.

References

Anonymous (1980) Mahale: study for the proposed Mahale Mountains National Park (final report, May 1980). Japan International Cooperation Agency, Tokyo

Assersohn C, Whiten A, Kiwede ZT, Tinka J, Karamagi J (2004) Use of leaves to inspect ectoparasites in wild chimpanzees: a third cultural variant? Primates 45:255–258

Azuma S, Toyoshima A (1961) Progress report of the survey of chimpanzees in their natural habitat, Kabogo Point area, Tanganyika. Primates 3:61–70

Boesch C (1995) Innovation in wild chimpanzees (*Pan troglodytes*). Int J Primatol 16:1–16

Boesch C (1996) Three approaches for assessing chimpanzee culture. In: Russon AE, Bard KA, Parker ST (eds) Reaching into thought: the minds of the great apes. Cambridge University Press, Cambridge, pp 404–429

Boesch C (1997) Evidence for dominant wild female chimpanzees investing more in sons. Anim Behav 54:811–815

Boesch C, Gone Bi ZB, Anderson D, Stahl D (2006) Food choice in Taï chimpanzees: are cultural differences present? In: Hohmann G, Robbins MM, Boesch C (eds) Feeding ecology in apes and other primates: ecological, physical and behavioral aspects. Cambridge University Press, New York, pp 183–201

Bonnie KE, de Waal FBM (2006) Affiliation promotes the transmission of a social custom: handclasp grooming among captive chimpanzees. Primates 47:27–34

Byrne RW (2007) Culture in great apes: using intricate complexity in feeding skills to trace the evolutionary origin of human technical prowess. Philos Trans R Soc Lond B 362:577–585 de Waal FBM, Seres M (1997) Propagation of handclasp grooming among captive chimpanzees. Am J Primatol 43:339–346

Emery Thompson M, Jones JH, Pusey AE, Brewer-Marsden S, Goodall J, Marsden D, Matsuzawa T, Nishida T, Reynolds V, Sugiyama Y, Wrangham RW (2007) Aging and fertility patterns in wild chimpanzees provide insights into the evolution of menopause. Curr Biol 17:2150–2156

Goodall J (1965) Chimpanzees of the Gombe Stream Reserve. In: De Vore I (ed) Primate behavior. Holt Reinhart and Winston, New York, pp 425–473

Goodall J (1986) The chimpanzees of Gombe: patterns of behavior. Harvard University Press, Cambridge, MA

Hanamura S, Kiyono M, Nakamura M, Sakamaki T, Itoh N, Zamma K, Kitopeni R, Matumula M, Nishida T (2006) A new code of observation employed at Mahale: prevention against a flu-like disease. Pan Afr News 13:1–16

Hanamura S, Kiyono M, Lukasik-Braum M, Mlengeya T, Fujimoto M, Nakamura M, Nishida T (2008) Chimpanzee deaths at Mahale caused by a flu-like disease. Primates 49:77–80

Hiraiwa-Hasegawa M, Hasegawa T, Nishida T (1984) Demographic study of a large-sized unit-group of chimpanzees in the Mahale Mountains, Tanzania: a preliminary report. Primates 25:401–413

Hosaka K (1995) A single flu epidemic killed at least 11 chimps. Pan Afr News 2:3–4

Hosaka K, Matsumoto-Oda A, Huffman MA, Kawanaka K (2000) Reactions to dead bodies of conspecifics by wild chimpanzees in the Mahale Mountains, Tanzania. Primate Res 16:1–15 [in Japanese with English summary]

Huffman MA, Kalunde MS (1993) Tool-assisted predation on a squirrel by a female chimpanzee in the Mahale Mountains, Tanzania. Primates 34:93–98

Ingold T (2001) The use and abuse of ethnography. Behav Brain Sci 24:337

Itani J (1979) Distribution and adaptation of chimpanzees in an arid area. In: Hamburg DA, McCown ER (eds) Perspectives on human evolution, vol 5, The great apes. Benjamin/Cummings Publishing, Menlo Park, CA, pp 55–71

Itani J (1993) When nature smiles. Heibon-sha, Tokyo [in Japanese]

Itoh N, Nishida T (2007) Chimpanzee grouping patterns and food availability in Mahale Mountains National Park, Tanzania. Primates 48:87–96

Itoh N, Sakamaki T, Hamisi M, Kitopeni R, Bunengwa M, Matumla M, Athumani K, Mwami M, Bunengwa H (1999) A new record of invasion by an unknown unit group into the center of M group territory. Pan Afr News 6:8–10

Itoh N, Nakamura M, Ihobe H, Uehara S, Zamma K, Pintea L, Seimon A, Nishida T (in press) Long-term changes in the social and natural environments surrounding the chimpanzees of the Mahale Mountains National Park. In: Plumptre AJ (ed) The ecological impact of long-term changes in Africa's rift valley. Nova Science, New York

Izawa K (1970) Unit groups of chimpanzees and their nomadism in the savanna woodland. Primates 11:1–46

Kahlenberg SM, Emery Thompson M, Muller MN, Wrangham RW (2008) Immigration costs for female chimpanzees and male protection as an immigrant counterstrategy to intrasexual aggression. Anim Behav 76:1497–1509

Kano T (1972) Distribution and adaptation of the chimpanzee on the eastern shore of Lake Tanganyika. Kyoto Univ Afr Stud 7:37–129

Kaur T, Singh J, Tong S, Humphrey C, Clevenger D, Tan W, Szekely B, Wang Y, Li Y, Muse EA, Kiyono M, Hanamura S, Inoue E, Nakamura M, Huffman MA, Jiang B, Nishida T (2008) Descriptive epidemiology of fatal respiratory outbreaks and detection of a human-related metapneumovirus in wild chimpanzees (*Pan troglodytes*) at Mahale Mountains National Park, Western Tanzania. Am J Primatol 70:755–765

Kiyono-Fuse M (2008) Use of wet hair to capture swarming termites by a chimpanzee in Mahale, Tanzania. Pan Afr News 15:8–12

Laland KN, Janik VM (2006) The animal cultures debate. Trends Ecol Evol 21:542–547

Marchant LF, McGrew WC (1999) Innovative behavior at Mahale: new data on nasal probe and nipple press. Pan Afr News 6:16–18

Matsusaka T (2007) Exploratory-threat behaviors in wild chimpanzees encountering a porcupine. Pan Afr News 14:29–31

Matsusaka T, Nishie H, Shimada M, Kutsukake N, Zamma K, Nakamura M, Nishida T (2006) Tool-use for drinking water by immature chimpanzees of Mahale: prevalence of an unessential behavior. Primates 47:113–122

McGrew WC (1992) Chimpanzee material culture: implications for human evolution. Cambridge University Press, Cambridge

McGrew WC (2004) The cultured chimpanzee: reflections on cultural primatology. Cambridge University Press, Cambridge

McGrew WC, Marchant LF (1998) Chimpanzee wears a knotted skin "necklace". Pan Afr News 5:8–9

McGrew WC, Tutin CEG (1978) Evidence for a social custom in wild chimpanzees? Man 13:234–251

McGrew WC, Marchant LF, Scott SE, Tutin CEG (2001) Intergroup differences in a social custom of wild chimpanzees: the grooming hand-clasp of the Mahale Mountains. Curr Anthropol 42:148–153

Nakamura M (1997) First observed case of chimpanzee predation on yellow baboons (*Papio cynocephalus*) at the Mahale Mountains National Park. Pan Afr News 4:9–11

Nakamura M (2002) Grooming-hand-clasp in Mahale M group chimpanzees: implication for culture in social behaviours. In: Boesch C, Hohmann G, Marchant LF (eds) Behavioural diversity in chimpanzees and bonobos. Cambridge University Press, Cambridge, pp 71–83

Nakamura M (2009) Aesthete in the forest? A female chimpanzee at Mahale collected and carried guineafowl feathers. Pan Afr News 16:17–19

Nakamura M, Itoh N (2001) Sharing of wild fruits among male chimpanzees: two cases from Mahale, Tanzania. Pan Afr News 8:28–31

Nakamura M, Itoh N (2005) Notes on the behavior of a newly immigrated female chimpanzee to the Mahale M group. Pan Afr News 12:20–22

Nakamura M, Itoh N (2008) Hunting with tools by Mahale chimpanzees. Pan Afr News 15:3–6

Nakamura M, Nishida T (2006) Subtle behavioral variation in wild chimpanzees, with special reference to Imanishi's concept of *kaluchua*. Primates 47:35–42

Nakamura M, Nishida T (2009) Chimpanzee tourism in relation to the viewing regulations at the Mahale Mountains National Park, Tanzania. Primate Conserv 24:85–90

Nakamura M, Uehara S (2004) Proximate factors of different types of grooming hand-clasp in Mahale chimpanzees: implications for chimpanzee social customs. Curr Anthropol 45:108–114

Nakamura M, McGrew WC, Marchant LF, Nishida T (2000) Social scratch: another custom in wild chimpanzees? Primates 41:237–248

Nishida T (1968) The social group of wild chimpanzees in the Mahale Mountains. Primates 9:167–224

Nishida T (1970) Social behavior and relationships among chimpanzees of Mahale Mountains. Primates 11:47–87

Nishida T (1979) The social structure of chimpanzees of the Mahale Mountains. In: Hamburg DA, McCown ER (eds) The great apes. Benjamin/Cummings, Menlo Park, CA, pp 73–121

Nishida T (1980) The leaf-clipping display: a newly-discovered expressive gesture in wild chimpanzees. J Hum Evol 9:117–128

Nishida T (1989) Social interactions between resident and immigrant female chimpanzees. In: Heltne PG, Marquardt LA (eds) Understanding chimpanzees. Harvard University Press, Cambridge, MA, pp 68–89

Nishida T (1990) A quarter century of research in the Mahale Mountains: an overview. In: Nishida T (ed) The chimpanzees of the Mahale Mountains: sexual and life history strategies. University of Tokyo Press, Tokyo, pp 3–35

Nishida T (1994) Review of recent findings on Mahale chimpanzees: implications and future research directions. In: Wrangham RW, McGrew WC, de Waal FBM, Heltne PG (eds) Chimpanzee cultures. Harvard University Press, Cambridge, MA, pp 373–396

Nishida T (1997) Sexual behavior of adult male chimpanzees of the Mahale Mountains National Park, Tanzania. Primates 38:379–398

Nishida T (1998) Deceptive tactic by an adult male chimpanzee to snatch a dead infant from its mother. Pan Afr News 5:13–15

Nishida T (2002) A self-medicating attempt to remove the sand flea from a toe by a young chimpanzee. Pan Afr News 9:5–6

Nishida T, Kawanaka K (1972) Inter-unit-group relationships among wild chimpanzees of the Mahali Mountains. Kyoto Univ Afr Stud 7:131–169

Nishida T, Nakamura M (2008) Long-term research and conservation in the Mahale Mountains, Tanzania. In: Wrangham RW, Ross E (eds) Science and conservation in African forests: the benefits of long-term research. Cambridge University Press, Cambridge, pp 173–183

Nishida T, Uehara S (1981) Kitongwe name of plants: a preliminary listing. Afr Stud Monogr 1:109–131

Nishida T, Wallauer WR (2003) Leaf-pile pulling: an unusual play pattern in wild chimpanzees. Am J Primatol 60:167–173

Nishida T, Wrangham RW, Goodall J, Uehara S (1983) Local differences in plant-feeding habits of chimpanzees between the Mahale Mountains and Gombe National Park, Tanzania. J Hum Evol 12:467–480

Nishida T, Hiraiwa-Hasegawa M, Hasegawa T, Takahata Y (1985) Group extinction and female transfer in wild chimpanzees in the Mahale National Park, Tanzania. Z Tierpsychol 67:284–301

Nishida T, Takasaki H, Takahata Y (1990) Demography and reproductive profiles. In: Nishida T (ed) The chimpanzees of the Mahale Mountains: sexual and life history strategies. University of Tokyo Press, Tokyo, pp 63–97

Nishida T, Kano T, Goodall J, McGrew WC, Nakamura M (1999) Ethogram and ethnography of Mahale chimpanzees. Anthropol Sci 107:141–188

Nishida T, Corp N, Hamai M, Hasegawa T, Hiraiwa-Hasegawa M, Hosaka K, Hunt KD, Itoh N, Kawanaka K, Matsumoto-Oda A, Mitani JC, Nakamura M, Norikoshi K, Sakamaki T, Turner L, Uehara S, Zamma K (2003) Demography, female life history, and reproductive profiles among the chimpanzees of Mahale. Am J Primatol 59:99–121

Nishida T, Mitani JC, Watts DP (2004) Variable grooming behaviours in wild chimpanzees. Folia Primatol 75:31–36

Nishida T, Matsusaka T, McGrew WC (2009) Emergence, propagation or disappearance of novel behavioral patterns in the habituated chimpanzees of Mahale: a review. Primates 50:23–36

Nishida T, Zamma K, Matsusaka T, Inaba A, McGrew WC (2010) Chimpanzee behavior in the wild: an audio-visual encyclopedia. Springer, Tokyo

Nishie H (2004) Increased hunting of yellow baboons (*Papio cynocephalus*) by M group chimpanzees at Mahale. Pan Afr News 11:10–12

Perry S, Godoy I, Lammers W (2012) The Lomas Barbudal Monkey Project: Two Decades of Research on *Cebus capucinus*. In: Kappeler PM (ed) Long-term field studies of primates. Springer, Heidelberg

Pruetz JD, Bertolani P (2007) Savanna chimpanzees, *Pan troglodytes verus*, hunt with tools. Curr Biol 17:412–417

Pusey AE, Murray C, Wallauer WR, Wilson ML, Wroblewski EE, Goodall J (2008) Severe aggression among female *Pan troglodytes schweinfurthii* at Gombe National Park, Tanzania. Int J Primatol 29:949–973

Rendell L, Whitehead H (2001) Culture in whales and dolphins. Behav Brain Sci 24:309–382

Sakamaki T (1998) First record of algae-feeding by a female chimpanzee at Mahale. Pan Afr News 5:1–3

Sakamaki T, Nakamura M (2007) Preliminary survey of unhabituated chimpanzees in the Mahale Mountains National Park, Tanzania: behavioral diversity across neighboring unit-groups and intergroup relationships. In: Formation of a strategic base for biodiversity studies. The 21st Century COE Program of Kyoto University, Kyoto, pp 278–280

Sakamaki T, Nakamura M, Nishida T (2007) Evidence of cultural differences in diet between two neighboring unit groups of chimpanzees in Mahale Mountains National Park, Tanzania. Pan Afr News 14:3–5

Shimada M (2002) Social scratch among the chimpanzees of Gombe. Pan Afr News 9:21–23

Shimada M, Matumula M (2004) Chimpanzee attitude toward a seriously weakened adolescent female at Mahale. Pan Afr News 11:6–8

Sugiyama Y (1981) Observations on the population dynamics and behavior of wild chimpanzees at Bossou, Guinea, in 1979–1980. Primates 22:435–444

Sugiyama Y (1989) Description of some characteristic behaviors and discussion on their propagation process among chimpanzees of Bossou, Guinea. In: Sugiyama Y (ed) Behavioral studies of wild chimpanzees at Bossou, Guinea. KUPRI, Inuyama, pp 43–47

Suzuki A (1969) An ecological study of chimpanzees in savanna woodland. Primates 10:103–148

Takahata Y, Hiraiwa-Hasegawa M, Takasaki H, Nyundo R (1986) Newly acquired feeding habits among the chimpanzees of the Mahale Mountains National Park, Tanzania. Hum Evol 1:277–284

Takasaki H (1983) Mahale chimpanzees taste mangoes: toward acquisition of a new food item? Primates 24:273–275

TANAPA (2006) Mahale Mountains National Park: general management plan 2006–2016. Tanzania National Parks, Arusha

Teleki G (1981) The omnivorous diet and eclectic feeding habits of chimpanzees in Gombe National Park, Tanzania. In: Harding RSO, Teleki G (eds) Omnivorous primates: gathering and hunting in human evolution. Columbia University Press, New York, pp 303–343

Tonooka R (2001) Leaf-folding behavior for drinking water by wild chimpanzees (*Pan troglodytes verus*) at Bossou, Guinea. Anim Cogn 4:325–334

Tsukahara T (1993) Lion eat chimpanzees: the first evidence of predation by lions on wild chimpanzees. Am J Primatol 29:1–11

Watts DP (2008) Tool use by chimpanzees at Ngogo, Kibale National Park, Uganda. Int J Primatol 29:83–94

Williams JM, Pusey AE, Carlis JV, Farm BP, Goodall J (2002) Female competition and male territorial behaviour influence female chimpanzees' ranging patterns. Anim Behav 63:347–360

Wrangham RW, de Waal FBM, McGrew WC (1994) The challenge of behavioral diversity. In: Wrangham RW, McGrew WC, de Waal FBM, Heltne PG (eds) Chimpanzee cultures. Harvard University Press, Cambridge, MA, pp 1–18

Zamma K (2002a) A chimpanzee trifling with a squirrel: pleasure derived from teasing? Pan Afr News 9:9–11

Zamma K (2002b) Leaf-grooming by a wild chimpanzee in Mahale. Primates 43:87–90

Zamma K, Fujita S (2004) Genito-genital rubbing among the chimpanzees of Mahale and Bossou. Pan Afr News 11:5–8

Chapter 16
Long-Term Studies of the Chimpanzees of Gombe National Park, Tanzania

Michael L. Wilson

Abstract The study of chimpanzees at Gombe National Park, which has continued for over 50 years, has yielded many discoveries that would have been impossible without long-term data. The basic data collection procedure has remained constant since the early 1970s, with a team of Tanzanian field assistants conducting all-day focal follows of individual chimpanzees in the Kasekela (1974–present) and Mitumba (1994–present) communities. Field assistants record map location and party composition at 15-min intervals and keep a continuous record of the focal subject's feeding behavior and narrative notes on behavior, including mating, agonistic interactions, tool use, hunting, boundary patrols, and intergroup interactions. Field assistants have also monitored the unhabituated Kalande community since 1999. A relational database developed from these data provides a powerful tool for analyzing long-term patterns. Studies using this database in combination with new technologies have deepened our understanding of chimpanzee pathogens, genetics, hormones, tool use, hunting, meat sharing, social relationships, habitat use, dispersal, life histories, and demography. This chapter focuses on life histories and demography, followed by a section that highlights findings on two topics for which long-term data have proven especially informative: intergroup aggression and disease ecology.

M.L. Wilson (✉)
Departments of Anthropology and Ecology, Evolution and Behavior, University of Minnesota, Minneapolis, MN, USA
e-mail: wilso198@umn.edu

P.M. Kappeler and D.P. Watts (eds.), *Long-Term Field Studies of Primates*, 357
DOI 10.1007/978-3-642-22514-7_16, © Springer-Verlag Berlin Heidelberg 2012

16.1 History and Infrastructure of Gombe Stream Research Centre

Long-term field studies are essential for understanding the lives of our closest relatives, including chimpanzees (*Pan troglodytes*). Jane Goodall began the first such study of chimpanzees at what was then Gombe Stream Game Reserve in 1960. Studies of captive chimpanzees by Wolfgang Köhler (1925) and Robert Yerkes (Yerkes and Yerkes 1929) had provided intriguing insights into chimpanzee cognition, but very little was known about chimpanzees in the wild. Nissen's short study of chimpanzees in French Guinea provided some of the first observations of feeding and social behavior (Nissen 1931). By 1960, technological advances such as jet travel and antimalarial drugs had made tropical field sites more accessible and medically safer for researchers from temperate zone countries, who had begun studying various primate species (e.g., Hanuman langurs (Jay 1962), savannah baboons (DeVore and Hall 1965), and mountain gorillas (Schaller 1963)), including chimpanzees (Kortlandt 1962).

16.1.1 Study Area

Gombe Stream Game Reserve was established in January 1943 (Thomas 1961) and upgraded to national park status in 1968 (Goodall 1986). Gombe National Park constitutes a 2–3.5-km-wide wedge-shaped strip of mountainous terrain rising from an elevation of 766 m along the shore of Lake Tanganyika to peaks ranging from 1,300 to 1,623 m along the crest of the rift escarpment in the east (Pintea et al. 2010; Fig. 16.1). Gombe's southern border is 15 km north of the town of Kigoma. The park extends roughly 14 km along the lake, with villages at its northern and southern ends and less densely settled land to the east (Fig. 16.2).

The park's area of 35.4 km^2 (13.7 square miles) consists of a series of steep ridges and valleys that descend westward to the lake. Roughly half of the 15 major valleys contain year-round streams; other streams flow seasonally. The vegetation varies from evergreen forest in valley bottoms, to thicket, vine tangle and semi-deciduous forest on the lower slopes; open woodland on the upper slopes; and grassland on the highest slopes (Clutton-Brock and Gillett 1979). Moist air from the lake combined with a north–south gradient of decreasing altitude creates a north–south gradient of decreasing rainfall. The north is thus more heavily forested, while the south is drier, with more woodland, and the density of chimpanzee food plants is substantially higher in the northern half of the park (Rudicell et al. 2010). In the 1960s, much of the park burned each year, creating an open habitat with little undergrowth. Fire has been suppressed since 1968, resulting in an increase in shrubs and vines and in overall greenness, especially in the north (Pintea et al. 2010). The increased undergrowth has not only likely improved the amount of food available to chimpanzees (Pintea et al. 2010), but has also reduced visibility and made following chimpanzees more difficult.

16 Long-Term Studies of the Chimpanzees of Gombe National Park, Tanzania

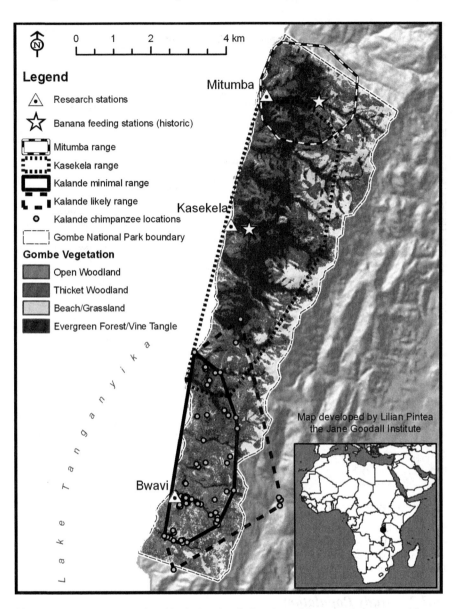

Fig. 16.1 Map of Gombe National Park, showing the locations of the research stations and former banana feeding stations, the 2007 ranges of the Mitumba and Kasekela communities, and the minimum and likely ranges of the Kalande community, based on sightings within the park and nest locations found near the park (2002–2009). The vegetation coverage within the park is based on classification from satellite images. The inset shows the location of Gombe within Tanzania. (Map based on Rudicell et al. (2010), Fig. 1, courtesy Lilian Pintea, the Jane Goodall Institute)

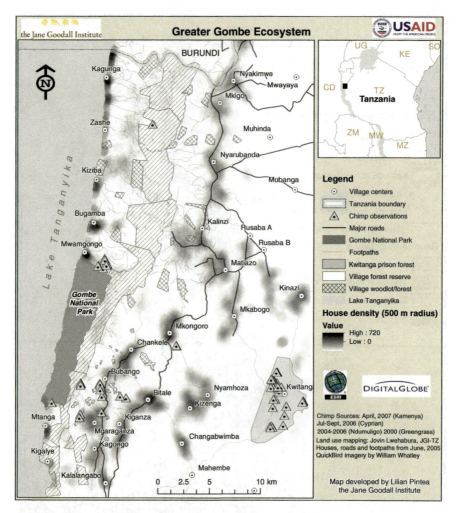

Fig. 16.2 Map of the greater Gombe ecosystem, showing Gombe National Park, the recently established village forest reserves, the location of chimpanzee sightings outside the park, and the density of houses, derived from 2005 QuickBird satellite imagery. (Map courtesy Lilian Pintea, the Jane Goodall Institute)

16.1.2 Study Population

Three chimpanzee communities have existed in Gombe throughout the duration of the study: Mitumba in the north, Kasekela[1] in the center, and Kalande in the south. As of January 2010, approximately 101–105 chimpanzees inhabited the park,

---

[1] Spelled "Kasakela" in earlier publications; "Kasekela" is the preferred local spelling.

including 25 in Mitumba, 61 in Kasekela, and perhaps 15–19 in Kalande. The Kahama community split from the Kasekela community in the early 1970s but survived only through 1977. Life history details are known for 292 chimpanzees in the Mitumba and Kasekela communities, with less detailed information available for another 40 or so in the Kalande community.

Most research has focused on the Kasekela community, which Goodall began studying in 1960. Efforts to habituate the Mitumba community began in 1985. It was originally intended to serve as a focus for tourism, but was subsequently excluded from tourism because of its small size (TANAPA 2005). Several efforts to habituate the Kalande community from 1968 to 1990 proved unsuccessful, but a monitoring program begun in 1999 continues to the present (Rudicell et al. 2010).

16.1.3 Data Collection Methods

Early efforts to observe wild chimpanzees faced challenges common to many other primate studies: chimpanzees use large ranges and generally flee from people. To overcome these challenges, Goodall employed an observational approach that differed greatly from previous attempts to study chimpanzees (e.g., both Nissen (1931) and Kortlandt had tried to watch chimpanzees from blinds), but is now nearly ubiquitous in primate studies. Like Carpenter (1934) had done in his pioneering field study, Goodall approached the animals in plain view, allowing them to become habituated to human presence gradually. Goodall searched for and watched chimpanzees from hilltops and other vantage points, and gradually observed them in more detail as habituation improved. By the end of her fourth month, Goodall had observed chimpanzees eating meat (Goodall 1963) and making and using tools to "fish" for termites (Goodall 1964).

During the 1960s, the study developed from a single researcher recording data *ad libitum* throughout the forest to a large research team focused on systematic collection of behavioral data in the vicinity of a banana provisioning area ("Camp") located near the center of the Kasekela community's range. Goodall began provisioning chimpanzees in 1963 to speed habituation and improve observation and filming conditions. She hired her first research assistant in 1964, and in 1965, she formally established the Gombe Stream Research Centre, which rapidly grew to host a large research team. Researchers kept daily records of chimpanzees seen in camp and of their interactions ("A-record"). Collection of these data continued until banana feeding ended in 2000. Starting in 1967, and then more regularly from 1970, chimpanzees were weighed in camp using a hanging spring balance baited with bananas (Pusey et al. 2005). Robert Hinde developed a check-sheet for recording mother–infant behavior in 1969, forming the basis of data collection that continues to the present. Banana feeding also attracted baboons, and by 1967 Goodall had established a long-term baboon study, which continues today (Ransom 1981; Collins et al. 1984; Packer et al. 1998).

Researchers began to follow chimpanzees away from Camp around 1968, for studies including hunting (Teleki 1973) and sexual behavior (McGinnis 1979). Tragically, in 1968, Ruth Davis fell to her death from a cliff while following chimpanzees in Kahama valley (Goodall 1986). Subsequently, researchers were required to work in teams of at least two for safety, and Goodall began hiring Tanzanian field assistants, starting with Hilali Matama in 1968. Initially, the Tanzanians simply accompanied the foreign students, but their excellent skill at tracking chimpanzees and observing their behavior led to the start of formal training in standardized data collection in 1970. In 1974, Larry Goldman and Donna Anderson began more intensive training of field assistants, introducing the use of check-sheets and reliability tests (Goodall 1986).

Focal follows, necessary for unbiased estimates of behavior rates (Altmann 1974), were first conducted in 1968, as students followed individual chimpanzees through the forest. Richard Wrangham initiated all-day focal sampling that involved following chimpanzees between consecutive night-nesting sites (Wrangham 1977). Focal data were called B-Record to distinguish them from the A-Record data. Field assistants conducted some focal follows in 1973 and have conducted daily all-day focal follows of individuals in the Kasekela community regularly since 1974. Comparable data collection for the Mitumba community began in 1994. Observers record map location, party composition, and female reproductive status at 15-min intervals ("travel and group check-sheets"); continuously record the focal subject's feeding behavior; note all occurrences of mating, agonistic interactions, grooming, and tool use by the focal subject; and take ad libitum narrative notes on other events, including hunting, boundary patrols, and intergroup interactions.

In the early 1970s, Goodall developed an affiliation with Stanford University through David Hamburg. From 1971 to 1975, Goodall was a visiting professor at Stanford. A series of Stanford undergraduates visited Gombe from 1972 to 1975, first receiving training in chimpanzee behavior at the Stanford Outdoor Primate Facility ("Gombe West") before spending 6 months at Gombe. Students also visited Gombe from the Zoology program at the University of Dar es Salaam, where Goodall was a guest lecturer.

The growing and productive international research effort at Gombe, including the partnership with Stanford, abruptly halted in 1975 with the kidnapping of four students by rebels from Zaire (Goodall 1986). All four students were eventually safely returned to their families, but Tanzanian authorities considered the park unsafe for foreign researchers, including Goodall, for many years following the kidnapping. The Tanzanian field assistants took over day-to-day operations at Gombe, which Goodall oversaw from her home base in Dar es Salaam. Thanks to the dedication of the Tanzanian research team, data collection was able to continue at Gombe throughout this time. To raise the funds necessary to continue long-term research, Goodall founded the Jane Goodall Institute in 1977, which continues to fund Gombe research while also expanding into a broader global mission.

In the 1980s, foreign researchers were gradually allowed to return to Gombe for increasing periods. Chris Boehm conducted studies of vocal communication and

16 Long-Term Studies of the Chimpanzees of Gombe National Park, Tanzania

behavior (1984–1990) and introduced the use of portable video cameras, which field assistants used to document behavior during B-record observations (Boehm 1989). Anthony Collins, who had studied baboons at Gombe in 1972–1975, returned in 1987 to oversee the baboon project, and since then has also been closely involved in management of the chimpanzee study.

By the 1980s, data accumulating from other sites made comparisons with Gombe possible. The first comparisons were between Gombe and Mahale and focused on diet (Nishida et al. 1983), habitat (Collins and McGrew 1988), and positional behavior (Hunt 1992). In 1990 and 1992, Christophe Boesch visited Gombe to compare hunting behavior at Gombe and Taï Forest, Côte d'Ivoire (Boesch 1994).

Foreign researchers returned to Gombe in greater numbers in the 1990s, including graduate students (e.g., Charlotte Uhlenbroek) and postdoctoral researchers (e.g., Craig Stanford). Several students started as volunteers habituating the Mitumba community, including Bill Wallauer, who subsequently began a 14-year-long video project to document chimpanzee behavior at Kasekela (Nishida and Wallauer 2003; Wilson et al. 2004; Pusey et al. 2008a). By the 2000s, Gombe once again hosted a regular contingent of graduate students from several universities. Many came from Anne Pusey's laboratory at the University of Minnesota and did work combining analysis of long-term data with new data collection. The 1990s and 2000s saw increased training of Tanzanian scientists, starting with Shadrack Kamenya, who studied red colobus monkeys (Kamenya 1997). In the 2000s, Tanzanian students conducted research for master's (Bakuza 2006; Ndimuligo 2007) and doctoral (Mjungu 2010) degrees.

While behavioral data collection continued using the same methods developed in the early 1970s, new technologies have greatly broadened and deepened our understanding of chimpanzees and their habitat. Advances in molecular technology permitted genetic analyses, including determination of kinship from noninvasively collected samples of hair (Morin et al. 1994) and feces (Constable et al. 2001; Wroblewski et al. 2009). Fecal samples also provided information for hormonal analyses (Emery Thompson et al. 2008), parasites (Gillespie et al. 2010), and virology (Keele et al. 2009). Hand-held video cameras have made possible more detailed analyses of behavior, including termite fishing (Lonsdorf et al. 2004) and meat sharing (Gilby 2006). Laser imaging technology has allowed creation of 3D models for analyzing skeletal materials (Kirchhoff 2010). The entry of demographic, ranging, feeding, and other behavioral data into a relational database has made decades worth of data available for analysis (e.g., Williams et al. 2002, 2004; Gilby et al. 2006; Murray et al. 2007; Wroblewski et al. 2009; Mjungu 2010). On a landscape scale, remote sensing images, from aerial photographs to satellite images, combined with Geographical Information Systems (GIS) software, have enabled the study of changes in habitat and land use in and around Gombe (Pintea et al. 2010). The phenology of key chimpanzee foods is regularly monitored in all three communities, and vegetation plots have been established and sampled throughout the park (Murray et al. 2006; Rudicell et al. 2010). Comparative studies using data from multiple field sites have been conducted on topics including demography (Hill et al. 2001; Emery Thompson et al. 2007), reproductive ecology (Emery Thompson 2005), and culture (Whiten et al. 1999; Nakamura et al. 2000; Nishida and Wallauer 2003; Schöning et al. 2008).

16.1.4 Infrastructure

Initially, Gombe research infrastructure consisted of a few simple tents, followed by prefabricated metal huts with thatched roofs. In 1971, George Dove helped upgrade the research infrastructure substantially, building breezeblock houses and offices along the lakeshore. The central office block still provides office and laboratory space, and an array of solar panels has provided electricity for lighting, computers, and other purposes, including a freezer for storing laboratory samples, since 2005. A small storage building has been converted into a necropsy laboratory. A herbarium cabinet stores plant samples for identification, and is now located in the JGI office in Kigoma for better protection from humidity. Breezeblock houses provide living quarters for research staff in Kasekela. Simple metal huts house research staff in Mitumba and in the Bwavi station in the Kalande community's range. To reduce risks of disease transmission, garbage pits and latrines are covered, and the entries to staff houses have been enclosed with wire mesh to provide "baboon-proof" outdoor living areas. Because of the rugged terrain, no roads reach the park, and Gombe is thus accessible only by foot or boat. The research center relies on wooden boats, with small ones stationed at Mitumba and Bwavi and larger ones at Kasekela.

16.1.5 Standardization and Management of Long-Term Data

The long-term study has produced many hundreds of thousands of pages of data. Analyzing these by hand is difficult and time consuming, and not really feasible for datasets that span many years, or for complicated analyses. Since 1990, Anne Pusey has worked with Goodall to develop an archive for housing all these datasheets, from which data are extracted and entered into a computer database. The Jane Goodall Institute's Center for Primate Studies operated at the University of Minnesota (1995–2009) and the archive has now moved to Duke University. Pusey has worked with computer scientists, including John Carlis, Shashi Shekhar, and their students, to develop a relational database, which keeps track of demographic data as well as behavioral data from the focal follows, including party composition, map location, and feeding, aggression, mating, and grooming. Data on ranging, party composition, and feeding are entered into the database from the maps and travel and group charts.

Extracting data from the B-record notes takes several steps, because the data are recorded in Swahili and written in a narrative instead of recorded on check-sheets.

(Several attempts have been made to design check-sheets to replace the narrative notes, but given the complexity of the data currently being collected, and the flexibility of the current system for recording both systematic data and descriptions of unusual events, designing check-sheets that are both sufficiently comprehensive and still practical to use in the field has been difficult.) The narrative notes are translated into English, and then entered into a database. Specific datasets such as mating, grooming, and aggression are extracted from the translated narrative notes and entered into tables, which are uploaded into an Access database; this allows efficient analysis of datasets that span many years (e.g., Williams et al. 2004; Gilby et al. 2006; Murray et al. 2006, 2007; Wroblewski et al. 2009; Rudicell et al. 2010). Work in progress includes the development of a searchable video database and computerizing of mother–infant data.

In the 1990s, the field maps used for recording the location of focal chimpanzees were redrawn from aerial photos rectified with the help of Global Positioning System (GPS) measurements. However, GPS signals are often difficult to obtain under heavy tree cover and in steep valley bottoms, so hand-drawn maps are still the main source of information on ranging. Comparison with GPS locations found that hand-drawn maps had a mean error of 133 m (Gilby et al. 2006). To document habitat change in and around the park, Pusey and Pintea have acquired remote sensing datasets, including aerial photographs and a series of satellite images of Gombe and the surrounding area, and have developed a digital elevation model for the park (Pintea et al. 2010).

16.2 Life Histories and Demography

16.2.1 Life Histories

Because chimpanzees can live nearly as long as humans, it has taken many years of study at Gombe and other sites (e.g., Mahale: Nishida et al. 2003; Bossou: Sugiyama 2004) to gain a clear picture of their life histories. During 50 years at Gombe, many individuals have been followed from birth to death. For example, researchers followed Goblin's life, from infancy (Goodall 1971, 1986, 1990), weaning (Pusey 1983), and adolescence (Pusey 1990) through his maturation into a politically savvy alpha male (Goodall 1986; Boehm 1992), followed by his fall from power in a brutal gang attack (Goodall 1992), his coalitionary behavior as a post-prime male (Gilby et al. 2009), and eventual old age and death (Williams et al. 2008; Terio et al. 2011). Necropsy revealed that Goblin had suffered from a multinodule infection of the nematode *Oesophagostomum* (Williams et al. 2008; Terio et al. 2011). Analysis of Goblin's skeleton found dental problems, including severely worn teeth and abscess drainage points (Kirchhoff 2010), which likely made it difficult for him to feed during the severe dry season of 2004, when he weakened and died.

Chimpanzees mature more rapidly and grow larger in captivity than in the wild, probably due to the abundance of food and protection from immunological and other stressors. Field data are thus necessary to obtain accurate measures of the timing of life history events in the context of the ecological constraints under which chimpanzee life history evolved. Moreover, life histories vary among sites; the following focuses on what has been learned from Gombe, especially from studies explicitly focused on development (e.g., Pusey 1983, 1990; Plooij 1984; Goodall 1986).

In the wild, chimpanzees grow slowly, being weaned at 50–86 months (Pusey 1983) and not reaching reproductive maturity until around 14–15 years for females and 15–16 years for males (Goodall 1986). At most chimpanzee study sites, the great majority of females transfer into new communities once they reach sexual maturity, whereas males universally stay in their natal communities (Pusey 1979; Stumpf et al. 2009; Nakamura and Nishida 2012). Females presumably disperse from their natal communities to avoid mating with close kin (Pusey 1979). About half of all females born into the Kaselela community, however, have stayed in their natal community (Pusey et al. 1997), perhaps due to limited dispersal opportunities. The proximate factors affecting dispersal and settlement decisions remain unclear, largely because data from neighboring habituated communities are limited. However, at least one Kasakela female that attempted to immigrate into Mitumba was repelled by aggressive attacks by resident females (Pusey et al. 2008a), suggesting that Mitumba is currently crowded, as far as female chimpanzees are concerned.

Once chimpanzees reach reproductive maturity, they focus their efforts on activities associated with reproductive success. Males and females both spend most of their time searching for and eating food, but important sex differences exist in reproductive strategies. Females must establish themselves in suitable areas, often in new communities, that have sufficient food to raise their young, are relatively safe from intergroup aggression, and have suitable mates (mature, unrelated males). Once established, females spend most of their time gestating, lactating, and otherwise caring for their offspring. For males, the main jobs in life are trying to attain high dominance rank, mating with estrous females, and joining with other males to defend and possibly expand the group territory.

By adulthood, male and female chimpanzees thus lead very different social lives. In general, males are more gregarious, often foraging in large parties, competing for dominance rank, grooming with allies, displaying at opponents, and mating with estrous females (Mitani 2009). Females are generally less gregarious than males (Pepper et al. 1999), perhaps because carrying infants makes mothers more sensitive to the costs of scramble competition associated with larger foraging parties (Wrangham 2000). Wrangham and Smuts (1980) found that Gombe mothers spent more than half of their time alone or with just their families. Females at some other sites, such as Taï and Ngogo, are more gregarious than Gombe females, likely due to greater abundance of food, but are still less gregarious than males (Lehmann and Boesch 2008; Wakefield 2008). When females have fully tumescent sexual swellings, they become more social, traveling with large parties of males or going off on consort with individual males (Goodall 1986).

16 Long-Term Studies of the Chimpanzees of Gombe National Park, Tanzania 367

Besides differing in overall gregariousness, the sexes differ in foraging patterns and range use. Females spend more time foraging for insects such as termites (Pandolfi et al. 2003), whereas males spend more time hunting and eating mammalian prey such as red colobus monkeys (Stanford et al. 1994). Males generally use the community's entire range, patrolling the borders and sometimes making deep incursions into neighboring ranges, whereas females generally use a smaller proportion of the community's total range (Williams et al. 2002). At Gombe, females spend much of their time in smaller "core areas" (Wrangham 1979; Wrangham and Smuts 1980). The extent to which males and females differ in their ranging patterns varies among sites (e.g., Lehmann and Boesch 2008), but evaluating these differences is complicated by the use of different definitions and methods of analysis across sites. At Gombe, female home ranges have been analyzed based on points from all-day follows in which females were "alone" (that is, unaccompanied by adults other than their female relatives) (Williams et al. 2002; Murray et al. 2006), a method not used at other sites. However, analysis of all ranging points indicates that Kanyawara females use different "neighborhoods" within the total range (Wilson 2001; Emery Thompson et al. 2007).

Early observations of these sex differences in space use suggested that the sexes might differ in community membership as well. Wrangham (1979) described three alternative models of chimpanzee social structure, in which females (1) ranged equally over the entire community with males, (2) ranged in smaller core areas but associated only with males of a particular community, or (3) ranged in smaller core areas distributed across the landscape independently of male ranging behavior. Studies at Gombe support the second model (Goodall 1983; Williams et al. 2002), while studies at Taï support the first model (Lehmann and Boesch 2008). In general, though, it appears that each female belongs to a specific community and that community memberships are stable. Exceptions to this rule occur when the number of males in a community declines precipitously, and parous females begin visiting or even transferring to neighboring communities (Nishida et al. 1985; Rudicell et al. 2010).

Both male and female chimpanzees compete for access to key resources. Early studies suggested that chimpanzees had completely egalitarian social relations (e.g., Reynolds and Reynolds 1965). Studies of habituated chimpanzees soon revealed that males compete vigorously for status (Bygott 1979). For females, though, decided agonistic interactions are infrequent and often subtle, leading to early speculations that dominance is not an important aspect of their social relationships (de Waal 1984; Wrangham 1980). Moreover, variation exists among sites; for example, Wakefield (2008) found no evidence for linear female dominance hierarchies at Ngogo. At Gombe, however, both males and females can be assigned dominance ranks based on the outcome of dyadic contests (Murray et al. 2006). At Gombe, higher dominance rank is associated with greater reproductive success for both males and females (Pusey et al. 1997; Constable et al. 2001; Wroblewski et al. 2009; Jones et al. 2010).

Rank appears to affect female reproductive success largely through access to better feeding areas, which in turn allows females to shorten their inter-birth intervals. In general, female chimpanzees reproduce slowly (Emery Thompson et al. 2007). Kasekela females had a median inter-birth interval of 4.9 years, including intervals where the previous infant died (Jones et al. 2010). High-ranking females at Gombe produce surviving infants at higher rates than low-ranking females and have daughters that mature at younger ages (Pusey et al. 1997). Vegetation plots conducted throughout the Kasekela community's range revealed that high-ranking females had smaller core areas with higher densities of preferred food trees (Murray et al. 2006). Fertility also varies with age (Emery Thompson et al. 2007; Jones et al. 2010), but considerable variation cannot be explained by either rank or age, suggesting that phenotypic quality is an important component of fertility (Jones et al. 2010). Phenotypic quality likely encompasses a broad range of traits, including maternal and other social skills plus immune system function.

Determining male reproductive success is challenging in chimpanzees, because females generally mate with multiple males. Goodall initially characterized the chimpanzee mating system as promiscuous (Goodall 1965). While it eventually became clear that high-ranking males tended to monopolize mating with available females, especially during the peri-ovulatory period (or "POP"; Tutin 1979), only with the advent of noninvasive genetic sampling did patterns of paternity become clear (Constable et al. 2001; Wroblewski et al. 2009). As it turned out, high-ranking males do have more offspring at Gombe (Constable et al. 2001; Wroblewski et al. 2009) and elsewhere (e.g., Taï: Boesch et al. 2006; Budongo: Newton-Fisher et al. 2010). However, by competing for access to peri-ovulatory females and by using alternative mating tactics (e.g., consortship) as well as mating with younger, less preferred females, lower-ranking males can achieve higher reproductive success than predicted by the priority-of-access model (Wroblewski et al. 2009).

Like other primates (Charnov and Berrigan 1993), chimpanzees live long lives compared to other mammals their size. Nonetheless, chimpanzees have higher age-specific mortality than human foragers, and appear to senesce more rapidly (Hill et al. 2001). A study compiling data from Gombe and several other sites found that mean life expectancy at birth was below 15 years for both males and females (Hill et al. 2001). The main causes of death were disease and conspecific aggression (Williams et al. 2008). Predation by leopards (Boesch 1991; Furuichi 2000) and lions (Inagaki and Tsukahara 1993) can be important sources of mortality at other sites, and probably were in the past at Gombe, before lions were extirpated and leopards reduced in numbers (Pierce 2009). Chimpanzees are considered "old" by their mid-30s (Goodall 1986), and by their late 30s, they often suffer from ailments of old age such as worn teeth and degenerative joint disease (Morbeck et al. 2002; Kirchhoff 2010). However, because exceptional individuals can live well beyond 50 years, even the 50-year-long Gombe study has not lasted long enough to document the longest chimpanzee lives fully. The oldest chimpanzee in Kasekela, Flo, was initially thought to be 41 ± 5 years old when she died (Goodall 1983). Flo's estimated age was later revised upward to 53 years, based on comparison with other chimpanzees (Williams et al. 2008). The oldest individual currently alive at Gombe, Sparrow, is approximately 52 years, and, with her full coat of hair and generally robust appearance, looks younger than Flo did at the end of her life.

Because Flo was already old in 1960 when the study started, and because Sparrow immigrated into the study community as an adolescent, neither individual's age is known precisely. Despite the longevity of these exceptional individuals, very few chimpanzees at Gombe have lived past 40 years. Among the individuals followed since infancy, the oldest female, Fifi, lived to be 46 years and appeared to be in good health when last seen. The oldest female followed since birth, Gremlin, is now 40 years. Several males are estimated to have lived to 40 years old or just beyond: Evered, Huxley, and McGregor. The oldest male followed since birth, Goblin, died 13 days before he would have turned 40 years.

Like most other mammals, but in contrast to humans, reproductive senescence in chimpanzee follows approximately the same schedule as somatic aging. The interval between births increases with age (Jones et al. 2010), and eventually some very old females may stop cycling altogether (Nishida et al. 2003; Nakamura and Nishida 2012), but females do not typically experience a post-reproductive life phase, or menopause (Emery Thompson et al. 2007).

16.2.2 Population Size

While Gombe is a small park, the park's total chimpanzee population remains uncertain. Estimates of the total population prior to about 2000 rely largely on guesswork. Goodall's estimate for the 1960s was 100–150 chimpanzees (van Lawick-Goodall 1968). Estimates made by extrapolating back from the known and estimated sizes of the Mitumba and Kalande communities yielded a similar range of 120–150 chimpanzees in the 1960s (Pusey et al. 2008b). Estimates of the population since about 2000 are more precise, thanks to the genotyping of fecal samples from the unhabituated Kalande community (Rudicell et al. 2010). Based on detailed demographic data for Mitumba and Kasekela and estimates for Kalande, the minimum population estimate ranged from 91 to 101 chimpanzees (median = 95) from 2000 to 2010, and the maximum estimate ranged from 100 to 109 chimpanzees (median = 103), with no clear upward or downward trend over time.

Despite the recent stability of the total population, the sizes of the individual communities have varied considerably. Although research on the Kasekela community began in 1960, it was not until 1966 that all individuals were known, at which time 60 individuals lived in Kasekela. By 1973, Kasekela had divided into two communities: Kasekela, with 40 chimpanzees, and Kahama, with at least 14 (and probably more, given the likelihood that Kahama also included unidentified, unhabituated females). Males from Kasekela killed at least five individuals from Kahama (Goodall 1986). By the end of 1977, the Kahama community no longer existed, as all the males had been killed or disappeared and the surviving females and offspring had rejoined Kasekela or dispersed elsewhere (Goodall 1986). Following a respiratory epidemic in 1987 that killed nine individuals, Kasekela reached its smallest recorded population size, 38, in 1989. Kasekela's population gradually recovered in the 1990s, and, thanks to immigration, grew rapidly in the 2000s, numbering 61 chimpanzees by the start of 2010.

The size of the Mitumba community has been documented only since 1985. Goodall (1986) estimated that Mitumba contained 50 individuals in the 1970s and early 1980s. During this time, the Mitumba successfully defended its boundaries against the Kasekela community, and even expanded its range at the expense of Kasekela in the early 1980s, suggesting that it was similar in size to Kasekela (which contained 51–58 individuals in 1978–1983). However, records from the first years of direct observation indicate that probably at least 20 individuals, but no more than 31, were alive in 1985 (Mjungu 2010). We do not know whether Mitumba's population declined in the early 1980s, whether some individuals were not recorded at that time, or if the pre-habituation estimates of the community's size were too high. However, by the mid-1990s, the chimpanzees were sufficiently well habituated, and observations sufficiently frequent, that all individuals were likely identified. A respiratory epidemic in 1996 caused Mitumba's population to drop from 25 to 20 (Mjungu 2010). The community's population declined to a minimum of 19 at the start of 1999, followed by a gradual recovery to 25 individuals at the start of 2010 (Mjungu 2010; Rudicell et al. 2010). Despite the overall increase, the number of adult males decreased from five in 1994 to only two from 2005 on (Mjungu 2010). The number of adult males in Mitumba may yet recover, though. The community has six males aged 8–13 years, suggesting that the Mitumba males might yet again pose a significant challenge to the Kasekela males in adulthood.

In 1969, Gale identified more than 20 individuals in the Kalande community and estimated that it totaled at least 40 (Rudicell et al. 2010) and might have been substantially larger. By the late 1990s, though, Kalande chimpanzees rarely met Kasekela chimpanzees, suggesting that Kalande had declined substantially. Estimates based on sightings, molecular genetics, and likely migration patterns indicate that Kalande contained 19–43 individuals in 1998, but declined to only 15–19 by the start of 2010, with a particularly severe loss in 2002 (Rudicell et al. 2010). During this time, Kalande had an unusually high prevalence of the virus SIVcpz, which is known to increase mortality in chimpanzees (Keele et al. 2009), suggesting that at least some of the decline resulted from the impact of SIVcpz infection (Rudicell et al. 2010).

As recently as the 1970s, forest and woodland habitat outside the park likely connected Gombe to other chimpanzee populations (Pintea et al. 2010). Intergroup encounters observed in 1975 to the extreme east of Kasekela's range suggest that an "Eastern" or "Rift" community might have persisted to the east of the park (Goodall et al. 1979), though those encounters might have simply involved individuals from Kalande moving north along the eastern margin of their rivals' range (Goodall 1986). By the late 1990s, however, rapid deforestation had turned Gombe into a largely isolated island of chimpanzee habitat (Pintea et al. 2010). Nonetheless, an unknown number of chimpanzees survive in forest fragments outside the park, including a small population (perhaps 15 weaned individuals) in Kwitanga Forest

Fig. 16.3 Goblin through the ages (**a**) juvenile in ~1969, (**b**) as a young adolescent in ~1972, (**c**) as alpha male in 1985, presiding over a red colobus carcass, (**d**) as an old male with severely worn teeth in 2003, (**e**) shortly before his death in August 2004, and (**f**) his skull in the Gombe skeleton collection. Photos ©David Bygott (**a, b**), ©Chris Boehm (**c**), ©Michael Wilson (**d, e**), and ©Claire Kirchhoff (**f**)

(Ndimuligo 2007), which may be close enough for chimpanzees to travel between there and Gombe. Chimpanzee nests have also been found near Zashe village, in the hills along the rift escarpment to the north of Gombe, near the border with Burundi (Fig. 16.3).

16.3 Research Highlights

16.3.1 Intergroup Dynamics

One topic for which Gombe is especially well known and that is relevant to many aspects of the long-term study is intergroup dynamics. Only long-term data provide information needed to understand the consequences of interactions between communities. As Goodall noted (1986:3), "Had my colleagues and I stopped after a mere 10 years, we should have had a very different picture of the Gombe chimpanzees than we do today. We would have observed many similarities in their behavior and ours, but we would have been left with the impression that chimpanzees were far more peaceable than humans".

Because chimpanzees live in fission–fusion societies, many years were required just to determine the boundaries of social groups at Gombe. Goodall initially supposed that the entire chimpanzee population interacted freely and peacefully (Goodall 1965). Reynolds formed a similar impression of "loose unstable groupings and apparent lack of group social organization" at Budongo (Reynolds and Reynolds 1965:422). In contrast, Itani and Suzuki (1967) inferred that chimpanzees lived in stable "large-sized groups." In 1965, Nishida began research in the Mahale Mountains, 150 km south of Gombe, with the specific goal of understanding the social organization of wild chimpanzees (Nishida 1979; Nakamura and Nishida 2012). By 1968, Nishida had inferred that chimpanzees lived in socially bounded fission–fusion societies. Observations at Gombe, and eventually other sites, confirmed that chimpanzees indeed live in socially bounded, mutually hostile groups called "communities" (Goodall et al. 1979).

Once observers began to follow Kasekela chimpanzees away from the feeding station in the late 1960s, they witnessed encounters with neighboring communities (Goodall et al. 1979). Around the same time, the Kasekela community began to fission. By 1970, two subgroups could be discerned. Six adult males, an adolescent male, and three adult females began avoiding the northern part of the range and spending more time in the south, while 8 adult males and 11 adult females occupied the north and avoided the south. The chimpanzees that ranged to the south were called the Kahama community, while those that stayed in the north retained the name Kasekela. Besides the three habituated females that became part of the Kahama community, an unknown number of unhabituated females also probably associated with the Kahama males (Goodall 1983). By 1973, the two communities were clearly distinct, and threatened each other when they met – with only occasional friendly contacts among some of the older males giving any hint that the two communities had once been one (Goodall 1983, 1986).

The first observed intergroup violence involved the killing of the infant of an unfamiliar female in 1971 (Bygott 1972). In 1973, observers following three males from the Kahama community found a freshly dead adult female, possibly a Kalande female killed by Kahama males (Wrangham 1975). Observers then witnessed a series of attacks from 1974 to 1977, during which Kasekela males fatally wounded four adult males and an adult female from Kahama; the freshly killed body of a fifth male was also found (Goodall 1986). By the end of 1977, all of Kahama's males had died or disappeared, and the remaining females either returned to Kasekela or joined other communities. Intergroup infanticides continued, with two in 1975 and one in 1979 (Goodall 1986). These attacks took place to the south of Kasekela's range, indicating that the infants' mothers were either unhabituated members of the Kahama community, or members of the Kalande community.

Goodall describes the extermination of the Kahama community as "The Four Year War" (Goodall and Berman 1999). While the rate of killings appears to have been particularly high during the 1970s, it also seems that chimpanzee communities are never fully at peace. In the late 1970s, having vanquished the Kahama community, the Kasekela chimpanzees experienced hostile encounters with their powerful neighbors to the north and south. By 1983, outside males were making deep incursions into the heart of Kasekela's range (Goodall 1986). Several otherwise healthy adult males disappeared, suggesting they might have been killed (*ibid.*), but no intergroup killings were directly observed in the 1980s. This may be at least partly due to the reduced number of observation hours per year compared to that in the preceding and following decades, resulting in fewer follows to border areas where intergroup interactions are most likely. The intergroup encounters that were observed were hostile, and severe attacks on females from neighboring communities were seen (Williams et al. 2004).

By the early 1990s, the Kasekela community had begun recovering in population size and range size. In March 1993, the Kasekela chimpanzees killed Rejea, an infant from the Mitumba community (Wilson et al. 2004). By the late 1990s, the expansion of the Kasekela range accelerated, largely at the expense of the Kalande community, which apparently started a steep population decline (Rudicell et al. 2010). On deep incursions into Kalande's range in 1998, Kasekela males attacked two infants, killing one, and severely attacked and likely killed an adolescent male (Wilson et al. 2004). During the 2000s, the Kasekela community maintained a large range, and intergroup violence continued intermittently. Kasekela males likely killed a Mitumba male, Rusambo, in 2002 (Wilson et al. 2004), and killed a Mitumba infant, Andromeda, in 2005 (Wrangham et al. 2006). In turn, Mitumba males killed Patti, an adult female from Kasekela (Wrangham et al. 2006; Williams et al. 2008).

Intergroup aggression has been a major source of mortality at Gombe (Williams et al. 2008). Intergroup killings have also been reported for all of the other long-term study sites, including Mahale, Kanyawara, Ngogo, and Budongo (summarized in Wrangham et al. 2006; Mitani et al. 2010), and Taï (Boesch et al. 2008). Killings have also occurred at some sites where studies have less time depth, such as Kalinzu (Hashimoto and Furuichi 2005) and Petit Loango (Boesch et al. 2007). Overall, the evidence strongly suggests that intergroup killing is a species-typical behavior.

16.3.1.1 Causes of Intergroup Aggression

Studies at Gombe and elsewhere have helped to clarify the causes of intergroup aggression. Males apparently benefit from defending and expanding territories by excluding rival males; by increasing the amount of food available for self, mates, and offspring; and by making territories more attractive to potential immigrants (Wilson and Wrangham 2003; Mitani et al. 2010). Females benefit from male territorial defense and expansion due to improved safety and improved food resources (Williams et al. 2004).

Male chimpanzees are hostile to foreign males, and genetic evidence supports the view that males are generally successful at preventing outside males from mating with their females. At Gombe, all infants tested in the Kasekela community had fathers from within their own community (Constable et al. 2001; Wroblewski et al. 2009). Paternity tests of Mitumba chimpanzees have likewise found no evidence of extra-community paternity (Wroblewski unpublished data). One female from the Kalande community apparently has an infant fathered by a Kasekela male (Rudicell et al. 2010). This conception might have occurred during a prolonged visit by that female to the Kasekela community (*ibid.*). No extra-group paternities have been reported for Budongo (Reynolds 2005), though extra-group paternity accounted for 7–11% of offspring born into communities at Taï (Vigilant et al. 2001; Boesch et al. 2006).

Several lines of evidence support the view that chimpanzees benefit by increasing the size of their territories. Analysis of 33 years of body mass data found that individuals in Kasekela were heavier when territory size was larger and population density was lower (Pusey et al. 2005). Analysis of 18 years of data for Kasekela found that females reproduced more quickly, individuals traveled in larger parties, and males encountered receptive females more often when the territory size was larger (Williams et al. 2004). Females dispersing from the declining Kalande community settled more often in the larger Kasekela territory than in the smaller Mitumba territory (Rudicell et al. 2010).

The quality of a territory may depend on other factors besides size, including food abundance and the density of competitors. When the number of adult males in Kalande was reduced to one, parous females began to visit the Kasekela community, and at least one emigrated permanently (Rudicell et al. 2010). This resembles the mass transfer of mothers from Mahale's K-group following the reduction of the community to a single adult male (Nishida et al. 1985) and suggests that females base their residence decisions on having enough males in the community to provide some or all of the following benefits: defense of food resources, protection from intergroup aggression, and providing sufficient mating partners. Investigations are currently under way to determine how food abundance and distribution affect the timing and location of intergroup encounters (Wilson et al. 2010).

Overall, the causes of intergroup aggression in chimpanzees are similar to those of many other group territorial species (Crofoot and Wrangham 2010). In most such species, though, intergroup conflict is rarely fatal. In contrast, among chimpanzees, humans, some social carnivores, and several social insects, intergroup conflict can account for a substantial portion of adult mortality. Manson and Wrangham (1991; cf. Wrangham 1999) argued that among mammals, fatal fighting is most likely in species with fission–fusion social systems, coalitionary bonds, and intergroup hostility, which together create opportunities for killing rivals at relatively low cost to the attackers (Manson and Wrangham 1991). Killing rivals, rather than merely chasing them off, benefits the killers by reducing the coalitionary strength of their opponents (Wrangham 1999). This "imbalance of power hypothesis," based on observations at Gombe, where the attackers generally greatly outnumbered the victims (Goodall 1986), has been supported by evidence from other sites, including the Kanyawara and Ngogo communities in Kibale National Park (Watts 2012). Playback experiments found that Kanyawara males were more likely to approach a simulated intruder the greater their numerical superiority (Wilson et al. 2001). Kanyawara males visited borders more often when in parties with more males (Wilson et al. 2007), and Ngogo males conducted boundary patrols more often when with many males (Mitani and Watts 2005).

16.3.2 Disease Ecology

Disease is the major cause of death for chimpanzees at Gombe (Williams et al. 2008), as at other sites (e.g., Mahale: Nishida et al. 2003 and Taï: Boesch and Boesch-Achermann 2000). Many deaths have occurred during epidemics, which included suspected polio, respiratory infections, and mange (Goodall 1986; Lonsdorf et al. 2006; Williams et al. 2008). Because chimpanzees and humans are so closely related, they share many of the same diseases. Molecular evidence indicates that at least some of the infections suffered by chimpanzees at research and tourism sites originate from humans (Köndgen et al. 2008). Humans may also contract diseases originating in chimpanzees (Wolfe et al. 1998). Understanding disease ecology is thus important for several reasons, such as guiding conservation management decisions, ensuring the health of people visiting and working in primate conservation areas, and understanding the origins and natural history of diseases that affect humans.

Various health data have been collected over the course of the study (Lonsdorf et al. 2006). Since March 2004, Lonsdorf and colleagues have been collecting systematic health data on the Kasekela and Mitumba chimpanzees (Lonsdorf et al. 2006; Travis et al. 2008). For each regular B-record follow, observers visually assess the health of the focal subject and collect a fecal sample for parasitological analysis. Gillespie and colleagues (2010) found that parasite prevalence was higher at Mitumba than Kasekela, perhaps due to the proximity of the Mitumba chimpanzees to the densely populated village of Mwamgongo. The health monitoring project has also provided tools and training to improve the recovery of samples and information during necropsies, which has proven invaluable in investigating the impact of another infection at Gombe: SIVcpz.

Current evidence indicates that HIV-1, the cause of the global AIDS pandemic, originated in the transmission of SIVcpz from chimpanzees to humans (Hahn et al. 2000). The strains of SIVcpz most closely related to HIV-1 occur in west-central Africa (Keele et al. 2006). Related strains of SIVcpz occur in the eastern Congo basin, but the virus is distributed unevenly and has not been detected in Uganda or Mahale (ibid.). Gombe is thus the only site with habituated chimpanzees at which some of the chimpanzees are naturally infected with SIVcpz (Santiago et al. 2002; Keele et al. 2006). SIVcpz infection was initially thought to be harmless (Silvestri 2008). However, detailed demographic data have revealed that it increases the mortality risk of infected chimpanzees by 10–16 times, and analysis of tissues collected from necropsies revealed that infected chimpanzees suffered AIDS-like damage to immune system tissues (Keele et al. 2009). Implications of this discovery for the future of Gombe's chimpanzees are sobering. The Kalande community, which has a much higher prevalence of the virus than Mitumba or Kasekela, has suffered a dramatic decline that may at be due at least partly to SIVcpz infection (Rudicell et al. 2010). Population models indicate that infected populations are likely to decline. Nonetheless, in small populations like Gombe, the virus may go extinct before the host population does. Moreover, factors such as immigration may help prevent population decline.

16.4 Discussion and Conclusions

In five decades of long-term research at Gombe, researchers have learned a great deal about the behavior and life histories of wild chimpanzees. While some striking behaviors such as tool use and hunting were observed early, only in the second decade of study did the outlines of chimpanzee social structure began to emerge, including group territorial defense, lethal aggression, and female transfer. The importance of female dominance rank was not clear until the study's fourth decade. Direct measures of male reproductive success were not obtained until the study's fifth decade, and we are still learning how female settlement patterns and ecology affect reproductive success. While many complete life histories have now been documented, the study has still not exceeded the lifespan of the longest lived chimpanzees. Gombe is small, but we are only now getting accurate estimates of the total chimpanzee population, and only two of the park's three communities are fully habituated. A wealth of data has been accumulated on intergroup dynamics, yet we are just starting to understand what influences frequencies of intergroup conflict. An integrated, multidisciplinary effort has begun to address disease, the major source of mortality, yet much work is needed to reduce the risk of disease transfer from humans to chimpanzees. Moreover, Gombe has unexpectedly emerged as a key site in the study of a virus that was unknown in 1960, but which has since claimed the lives of millions of people around the world. The presence of SIVcpz in Gombe presents both a fascinating opportunity for research and a challenging threat to the population.

Meanwhile, other field studies across Africa have broadened our understanding of chimpanzees and have confirmed that many patterns of behavior observed at Gombe are species typical. These studies have also found important differences among chimpanzee societies, revealing a diversity of social customs and raising questions about the extent to which behavioral differences depend on genetic differences, ecology, or social learning.

Many questions remain, even within a specific topic, such as aggression. For example: Will the Kasekela males kill the remaining Mitumba and Kalande males? If they eliminate their rivals, will they control a single super community, or will the community fission? Instead, if the adolescent males of Mitumba survive into adulthood, will they take back territory from Kasekela?

In the coming decades, molecular analyses of noninvasively collected samples will no doubt continue to yield new ways of detecting and analyzing viruses, parasites, genes, hormones, and more. The analysis of entire genomes will become practical. Molecular studies will continue to complement behavioral work in important ways, helping to confirm the identities of individuals as they move among communities and identifying likely maternal and paternal relationships, even among unhabituated chimpanzees (Rudicell et al. 2010). Yet however much technology advances, the core of the research must remain the individual chimpanzees, with their distinctive personalities and sometimes unexpected behavior. The more we learn about these individuals and their relationships, the more potential there is to ask interesting questions, including many that we have not yet thought of.

One of the most pressing questions is whether chimpanzees will survive in the wild. As human populations grow and forests become fragmented by logging and conversion to agriculture, many chimpanzee populations face a future similar to Gombe: living in ecological islands, surrounded by people, and heavily impacted by human activities and climate change (see also Jolly 2012). Gombe provides a field laboratory for promoting the survival of chimpanzees across Africa. Current efforts seek to address two major threats: disease risk, and habitat loss and subsequent population isolation. The health monitoring project is collecting baseline data essential for assessing chimpanzee health. Efforts have been made to improve sanitation around human settlements and minimize contact between chimpanzees and humans, including researchers, park staff, and tourists. Improvements in sanitation and enforcement of existing rules are still needed. The Jane Goodall Institute has been working with local villages to establish an interconnected network of village forest reserves around Gombe (Fig. 16.2). These could provide benefits both to villagers, in terms of protected watersheds, reduced erosion, and better conservation of forest products, and to chimpanzees, by providing a buffer and by enabling them to move more easily between Gombe and other remnant populations in the area (Pintea et al. 2010). If chimpanzees begin moving through these forest reserves, efforts will be needed to maintain local good will and prevent poaching.

Despite many threats, the Gombe chimpanzee population has remained stable over the past 10 years, giving some hope that it will survive well into the future.

Moreover, the government of Tanzania continues to demonstrate that it regards Gombe as one of the nation's natural treasures. In 2010, President Kikwete announced that the Tanzanian government would nominate Gombe for consideration as a World Heritage Site. Moreover, plans are underway to extend the park boundaries 1.5 km into the lake to protect freshwater species, resulting in a total park area of over 56 km^2 (Kayanda 2010). Gombe will need continued national and international support to ensure that its chimpanzees continue to survive, and that we will continue to learn from them.

Acknowledgments Research at Gombe is conducted with the permission and support of the Tanzania Commission for Science and Technology, the Tanzania Wildlife Research Institute, and the Tanzania National Parks. Over the past 50 years, numerous sources have provided funding for this research, including the Wilkie Foundation, the National Geographic Society, the Leakey Foundation, the William T. Grant Foundation, and the Jane Goodall Institute. The development of the relational database and studies of genetics and intergroup aggression have been supported by the National Science Foundation (DBS-9021946, SBR-9319909, BSC-0452315, IIS-0431141, and BSC-0648481), the Harris Steel Group, and the University of Minnesota. Research on SIVcpz was supported by grants from the National Institutes of Health (R01 AI50529, R01 AI58715, T32 GM008111, and K01HD051494) and the UAB Center for AIDS Research (P30 AI 27767). The Solar Electric Light Fund and the US Fish and Wildlife Service have contributed to infrastructure improvements. The health monitoring project has been supported by the US Fish and Wildlife Service, The Lester E. Fisher Center for the Study and Conservation of Apes, and the Davee Center for Epidemiology and Endocrinology of Lincoln Park Zoo. Conservation efforts in the Greater Gombe Ecosystem have been funded by the United States Agency for International Development (USAID). Becky Sun assisted with artwork. Anne Pusey and Anthony Collins provided helpful comments on earlier versions of this manuscript. Above all, the long-term study depends on the hard work of the field staff and visiting researchers at the Gombe Stream Research Centre, who have contributed countless hours to this enormously collaborative project.

References

Altmann J (1974) Observational study of behavior: sampling methods. Behaviour 49:227–266
Bakuza JS (2006) Ecological parasitology of chimpanzees in Gombe National Park, Tanzania. MSc thesis, University of Dar es Salaam, Tanzania
Boehm C (1989) Methods for isolating chimpanzee vocal communication. In: Heltne PG, Marquardt LA (eds) Understanding chimpanzees. Harvard University Press, Cambridge/MA, pp 38–59
Boehm C (1992) Segmentary 'warfare' and the management of conflict: comparison of East African chimpanzees and patrilineal-patrilocal humans. In: Harcourt AH, de Waal FBM (eds) Coalitions and alliances in humans and other animals. Oxford University Press, Oxford, pp 137–173
Boesch C (1991) The effects of leopard predation on grouping patterns in forest chimpanzees. Behaviour 117:220–242
Boesch C (1994) Cooperative hunting in wild chimpanzees. Anim Behav 48:653–667
Boesch C, Boesch-Achermann H (2000) The chimpanzees of the Taï Forest: behavioural ecology and evolution. Oxford University Press, Oxford
Boesch C, Kohou G, Néné H, Vigilant L (2006) Male competition and paternity in wild chimpanzees of the Taï forest. Am J Phys Anthropol 130:103–115

Boesch C, Head J, Tagg N, Arandjelovic M, Vigilant L, Robbins MM (2007) Fatal chimpanzee attack in Loango National Park, Gabon. Int J Primatol 28:1025–1034

Boesch C, Crockford C, Herbinger I, Wittig RM, Moebius Y, Normand E (2008) Intergroup conflicts among chimpanzees in Taï National Park: lethal violence and the female perspective. Am J Primatol 70:519–532

Bygott JD (1972) Cannibalism among wild chimpanzees. Nature 238:410–411

Bygott JD (1979) Agonistic behaviour and dominance among wild chimpanzees. In: Hamburg DA, McCown ER (eds) The great apes. Benjamin/Cummings, Menlo Park/CA, pp 405–427

Carpenter CR (1934) A field study of the behavior and social relations of howling monkeys (*Alouatta palliata*). Comp Psychol Monogr 10:1–168

Charnov EL, Berrigan D (1993) Why do female primates have such long lifespans and so few babies? or life in the slow lane. Evol Anthropol 1:191–194

Clutton-Brock TH, Gillett JB (1979) A survey of the forest composition in the Gombe National Park, Tanzania. Afr J Ecol 17:131–158

Collins DA, McGrew WC (1988) Habitats of three groups of chimpanzees (*Pan troglodytes*) in western Tanzania compared. J Hum Evol 17:553–574

Collins DA, Busse CD, Goodall J (1984) Infanticide in two populations of savanna baboons. In: Hausfater G, Hrdy SB (eds) Infanticide: comparative and evolutionary perspectives. Aldine, New York, pp 193–215

Constable JL, Ashley MV, Goodall J, Pusey AE (2001) Noninvasive paternity assignment in Gombe chimpanzees. Mol Ecol 10:1279–1300

Crofoot MC, Wrangham RW (2010) Intergroup aggression in primates and humans: the case for a unified theory. In: Kappeler PM, Silk JB (eds) Mind the gap. Springer, Berlin, pp 171–195

DeVore I, Hall KRL (1965) Baboon ecology. In: DeVore I (ed) Primate behavior: field studies of monkeys and apes. Holt Rienehart and Winston, New York, pp 20–52

de Waal FBM (1984) Sex differences in the formation of coalitions among chimpanzees. Ethol Sociobiol 5:239–255

Emery Thompson M (2005) Reproductive endocrinology of wild female chimpanzees (*Pan troglodytes schweinfurthii*): methodological considerations and the role of hormones in sex and conception. Am J Primatol 67:137–158

Emery Thompson M, Jones JH, Pusey AE, Brewer-Marsden S, Goodall J, Marsden D, Matsuzawa T, Nishida T, Reynolds V, Sugiyama Y, Wrangham RW (2007) Aging and fertility patterns in wild chimpanzees provide insights into the evolution of menopause. Curr Biol 17:2150–2156

Emery Thompson M, Wilson ML, Gobbo G, Muller MN, Pusey AE (2008) Hyperprogesteronemia in response to *Vitex fischeri* consumption in wild chimpanzees (*Pan troglodytes schweinfurthii*). Am J Primatol 70:1064–1071

Furuichi T (2000) Possible case of predation on a chimpanzee by a leopard in the Petit Loango Reserve, Gabon. Pan Afr News 7:21–23

Gilby IC (2006) Meat sharing among the Gombe chimpanzees: harassment and reciprocal exchange. Anim Behav 71:953–963

Gilby IC, Eberly LE, Pintea L, Pusey AE (2006) Ecological and social influences on the hunting behaviour of wild chimpanzees, *Pan troglodytes schweinfurthii*. Anim Behav 72:169–180

Gilby IC, Wilson ML, Mohlenhoff KA, Pusey AE (2009) The function of coalitionary aggression among wild chimpanzees. 78th Annual Meeting of the American Association of Physical Anthropologists. Am J Phys Anthropol 138 (S48): 133

Gillespie TR, Lonsdorf EV, Canfield EP, Meyer DJ, Nadler Y, Raphael J, Pusey AE, Pond J, Pauley J, Mlengeya T, Travis DA (2010) Demographic and ecological effects on patterns of parasitism in eastern chimpanzees (*Pan troglodytes schweinfurthii*) in Gombe National Park, Tanzania. Am J Phys Anthropol 143:534–544

Goodall J (1963) Feeding behaviour of wild chimpanzees. A preliminary report. Symp Zool Soc Lond 10:39–47

Goodall J (1964) Tool-using and aimed throwing in a community of free-living chimpanzees. Nature 201:1264–1266

Goodall J (1965) Chimpanzees of the Gombe Stream Reserve. In: DeVore I (ed) Primate behavior. Holt Rinehart and Winston, New York, pp 425–473

Goodall J (1971) In the shadow of man. Houghton Mifflin, Boston

Goodall J (1983) Population dynamics during a 15 year period in one community of free-living chimpanzees in the Gombe National Park, Tanzania. Z Tierpsychol 61:1–60

Goodall J (1986) The chimpanzees of Gombe: patterns of behavior. Belknap, Cambridge, MA

Goodall J (1990) Through a window: my thirty years with the chimpanzees of Gombe. Houghton Mifflin, Boston

Goodall J (1992) Unusual violence in the overthrow of an alpha male chimpanzee at Gombe. In: Nishida T, McGrew WC, Marler P, Pickford M, de Waal FBM (eds) Topics in primatology, vol 1, Human origins. University of Tokyo Press, Tokyo, pp 131–142

Goodall J, Berman P (1999) Reason for hope: a spiritual journey. Warner Books, New York

Goodall J, Bandora A, Bergmann E, Busse C, Matama H, Mpongo E, Pierce A, Riss D (1979) Intercommunity interactions in the chimpanzee population of the Gombe National Park. In: Hamburg DA, McCown ER (eds) The great apes. Benjamin/Cummings, Menlo Park/CA, pp 13–53

Hahn BH, Shaw GM, De Cock KM, Sharp PM (2000) AIDS as a zoonosis: scientific and public health implications. Science 287:607–614

Hashimoto C, Furuichi T (2005) Possible intergroup killing in chimpanzees in the Kalinzu Forest, Uganda. Pan Afr News 12:3–5

Hill K, Boesch C, Goodall J, Pusey AE, Williams JM, Wrangham RW (2001) Mortality rates among wild-living chimpanzees. J Hum Evol 40:437–450

Hunt KD (1992) Positional behavior of *Pan troglodytes* in the Mahale Mountains and Gombe Stream National Parks, Tanzania. Am J Phys Anthropol 87:83–105

Inagaki H, Tsukahara T (1993) A method of identifying chimpanzee hairs in lion feces. Primates 34:109–112

Itani J, Suzuki A (1967) The social unit of chimpanzees. Primates 8:355–381

Jay PC (1962) Aspects of maternal behavior among langurs. Ann NY Acad Sci 102:468–476

Jolly A (2012) Berenty Reserve, Madagascar: A long time in a small space. In: Kappler PM, Watts DP (eds) Long-Term Field Studies of Primates. Springer, Heidelberg

Jones JH, Wilson ML, Murray CM, Pusey AE (2010) Phenotypic quality influences fertility in Gombe chimpanzees. J Anim Ecol 79:1262–1269

Kamenya S (1997) Changes in land use patterns and their impacts on red colobus monkeys' behavior: implications for primate conservation in Gombe National Park, Tanzania. PhD thesis, Boulder, University of Colorado

Kayanda A (2010) Gombe reserve set to extend its boundaries. The Citizen: 22 July 2010

Keele BF, Van Heuverswyn F, Li Y, Bailes E, Takehisa J, Santiago ML, Bibollet-Ruche F, Chen Y, Wain LV, Liegeois F, Loul S, Mpoudi Ngole E, Bienvenue Y, Delaporte E, Brookfield JFY, Sharp PM, Shaw GM, Peeters M, Hahn BH (2006) Chimpanzee reservoirs of pandemic and nonpandemic HIV-1. Science 313:523–526

Keele BF, Jones JH, Terio KA, Estes JD, Rudicell RS, Wilson ML, Li Y, Learn GH, Beasley TM, Schumacher-Stankey JC, Wroblewski EE, Mosser A, Raphael J, Kamenya S, Lonsdorf EV, Travis DA, Mlengeya T, Kinsel MJ, Else JG, Silvestri G, Goodall J, Sharp PM, Shaw GM, Pusey AE, Hahn BH (2009) Increased mortality and AIDS-like immunopathology in wild chimpanzees infected with SIVcpz. Nature 460:515–519

Kirchhoff CA (2010) From birth to bones: skeletal evidence for health, disease, and injury in the Gombe chimpanzees. PhD thesis, University of Minnesota, Minneapolis

Köhler W (1925) The mentality of apes. Harcourt, New York

Köndgen S, Kühl H, N'Goran PK, Walsh PD, Schenk S, Ernst N, Biek R, Formenty P, Mätz-Rensing K, Schweiger B, Junglen S, Ellerbrok H, Nitsche A, Briese T, Lipkin WI, Pauli G, Boesch C, Leendertz FH (2008) Pandemic human viruses cause decline of endangered great apes. Curr Biol 18:260–264

Kortlandt A (1962) Chimpanzees in the wild. Sci Am 206:128–138

Lehmann J, Boesch C (2008) Sexual differences in chimpanzee sociality. Int J Primatol 29:65–81

Lonsdorf EV, Eberly LE, Pusey AE (2004) Sex differences in learning in chimpanzees. Nature 428:715–716

Lonsdorf EV, Travis D, Pusey AE, Goodall J (2006) Using retrospective health data from the Gombe chimpanzee study to inform future monitoring efforts. Am J Primatol 68:897–908

Manson JH, Wrangham RW (1991) Intergroup aggression in chimpanzees and humans. Curr Anthropol 32:369–390

McGinnis PR (1979) Sexual behavior in free-living chimpanzees: consort relationships. In: Hamburg DA, McCown ER (eds) The great apes. Benjamin/Cummings, Menlo Park, CA, pp 429–439

Mitani JC (2009) Cooperation and competition in chimpanzees: current understanding and future challenges. Evol Anthropol 18:215–227

Mitani JC, Watts DP (2005) Correlates of territorial boundary patrol behaviour in wild chimpanzees. Anim Behav 70:1079–1086

Mitani JC, Watts DP, Amsler SJ (2010) Lethal intergroup aggression leads to territorial expansion in wild chimpanzees. Curr Biol 20:R507–R508

Mjungu DC (2010) Dynamics of intergroup competition in two neighboring chimpanzee communities. PhD thesis, University of Minnesota, St. Paul

Morbeck ME, Galloway A, Richman Sumner D (2002) Getting old at Gombe: skeletal aging in wild-ranging chimpanzees. In: Erwin JM, Hof PR (eds) Aging in nonhuman primates. Karger, Basel, pp 48–62

Morin PA, Moore JJ, Chakraborty R, Jin L, Goodall J, Woodruff DS (1994) Kin selection, social structure, gene flow, and the evolution of chimpanzees. Science 265:1193–1201

Murray CM, Eberly LE, Pusey AE (2006) Foraging strategies as a function of season and rank among wild female chimpanzees (*Pan troglodytes*). Behav Ecol 17:1020–1028

Murray CM, Mane SV, Pusey AE (2007) Dominance rank influences female space use in wild chimpanzees, *Pan troglodytes*: towards an ideal despotic distribution. Anim Behav 74:1795–1804

Nakamura M, McGrew WC, Marchant LF, Nishida T (2000) Social scratch: another custom in wild chimpanzees? Primates 41:237–248

Nakamura M, Nishida T (2012) Long-term field studies of chimpanzees at Mahale Mountains National Park, Tanzania. In: Kappler PM, Watts DP (eds) Long-term field studies of primates. Springer, Heidelberg

Ndimuligo SA (2007) Assessment of chimpanzee (*Pan troglodytes*) population and habitat in Kwitanga Forest, western Tanzania. MSc thesis, University of Witwatersrand, Johannesburg, South Africa

Newton-Fisher NE, Emery Thompson M, Reynolds V, Boesch C, Vigilant L (2010) Paternity and social rank in wild chimpanzees (*Pan troglodytes*) from the Budongo Forest, Uganda. Am J Phys Anthropol 142:417–428

Nishida T (1968) The social group of wild chimpanzees in the Mahali Mountains. Primates 9:167–224

Nishida T (1979) The social structure of chimpanzees of the Mahale Mountains. In: Hamburg DA, McCown ER (eds) The great apes. Benjamin/Cummings, Menlo Park, CA, pp 73–121

Nishida T, Wallauer WR (2003) Leaf-pile pulling: an unusual play pattern in wild chimpanzees. Am J Primatol 60:167–173

Nishida T, Wrangham RW, Goodall J, Uehara S (1983) Local differences in plant-feeding habits of chimpanzees between the Mahale Mountains and Gombe National Park, Tanzania. J Hum Evol 12:467–480

Nishida T, Hiraiwa-Hasegawa M, Hasegawa T, Takahata Y (1985) Group extinction and female transfer in wild chimpanzees in the Mahale National Park, Tanzania. Z Tierpsychol 67:284–301

Nishida T, Corp N, Hamai M, Hasegawa T, Hiraiwa-Hasegawa M, Hosaka K, Hunt KD, Itoh N, Kawanaka K, Matsumoto-Oda A, Mitani JC, Nakamura M, Norikoshi K, Sakamaki T, Turner L,

Uehara S, Zamma K (2003) Demography, female life history, and reproductive profiles among the chimpanzees of Mahale. Am J Primatol 59:99–121

Nissen H (1931) A field study of the chimpanzee: observations of chimpanzee behavior and environment in western French Guinea. Comp Psychol Monogr 8:1–122

Packer C, Tatar M, Collins A (1998) Reproductive cessation in female mammals. Nature 392:807–811

Pandolfi SS, van Schaik CP, Pusey AE (2003) Sex differences in termite fishing among Gombe chimpanzees. In: de Waal FBM, Tyack PL (eds) Animal social complexity: intelligence, culture, and individualized societies. Harvard University Press, Cambridge/MA, pp 414–418

Pepper JW, Mitani JC, Watts DP (1999) General gregariousness and specific social preferences among wild chimpanzees. Int J Primatol 20:613–632

Pierce A (2009) An encounter between a leopard and a group of chimpanzees at Gombe National Park. Pan Afr News 16:22–24

Pintea L, Pusey AE, Wilson ML, Gilby IC, Collins DA, Kamenya S, Goodall J (2010) Long-term changes in the ecological factors surrounding the chimpanzees of Gombe National Park. In: Plumptre AJ (ed) Long term changes in Africa's Rift Valley: impacts on biodiversity and ecosystems. Nova, New York, pp 194–210

Plooij FX (1984) The behavioral development of free-living chimpanzee babies and infants. Ablex Publishing Corporation, Norwood, NJ

Pusey AE (1979) Intercommunity transfer of chimpanzees in Gombe National Park. In: Hamburg DA, McCown ER (eds) The great apes. Benjamin/Cummings, Menlo Park, CA, pp 464–479

Pusey AE (1983) Mother-offspring relationships in chimpanzees after weaning. Anim Behav 31:363–377

Pusey AE (1990) Behavioural changes at adolescence in chimpanzees. Behaviour 115:203–246

Pusey AE, Williams JM, Goodall J (1997) The influence of dominance rank on the reproductive success of female chimpanzees. Science 277:828–831

Pusey AE, Oehlert GW, Williams JM, Goodall J (2005) Influence of ecological and social factors on body mass of wild chimpanzees. Int J Primatol 26:3–31

Pusey AE, Murray C, Wallauer WR, Wilson ML, Wroblewski EE, Goodall J (2008a) Severe aggression among female *Pan troglodytes schweinfurthii* at Gombe National Park, Tanzania. Int J Primatol 29:949–973

Pusey AE, Wilson ML, Collins DA (2008b) Human impacts, disease risk, and population dynamics in the chimpanzees of Gombe National Park, Tanzania. Am J Primatol 70:738–744

Ransom TW (1981) Beach troop of the Gombe. Associated University Press, London

Reynolds V (2005) The chimpanzees of the Budongo Forest: ecology, behaviour, and conservation. Oxford University Press, Oxford

Reynolds V, Reynolds F (1965) Chimpanzees of the Budongo Forest. In: DeVore I (ed) Primate behavior: field studies of monkeys and apes. Holt Rinehart and Winston, New York, pp 368–424

Rudicell RS, Jones JH, Wroblewski EE, Learn GH, Li Y, Robertson JD, Greengrass E, Grossmann F, Kamenya S, Pintea L, Mjungu DC, Lonsdorf EV, Mosser A, Lehman C, Collins DA, Keele BF, Goodall J, Hahn BH, Pusey AE, Wilson ML (2010) Impact of simian immunodeficiency virus infection on chimpanzee population dynamics. PLoS Pathogens 6:e1001116. doi:10.1371/journal.ppat.1001116

Santiago ML, Rodenburg CM, Kamenya S, Bibollet-Ruche F, Gao F, Bailes E, Meleth S, Soong S-J, Kilby JM, Moldoveanu Z, Fahey B, Muller MN, Ayouba A, Nerrienet E, McClure HM, Heeney JL, Pusey AE, Collins DA, Boesch C, Wrangham RW, Goodall J, Sharp PM, Shaw GM, Hahn BH (2002) SIVcpz in wild chimpanzees. Science 295:465

Schaller GB (1963) The mountain gorilla – ecology and behavior. University of Chicago Press, Chicago

Schöning C, Humle T, Möbius Y, McGrew WC (2008) The nature of culture: technological variation in chimpanzee predation on army ants revisited. J Hum Evol 55:48–59

Silvestri G (2008) Immunity in natural SIV infections. J Intern Med 265:97–109

Stanford CB, Wallis J, Matama H, Goodall J (1994) Patterns of predation by chimpanzees on red colobus monkeys in Gombe National Park, 1982–1991. Am J Phys Anthropol 94:213–228

Stumpf RM, Emery Thompson M, Muller MN, Wrangham RW (2009) The context of female dispersal in Kanyawara chimpanzees. Behaviour 146:629–656

Sugiyama Y (2004) Demographic parameters and life history of chimpanzees at Bossou, Guinea. Am J Phys Anthropol 124:154–165

TANAPA (2005) Gombe National Park: general management plan 2005–2015. Tanzania National Parks, Arusha

Teleki G (1973) The predatory behavior of wild chimpanzees. Bucknell University Press, Lewisburg/PA

Terio KA, Kinsel MJ, Raphael J, Mlengeya T, Lipende I, Kirchhoff C, Gilagiza B, Wilson ML, Kamenya S, Estes JD, Keele BF, Rudicell RS, Liu W, Patton S, Collins DA, Hahn BH, Travis DA, Lonsdorf EV (2011) Pathological lesions in chimpanzees (Pan troglodytes schweinfurthii) from Gombe National Park, Tanzania, 2004–2010. J Zoo and Wildlife Medicine 42(4):597–607

Thomas DK (1961) The Gombe stream game reserve. Tanganyika Notes Records 56:34–39

Travis DA, Lonsdorf EV, Mlengeya T, Raphael J (2008) A science-based approach to managing disease risks for ape conservation. Am J Primatol 70:745–750

Tutin CEG (1979) Mating patterns and reproductive strategies in a community of wild chimpanzees (Pan troglodytes schweinfurthii). Behav Ecol Sociobiol 6:29–38

van Lawick-Goodall J (1968) The behaviour of free-living chimpanzees in the Gombe Stream Area. Anim Behav Monogr 1:161–311

Vigilant L, Hofreiter M, Siedel H, Boesch C (2001) Paternity and relatedness in wild chimpanzee communities. Proc Natl Acad Sci USA 98:12890–12895

Watts DP (2012) Long-term research on chimpanzee behavioral ecology in Kibale National Park, Uganda. In: Kappler PM, Watts DP (eds) Long-Term Field Studies of Primates. Springer, Heidelberg

Wakefield ML (2008) Grouping patterns and competition among female Pan troglodytes schweinfurthii at Ngogo, Kibale National Park, Uganda. Int J Primatol 29:907–929

Whiten A, Goodall J, McGrew WC, Nishida T, Reynolds V, Sugiyama Y, Tutin CEG, Wrangham RW, Boesch C (1999) Chimpanzee cultures. Nature 399:682–685

Williams JM, Pusey AE, Carlis JV, Farm BP, Goodall J (2002) Female competition and male territorial behaviour influence female chimpanzees' ranging patterns. Anim Behav 63:347–360

Williams JM, Oehlert GW, Carlis JV, Pusey AE (2004) Why do male chimpanzees defend a group range? Anim Behav 68:523–532

Williams JM, Lonsdorf EV, Wilson ML, Schumacher-Stankey JC, Goodall J, Pusey AE (2008) Causes of death in the Kasekela chimpanzees of Gombe National Park, Tanzania. Am J Primatol 70:766–777

Wilson ML (2001) Imbalances of power: how chimpanzees respond to the threat of intergroup aggression. PhD thesis, Harvard University Press, Cambridge, MA

Wilson ML, Wrangham RW (2003) Intergroup relations in chimpanzees. Annu Rev Anthropol 32:363–392

Wilson ML, Hauser MD, Wrangham RW (2001) Does participation in intergroup conflict depend on numerical assessment, range location, or rank for wild chimpanzees? Anim Behav 61:1203–1216

Wilson ML, Wallauer WR, Pusey AE (2004) New cases of intergroup violence among chimpanzees in Gombe National Park, Tanzania. Int J Primatol 25:523–549

Wilson ML, Hauser MD, Wrangham RW (2007) Chimpanzees (Pan troglodytes) modify grouping and vocal behaviour in response to location-specific risk. Behaviour 144:1621–1653

Wilson ML, Mjungu DC, Pusey AE (2010) Causes of intergroup aggression among chimpanzees at Gombe. Abstract: XXIII Congress of the International Primatological Society, Kyoto, Japan

Wolfe ND, Escalante AA, Karesh WB, Kilbourn A, Spielman A, Lal AA (1998) Wild primate populations in emerging infectious disease research: the missing link? Emerg Infect Dis 4:149–158

Wrangham RW (1975) The behavioural ecology of chimpanzees in Gombe National Park, Tanzania. PhD thesis, Cambridge University

Wrangham RW (1977) Feeding behaviour of chimpanzees in Gombe National Park, Tanzania. In: Clutton-Brock TH (ed) Primate ecology: studies of feeding and ranging behaviour in lemurs, monkeys and apes. Academic, New York, pp 503–538

Wrangham RW (1979) Sex differences in chimpanzee dispersion. In: Hamburg DA, McCown ER (eds) The great apes. Benjamin/Cummings, Menlo Park, CA, pp 481–489

Wrangham RW (1980) An ecological model of female-bonded primate groups. Behaviour 75:262–300

Wrangham RW (1999) Evolution of coalitionary killing. Yearbk Phys Anthropol 42:1–30

Wrangham RW (2000) Why are male chimpanzees more gregarious than mothers? A scramble competition hypothesis. In: Kappeler PM (ed) Primate males: causes and consequences of variation in group composition. Cambridge University Press, Cambridge, pp 248–258

Wrangham RW, Smuts BB (1980) Sex differences in the behavioural ecology of chimpanzees in the Gombe National Park, Tanzania. J Reprod Fert 28(Suppl):13–31

Wrangham RW, Wilson ML, Muller MN (2006) Comparative rates of violence in chimpanzees and humans. Primates 47:14–26

Wroblewski EE, Murray CM, Keele BF, Schumacher-Stankey JC, Hahn BH, Pusey AE (2009) Male dominance rank and reproductive success in chimpanzees, *Pan troglodytes schweinfurthii*. Anim Behav 77:873–885

Yerkes RM, Yerkes AW (1929) The great apes: a study of anthropoid life. Yale University Press, New Haven

Chapter 17
Long-Term Research on Grauer's Gorillas in Kahuzi-Biega National Park, DRC: Life History, Foraging Strategies, and Ecological Differentiation from Sympatric Chimpanzees

Juichi Yamagiwa, Augustin Kanyunyi Basabose, John Kahekwa, Dominique Bikaba, Chieko Ando, Miki Matsubara, Nobusuke Iwasaki, and David S. Sprague

Abstract We have conducted long-term research on sympatric gorillas and chimpanzees in Kahuzi-Biega National Park since 1987. The demographic history of habituated gorillas has provided insights into their reproductive strategies. Infanticide by male gorillas, which has occurred frequently in the Virunga mountain gorilla population, had not been reported in Kahuzi for more than 20 years. However, soon after the large-scale killing of gorillas during a war in the late 1990s, it occurred three times within a few months. The infanticidal male might have discriminated between infants who were not his offspring and an infant whom he presumably sired based on past interactions with their mothers. At Kahuzi, births occurred most frequently during the period of ripe fruit abundance, and female Grauer's gorillas show longer inter-birth interval than female mountain gorillas in the Virungas. A comparison of reproductive strategies among different gorilla populations suggests that seasonal fluctuation in food abundance may lead to slow reproduction, whereas the potential pressure of infanticide may promote rapid reproduction. The reduced ranges and increased encounters between unfamiliar groups induced by large human disturbance, such as wars or conversion of their habitat to farmland, might have produced conditions leading to infanticide.

J. Yamagiwa (✉) • C. Ando
Department of Zoology, Kyoto University, Kyoto, Japan
e-mail: yamagiwa@jinrui.zool.kyoto-u.ac.jp

A.K. Basabose
International Gorilla Conservation Program, Mont Goma, Democratic Republic of Congo

J. Kahekwa • D. Bikaba
Pole Pole Foundation, Miti, Democratic Republic of Congo

M. Matsubara
Primate Research Institute, Kyoto University, Inuyama, Aichi, Japan

N. Iwasaki • D.S. Sprague
National Institute for Agro-Environmental Sciences, Tsukuba, Ibaragi, Japan

P.M. Kappeler and D.P. Watts (eds.), *Long-Term Field Studies of Primates*,
DOI 10.1007/978-3-642-22514-7_17, © Springer-Verlag Berlin Heidelberg 2012

Long-term data on diet composition indicate extensive overlap in fruit foods between sympatric gorillas and chimpanzees. However, their ranging and fallback strategies differed. The gorillas tended to use a large area evenly, whereas the chimpanzees visited small areas repeatedly. When preferred fruits were scarce, gorillas increased consumption of vegetative foods including leaves, pith, and barks, and chimpanzees ate more leaves, pith, and animal foods. These differences are tied to differences in food processing and digestive abilities, positional behavior, and cognitive abilities that mitigate feeding competition between the two species. The growing human population and political instability caused by the recent war have increased pressures on wildlife in and around the park. Conflicts between the park authorities and unemployed local people could be an important additional factor that predisposes individuals to hunt gorillas. The role of local conservation NGOs has become increasingly important for mitigating such conflicts and reducing destructive activities. We should use our knowledge of gorillas and chimpanzees gained by long-term studies for creating appropriate conservation measures.

17.1 Research History of Gorillas and Chimpanzees in Kahuzi-Biega National Park

Most primates have relatively long life spans and slow reproduction (Ross 1998; Kappeler et al. 2003). The great apes, in particular, have the slowest reproduction and the longest period of immaturity (Goodall 1986; Watts 1990; Nishida et al. 2003; Wich et al. 2004; Harcourt and Stewart 2007). Gorillas and chimpanzees are gregarious, and association with other individuals in social groups and individual transfer between groups greatly influence their survival and reproductive success (Fossey 1983; Goodall 1986; Harcourt and Stewart 2007). Therefore, long-term studies of demography, social changes, and reproduction are necessary to obtain the life history data crucial for understanding social and behavioral evolution in these species. In addition, long-term studies allow us to analyze (a) how apes and other nonhuman primates make behavioral adjustments to environmental changes, (b) how dispersal influences social structure, (c) how demographic changes influence intergroup relationships, and (d) how niche divergence allows primate species to live sympatrically.

Long-term research projects on primates living in natural habitats should also involve conservation efforts in response to increasing human disturbance such as deforestation, mining, and poaching. Ecotourism has been promoted as part of conservation strategies in many protected areas. Kahuzi-Biega National Park (KBNP), in the eastern Democratic Republic of Congo (DRC), has been a suitable site for both long-term research and ecotourism. Several groups of Grauer's gorillas (*Gorilla beringei graueri*) were habituated for research and tourism, and their demography has been recorded for more than 20 years. Eastern chimpanzees (*Pan troglodytes schweinfurthii*) are sympatric with Grauer's gorillas, and a group of chimpanzees was also habituated for research. In this chapter, we describe some of the major results of our long-term research on these gorillas and chimpanzees and of conservation efforts in Kahuzi.

Grauer's gorillas are only found in eastern DRC. They occur in two national parks (Maiko and Kahuzi-Biega) and two national reserves (Itombwe and Tanya), with the largest population in Kahuzi-Biega (Hall et al. 1998). Eastern chimpanzees also occur in these habitats. Kahuzi is divided into a lowland sector (5,400 km^2 at an altitude of 600–1,800 m) covered with lowland moist evergreen forest and a highland sector (600 km^2 at an altitude of 1,800–3,308 m) covered with montane forest. The two sectors are connected by a corridor that is 4 km wide and 20 km long (Fig. 17.1). The highland sector was gazetted as a national park in 1970, mainly for the protection of Grauer's gorillas (Mankoto 1988). The park was extended into the lowland sector in 1975, and the whole park was inscribed in the World Heritage List in 1980.

A team from Kyoto University and Centre de Recherche en Sciences Naturelles (CRSN) carried out preliminary research on the Kahuzi gorillas and chimpanzees from 1987 to 1991 and estimated population densities for both species. This team found extensive overlap in diet and ranging, documented frugivory and insectivory by the gorillas, and found that frugivory had a positive effect on gorilla daily path length (Yamagiwa et al. 1989, 1991, 1994; Yamagiwa and Mwanza 1994). Population surveys were also conducted in the lowland sector and adjacent forest by WCS in 1994 (Hall 1994; Hall et al. 1998). The density of apes in the lowland sector was estimated to be 0.27–0.33 gorillas/km^2 and 0.27–0.32 chimpanzees/km^2 by nest counts along reconnaissance walks (Yamagiwa et al. 1989), and 0.26–0.58 gorillas/km^2 and 0.81–1.78 chimpanzees/km^2 by nest counts along transects (Hall et al. 1998).

In the highland sector, four groups of gorillas were habituated for a gorilla tourism program (Fig. 17.2). The Mushamuka Group and the Maeshe Group were habituated in the early 1970s, and demographic changes in these groups have been recorded since the late 1970s (Yamagiwa 1983; Mankoto et al. 1994). The Mubalala Group was habituated in 1987 (Mankoto 1988), and the Nindja Group formed in 1989, when several females and immatures from the Mushamuka and Maeshe Groups joined a young silverback who had been born in the Mushamuka Group. Research on the Mushamuka and Maeshe Groups was conducted in the 1970s and focused on ranging patterns, feeding ecology and nutrition, social relationships, and relationships between groups (Casimir and Butenandt 1973; Casimir 1975; Goodall 1977; Yamagiwa 1983). The first gorilla census in the highland sector was conducted using nest counts in 1978, and the population was estimated at 223 gorillas (Table 17.1; Murnyak 1981).

In 1985, the Institut Zairois pour la Conservation de la Nature (IZCN, now ICCN) launched a conservation project in cooperation with the German Technical Cooperation Agency (GTZ) to train guides and regulate the gorilla tourism program (von Richter 1991). Gorilla tourism was well organized during this time, and it generated significant revenue. From 1989 to 1993, annual revenue was about US$ 210,000 (Butynski and Kalina 1998). A second census, conducted in 1990

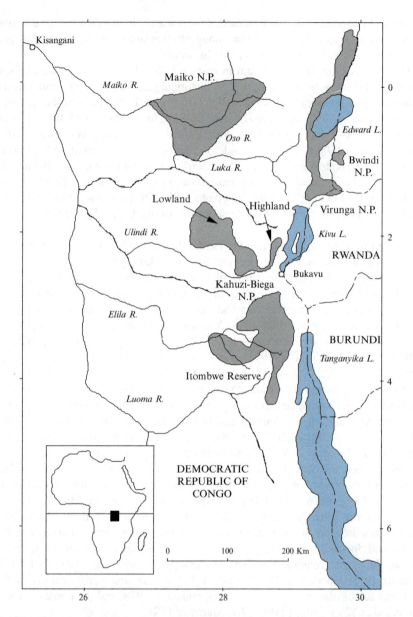

Fig. 17.1 Map showing the location of Kahuzi-Biega National Park. Gray areas represent a national park or forest reserve. *Blue* areas represent a lake

by Kyoto University, GTZ, CRSN, and IZCN, found about 258 gorillas (Yamagiwa et al. 1993). The first census of chimpanzees in the highland sector was conducted simultaneously, and three chimpanzee groups (communities) were confirmed by nest counts. Their nests were found in the small areas covered with primary forest,

17 Long-Term Research on Grauer's Gorillas in Kahuzi-Biega National Park, DRC 389

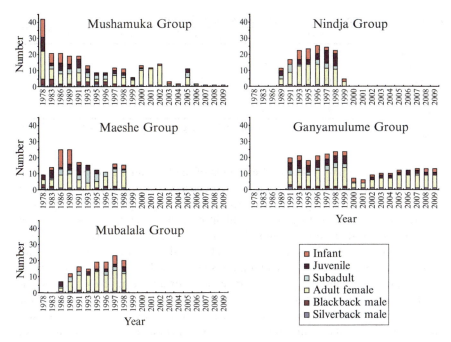

Fig. 17.2 Annual change in group size and composition of habituated gorilla groups at Kahuzi

and their density was estimated to be 0.13 chimpanzees/km^2 (Yamagiwa et al. 1992). During this survey, we found extensive overlap between one group of chimpanzees and four groups of gorillas, and decided to habituate these groups for long-term studies (Yamagiwa et al. 1996). By 1994, we had habituated the Ganyamulume Group (21 gorillas) and the Kaboko Group (22 chimpanzees) for research on diet and ranging (Yamagiwa and Basabose 2006a, b).

However, political instability in the Kivu region interrupted our research many times in the 1990s. The problems started with the outbreak of riots in Kinshasa in 1991 and were exacerbated in 1994 when the Rwandan genocide provoked the influx of about 450,000 refugees into the area around the park (Hall 1994). During the subsequent war from 1996 to 1999, these refugees were driven from the area, and many people died. Insecurity has continued on and off to the present day, associated with the presence of *Interahamwe* who caused the genocide in Rwanda and of other armed groups. Most of the elephants and half the gorilla population in the highland sector were lost during the war when park rangers were disarmed and prohibited by the rebel government from patrolling the park (Yamagiwa 1999). Armed militia groups frequently camped in the lowland sector and poaching there was probably more intense than in the highland sector (Hart et al. 2007). Three population surveys did not show a marked decline in the population of chimpanzees, but clearly indicated a sudden decrease in the number of gorilla groups during the war and a gradual increase in group size and in the proportion of infants (Table 17.1).

Table 17.1 Population census in the highland sector of KBNP

	1978[1]	1990[2]	1996[3]	2000[4]	2004[5]
Population size	223	258	247	130	163
# Groups	14	25	25	13	15
Mean group size	15.6	10.8	9.8	9.6	17.3
% Infant	17.0	8.4	12,7	9.3	15.9
# Solitary males	5	9	2	4	2

Source: 1, Murnyak, 1981; 2, Yamagiwa et al., 1993; 3, Inogwabini et al., 2000; 4, 5, Amsini et al., 2007.

Since 1994, we have followed fresh trails of the Ganyamulume Group and the Kaboko Group (up to one-day old) between consecutive nest sites and used GPS data to record their ranging. We confirmed estimates of group size and composition by direct observation and by counting night nests and measuring the feces they contained. We estimated diet composition from direct observations, evaluation of feeding remains along fresh trails, and fecal analysis. Fresh feces were collected mainly at nest sites, washed in 1-mm mesh sieves, dried in sunlight, and stored in plastic bags. The contents of each sample were examined macroscopically and listed as seeds, fruit skins, fiber, leaves, fragments of insects, and other matter. Fruit seeds and skin were identified at the species level macroscopically. We also kept track of changes in the composition of the groups habituated for tourism and used data on these groups to investigate gorilla life history strategies.

The home ranges of the Ganyamulume Group and the Kaboko Group were composed mainly of bamboo (*Sinarundinaria alpina*) forest, primary montane forest, secondary montane forest, and *Cyperus* (*Cyperus latifolius*) swamps. We set up a belt transect 5,000 m long and 20 m wide at an altitude of 2,050–2,350 m within the Ganyamulume Group's home range in 1994, and recorded 2,033 individual trees and woody vines above 10 cm in diameter at breast height (DBH) belonging to 49 species. The diversity of tree species in the highland sector is distinctly lower than that in the lowland sector, where we identified 6,922 trees and vines belonging to at least 150 species found in a belt transect 8,000 m long and 10 m wide. To estimate fruit abundance, we have monitored 28 species of trees and shrubs twice each month since 1994; chimpanzees and gorillas eat fruit from 24 of these and the other four are not ape food species. For each species, fruits (ripe and unripe) of at least ten reproductively mature trees were monitored. The monthly datum of the presence of fruit was the average of the two records. To estimate fruit abundance (biomass and number) of tree species, we used DBH (Chapman et al. 1992). We calculated a monthly fruit index (F_m) as

$$F_m = \sum_{k=1}^{s} P_{km} B_k,$$

where P_{km} denotes the proportion of the number of trees in fruit for species k in month m, and B_k denotes the total basal area per ha for species k in the line transect. The ripe fruit index was calculated from six fruit species preferred by gorillas and 12

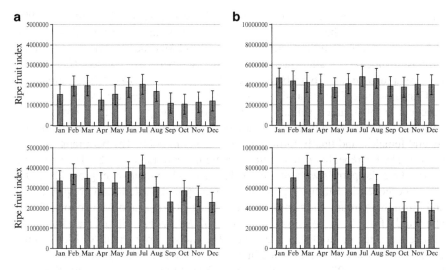

Fig. 17.3 Monthly fluctuation in the ripe fruit index for gorillas (**a**) and chimpanzees (**b**) in the primary (*upper*) and the secondary (*lower*) forests

preferred by chimpanzees, where preferred fruit species were defined as those ranked in the top three fruit species by fecal analysis in a month (Yamagiwa et al. 2008).

Meteorological data are available from the Meteorological Station at CRSN (1,600 m above sea level), which is located 4 km from the study area. The mean annual rainfall from 1994 to 2009 was 1,607 mm (range: 1,316–2,180 mm), with a distinct dry season in June, July, and August, in which the mean monthly rainfall was below 100 mm. The mean monthly temperature was 20.1°C (mean maximum: 26.5°C; mean minimum: 13.8°C).

In the lowland tropical forests that gorillas and chimpanzees inhabit sympatrically, the dry season is regarded as a period of fruit scarcity (Lope, Gabon: Tutin and Fernandez 1993; Ndoki, Congo: Kuroda et al. 1996; Moukalaba, Gabon: Takenoshita et al. 2008). In contrast, more ripe fruits are available during the dry season in Kahuzi (Fig. 17.3).

17.2 Life History of Kahuzi Gorillas

Starting in 1983, we named each individual and used the names for systematic recording of demographic changes. J. Kahekwa worked as a guide for gorilla tourism and named all individuals of the habituated groups, and J. Yamagiwa confirmed demographic changes with J. Kahekwa every year. Although the exact dates of birth, death, immigration, and emigration of some individuals were recorded, most demographic data were only collected weekly or monthly, due to the limits on the time that tourists were allowed to contact the gorillas per day.

Unfortunately, large-scale poaching when the war resumed in 1998 and 1999 decimated half of the gorilla population in the highland sector (Table 17.1). Habituated groups were the focus of poaching, and Maeshe, Mubalala, and Nindja Groups disintegrated after the majority of members were killed (Fig. 17.2). In Mushamuka Group, a blackback male (11 years old in 1998) led the 11 survivors. In Ganyamulume Group, a silverback male and three females were killed by poachers in 1998, and a solitary silverback joined the group after 6 months as a new leader. However, a blackback male started regularly forming a subgroup with several females, and finally formed another group in 2000. The park rangers were disarmed by the rebel force and could not enter the park for 9 months. The park staff resumed regular patrols without arms and started to habituate other groups in 1999. They succeeded in habituating Mishebere Group and Mufanzala Group in 2001, and Birindwa Group, Chimanuka Group, Langa Group, and Mankokoto Group afterward (Fig. 17.4). A silverback male of Mishebere Group was killed by poachers and the group disintegrated in 2003, and the young male of Mushamuka Group recently lost all females and immature to become solitary, but other groups have survived until 2009. Data collection on the habituated groups has continued so far, but life history data are incomplete due to frequent movements of individuals between habituated and unhabituated groups.

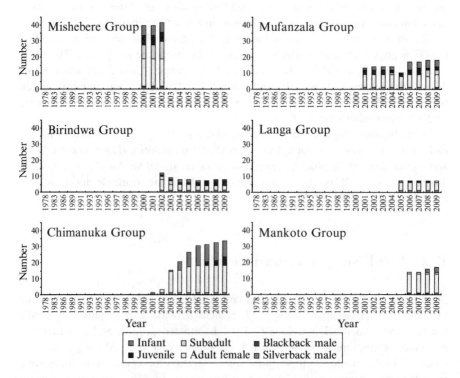

Fig. 17.4 Annual change in group size and composition of newly habituated gorilla groups at Kahuzi

We reconstructed the demographic history of habituated groups from fragmented records on each individual, mainly from data before the large-scale killing in 1998, and compared them with those of mountain gorillas in the Virungas (Table 17.2). Most previous information on gorilla social organization and life histories has come from the long-term data on mountain gorillas (*G. beringei*) gathered by the Karisoke Research Center. Mountain gorillas are year-round breeders, and no evidence of birth seasonality exists (Watts 1991). They form cohesive groups, and the home range of neighboring groups overlaps extensively (Schaller 1963; Harcourt 1978). Low variation in habitat quality and high home range overlap reduce the ecological cost of dispersal, while reproductive strategies strongly influence patterns of female transfer (Watts 1990, 1996; Robbins 1995; Harcourt and Stewart 2007).

Infanticide by males is a major source of infant mortality in the Virunga population (Fossey 1984; Watts 1989, 1991; Robbins et al. 2009). The contexts in which it occurs and its effects on female transfers and reproductive rates are consistent with the sexual selection hypothesis (Hrdy 1979; Sommer 1994; van Schaik 2000). Female mountain gorillas might respond to the threat of infanticide by seeking effective male protection. For example, they tend to transfer into multi-male groups instead of to single-male groups or solitary males, presumably because defense by multiple males is more effective and because the presence of multiple males reduces that chance that male death leaves a female with a dependent infant without any protection (Watts 1989, 1996, 2000). Multi-male groups attract more

Table 17.2 Reproduction and life history features of gorillas in Kahuzi and Virungas

Life history parameter	N	Kahuzi	N	Virunga
Seasonality in birth	47	$p < 0.05$	65	N.S.
Sex ratio at birth (number of males/number of females	64	0.94	59	0.79
Infant mortality 1 Primiparous	21	33.3%	14	42.9%
Parous	25	20.0%	45	17.8%
Infant mortality 2 First year	46	19.6%	65	26.2%
Second year	46	6.5%	65	7.7%
Minimum age at first observed copulation		5.2 years		5.8 years
Number of producing their first infant in natal group	18	5	16	7
Number of producing their first infant in non-natal group	18	9	16	9
Number of females emigrating from their natal group after the first birt	5	4	7	4
Age at first parturition	6	10.6 years (9.1-12.1)	8	10.1 years (8.7-12.8)
Interval between surviving births	9	4.6 years (3.4-6.6)	26	3.9 years (3.0-7.3)
Age at male emigration from natal group	6	12.7 years (9.6-14.4)	9	13.5 years (12-15)
Maximum number of silverback males within a group		2		7
Proportion of multi-male group in the population		8%		40%

Source: Watts (1990, 1991); Robbins (1995), Yamagiwa & Kahekwa (2001) Yamagiwa et al (2003); Robbins et al (2007)

females than single-male groups; this in turn might prevent maturing males from leaving their natal groups because they have good chances of mating within those groups (Robbins 1995, 2001; Robbins and Robbins 2005; Harcourt and Stewart 2007). This could explain the high proportion of multi-male groups (about 40%) in the Virunga population (Weber and Vedder 1983; Robbins 1995).

In contrast, births at Kahuzi are significantly clumped in May and June – the early dry season (Fig. 17.5; $X^2 = 23.6$, $df = 11$, $p < 0.05$), at which time the abundance of ripe fruit increases (Fig. 17.3a). Infant mortality was lower at Kahuzi than in the Virungas, especially that of infants born to primiparous females and mortality in the first year after birth (Table 17.2). Infanticide might have caused this difference. In Kahuzi where infanticide had not been observed until recently, females tended to transfer with immatures, and 9 cases of females with infants were observed between 1989 and 1997 (Yamagiwa and Kahekwa 2001). After the death of silverback males, females did not disperse but maintained cohesive groups without any mature males for prolonged periods (Maeshe Group for 29 months, Mushamuka Group for 15 months, and Nindja Group for 9 months). In the Virungas, single-male groups of mountain gorillas always disintegrated after the death of silverback males, and females joined neighboring groups or solitary males within a few months. The absence of infanticide might have facilitated natal dispersal by maturing males either alone or with other individuals before maturity. The high probability of co-transfer by multiple females may make it relatively easy for solitary males to form new groups and for males to form new groups via fissioning of existing ones; this might explain the small proportion of multi-male group (8%) in the Kahuzi population (Yamagiwa et al. 1993, 2003).

We thought that male Grauer's gorillas were not infanticidal (Yamagiwa and Kahekwa 2001), but social and demographic changes following the war showed this to be wrong (Yamagiwa and Kahekwa 2004). The first case of infanticide at Kahuzi was observed in 2003. When a female transferred with other females from the Mugaruka

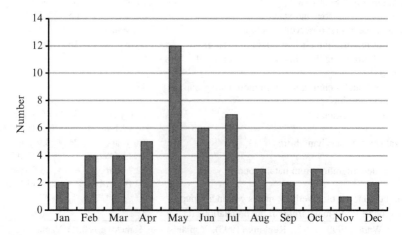

Fig. 17.5 Monthly fluctuation in the number of births observed in the habituated groups

(the former Mushamuka) Group to the Chimanuka Group, her newborn infant was killed by Chimanuka, the silverback male of that group. The victims in the second and the third cases were newborn infants whose mothers transferred into Chimanuka Group 1 and 2 months before giving birth, respectively. But another newborn infant whose mother transferred into the same group more than 1 year before was not killed by Chimanuka. These observations suggest that the infanticidal male discriminated between infants who were not his offspring and an infant whom he had presumably sired, probably based on past interactions with their mothers (Yamagiwa et al. 2009).

The occurrence of infanticide might have changed patterns of female association. A female transferred from the Mugaruka Group to the Chimanuka Group 1 month after the last infanticide, and she left a dependent infant (31-month old). She might have learned from the first case of infanticide that she witnessed and effectively responded to the threat by abandoning her infant at transfer. Since the first infanticide occurred, female transfer with dependent infants or formation of all-female groups have never been seen again.

Although infanticide did not promote the formation of multi-male groups at Kahuzi, it might have accelerated female reproduction. In the Chimanuka Group, many females gave birth soon after the last case of infanticide, and the number of infants rapidly increased (Fig. 17.4). Between 2005 and 2010, four twins were born in the Chimanuka Group, and two pairs have survived until present. Age at first birth is slightly older at Kahuzi than in the Virungas, and inter-birth intervals are slightly longer (Table 17.2). Both ecological and social factors are responsible. Seasonal fluctuation in food abundance may lead to slow reproduction. Greater frugivory, stronger habitat seasonality, lower densities of herbaceous foods, and lower abundance of weaning foods may promote slower development and reproduction in western lowland gorillas (Doran and McNeilage 2001; Nowell and Fletcher 2008), which mature more slowly and have a later age at first parturition and longer inter-birth intervals than mountain gorillas (Robbins et al. 2004; Breuer et al. 2009). In contrast, the potential pressure of infanticide may promote rapid reproduction in the Virungas (Fig. 17.6). Although no difference is found in the mean group size across gorilla populations in different habitats, the maximum group size is larger in eastern gorillas than western gorillas, and the proportion of multi-male groups in the population is larger in mountain gorillas than in Grauer's gorillas and western gorillas (Yamagiwa et al. 2003). The more frugivorous diets of western gorillas may set lower limits on maximum group size because they lead to stronger within-group feeding competition, which in turn makes it easier for single males to monopolize females and prevents the formation of multi-male groups (Harcourt and Stewart 2007; Breuer et al. 2009).

The probability of infanticide may be increased by rapid changes in density of gorilla social units and associated changes in relationships between groups like those that followed large-scale poaching at Kahuzi. The leading silverbacks in all five habituated groups were killed by poachers, and group disintegrations and frequent female transfers occurred after their deaths, although females continued to associate with each other and without silverbacks for several months in some groups. Young silverbacks who had been solitary males survived during the war and joined these female groups as the new leading males. Among eight groups newly monitored by the

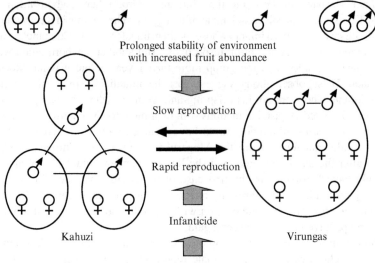

Fig. 17.6 Two types of social structure shifting to each direction by environmental changes

park and by us after large-scale poaching, seven groups were led by young silverbacks estimated to be less than 20 years old (Yamagiwa et al. 2009). Chimanuka, the infanticidal male, was 17 years old when he killed three newborn infants. Mugaruka, who lost all his females to Chimanuka, was one year younger than Chimanuka. Both males were born in different habituated groups (Maeshe and Mushamuka Groups) and were presumably unrelated. These maturing silverbacks are strongly motivated to attract females in order to establish their own groups (Harcourt 1978; Yamagiwa 1986), and they show a greater potential to commit infanticide than do group males in the Virungas (Watts 1989). The sudden changes in the population of mountain gorillas following the large conversion (40%) of their habitat to farmland might also have promoted infanticide in the Virunga region. The gorilla population was reduced by half after the conversion, as was observed in Kahuzi. The reduced ranges and increased encounters between unfamiliar groups induced by such disturbances might have produced conditions leading to infanticide.

17.3 Research on Sympatric Populations of Gorillas and Chimpanzees

The pioneering studies of sympatric populations of gorillas and chimpanzees in the 1950s and 1960s suggested that several aspects of niche differentiation, including differences in diet, habitat choice, and use of forest strata, allow coexistence of gorillas and chimpanzees (Schaller 1963; Jones and Sabater Pi 1971). Field studies on allopatric populations of gorillas and chimpanzees have also reported distinct differences in their socioecology. Gorillas are more folivorous, and the continuous availability and relatively uniform quality of their terrestrial food resources reduce the potential for feeding competition and enable them to form cohesive groups in which females have egalitarian social relationships. Gorillas are not territorial and groups show weak site fidelity (Stewart and Harcourt 1987; Watts 1991, 1996, 1998a,b; Harcourt 1992). In contrast, pronounced frugivorous diets and frequent use of depletable high-quality fruit patches prevent chimpanzees from forming cohesive groups and promote a fission–fusion sociality as a way to reduce feeding competition (Nishida 1968; Goodall 1986; Chapman et al. 1995; Wrangham et al. 1996). Chimpanzees form communities (unit groups) that have stable memberships. They defend territories against members of neighboring communities, and male chimpanzees occasionally patrol the peripheral parts of their ranges, where lethal intercommunity aggression can occur (Goodall et al. 1979; Nishida et al. 1985; Watts and Mitani 2001; Wilson and Wrangham 2003; Boesch et al. 2007; Watts 2012; Wilson 2012).

However, most of the data on gorillas have come from long-term studies on a single population of mountain gorillas in the Virunga volcanoes living at the upper altitudinal extreme of the species' geographic distribution. Research on western lowland and Grauer's gorillas, and that on mountain gorillas inhabiting the low altitudinal forest of Bwindi show that they are considerably more frugivorous than Virunga gorillas and often feed and nest in trees (Tutin and Fernandez 1984; Williamson et al. 1990; Yamagiwa et al. 1994; Kuroda et al. 1996; Remis et al. 2001; Robbins and McNeilage 2003; Stanford and Nkurunungi 2003; Rogers et al. 2004). Nevertheless, they also tend to form cohesive groups like mountain gorillas (Harcourt et al. 1981; Tutin 1996; Doran and McNeilage 1998). Average group size and home range size of gorillas vary little among habitats, although the proportion of multi-male groups is higher in mountain gorillas than in eastern and western lowland gorillas (Watts 1996; Doran and McNeilage 2001; Yamagiwa et al. 2003; Yamagiwa and Basabose 2006b).

In contrast, the members of a chimpanzee community form temporary parties that frequently change size and composition in association with variation in fruit availability and with social factors such as the presence of estrous females and conflict among males, and their annual range size varies among habitats (Chapman et al. 1995; Matsumoto-Oda et al. 1998; Yamagiwa 1999; Anderson et al. 2002; Mitani et al. 2002; Basabose 2004). Thus, gorillas and chimpanzees have great social and ecological flexibility, but they respond differently to variation in food abundance and distribution. Both gorillas and chimpanzees have unspecialized digestive systems, and they are less able to digest unripe fruits and mature leaves than Old World monkeys (Chivers and Hladik 1980; Fleagle 1984; Remis 2000). Consequently, their diet is highly diverse and includes a broad range of non-fruit foods (van Schaik et al. 2004; Yamagiwa 2004). Their behavioral and social flexibilities have possibly evolved to cope with these dietary constraints.

Niche separation between sympatric primate species becomes more pronounced during periods of food shortage (Ungar 1996; Tan 1999; Powzyk and Mowry 2003). Fallback foods, which are used more when preferred foods are scarce, might have had particularly important evolutionary influences on the morphology and behavior of the African apes (Lambert 2007; Marshall and Wrangham 2007). Research on sympatric populations is the best way to elucidate the impact of fallback foods and other environmental factors on morphological and behavioral differentiation between gorillas and chimpanzees.

In Kahuzi, long-term data on diet composition indicate extensive overlap in fruit foods between gorillas and chimpanzees (Yamagiwa and Basabose 2006a). However, patterns of frugivory were different. For gorillas, the monthly abundance of ripe fruit from preferred species was positively correlated with the diversity of fruit in the monthly diet (the monthly mean number of fruit species per fecal sample: $R^2 = 0.436, p < 0.0001$) and with the monthly mean proportion of fruit remains per fecal sample ($R^2 = 0.244, p < 0.001$). For the chimpanzees, the abundance of preferred fruit was positively correlated with the monthly proportion of fruit in fecal samples ($R^2 = 0.197, p < 0.01$), but the diversity of fruits consumed was not significantly related to the abundance of ripe fruit ($R^2 = 0.005, p = 0.623$). Among fruit species preferred by both apes, both consumed ripe fruit of *Myrianthus holstiii* in proportion to its abundance ($R^2 = 0.482, p < 0.001$ for gorillas; $R^2 = 0.281, p < 0.001$). However, consumption of figs by both apes was not significantly correlated with the abundance of ripe fruits, and chimpanzees always consumed more figs than gorillas.

These differences in feeding patterns are reflected in differences in ranging patterns. We plotted daily movements of the study groups on a grid of 250×250 m quadrats superimposed on a 1:25,000 vegetation map of the study area (Yamagiwa and Basabose 2006b). Based on the number of grid squares visited by the study groups, the total home range of the gorilla group (42.3 km^2 for 92 months) was three times larger than that of the chimpanzee group (15.7 km^2 for 82 months). Both annual range (average 15.5 km^2) and core area (average 3.4 km^2) of the gorilla group were about twice as large as those of chimpanzees (7.1 km^2 and 1.5 km^2, respectively). The gorillas tended to use a large area evenly, whereas the chimpanzees visited small areas repeatedly. The proportion of the total home range reused within 1 month was larger for chimpanzees (55%) than for gorillas (25%). The gorillas tended to shift their range monthly and yearly, while the chimpanzees showed high site fidelity. The number of new grid squares visited by the gorilla group tended to increase every year, while that of chimpanzees did not increase substantially after the first 2 years (Yamagiwa and Basabose 2006b). The chimpanzees covered more than half their total home range within the first year (Basabose 2005).

The ranging patterns of the gorillas were similar to those of mountain gorillas in the Virunga region, and reflect a strategy to seek high-quality herbaceous vegetation and to avoid previously used areas that contained trampled vegetation (Vedder 1984; Watts 1987, 1998b). In contrast, the ranging patterns of the chimpanzee study group are similar to those of chimpanzees inhabiting lowland forests in Taï (Lehmann and Boesch 2003) and Gombe (Williams et al. 2004), and reflect a strategy to harvest fruit efficiently by adjusting search paths and patch revisitation to the current distribution of fruit (Waser 1984; Garber 1989).

Table 17.3 Range overlap between the gorilla and chimpanzee groups

	Total range	Core 75%	Core 50%
Overlap area	11.0 km^2	2.1 km^2	0.25 km^2
Primary forest	21.5%	42.3%	62.5%
Secondary forest	65.8%	51.3%	35.6%
*Cyperus*swamp	1.9%	1.9%	2.2%
Bamboo forest	10.7%	4.5%	0%
% in gorilla range	25.9%	14.9%	4.0%
% in chimpanzee range	69.7%	41.3%	9.8%

Most (70%) of the chimpanzee study group's home range overlapped with that of the gorilla study group (Table 17.3). However, the small core area overlap (10% from the chimpanzee perspective) suggests that the chimpanzee avoided the core area of the gorillas, which included primary forests, secondary forests, and swamps. In contrast, that of the chimpanzees consisted mainly of primary forest fragments. Several encounters between gorillas and chimpanzees were observed in the primary forest (Yamagiwa et al. 1996). All lasted for only a few minutes and ended with mutual avoidance. Chimpanzees possibly avoided visiting fruiting patches that the gorillas also used, especially when they were in small parties (Basabose 2002). Although high overlap in fruit diets predicts frequent encounters between gorillas and chimpanzees at fruiting crops, their different foraging strategies may reduce the opportunities for encounters. They may also avoid direct contact where their preferred fruits are available. Co-feeding between gorillas and chimpanzees has been observed in the lowland tropical forests at Ndoki and Goualago (Kuroda et al. 1996; Morgan and Sanz 2006), and aggressive interactions were observed in the montane forest at Bwindi (Stanford and Nkurunungi 2003).

September to January is a period of fruit scarcity in Kahuzi. The abundance of ripe fruits for chimpanzees decreased distinctly in the secondary forest that dominated their home range (Fig. 17.4). Chimpanzees possibly need to use the small patches of primary forest to seek fruit during this period. Gorillas usually shifted to the bamboo forest and fed on bamboo shoots (Casimir and Butenandt 1973; Goodall 1977). However, from 1994 to 1997, many bamboo stands withered and died after flowering, and the gorillas only occasionally visited the bamboo forest, probably to search bamboo shoots, and returned to the previous ranging area in the primary or the secondary forest after a few days (Fig. 17.7). When bamboo shoots appeared in 1998, the gorillas used the bamboo forest more.

When fruit was scarce, gorillas and chimpanzees relied on different fallback foods in Kahuzi. We examined their fallback foods using the definition of staple/filler fallback foods by Marshall and Wrangham (2007). Fecal analysis showed that gorillas regularly consumed vegetative foods (leaves, pith, and bark) in every month, and in certain months, all gorilla fecal samples (100%) were composed entirely of vegetative plant parts (Yamagiwa and Basabose 2009). The mean proportion of vegetative foods per fecal sample of gorillas was negatively

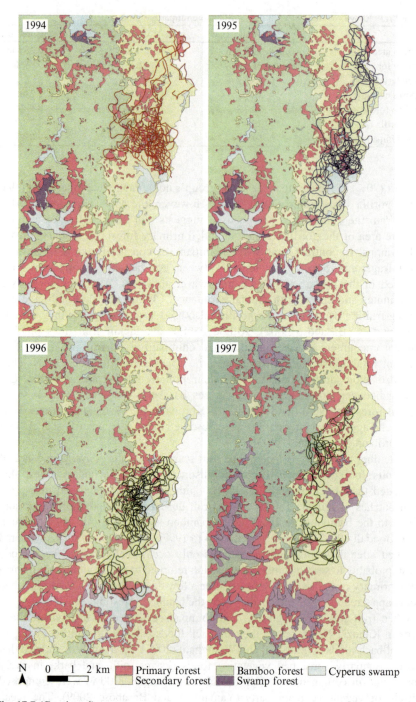

Fig. 17.7 (Continued)

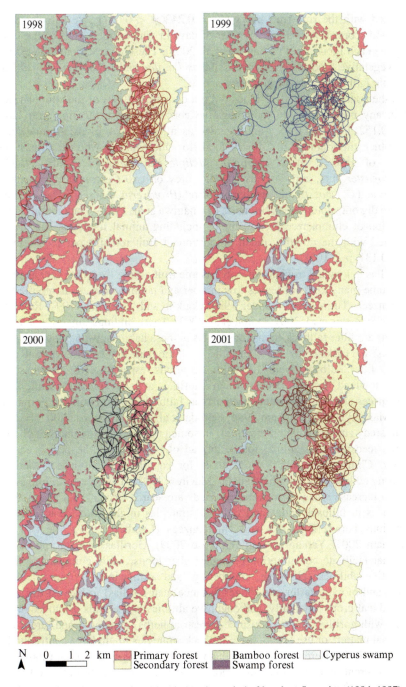

Fig. 17.7 Daily travel routes of gorillas during the period of bamboo flowering (1994–1997) and after bamboo shoots become available (1998–2001)

correlated with the ripe fruit index ($R^2 = 0.242$, $p < 0.001$). We calculated the daily food scores (DFS) as the number of days in which the gorillas of the study group ate each food item (Yamagiwa et al. 2005). These observations suggest that some vegetative foods, such as leaves of *Basella alba* and *Jasminium abyssinicus*, are staple fallback foods for gorillas.

On the contrary, vegetative foods did not constitute 100% of the chimpanzee's diet in any month. Due to their negative correlation with the ripe fruit index ($R^2 = 0.152$, $p < 0.01$), they are regarded as filler fallback foods. Animal foods are another important set of fallback foods for chimpanzees, who eat at least five species of insects (*Apis mellifera*, *Meliplebeia tanganyikae* aff. *Nigrita*, *Crematogaster* spp., and *Dorylus* sp.) and prey on two *Cercopithecus* monkeys (*C. mitis* and *C. l'hoesti*) and the giant squirrel (*Protoxerus stangeri*). They also use sticks to dig out subterranean bee nests (Yamagiwa et al. 1988). The monthly mean proportion of chimpanzee fecal samples including animal foods was negatively correlated with the monthly mean proportion of fruit remains per fecal sample ($R^2 = 0.131$, $p < 0.01$).

Gorillas in lowland tropical forests consume fruit during the entire year and tend to consume many kinds of leaves low in fiber and tannin contents, as reported for chimpanzees (Tutin and Fernandez 1993; Conklin and Wrangham 1994; Reynolds et al. 1998; Remis et al. 2001; Doran et al. 2002). These findings suggest that leaves and piths are filler fallback foods for gorillas in these habitats, while they are staple fallbacks, and bark serves as a filler fallback, in montane habitats. Leaves and piths are also filler fallback foods for chimpanzeesans, especially, for bonobos that inhabit lowland tropical forest where gorillas are not present, while both chimpanzees and bonobos rarely use barks as fallback foods (Tutin and Fernandez 1993; Malenky and Wrangham 1994; Kuroda et al. 1996; Basabose 2002).

Our study suggests that chimpanzees in some populations or habitats, including Kahuzi, feed more on animal foods, instead of bark, during the period of fruit scarcity. Chimpanzees use various tools for capturing social insects and for extracting edible parts from hard shells (McGrew 1992). Tool use by chimpanzees tends to increase when their preferred foods are scarce (Yamakoshi 1998). Such differences in fallback foods and fallback strategies might have led to different adaptations between gorillas and chimpanzees (Lambert 2007; Marshall and Wrangham 2007; Yamagiwa and Basabose 2009). Gorillas have morphological traits that facilitate consumption of higher fiber foods, while chimpanzees have locomotive abilities (climbing, clambering, and arm-swinging from branch to branch, and moving faster and longer distance than gorillas) to harvest dispersed fruit and leaf crops efficiently, and cognitive abilities to use tools. Bonobos do not coexist with gorillas, tend to consume terrestrial herbs frequently, and rarely feed on animal matters or use tools. These fallback strategies might have contributed to socioecological differences between them (Fig. 17.8). Sympatry might have promoted different fallback strategies in gorillas and chimpanzees to mitigate feeding competition; these differences in turn allowed range expansion into different habitats (gorillas to higher montane forest, and chimpanzees to more open and drier habitat).

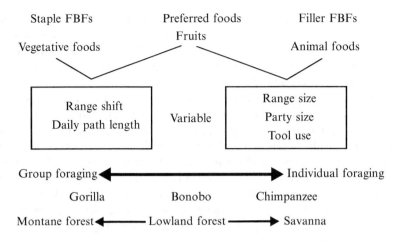

Fig. 17.8 Fallback strategies of the African apes. See text for explanation

17.4 Human Population Pressure and Conservation Measures

The South Kivu region, where the KBNP is located, has a high human density (over 300 people/km^2) and an annual population growth rate of 4%. People have extensively cultivated the highland region, which has highly fertile volcanic soils, and no natural forest remains outside the national park. Since the 1980s, people have suffered a severe economic crisis and faced political instabilities. As of the early 1990s, various projects funded by the World Bank and other international NGOs had improved infrastructure, and gorilla tourism was generating significant revenues, which helped boost support for gorilla conservation. However, the number of tourists has drastically declined and most foreign aid and cooperation has been suspended since 1994. Most of the elephants and half of the gorilla population were lost in the highland sector of the park during the war, and the situation is worse in the lowland sector where armed militia groups are still poaching wildlife (Yamagiwa 2003; Hart et al. 2007).

Deforestation, mining, charcoal production, cattle and goat encroachment, and slash-and-burn agriculture are the main destructive activities in the highland sector of the park. In 2000 and 2001, approximately 15,000 people rushed into the park to mine coltan when its price suddenly increased by more than ten times (Kasereka et al. 2006). In 2004, the military forces left the park and the park rangers resumed regular patrols, but more than 8,000 people were still mining coltan, cassiterite, and gold (Iyomi and Schuler 2004) and camped inside the park. Their subsistence activities (farming and hunting) also caused permanent damages to vegetation and wild animals. The human pressure was strongest in the corridor linking the highland and lowland sectors of KBNP. By 2008, 60% of the corridor had been degraded due to farming and mining, and the park authority has declared an urgent need to protect this area (Nishuli 2008). Recent damage to vegetation by humans is serious. Plants are collected for construction, food, fuel, and medicine. Sivha (1999) interviewed 213 people living around the highland sector of KBNP and reported that 93% of 249 wild plants they used came directly from the park. Suspension of public services such as clinics and transportation has accelerated exploitative activities of local people after the war.

The growing population has increased pressures on wildlife in and around the park. Various conservation activities in cooperation with international and national NGOs have been undertaken to support the park (Bikaba 2007). The IZCN-GTZ project funded by the World Bank, the World Wide Fund for Nature, and other international NGOs supported the wages and infrastructure of KBNP, improved public health, and helped lessen the workload of local women by providing water piped from natural sources (Steinhauer-Burkart et al. 1995). Park authorities decided to devote 40% of park revenues to park management and community development. The Gorilla Organization initiated a campaign to minimize the impact of mining, Partners In Conservation (PIC-Columbus Zoo) funded an animal breeding program that enabled 147 households around the park to obtain a goat, and the Rhodes Scholar's Southern Africa Forum (RSSAF/Oxford University) has funded a farming project that distributed agricultural tools and seeds to households around the park.

Starvation caused by the recent economic and political crisis is the main factor that led the people to destructive activities in the park. Conflict between the park authorities and unemployed local people could be an important additional factor that predisposes individuals to hunt gorillas. For many local residents, the park had been a long-standing source of resentment and conflict. When the national government created the park in the 1970s, it required many local villagers to abandon their lands and refrain from using the natural resources of the new reserve. They were also prohibited from shooting the elephants that frequently raided their crops. Furthermore, local villages were ordered to absorb the people who were evicted from the new reserve. To mitigate these growing conflicts, promote local conservation knowledge, and improve the quality of life in the area's communities, we established the Pole Pole Foundation (POPOF), a local NGO, in 1992. A few park rangers and some inhabitants of villages adjacent to the park constitute the bulk of the POPOF members, and researchers joined as advisers. POPOF sought to promote local community development through a variety of projects that would simultaneously benefit conservation in the park and neighboring lands. These included the establishment of a handicraft center, a tree nursery, and a school for environment education that employed local people, including former poachers. POPOF is currently working with 21 schools and running 11 adult literacy centers. In 2007, POPOF built a secondary school for environmental education, which has six class rooms and currently hosts 213 pupils. About 100,000 tree seedlings have been produced in the tree nursery center and distributed to villagers around the park. PIC, the Gorilla Organization, and POPOF-Japan have supported these projects.

Recently, we conducted a socioeconomic survey in two administrative sectors surrounding the highland sector of the park to understand the human community reliance on park's natural resources and to try implementing alternative subsistence measures that would reduce human impacts on the park. We selected five households in each sector and monitored foods and materials for subsistence they harvested, got, gave, sold, or purchased every day for 13 months in 2007 and 2008. The results showed that the local people obtained barely enough grains and vegetables necessary for subsistence and got hardly animal protein and not enough fuel (Fig. 17.9). Wood is their main source of fuel, supplemented with charcoal and kerosene obtained mainly by purchase or barter. Cooking heavily depends on palm oil, transported from lowland forests areas and usually purchased with cash. Income declines caused by the economic collapse triggered by the wars have increased such needs and have stimulated the local people to collect natural resources in the park for subsistence. The following alternative measures are proposed.

1. Promote animal breeding projects and fish farming to increase the availability of meat and milk.
2. Resume and strengthen the trees planting projects around the park to supply communities with wood and other natural resources to be managed rationally.
3. Promote farming projects to improve food security around the park, including sustainable land management.
4. Promote projects that use appropriate sustainable technologies to meet fuel needs and rational management strategies for using farmland, such as production of vegetable oil from locally grown crops such as peanuts and sunflowers.
5. Promote socioeconomic and income-generating activities. Development of a fair trade system accompanied by ecotourism is one of our recommended activities.

The role of local NGOs in conservation of natural resources has become increasingly important in the areas with political instability, such as Kahuzi. With weak enforcement of rules and regulations and national institutions in disarray, local people in this region have made decisions according to short-term personal or local interests (Kasereka et al. 2006; Bikaba 2007). Park authorities are powerless to stop local people from using protected areas for agricultural production and mining. The local

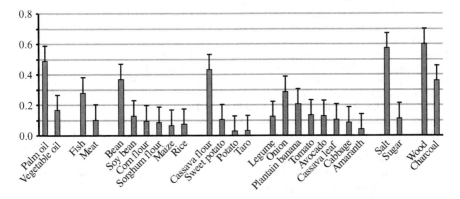

Fig. 17.9 Mean frequency of usage for each material of subsistence per meal by ten families inhabiting near the Kahuzi-Biega National Park

NGOs, such as POPOF, should cooperate with international NGOs to reinforce conservation knowledge among local people and provide alternatives to destructive activities. Scientists should also cooperate with them to use scientific knowledge for creating appropriate conservation measures. We should use our understanding of the ecological and social flexibility of gorillas and chimpanzees gained by long-term studies to help with management of the park, and we urgently need to inspire positive participation by local people in research and conservation activities.

Acknowledgments This paper was originally prepared for the Symposium on "Long Term Studies of Primates," which was held in Göttingen, Germany, on December 9–12, 2009. We would like to express our hearty thanks to Prof. Peter Kappeler for giving us the opportunity to present our work at the symposium. This study was financed by Ministry of Education, Culture, Sports, Science and Technology, Japan (No.162550080, No.19107007 to J. Yamagiwa); by Japanese Ministry of Environment (Global Environmental Research Fund F-061 to T. Nishida); and by Kyoto University Global COE Program (Formation of a Strategic Base for Biodiversity and Evolutionary Research), and was conducted in cooperation with CRSN and ICCN. We thank Dr. S. Bashwira, Dr. B. Baluku, Mr. M.O. Mankoto, Mr. B. Kasereka, Mr. L. Mushenzi, Ms. S. Mbake, Mr. B.I. Iyomi, Mr. C. Schuler, and Mr. R. Nishuli for their administrative help. We are also greatly indebted to Mr. M. Bitsibu, Mr. S. Kamungu, and all of the guides, guards, and field assistants in the Kahuzi-Biega National Park for their technical help and hospitality throughout the fieldwork.

References

Anderson DP, Nordheim EV, Boesch C, Moermond TC (2002) Factors influencing fission-fusion grouping in chimpanzees in the Taï National Park, Cote d'Ivoire. In: Boesch C, Hohmann G, Marchant LF (eds) Behavioral diversity in chimpanzees and bonobos. Cambridge University Press, Cambridge, pp 90–101

Basabose AK (2002) Diet composition of chimpanzees inhabiting the montane forest of Kahuzi, Democratic Republic of Congo. Am J Primatol 58:1–21

Basabose AK (2004) Fruit availability and chimpanzee party size at Kahuzi montane forest, Democratic Republic of Congo. Primates 45:211–219

Basabose AK (2005) Ranging patterns of chimpanzees in a montane forest of Kahuzi, Democratic Republic of Congo. Int J Primatol 26:33–54

Bikaba DZ (2007) Post-war Kahuzi-Biega National Park. Gorilla Gazette 20:13–15

Boesch C, Head J, Tagg N, Arandjelovic M, Vigilant L, Robbins MM (2007) Fatal chimpanzee attack in Loango National Park, Gabon. Int J Primatol 28:1025–1034

Breuer T, Breuer-Ndoundou Hockemba M, Olejniczak C, Parnel RJ, Stokes EJ (2009) Physical maturation, life-history classes and age estimates of free-ranging western gorillas - insights from Mbeli Bai, Republic of Congo. Am J Primatol 71:106–119

Butynski TM, Kalina J (1998) Gorilla tourism: a critical look. In: Milner-Gulland EJ, Mace R (eds) Conservation of biological resources. Blackwell Science Ltd, Oxford, pp 280–300

Casimir MJ (1975) Feeding ecology and nutrition of an eastern gorilla group in the Mt. Kahuzi region (République du Zaïre). Folia Primatol 24:81–136

Casimir MJ, Butenandt E (1973) Migration and core area shifting in relation to some ecological factors in a mountain gorilla group (*Gorilla gorilla beringei*) in the Mt. Kahuzi region (République du Zaïre). Z Tierpsychol 33:514–522

Chapman CA, Chapman LJ, Wrangham RW, Hunt K, Gebo D, Gardner L (1992) Estimators of fruit abundance of tropical trees. Biotropica 24:527–531

17 Long-Term Research on Grauer's Gorillas in Kahuzi-Biega National Park, DRC

Chapman CA, Wrangham RW, Chapman LJ (1995) Ecological constraints on group size: an analysis of spider monkey and chimpanzee subgroups. Behav Ecol Sociobiol 36:59–70

Chivers DJ, Hladik CM (1980) Morphology of the gastrointestinal tract in primates: comparisons with other mammals in relation to diet. J Morphol 116:337–386

Conklin NL, Wrangham RW (1994) The value of figs to hind-gut fermenting frugivore: a nutritional analysis. Biochem Syst Ecol 22:137–151

Doran DM, McNeilage A (1998) Gorilla ecology and behavior. Evol Anthropol 6:120–131

Doran DM, McNeilage A (2001) Subspecific variation in gorilla behavior: the influence of ecological and social factors. In: Robbins MM, Sicotte P, Stewart KJ (eds) Mountain gorillas: three decades of research at Karisoke. Cambridge University Press, Cambridge, pp 123–149

Doran DM, McNeilage A, Greer D, Bocian C, Mehlman P, Shah N (2002) Western lowland gorilla diet and resource availability: new evidence, cross-site comparisons, and reflections on indirect sampling methods. Am J Primatol 58:91–116

Fleagle JG (1984) Primate locomotion and diet. In: Chivers DJ, Wood BA, Bilsborough A (eds) Food acquisition and processing in primates. Plenum, New York, pp 105–117

Fossey D (1983) Gorillas in the mist. Houghton Mifflin, Boston

Fossey D (1984) Infanticide in mountain gorillas (*Gorilla gorilla beringei*) with comparative notes on chimpanzees. In: Hausfater G, Hrdy SB (eds) Infanticide: comparative and evolutionary perspectives. Aldine, Hawthorne/NY, pp 217–235

Garber PA (1989) Role of spatial memory in primate foraging patterns: *Saguinus mystax* and S*aguinus fuscicollis*. Am J Primatol 19:203–216

Goodall AG (1977) Feeding and ranging behavior of a mountain gorilla group (*Gorilla gorilla beringei*) in the Tshibinda-Kahuzi resion (Zaïre). In: Clutton-Brock TH (ed) Primate ecology: studies of feeding and ranging behavior in lemurs, monkeys and apes. Academic, New York, pp 450–479

Goodall J (1986) The chimpanzees of Gombe: patterns of behavior. Belknap, Cambridge/MA

Goodall J, Bandora A, Bergmann E, Busse C, Matama H, Mpongo E, Pierce A, Riss D (1979) Intercommunity interactions in the chimpanzee population of the Gombe National Park. In: Hamburg DA, McCown ER (eds) The great apes. Benjamin/Cummings, Menlo Park/CA, pp 13–53

Hall JS (1994) Counting gorillas in Zaïre's Kahuzi-Biega NP. African Wildlife Update, Nov–Dec:6

Hall JS, White LJT, Inogwabini B-I, Omari I, Morland HS, Williamson EA, Saltonstall K, Walsh P, Sikubwabo C, Bonny N, Kaleme KP, Vedder AL, Freeman K (1998) Survey of Grauer's gorillas (*Gorilla gorilla graueri*) and eastern chimpanzees (*Pan troglodytes schweinfurthii*) in the Kahuzi-Biega National Park lowland sector and adjacent forest in eastern Democratic Republic of Congo. Int J Primatol 19:207–235

Harcourt AH (1978) Strategies of emigration and transfer by primates, with particular reference to gorillas. Z Tierpsychol 48:401–420

Harcourt AH (1992) Coalitions and alliances: are primates more complex than non-primates? In: Harcourt AH, de Waal FBM (eds) Coalitions and alliances in humans and other animals. Oxford University Press, Oxford, pp 445–472

Harcourt AH, Stewart KJ (2007) Gorilla society: conflict, compromise, and cooperation between the sexes. University of Chicago Press, Chicago

Harcourt AH, Fossey D, Sabater Pi J (1981) Demography of *Gorilla gorilla*. J Zool Lond 195:215–233

Hart J, Carbo M, Amsini F, Grossmann F, Kibambe C (2007) Parc National de Kahuzi-Biega Secteur de Basse Altitude: Inventaire préliminaire de la grande faune avec une évaluation de l'impact des activités humaines et la situation sécuritaire 2004–2007. WCS/RDC, IMU Technical Report No 8

Hrdy SB (1979) Infanticide among animals: a review, classification, and examination of the implications for the reproductive strategies of females. Ethol Sociobiol 1:13–40

Iyomi IB, Schuler C (2004) News from the Kahuzi-Biega National Park. Gorilla J 31:3

Jones C, Sabater Pi J (1971) Comparative ecology of *Gorilla gorilla* (Savage and Wyman) and *Pan troglodytes* (Blumenbach) in Rio Muni, West Africa. Bibl Primatol 13:1–96

Kappeler PM, Pereira ME, van Schaik CP (2003) Primate life histories and socioecology. In: Kappeler PM, Pereira ME (eds) Primate life histories and socioecology. Chicago University Press, Chicago, pp 1–23

Kasereka B, Muhigwa J-BB, Shalukoma C, Kahekwa JM (2006) Vulnerability of habituated Grauer's gorilla to poaching in the Kahuzi-Biega National Park, DRC. Afr Stud Mongr 27: 15–26

Kuroda S, Nishihara T, Suzuki S, Oko RA (1996) Sympatric chimpanzees and gorillas in the Ndoki Forest, Congo. In: McGrew WC, Marchant LF, Nishida T (eds) Great ape societies. Cambridge University Press, Cambridge, pp 71–81

Lambert JE (2007) Seasonality, fallback strategies, and natural selection: a chimpanzee and cercopithecoid model for interpreting the evolution of hominin diet. In: Ungar PS (ed) Evolution of the human diet: the known, the unknown and the unknowable. Oxford University Press, Oxford, pp 324–343

Lehmann J, Boesch C (2003) Social influences on ranging patterns among chimpanzees (*Pan troglodytes verus*) in the Taï National Park, Côte d'Ivoire. Behav Ecol 14:642–649

Malenky RK, Wrangham RW (1994) A quantitative comparison of terrestrial herbaceous food consumption by *Pan paniscus* in the Lomako Forest, Zaïre, and *Pan troglodytes* in the Kibale Forest, Uganda. Am J Primatol 32:1–12

Mankoto MO (1988) La géstion du Parc National du Kahuzi-Biega (Zaïre). Cah Ethol Appl 8:447–450

Mankoto MO, Yamagiwa J, Steinhauer-Burkart B, Mwanza N, Maruhashi T, Yumoto T (1994) Conservation of Grauer's gorillas in the Kahuzi-Biega National Park, Zaïre. In: Thierry B, Anderson JR, Roeder JJ, Herrenschmidt N (eds) Current primatology, vol I, Ecology and evolution. University of Louis Pasteur, Strasbourg, pp 113–122

Marshall AJ, Wrangham RW (2007) Evolutionary consequences of fallback foods. Int J Primatol 28:1219–1235

Matsumoto-Oda A, Hosaka K, Huffman MA, Kawanaka K (1998) Factors affecting party size in chimpanzees of the Mahale Mountains. Int J Primatol 19:999–1011

McGrew WC (1992) Chimpanzee material culture: implications for human evolution. Cambridge University Press, Cambridge

Mitani JC, Watts DP, Pepper JW, Merriwether DA (2002) Demographic and social constraints on male chimpanzee behaviour. Anim Behav 64:727–737

Morgan D, Sanz C (2006) Chimpanzee feeding ecology and comparisons with sympatric gorillas in the Goualougo Triangle, Republic of Congo. In: Hohmann G, Robbins MM, Boesch C (eds) Feeding ecology in apes and other primates: ecological, physiological and behavioural aspects. Cambridge University Press, Cambridge, pp 97–122

Murnyak DF (1981) Censusing the gorillas in Kahuzi-Biega National Park. Biol Conserv 21:163–176

Nishida T (1968) The social group of wild chimpanzees in the Mahali Mountains. Primates 9:167–224

Nishida T, Hiraiwa-Hasegawa M, Hasegawa T, Takahata Y (1985) Group extinction and female transfer in wild chimpanzees in the Mahale National Park, Tanzania. Z Tierpsychol 67:284–301

Nishida T, Corp N, Hamai M, Hasegawa T, Hiraiwa-Hasegawa M, Hosaka K, Hunt KD, Itoh N, Kawanaka K, Matsumoto-Oda A, Mitani JC, Nakamura M, Norikoshi K, Sakamaki T, Turner L, Uehara S, Zamma K (2003) Demography, female life history, and reproductive profiles among the chimpanzees of Mahale. Am J Primatol 59:99–121

Nishuli R (2008) A new gorilla group. Gorilla J 37:3

Nowell AA, Fletcher AW (2008) The development of feeding behaviour in wild western lowland gorillas (*Gorilla gorilla gorilla*). Behaviour 145:171–193

17   Long-Term Research on Grauer's Gorillas in Kahuzi-Biega National Park, DRC   409

Powzyk JA, Mowry CB (2003) Dietary and feeding differences between sympatric *Propithecus diadema diadema* and *Indri indri*. Int J Primatol 24:1143–1162

Remis MJ (2000) Initial studies on the contribution of body size and gastrointestinal passage rates to dietary flexibility among gorillas. Am J Phys Anthropol 112:171–180

Remis MJ, Dierenfeld ES, Mowry CB, Carroll RW (2001) Nutritional aspects of western lowland gorilla (*Gorilla gorilla gorilla*) diet during seasons of fruit scarcity at Bai Hokou, Central African Republic. Int J Primatol 22:807–836

Reynolds V, Plumptre AJ, Greenham J, Harborne J (1998) Condensed tannins and sugars in the diet of chimpanzees (*Pan troglodytes schweinfurthii*) in the Budongo Forest, Uganda. Oecologia 115:331–336

Robbins MM (1995) A demographic analysis of male life history and social structure of mountain gorillas. Behaviour 132:21–47

Robbins MM (2001) Variation in the social system of mountain gorillas: the male perspective. In: Robbins MM, Sicotte P, Stewart KJ (eds) Mountain gorillas: three decades of research at Karisoke. Cambridge University Press, Cambridge, pp 29–58

Robbins MM, McNeilage A (2003) Home range and frugivory patterns of mountain gorillas in Bwindi Impenetrable National Park, Uganda. Int J Primatol 24:467–491

Robbins AM, Robbins MM (2005) Fitness consequences of dispersal decisions for male mountain gorillas (*Gorilla gorilla beringei*). Behav Ecol Sociobiol 58:295–309

Robbins MM, Bermejo M, Cipolletta C, Magliocca F, Parnell RJ, Stokes E (2004) Social structure and life-history patterns in western gorillas (*Gorilla gorilla gorilla*). Am J Primatol 64:145–159

Robbins AM, Stoinski TS, Fawcett KA, Robbins MM (2009) Socioecological influences on the dispersal of female mountain gorillas – evidence of a second folivore paradox. Behav Ecol Sociobiol 63:477–489

Rogers ME, Abernethy K, Bermejo M, Cipolletta C, Doran D, McFarland K, Nishihara T, Remis M, Tutin CEG (2004) Western gorilla diet: a synthesis from six sites. Am J Primatol 64:173–192

Ross C (1998) Primate life histories. Evol Anthropol 6:54–63

Schaller GB (1963) The mountain gorilla: ecology and behavior. University of Chicago Press, Chicago

Sivha M (1999) Conservation of resources in Kahuzi-Biega. Gorilla J 19:6–7

Sommer V (1994) Infanticide among the langurs of Jodhpur: testing the sexual selection hypothesis with a long-term record. In: Parmigiani S, Vom Saal FS (eds) Infanticide and parental care. Harwood Academic Publishers, Chur, pp 155–198

Stanford CB, Nkurunungi JB (2003) Behavioral ecology of sympatric chimpanzees and gorillas in Bwindi Impenetrable National Park, Uganda: diet. Int J Primatol 24:901–918

Steinhauer-Burkart B, Mühlenberg M, Slowik J (1995) Kahuzi-Biega National Park. The IZCN/ GTZ Project Integrated Nature Conservation in East Zaïre

Stewart KJ, Harcourt AH (1987) Gorillas: variation in female relationships. In: Smuts BB, Cheney DL, Seyfarth RM, Wrangham RW, Struhsaker TT (eds) Primate societies. University of Chicago Press, Chicago, pp 155–164

Takenoshita Y, Ando C, Iwata Y, Yamagiwa J (2008) Fruit phenology of the great ape habitat in the Moukalaba-Doudou National Park, Gabon. Afr Stud Monogr 39(suppl):23–39

Tan CL (1999) Group composition, home range size, and diet of three sympatric bamboo lemur species (Genus *Hapalemur*) in Ranomafana National Park, Madagascar. Int J Primatol 20:547–566

Tutin CEG (1996) Ranging and social structure of lowland gorillas in the Lopé Reserve, Gabon. In: McGrew WC, Marchant LF, Nishida T (eds) Great ape societies. Cambridge University Press, Cambridge, pp 58–70

Tutin CEG, Fernandez M (1984) Nationwide census of gorilla (*Gorilla g. gorilla*) and chimpanzee (*Pan t. troglodytes*) populations in Gabon. Am J Primatol 6:313–336

Tutin CEG, Fernandez M (1993) Composition of the diet of chimpanzees and comparisons with that of sympatric lowland gorillas in the Lopé Reserve, Gabon. Am J Primatol 30:195–211

Ungar PS (1996) Relationship of incisor size to diet and anterior tooth use in sympatric Sumatran anthropoids. Am J Primatol 38:145–156

van Schaik CP (2000) Infanticide by male primates: the sexual selection hypothesis revisited. In: van Schaik CP, Janson CH (eds) Infanticide by males and its implication. Cambridge University Press, Cambridge, pp 27–60

van Schaik CP, Preuschoft S, Watts DP (2004) Great ape social systems. In: Russon AE, Begun DR (eds) The evolution of thought: evolutionary origins of great ape intelligence. Cambridge University Press, Cambridge, pp 190–209

Vedder AL (1984) Movement patterns of a group of free-ranging mountain gorillas (*Gorilla gorilla beringei*) and their relation to food availability. Am J Primatol 7:73–88

von Richter W (1991) Problems and limitations of nature conservation in developing countries: a case study in Zaïre. In: Erdelen W, Ishwaran N, Muller P (eds) Tropical ecosystems. Margraf Scientific Books, Weikersheim, pp 185–194

Waser PM (1984) Ecological differences and behavioral contrasts between two mangabey species. In: Rodman PS, Cant JGH (eds) Adaptations for foraging in nonhuman primates. Columbia University Press, New York, pp 195–216

Watts DP (1987) Effects of mountain gorilla foraging activities on the productivity of their food plant species. Afr J Ecol 25:155–163

Watts DP (1989) Infanticide in mountain gorillas: new cases and a reconsideration of the evidence. Ethology 81:1–18

Watts DP (1990) Mountain gorilla life histories, reproductive competition, and sociosexual behavior and some implication for captive husbandry. Zoo Biol 9:185–200

Watts DP (1991) Strategies of habitat use by mountain gorillas. Folia Primatol 56:1–16

Watts DP (1996) Comparative socio-ecology of gorillas. In: McGrew WC, Marchant LF, Nishida T (eds) Great ape societies. Cambridge University Press, Cambridge, pp 16–28

Watts DP (1998a) Long-term habitat use by mountain gorillas (*Gorilla gorilla beringei*). 1. Consistency, variation, and home range size and stability. Int J Primatol 19:651–680

Watts DP (1998b) Long-term habitat use by mountain gorillas (*Gorilla gorilla beringei*). 2. Reuse of foraging areas in relation to resource abundance, quality, and depletion. Int J Primatol 19:681–702

Watts DR (2000) Mountain gorilla habitat use strategies and group movements. In: Boinski S, Garber PA (eds) On the move: how and why animals travel in groups. University of Chicago Press, Chicago, pp 351–374

Watts DP (2012) Long-term research on Chimpanzee behavioral ecology in Kibale National Park, Uganda. In: Kappler PM, Watts DP (eds) Long-term field studies of primates. Springer, Heidelberg

Watts DP, Mitani JC (2001) Boundary patrols and intergroup encounters in wild chimpanzees. Behaviour 138:299–327

Weber AW, Vedder AL (1983) Population dynamics of the Virunga gorillas: 1959–1978. Biol Conserv 26:341–366

Wich SA, Utami-Atmoko SS, Mitra Setia T, Rijksen HD, Schürmann C, van Hooff JARAM, van Schaik CP (2004) Life history of wild Sumatran orangutans (*Pongo abelii*). J Hum Evol 47:385–398

Williams JM, Oehlert GW, Carlis JV, Pusey AE (2004) Why do male chimpanzees defend a group range? Anim Behav 68:523–532

Williamson RA, Tutin CEG, Rogers ME, Fernandez M (1990) Composition of the diet of lowland gorillas at Lopé in Gabon. Am J Primatol 21:265–277

Wilson ML (2012) Long-term studies of the Chimpanzees of Gombe National Park, Tanzania. In: Kappler PM, Watts DP (eds) Long-term field studies of primates. Springer, Heidelberg

Wilson ML, Wrangham RW (2003) Intergroup relations in chimpanzees. Annu Rev Anthropol 32:363–392

Wrangham RW, Chapman CA, Clark-Arcadi AP, Isabirye-Basuta G (1996) Social ecology of Kanyawara chimpanzees: implications for understanding the costs of great ape groups.

In: McGrew WC, Marchant LF, Nishida T (eds) Great ape societies. Cambridge University Press, Cambridge, pp 45–57

Yamagiwa J (1983) Diachronic changes in two eastern lowland gorilla groups (*Gorilla gorilla graueri*) in the Mt. Kahuzi region, Zaïre. Primates 24:174–183

Yamagiwa J (1986) Activity rhythm and the ranging of a solitary male mountain gorilla (*Gorilla gorilla beringei*). Primates 27:273–282

Yamagiwa J (1999) Socioecological factors influencing population structure of gorillas and chimpanzees. Primates 40:87–104

Yamagiwa J (2003) Bushmeat poaching and the conservation crisis in Kahuzi-Biega National Park, Democratic Republic of the Congo. J Sustain Forest 16:115–135

Yamagiwa J (2004) Diet and foraging of the great apes: ecological constraints on their social organizations and implications for their divergence. In: Russon AE, Begun DR (eds) The evolution of thought: evolutionary origins of great ape intelligence. Cambridge University Press, Cambridge, pp 210–233

Yamagiwa J, Basabose AK (2006a) Diet and seasonal changes in sympatric gorillas and chimpanzees at Kahuzi-Biega National Park. Primates 47:74–90

Yamagiwa J, Basabose AK (2006b) Effects of fruit scarcity on foraging strategies of sympatric gorillas and chimpanzees. In: Hohmann G, Robbins MM, Boesch C (eds) Feeding ecology in apes and other primates. Cambridge University Press, Cambridge, pp 73–96

Yamagiwa J, Basabose AK (2009) Fallback foods and dietary partitioning among *Pan* and *Gorilla*. Am J Phys Anthropol 140:739–750

Yamagiwa J, Kahekwa J (2001) Dispersal patterns, group structure, and reproductive parameters of eastern lowland gorillas at Kahuzi in the absence of infanticide. In: Robbins MM, Sicotte P, Stewart KJ (eds) Mountain gorillas: three decades of research at Karisoke. Cambridge University Press, London, pp 89–122

Yamagiwa J, Kahekwa J (2004) First observations of infanticides by a silverback in Kahuzi-Biega. Gorilla J 29:6–9

Yamagiwa J, Mwanza N (1994) Day-journey length and daily diet of solitary male gorillas in lowland and highland habitats. Int J Primatol 15:207–224

Yamagiwa J, Yumoto T, Mwanza N, Maruhashi T (1988) Evidence of tool-use by chimpanzees (*Pan troglodytes schweinfurthii*) for digging out a bee-nest in the Kahuzi-Biega National Park, Zaïre. Primates 29:405–411

Yamagiwa J, Maruhashi T, Yumoto T, Mwanza N (1989) A prelimary survey on sympatric populations of *Gorilla gorilla graueri* and *Pan troglodytes schweinfurthii* in eastern Zaïre. Grant-in-Aid for Overseas Scientific Research Report: Interspecies Relationships of Primates in the Tropical and Montane Forests 1:23–40

Yamagiwa J, Mwanza N, Yumoto T, Maruhashi T (1991) Ant eating by eastern lowland gorillas. Primates 32:247–253

Yamagiwa J, Mwanza N, Spangenberg A, Maruhashi T, Yumoto T, Fischer A, Steinhauer-Burkart B, Refisch J (1992) Population density and ranging pattern of chimpanzees in Kahuzi-Biega National Park, Zaïre: comparison with a sympatric population of gorillas. Afr Stud Monogr 13:217–230

Yamagiwa J, Mwanza N, Spangenberg A, Maruhashi T, Yumoto T, Fischer A, Steinhauer-Burkart B (1993) A census of the eastern lowland gorillas *Gorilla gorilla graueri* in Kahuzi-Biega National Park with reference to mountain gorillas *G. g. beringei* in the Virunga region, Zaïre. Biol Conserv 64:83–89

Yamagiwa J, Mwanza N, Yumoto Y, Maruhashi T (1994) Seasonal change in the composition of the diet of eastern lowland gorillas. Primates 35:1–14

Yamagiwa J, Maruhashi T, Yumoto T, Mwanza N (1996) Dietary and ranging overlap in sympatric gorillas and chimpanzees in Kahuzi-Biega National Park, Zaïre. In: McGrew WC, Marchant LF, Nishida T (eds) Great ape societies. Cambridge University Press, Cambridge, pp 82–98

Yamagiwa J, Kahekwa J, Basabose AK (2003) Intra-specific variation in social organization of gorillas: implications for their social evolution. Primates 44:359–369

Yamagiwa J, Basabose AK, Kaleme K, Yumoto T (2005) Diet of Grauer's gorillas in the montane forest of Kahuzi, Democratic Republic of Congo. Int J Primatol 26:1345–1373

Yamagiwa J, Basabose AK, Kaleme KP, Yumoto T (2008) Phenology of fruits consumed by a sympatric population of gorillas and chimpanzees in Kahuzi-Biega National Park, Democratic Republic of Congo. Afr Stud Monogr 39(suppl):3–22

Yamagiwa J, Kahekwa J, Basabose AK (2009) Infanticide and social flexibility in the genus *Gorilla*. Primates 50:293–303

Yamakoshi G (1998) Dietary responses to fruit scarcity of wild chimpanzees at Bossou, Guinea: possible implications for ecological importance of tool use. Am J Phys Anthropol 106:283–295

Chapter 18
Long-Term Studies on Wild Bonobos at Wamba, Luo Scientific Reserve, D. R. Congo: Towards the Understanding of Female Life History in a Male-Philopatric Species

Takeshi Furuichi, Gen'ichi Idani, Hiroshi Ihobe, Chie Hashimoto, Yasuko Tashiro, Tetsuya Sakamaki, Mbangi N. Mulavwa, Kumugo Yangozene, and Suehisa Kuroda

Abstract Long-term studies on wild bonobos began at Wamba, in the current Luo Scientific Reserve, in 1973. Except for several interruptions due to political instability and civil war, we have been conducting studies of identified individual bonobos over 35 years, providing valuable data on their population dynamics and life history. Although the number of groups and number of individuals in the northern section of the reserve decreased by half during the interruptions of the study, the number of members of the main study group has steadily increased since 2002 when we resumed the study. Our long-term data demonstrated the male-philopatric structure of the group. There is no confirmed case of emigration of males from the study group, and no case of immigration of males into the group. On the other hand, all females born into the study group disappeared by the age of 10 years, and females with estimated ages of 6–13 years immigrated into the study group. These ages of intergroup transfer are much earlier than those reported for chimpanzees. Exceptional cases of immigration of two adult males and two adult

---

T. Furuichi (✉) • C. Hashimoto • T. Sakamaki
Primate Research Institute, Kyoto University, Inuyama, Aichi, Japan
e-mail: furuichi@pri.kyoto-u.ac.jp

G. Idani
Wildlife Research Center, Kyoto University, Kyoto, Japan

H. Ihobe
School of Human Sciences, Sugiyama Jogakuen University, Nisshin, Aichi, Japan

Y. Tashiro
Hayashibara Great Ape Research Institute, Tamano, Okayama, Japan

M.N. Mulavwa • K. Yangozene
Research Center of Ecology and Forestry, Ministry of Scientific Research, Bikolo, Equateur, Democratic Republic of Congo

S. Kuroda
School of Human Cultures, The University of Shiga Prefecture, Hikone, Shiga, Japan

---

P.M. Kappeler and D.P. Watts (eds.), *Long-Term Field Studies of Primates*,
DOI 10.1007/978-3-642-22514-7_18, © Springer-Verlag Berlin Heidelberg 2012

females with offspring occurred right after the war. It is likely that remnants of extinct groups joined the study group. Such integration of members of foreign groups highlights the peaceful nature of bonobo society. The study group is characterized by an extremely high tendency for female aggregation. Various factors, including high density of food patches, female initiative in ranging, prolonged estrus of females, and high social status of females, seem to be responsible for the high attendance ratio of females in mixed-sex parties. Our long-term observations therefore provided evidence for interesting behavioral contrasts with chimpanzees.

18.1 Introduction

The so-called last ape, or bonobo (*Pan paniscus*), which lives only in areas bordered by the Congo River, was first described as a subspecies of chimpanzee in 1929 (Schwarz 1929). It attracted the interest of many anthropologists and primatologists because of its morphological similarity to human ancestors and unique sexual behaviors (Susman 1984; Kano 1992). Long-term studies on wild bonobos began in the 1970s at two study sites, Wamba and Lomako in the Republic of Zaïre. However, mainly due to the country's political instability, it was very difficult to continue these studies. The economic situation worsened through the 1980s, and the first violent riot arose in the capital city of Kinshasa in 1991, when many researchers had to evacuate their study sites. Though studies gradually resumed at various sites, a civil war occurred in 1996 and a new regime was established, bringing with it a new name for the country: Democratic Republic of Congo. Security conditions improved somewhat and researchers returned or attempted to return to their field sites around 1998, but again a war involving several African countries occurred in August of the same year. Researchers could do nothing until around 2002, when a ceasefire was agreed upon.

Because of these difficulties, there are even now only a few long-term study sites at which researchers are carrying out behavioral studies of wild bonobos. Studies in Lomako were interrupted by the wars that occurred in 1996 and 1998. The Lomako-Yokokala Faunal Reserve was created in 2006, and studies of bonobos are currently being undertaken (Dupain et al. 2000; White et al. 2008), but not mainly on the same study groups or identified individuals. Very active research programs studying wild bonobos have been undertaken at Lui Kotale since 2002 (Hohmann and Fruth 2003), but even this study remains shorter than those at many chimpanzee study sites. Therefore, we have much to learn from studies at Wamba, which have continued for more than 35 years except during periods of political instability and war, especially concerning the life history of wild bonobos.

In this chapter, we will first provide an overview of the history of our research and conservation activities at Wamba. We will describe changes in the bonobo population structure during times of economic and political instability, focusing on the general effects of such instability. Second, we will summarize the outcome of our studies concerning the life history of bonobos at Wamba, using the data available for the main study group. Bonobos form semiclosed social groups, called unit groups (Nishida 1979) or communities (Goodall 1983), which we refer to as "groups" in this chapter. Groups of bonobos are male-philopatric, and females transfer between them before the onset of sexual maturation (Furuichi 1989; Kano 1992). Therefore, long-term data are indispensable for understanding the life history of females. We will also briefly review our findings pertaining to female sexuality and social behavior. Though bonobos are close relatives of chimpanzees, there are many interesting differences between the sexual and social behaviors of these two species, particularly among females. Therefore, the investigation of female life histories and behavior should greatly improve our understanding of what bonobos can tell us about the evolution of hominoids.

18.2 History of Research and Conservation Activities

18.2.1 Onset of Studies at Wamba and the Creation of the Luo Scientific Reserve

After the Congo crisis of 1960–1966, Toshisada Nishida carried out a preliminary survey of wild bonobos in an area adjacent to Lake Tumba in 1971, and confirmed that bonobos still survived in the Congo Basin (Nishida 1972). Using the information provided by this survey, Takayoshi Kano conducted a survey over a wider area to find an appropriate site to begin observation of wild bonobos in 1973 (Kano 1992). This trip was extremely difficult, partly because he had to roam an enormous area in the Congo basin on a bicycle, and partly because the indigenous people were not accustomed to non-African visitors at that time. People of some villages were afraid of him and did not treat him with much respect, forcing him to continue his journey without staying at their villages. When he reached a village called Wamba, however, the people welcomed him. He also heard vocalizations of bonobos from the village. Thus, he decided to choose this village for his field research.

In 1974, Kano sent Suehisa Kuroda to Wamba to begin the study of bonobos there. Kuroda first tried to habituate the bonobos ranging around the Bokela stream by artificial provisioning, but was unsuccessful. He then changed his target to another group that sometimes visited agricultural fields to feed on sugar cane. He named this group "E group," after "elanga," which refers to an agricultural field in the local language. He successfully habituated most bonobos in E group and, with Kano, identified its members by 1976. Bonobos of E group came to regularly visit the sugar cane fields in 1977, and the first artificial provisioning site was opened in the forest in 1978 (Kano 1992; Kano et al. 1996).

Since then, many researchers of the team headed by Kano have visited Wamba and carried out research on bonobo ecology and behavior. Bonobos used the forest around the village and had little fear of local people, who were tolerant of them because of their traditional belief that they and the bonobos both descended from the same family of bonobos that had lived in the forest in the past. According to this belief, an older brother in a family of bonobos held to their traditional lifestyle and his descendants thus remained in the forest as bonobos. However, his younger brother was tired of eating raw foods. Once upon a time, he was roaming in the forest, crying, a spirit of the forest taught him to make fire, after which he left the forest and began eating cooked food. His descendants became humans. Therefore, the village people consider the bonobos akin to distant brothers and do not kill or eat them.

In fact, there were no suspected cases of poaching before 1983, but during the absence of researchers in 1984, one of the young adult males of E1 group was killed by a poacher from outside of Wamba who tried to sell its body as meat. A second poaching incident that occurred during the absence of researchers in 1987 was more serious. A military troop ordered village people to capture two infants, and their mothers and some other adults of the study groups were killed during this attempt. According to an unconfirmed report, the two infants were carried to the capital Kinshasa and were given by the government as a gift to the King of Belgium. These incidents, together with a growing awareness of the importance of bonobos at Wamba for conservation and research purposes, encouraged us to create a reserve in the Wamba area (Kano et al. 1996).

Following discussions with our research team, our counterpart organization (Research Center of Natural Science, or CRSN, of the Republic of Zaïre) decided to create a reserve for scientific research, and we agreed to provide technical and financial aid to help with its creation. Delimitation of the proposed reserve area started in 1987, and in 1988, CRSN and the local communities signed agreements that created a reserve named the Reserve Scientifique de Luo (called "Luo Reserve" hereafter; Fig. 18.1). The reserve was officially recognized by the Ministry of the Environment in March 1990.

After some argument about what status the reserve should have, we chose to create a special reserve for scientific research instead of a national park, the creation of which would have required local people to leave. Though there were only small hunting camps in the southern section of the reserve, there were six homesteads and more than 1,000 people living in the northern section where we had been carrying out research. We therefore made efforts to help local people maintain their traditional ways of coexistence with the bonobos. Hunting primates and the use of guns or wire snares were prohibited, but hunting other animals with traditional techniques was allowed. Also, clearing of new fields in primary or old secondary forest to grow cash crops was prohibited, but reusing or even creating new fields for cassava or other subsistence crops was allowed within limited areas, mostly within 1 km of the road. To compensate people for these limits on forest use, the CRSN promised to support local development, including provision of support for local schools and medical services. Our research team has since tried to help CRSN, now CREF (Research Center of Ecology and Forestry), to provide such support.

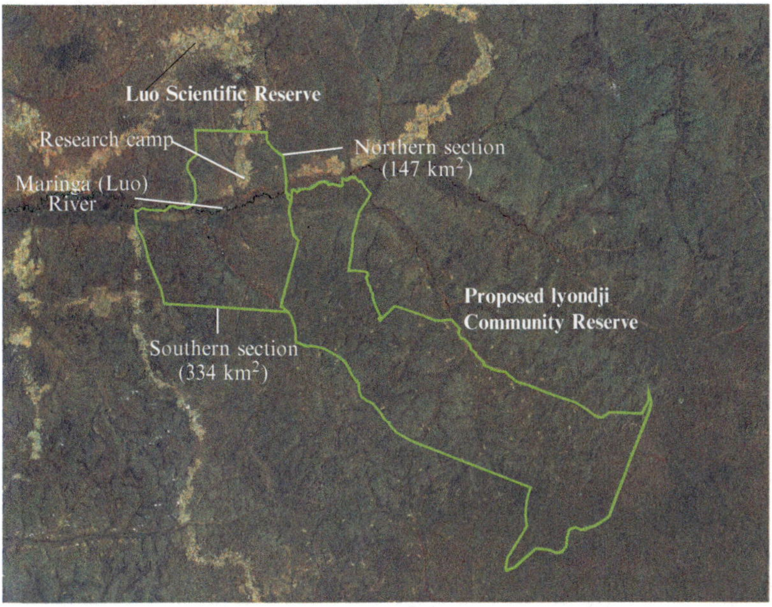

Fig. 18.1 The Luo Scientific Reserve and the proposed Iyondji Community Reserve. Image: NASA Landsat Program, c. 1998–2002 image composite. The *pale colorations* show roads, homesteads, agricultural fields, and young secondary forest

These efforts were initially fairly successful. However, promoting both conservation and the well-being of local people long term is extremely difficult, particularly when the conditions of human societies change rapidly. At Wamba, requests by local people for support became very serious when their lives were badly affected by political instability and war in the 1990s. Furthermore, requests for assistance by local people increased when the globalization of information made them more aware of the importance of bonobos to the conservation community and of the gaps between their standard of living and those of people in the developed world. Previously successful strategies to balance nature conservation with the local peoples' quality of life needed modification in subsequent years. The trials of achieving coexistence between the ever-changing needs of humans with unchanging nature seem to be endless.

18.2.2 Changes in Bonobo Population Structure in the Reserve

The Luo Reserve is divided into two sections by the Maringa (or Luo) River (Fig. 18.1), which bonobos cannot cross. A large portion of the northern section, where most of our research activities have been carried out, consists of young secondary forest and both abandoned and currently used homesteads and fields.

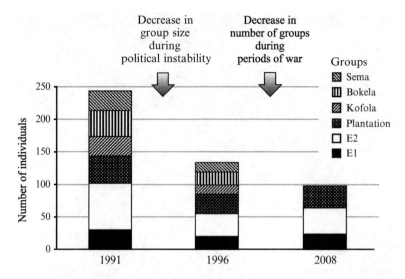

Fig. 18.2 Changes in the bonobo population structure in the northern section of the Luo Reserve. Figure modified from Hashimoto et al. (2008) using data from 1991 to 2008

In the beginning of 1991, six groups of bonobos with about 250 individuals used the northern section (Fig. 18.2). However, the population decreased by half during our absence, which began in 1991 because of political instability. When we resumed our studies in 1994, the number of bonobos estimated by direct observations was about 130, and their density estimated by nest counts was 0.54 independent animals per km^2 in 1996 (Hashimoto and Furuichi 2001). We were again forced to leave the site in 1996 due to the civil war. When we returned to the site in 2002, we could not find three of the six groups, and only about 100 bonobos remained in the northern section.

In contrast to the northern section, most of the southern section is primary or old secondary forest, and there are fewer villages, although many small hunting camps exist. Unfortunately however, the situation in this section seems to be even worse than in the north. The density of bonobos in the western area of the southern section was estimated to be 0.28 in 1996 (Hashimoto and Furuichi 2001). However, there were almost no bonobos in the western area in 2007 (Mulavwa unpublished data). Apparently, the population of bonobos in the area near the road was significantly reduced during the war. Though a good number of bonobos seems to remain in the eastern area of the southern section, an intensive survey in this section is urgently needed.

The main cause of the loss of bonobos during the war seemed to be hunting, especially by or on the orders of soldiers. Because many of the soldiers deployed in the Luo Reserve were from other areas of the country and many did not have a taboo against killing and eating bonobos, they sometimes hunted or asked villagers to hunt these animals for them. The village people might also have hunted the bonobos spontaneously to sell the meat as a means of surviving the war. Vegetation in the

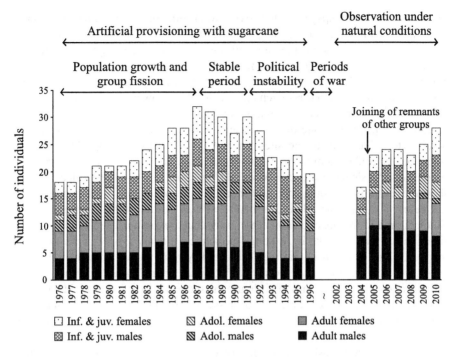

Fig. 18.3 Changes in the number of bonobos in the main study group. Figure modified from Hashimoto et al. (2008) using data from 1976 to 2010

primary forest was also damaged during the war. Many people fled from the villages along the road and lived instead in hunting camps within the forest, making cassava fields in the areas surrounding their encampments. Thus, political instability and war considerably damaged the fauna and flora in the reserve, although no fighting actually occurred in this area.

The number of bonobos in the main study group was also strongly influenced by the political conditions. The main study group, E, split into two groups, E1 and E2, by 1983 (Furuichi 1987). Figure 18.3 shows changes in the number of bonobos in the subgroup of E that eventually became E1 before 1983, and the number of bonobos in E1 since 1984. The group increased steadily in size until the split, and continued to increase thereafter. However, these increases stopped in 1987, and the group rapidly decreased in size during the period of political instability beginning in 1991. Though we lack data on group size during the war that started in 1996, E1 had 17 confirmed members in the beginning of 2004. Thus, the number of bonobos in E1 group did not decrease significantly during the 1996–2004 period.

Although we had been employing artificial provisioning for our observations before 1996, we decided to resume our research without this practice to avoid disease transmission and to observe bonobos under more natural conditions. Starting in 2004, the size of E1 group rapidly increased, seemingly because remnant individuals of other disappearing groups joined them (see below). Since then, the number of bonobos of E1 has steadily increased, attaining a current number close to that of 1991.

It was surprising that, in spite of the disappearance of three neighboring groups, the size of E1 group did not significantly decrease during our war-induced absence. It may seem that this occurred because the group was ranging in the middle of the northern section where our research camp is located. However, the risk of hunting was likely higher in this central area because a troop of soldiers was staying at our research camp. In fact, one of our research assistants was repeatedly ordered by soldiers to guide them to the sites where E1 group was sleeping. Although he intentionally guided them to the wrong sites several times, he was finally forced to guide them to a sleeping site after being told "guide us, or you will be killed." When they arrived at the sleeping site, soldiers asked him which one to shoot. He could not help but point to an adolescent male that was the first to appear from a bed. Soldiers fired at the bonobo from close range, but they still needed to fire several times before finally killing it. Although the soldiers did take it back and eat it, they stopped killing bonobos in that area because they were afraid that some "magical power" had prevented their bullets from hitting the animal. The research assistant hid one of the bonobo's bones in his house; he showed it to us late at night when he reported these events after our return to the site in 2002. Probably the biggest reason that this group survived the war was that our assistants and other local people tried to protect it, in the hope that the researchers would return after the war.

One of the more positive outcomes of our conservation activities is that the people of an adjacent village, Iyondje, proposed to create a community reserve for the conservation of bonobos in 2007 (Fig. 18.1). They wanted to stimulate research activities and tourism in the area and to protect their forest against intrusion by farmers from other villages. This reserve would be connected to the Luo reserve and would triple its size. Our research team is now working to help local people achieve this goal, in collaboration with the US Fish and Wildlife Service and the African Wildlife Foundation.

18.3 Life History of Bonobos at Wamba and Characteristics of Female Behavior

18.3.1 Intergroup Transfer of Individuals

Like chimpanzees, bonobos were thought to form male-philopatric groups, among which only females transferred (Nishida 1979; Wrangham 1986; Furuichi 1989; Kano 1992). This assumption was supported by the fact that the genetic distances between male bonobos of the same group were shorter than those between females (Hashimoto et al. 1996; Gerloff et al. 1999). However, there are many reports of exceptions to male philopatry and female transfer in chimpanzees. For example, some females at Gombe remain in their natal group for entire lives (Goodall 1986).

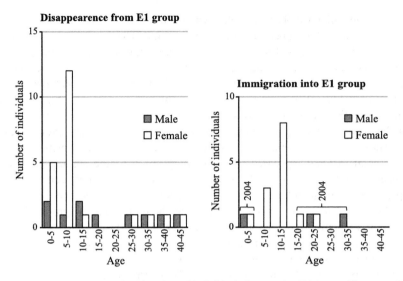

Fig. 18.4 Disappearance and immigration of individuals from and to E1 group. Figure modified from Hashimoto et al. (2008) using data from 1976 to 2009

At Bossou, Guinea, both males and females disappeared from their natal group, and some extra-group males were observed to visit the study group and stay temporarily (Sugiyama 1999, 2004). Therefore, the patterns of transfer of individuals need to be examined through long-term studies of specific groups.

Figure 18.4 shows the number of individuals of different age classes that disappeared from, or immigrated into, the study group at Wamba. Many females disappeared between the ages of 6 and 10 and presumably emigrated (Hashimoto et al. 2008). Only a few individuals in other age classes disappeared; their disappearances might have been deaths. If we exclude mergers between the remnants of adjacent groups in 2004, only females at estimated ages of 6–13 years old immigrated into E1 group. Females who immigrated into the study group under the age of 10 years disappeared shortly thereafter. However, most females who appeared in E1 at the age of 10 or older eventually had offspring, and settled into the group permanently.

Based on this information, we can infer that females leave their natal groups at the age of 6–10, and settle into new groups at the age of 10–13, when they start to reproduce. Because there is only one case of a supposed emigration by a female at 6 years old, this might in fact be a disappearance due to death rather than emigration. However, all the other females born into the study group disappeared between 7 and 10 years of age (Hashimoto et al. 2008), which is much younger than the age of first transfer among female chimpanzees; for example, females leave their natal groups between the ages of 8 and 16 at Mahale, Tanzania (Nishida et al. 2003). One of the unanswered questions concerns what females are doing during the years after leaving their natal groups and before settling down in new groups. To date, there are many cases in which we observed unknown females of 10 years or younger visiting our study group and disappearing after a short period. Therefore, it seems possible that young adolescent females who have left their natal groups move from group to group before finally immigrating permanently into a group in late adolescence.

Disappearance of males occurred almost evenly across the different age classes, and this probably reflects their deaths. By contrast, except for mergers of the remnants of adjacent groups in 2004, there were no observed cases of immigration by males. This strongly suggests that bonobos form male-philopatric groups (Kano 1982; Hashimoto et al. 1996). There were, however, some cases in which adolescent or adult males disappeared from the study group before coming back after absences of several months. Males ranging alone were sometimes seen by village people, albeit infrequently. In one case, an adult male that had been absent for several months had a number of new scars when he returned to the group. We still have little indication of what these males are doing during their absence, in particular whether they stay on their own within the ranges of their own groups or visit other areas or other groups temporarily.

One exception regarding the transfer of males occurred in 2004 (Hashimoto et al. 2008). Researchers sometimes observed two strange females with infants, and two strange adult males traveling with the study group. In the beginning, this happened only when the study group was in the eastern area of its home range, near the home range of the extinct K group. The north-eastern area was undoubtedly in the home range of K group, and E1 was never observed to use this area before the wars. It is likely that E1 group expanded their home range into this newly vacant area, and the remnants of the extinct K group subsequently joined them. By 2006, each of those six individuals was completely integrated into E1 group, even ranging with other E1 members into western parts of the group's home range, far from the range of their likely original group, K. The two adult males were usually found in the central part of mixed-sex parties, and we could not distinguish their behavioral tendencies from those of the original E1 males.

Another example of transfer of male bonobos was reported from Lomako (Hohmann 2001). When Hohmann resumed observation in 1997 after the first war, he found two strange males in the range of the study group, Eyengo. They were severely attacked by resident males and females, and one of them was no longer observed after 6 months. However, the other male established affiliative relations with resident members and remained with the group for at least 11 months, after which observations were interrupted by the second war. Though possibly a coincidence, it is noteworthy that these cases at Wamba and Lomako both occurred immediately following a period of war.

Such facultative intergroup transfers have also been observed under similar circumstances in Japanese macaques (*Macaca fuscata*), which form female-philopatric groups among which only males typically transfer (Suzuki et al. 1998). In one case, a group gradually decreased in size, and when only two females remained, they finally immigrated into a neighboring group (Takahata et al. 1994). Such a transfer of individuals of the normally nondispersing sex may occur in unusual situations involving group extinctions. In addition, these incidents in bonobos also reflect their relatively peaceful nature. Unlike the intensely hostile intergroup relationships among chimpanzees, when two groups of bonobos at Wamba meet, they frequently merge temporarily and travel together while engaging in affiliative social and sexual interactions (Idani 1990; Kano 1992). Although some aggressive interactions occur in the beginning of encounters, they are mostly displays without physical attacks. Although Hohmann reported that encounters between neighboring groups at Lomako were often agonistic, he also noted that aggressive behaviors were mainly restricted to displays (Hohmann et al. 1999). Such tolerance of members of different groups might allow for the immigration of adult males under conditions such as those outlined here.

18.3.2 Cohesive Grouping of bonobos

Both chimpanzees and bonobos have fission–fusion social systems, which are characterized by group members splitting into smaller subgroups, called parties, which change flexibly in membership over time (Kuroda 1979; Kano 1982; Nishida 1979; Wrangham 1979; Chapman et al. 1994; van Elsacker et al. 1995). Early studies on the grouping of bonobos reported that their party sizes were larger than those of chimpanzees (Kuroda 1979; Kano 1986), and various hypotheses to explain this difference were presented (White and Wrangham 1988; Kano 1992; Furuichi 2009). However, more recent analyses showed that party size varies considerably among sites in both species, leading to uncertainty about the differences between species (Furuichi 2009). If we compare the data obtained using similar definitions of a "party," the number of individuals ranges from 4.0 to 10.3 for chimpanzees and from 4.9 to 14 for bonobos in studies including independent individuals only, while it ranges from 4.0 to 10.0 for chimpanzees and from 8.5 to 22.7 for bonobos in studies including all individuals. Thus, there is no statistically significant difference in party size between the two species. However, relative party size (Boesch 1996) or attendance ratio (Furuichi et al. 2008), defined as party size relative to the size of the whole group, differs significantly. This value ranges from 9 to 30% for chimpanzees but from 27 to 51% for bonobos. This might partly be explained by the larger size of chimpanzee groups, but it also seems to reflect real differences in behavior.

Fission–fusion events typically occur at high frequencies in chimpanzees (e.g., Mahale: Nishida 1968). Different chimpanzee parties may simultaneously use distant areas within the group home range, and it is therefore very rare that a researcher can observe all members of the group within a single day. In Kalinzu Forest, Uganda, where we have been studying the ecology of chimpanzees since 1992, we could not see some females for as long as 4–5 years, probably because they ranged alone or in small parties while pregnant and lactating. In contrast to chimpanzees, different bonobo parties at Wamba usually range in similar areas while keeping track of each other's position through vocal exchanges. As a result, especially during the high-fruiting season when these bonobos form large parties

(Mulavwa et al. 2008), all members of E1 group are often observed on a single day's observation. For example, all of the group members were observed on 35 of the 125 days on which E1 group was observed from sleeping site to sleeping site in 2008. Thus, the grouping pattern of bonobos at Wamba is considerably different from that of chimpanzees, although the term "fission and fusion" is often used for the grouping patterns of both species.

When we follow bonobos in the forest, we are impressed by some behavioral tendencies that we rarely or never observe in chimpanzees. Though the daily travel distances of bonobos tend to be shorter than those of chimpanzees (Furuichi et al. 2008), bonobos sometimes travel long distances to shift between different areas in the home range. During such long-distance travel, they sometimes split into several parties that often exchange vocalizations, and some bonobos in the leading party sit on the ground and look back, awaiting the arrival of those left behind. Another interesting contrast with chimpanzees is that the first individuals to arrive at big fruiting trees usually do not climb up immediately, but instead wait for the arrival of other members of the party. When these other members arrive, they make a chorus of soft vocalizations, and then all climb the tree together. Our observations at Kalinzu suggest that chimpanzees sometimes give loud calls to attract other members to feeding trees, but they do not wait for other members to arrive before climbing up.

The behavior of bonobos in the evening also warrants consideration. When the time to sleep approaches, parties that are travelling separately but in the same vicinity start calling to each other. We call these vocalizations "sunset calls." Parties sometimes approach each other while exchanging sunset calls, and begin to make nests after they join. This happens more frequently during the high-fruiting season. Bonobos usually forage in large parties during the daytime and aggregate to form even larger parties in the evening, before again splitting into several parties the next morning to forage (Kuroda 1979; Fruth and Hohmann 1994; Mulavwa et al. 2010). We do not hear such typical calls in the evening among chimpanzees at Kalinzu, and there is a report from Budongo, Uganda, that the party sizes of chimpanzees decrease towards the evening (Reynolds 2005). All these behaviors seem to show that bonobos are strongly motivated to maintain as much cohesion as ecological circumstances allow.

18.3.3 Aggregation of Females

Unlike the tendency for female chimpanzees to range alone or in small parties, female bonobos tend to attend parties more frequently than do males, which contributes greatly to their group's overall cohesion (Kuroda 1979; Kano 1982, 1992; Furuichi 1987; White 1988; Furuichi et al. 2008). Even at Taï, Côte d'Ivoire, where chimpanzees form larger parties than at other sites, females have a lower tendency than males to attend such parties (Boesch 1996; Boesch and Boesch-Achermann 2000). As an example, this difference can be illustrated by comparing

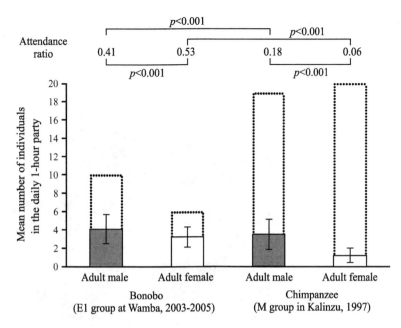

Fig. 18.5 Comparison of the group size and the mean 1-h party size between bonobos at Wamba and chimpanzees in the Kalinzu Forest. The length of the *bars with solid lines* represents the average numbers of individuals found in the 1-h party, and the length of the *bars with dotted lines* represents the number of all individuals belonging to the unit-group. The 1-h party is a method developed for comparison of party sizes between different species and sites (Hashimoto et al. 2001). Figure modified from Mulavwa et al. (2008)

party sizes relative to group sizes for bonobos of E1 group with chimpanzees at Kalinzu (Fig. 18.5). Although there are various definitions for party size, we employed the same 1-h party size method (Hashimoto et al. 2001) at both sites for a direct comparison. The numbers of individuals in the 1-h parties were not substantially different between the two species. However, while almost half of the group members were found in the 1-h parties of bonobos on average, a smaller proportion of the group members were found in parties of chimpanzees. In particular, we found a marked difference between species in female attendance in parties. While less than one-tenth of the females were found in parties of chimpanzees on average, almost two-thirds of the females were found in parties of bonobos. Female bonobos do not only attend parties frequently, but usually aggregate in the central part of the parties (Fig. 18.6).

The tendency for female chimpanzees to range alone or in small parties has been explained by various hypotheses. Due to their subordinate status relative to males, female chimpanzees may be subject to larger costs from contest competition at feeding sites than males are. Furthermore, foraging in a larger party may increase the frequency of shifts between food patches, which increases the cost from scramble competition for the slower moving females because they require longer travelling times between food patches and then arrive late to feeding sites

Fig. 18.6 Aggregation of females and their offspring in the central part of a mixed-sex party of bonobos

(Chapman et al. 1995; Wrangham et al. 1996; Wrangham 2000; Furuichi 2009). If true, the difference in the attendance ratio between males and females should be greater in larger parties because females should avoid attending such large parties. In fact, such a tendency has been confirmed in chimpanzees at Kanywara in Kibale National Park, Uganda (Wrangham 2000). However, this prediction was not verified in bonobos (Furuichi et al. 2008). As seen in Fig. 18.7, the attendance ratio of female bonobos was always higher than that of males, irrespective of party size. This suggests that influences of participation in parties differ quantitatively and/or qualitatively for female bonobos and female chimpanzees.

There are four major factors that may be responsible for the difference between the two species (Fig. 18.8; Furuichi 2011). The first factor is the difference in the density of food patches, which may typically be higher in bonobo habitats. Though we have not yet analyzed the data quantitatively, we encounter fruiting trees or food remains at much shorter intervals when following bonobos at Wamba than when following chimpanzees in the Kalinzu Forest. These food sources are not necessarily large fruiting trees, but include terrestrial herbs and the fruit and leaves of small trees that are more abundant in bonobo habitats (White and Wrangham 1988; Malenky and Wrangham 1994). The existence of such food sources between

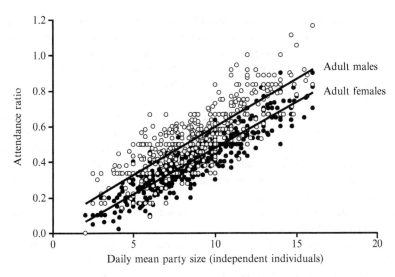

Fig. 18.7 Attendance ratio of males and females. Each *dot* in this figure shows the daily mean attendance ratio, which is the mean probability that each male or female was observed in a 1-h party. The *x* axis is the daily mean 1-h party size. Figure cited from Furuichi et al. (2008)

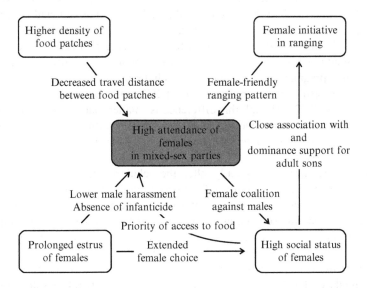

Fig. 18.8 Factors related to the aggregation of females. Figure cited from Furuichi (2011)

large food patches may decrease both the rate and distance of travel between the large food patches, thus reducing the costs for the slower moving females (Wrangham 2000).

The second factor to consider is the extent of female initiative in ranging (Furuichi et al. 2008). At Wamba, party movements typically occur when members descend from tall fruit trees and take short breaks in lower layers of the forest. They observe one another's movements as if to confirm which way others want to travel. Sitting on low trees seems to provide better visibility than sitting on the ground. Some individuals, most frequently males, climb down and perform branch-dragging displays, seemingly to propose a direction of movement. However, the entire party does not move until high-ranking females climb down and initiate movement in a direction of their own choice. Parties of E1 group usually range in one area for up to several weeks and then shift to another area. They typically move only short distances per day while in particular areas. Even when males travel farther, seemingly to shift to another area, they usually abandon these attempts if the females do not follow them, at which point they return to the original area to rejoin the females by evening (Furuichi et al. 2008). If females can control the direction and distance of travel, they can avoid incurring large costs associated with ranging in large mixed-sex parties.

The third factor is the prolonged estrous periods of females (Savage-Rumbaugh and Wilkerson 1978; Thompson-Handler et al. 1984; Furuichi 1987; Kano 1992; Furuichi and Hashimoto 2002). Female chimpanzees usually do not resume cyclic estrus until 4–5 years after parturition when their infants survive, and they stop showing signs of estrus by two months after conception. Thus, if we extrapolate from reproductive parameters recorded at Gombe and Mahale, female chimpanzees are in estrus on only about 5% of days in their adult life (Furuichi and Hashimoto 2002). By contrast, female bonobos resume swelling cycles only 1 year after parturition. Because these females are still experiencing postpartum amenorrhea, this is a pseudo-estrus without possibility of conception. They also show pseudo-estrus during pregnancy until about 1 month before parturition. Thus, female bonobos show estrus for as much as about 27% of their adult lives (Furuichi and Hashimoto 2002). Though this difference is smaller than Wrangham's (1993) estimate of 36 estrus (or pseudo-estrus) days between births for chimpanzees vs. 497 days for bonobos, bonobos still show estrus on more than five times as many days as do chimpanzees.

Such a difference must greatly alter the relationships between males and females. Male chimpanzees compete for a very limited number of estrous females, and alpha males usually have priority of access to them. Although females sometimes perform opportunistic mating or form consortships with lower ranking males, females cannot usually refuse sexual solicitation by alpha males. In some large groups with many males, such as at Ngogo and Kalinzu, opportunistic mating occurs more frequently than in smaller groups but females are sometimes cooperatively herded by a few high-ranking males, and do not usually refuse solicitation by those males. Females are subject to aggressive behaviors during the solicitation of copulations, and female choice for mating partners is very limited (Tutin 1979; de Waal 1982; Hasegawa and Hiraiwa-Hasegawa 1983; Watts 1998; Stumpf and Boesch 2005; Boesch et al. 2006; Hashimoto and Furuichi 2006a, b; Muller et al. 2007).

Among bonobos, in contrast, there are typically many estrous females in a group, and even high-ranking males cannot monopolize all of them. Therefore, males do not frequently fight with each other over access to estrous females, but instead devote greater effort to soliciting females for copulation. Copulation success is therefore largely dependent upon whether females accept such sexual solicitations. Under such circumstances, males rarely attack females, and females can easily ignore even solicitations of the alpha male. Thus, female choice of mating partners is enhanced in bonobos. In addition to reducing sexual coercion, the prolonged estrus of female bonobos may also contribute to paternity confusion, which may help to prevent infanticide (Kano 1992; Wrangham 1993; de Waal 1995; Furuichi 1997; Furuichi and Hashimoto 2002, 2004).

The fourth and final factor discussed here is the relatively high social status of female bonobos. Although there has been dispute regarding whether female bonobos are dominant to males or whether both males and females share equal status, the balance of power between the sexes certainly differs between the two species (Kano 1992; Parish 1996; Furuichi 1997; Vervaecke et al. 2000; Stevens et al. 2007). The high social status of females may contribute to the aggregation of female bonobos in three ways. First is that the contest competition with males over food is not a constraint for females to attend the mixed-sex parties because females have priority of access to food (Kano 1992; Parish 1994; Furuichi 1997; Vervaecke et al. 2000). Second is that females can avoid harassment and infanticide by males because of their high status (Parish 1996; Furuichi 1997; Kano 2001). Third is that they can take the initiative in ranging through the close association with their adult sons. Males usually range with their mothers and frequently exchange grooming with them, and females support their sons in the agonistic interactions between males and help them to attain higher dominance rank (Furuichi 1989, 1997, in press; Kano 1992; Furuichi and Ihobe 1994; Surbeck et al. 2011). Reciprocally, the high degree of aggregation among females may contribute to their high social status through the formation of coalitions (Parish 1996; Furuichi 1997). Furthermore, the social status of females may be enhanced by the extended female choice in mating resulting from the prolonged estrus.

Thus, many factors seem to synergistically contribute to the aggregation of female bonobos. Although we do not yet know how this entire system evolved, Fig. 18.8 may suggest that the evolution of physiological traits concerning prolonged estrus may be one of the key candidates capable of initiating this system. For a better understanding of the evolutionary significance of sexual, behavioral, and sociological divergence between our closest living relatives, we undoubtedly need more information concerning the life history of females. We hope that our long-term studies at Wamba will continue to provide such information.

Acknowledgments All research and conservation activities reported here have been carried out with Drs. Takayoshi Kano, Kohji Kitamura, Akio Mori, Evelyn Ono-Vineberg, Tomoo Enomoto, Naobi Okayasu, Ellen J. Ingmanson, Hiroyuki Takemoto, and other researchers who visited Wamba for short periods. Mr. Ekam Wina, Dr. Mwanza N. Ndunda, Mr. Mikwaya Yamba-Yamba, Mr. Balemba Motema-Salo of the Research Center for Ecology and Forestry, Ministry of Scientific Research, D.R. Congo also carried out research with us. Though we could not include their names as coauthors, we'd like to give special thanks for their great contributions to this work. We also thank Drs. Toshisada Nishida, Juichi Yamagiwa, and Tetsuro Matsuzawa for their continued support of our studies, and members of the Laboratory of Human Evolution Studies and Primate Research Institute of Kyoto University for providing valuable suggestions. This study was mainly supported by the Japan Society for the Promotion of Science (JSPS) Grants-in-aid for Scientific Research (X491036, X501069, X501032, X511074, X511034, X521081, X521033, X00160-304330, X00160-304135, X00160-404337, 59041053, 60043055, 61041071, 62043067, 63041078, 02041049, 06041064, 09041160, 10640613 to Kano; X00160-404120, X00160-504319, 56041018, 57043014, 58041025, 59043022, 60041020, 62041021 to Nishida; 10CE2005 to Takenaka; 12575017, 17255005, 22255007 to Furuichi; 17570193, 19405015 to Hashimoto; 19107007 to Yamagiwa; 21255006 to Ihobe), the National Geographic Fund for Research and Exploration (7511–03 to Furuichi), JSPS Core-to-Core program (15001 to Primate Research Institute), Toyota Foundation (D04-B-285 to Furuichi), JSPS International Training Program (2009–8 to Primate Research Institute), JSPS Asia-Africa Science Platform Program (2009–8 to Furuichi), JSPS Institutional Program for Young Researcher Overseas Visits (2009–37 to Primate Research Institute), Japan Ministry of the Environment Global Environment Research Fund (F-061 to Nishida), Africa Support Fund (2010 to Support for Conservation of Bonobos), and Japan Ministry of the Environment Research and Technology Development Fund (D-1007 to Furuichi).

References

Boesch C (1996) Social grouping in Taï chimpanzees. In: McGrew WC, Marchant LF, Nishida T (eds) Great ape societies. Cambridge University Press, Cambridge, pp 101–113
Boesch C, Boesch-Achermann H (2000) The chimpanzees of the Taï forest: behavioural ecology and evolution. Oxford University Press, New York
Boesch C, Kohou G, Néné H, Vigilant L (2006) Male competition and paternity in wild chimpanzees of the Taï forest. Am J Phys Anthropol 130:103–115
Chapman CA, White FJ, Wrangham RW (1994) Party size in chimpanzees and bonobos. In: Wrangham RW, McGrew WC, de Waal FBM, Heltne PG (eds) Chimpanzee cultures. Harvard University Press, Cambridge, MA, pp 41–57
Chapman CA, Wrangham RW, Chapman LJ (1995) Ecological constraints on group size: an analysis of spider monkey and chimpanzee subgroups. Behav Ecol Sociobiol 36:59–70
de Waal FBM (1982) Chimpanzee politics: power and sex among apes. Jonathan Cape, London
de Waal FBM (1995) Bonobo sex and society. Sci Am 272:82–88
Dupain J, van Krunkelsven E, van Elsacker L, Verheyen RF (2000) Current status of the bonobo (*Pan paniscus*) in the proposed Lomako Reserve (Democratic Republic of Congo). Biol Conserv 94:265–272
Fruth B, Hohmann G (1994) Comparative analyses of nest building behavior in bonobos and chimpanzees. In: Wrangham RW, McGrew WC, de Waal FBM, Heltne PG (eds) Chimpanzee cultures. Harvard University Press, Cambridge, MA, pp 109–128
Furuichi T (1987) Sexual swelling, receptivity, and grouping of wild pygmy chimpanzee females at Wamba, Zaïre. Primates 28:309–318
Furuichi T (1989) Social interactions and the life history of female *Pan paniscus* in Wamba, Zaïre. Int J Primatol 10:173–197
Furuichi T (1997) Agonistic interactions and matrifocal dominance rank of wild bonobos (*Pan paniscus*) at Wamba. Int J Primatol 18:855–875
Furuichi T (2009) Factors underlying party size differences between chimpanzees and bonobos: a review and hypotheses for future study. Primates 50:197–209
Furuichi T (2011) Female contributions to the peaceful nature of bonobo society. Evol Anthropol
Furuichi T, Hashimoto C (2002) Why female bonobos have a lower copulation rate during estrus than chimpanzees. In: Boesch C, Hohmann G, Marchant LF (eds) Behavioural diversity in chimpanzees and bonobos. Cambridge University Press, New York, pp 156–167

18 Long-Term Studies on Wild Bonobos at Wamba, Luo Scientific Reserve

Furuichi T, Hashimoto C (2004) Sex differences in copulation attempts in wild bonobos at Wamba. Primates 45:59–62

Furuichi T, Ihobe H (1994) Variation in male relationships in bonobos and chimpanzees. Behaviour 130:211–228

Furuichi T, Mulavwa MN, Yangozene K, Yamba-Yamba M, Motema-Salo B, Idani G, Ihobe H, Hashimoto C, Tashiro Y, Mwanza NN (2008) Relationships among fruit abundance, ranging rate, and party size and composition of bonobos at Wamba. In: Furuichi T, Thompson J (eds) The bonobos: behavior, ecology, and conservation. Springer, New York, pp 135–149

Gerloff U, Hartung B, Fruth B, Hohmann G, Tautz D (1999) Intracommunity relationships, dispersal pattern and paternity success in a wild living community of bonobos (*Pan paniscus*) determined from DNA analysis of faecal samples. Proc R Soc Lond B 266:1189–1195

Goodall J (1983) Population dynamics during a 15 year period in one community of free-living chimpanzees in the Gombe National Park, Tanzania. Z Tierpsychol 61:1–60

Goodall J (1986) The chimpanzees of Gombe: patterns of behavior. Harvard University Press, Cambridge, MA

Hasegawa T, Hiraiwa-Hasegawa M (1983) Opportunistic and restrictive matings among wild chimpanzees in the Mahale Mountains, Tanzania. J Ethol 1:75–85

Hashimoto C, Furuichi T (2001) Current situation of bonobos in the Luo Reserve, Equateur, Democratic Republic of Congo. In: Galdikas BMF, Erickson Briggs N, Sheeran LK, Shapiro GL, Goodall J (eds) All apes great and small. Vol 1: African apes. Kluwer Academic/Plenum, New York, pp 83–90

Hashimoto C, Furuichi T (2006a) Comparison of behavioral sequence of copulation between chimpanzees and bonobos. Primates 47:51–55

Hashimoto C, Furuichi T (2006b) Frequent copulations by females and high promiscuity in chimpanzees in the Kalinzu Forest, Uganda. In: Newton-Fisher NE, Notman H, Paterson JD, Reynolds V (eds) Primates of western Uganda. Springer, New York, pp 247–257

Hashimoto C, Furuichi T, Takenaka O (1996) Matrilineal kin relationship and social behavior of wild bonobos (*Pan paniscus*): sequencing the D-loop region of mitochondrial DNA. Primates 37:305–318

Hashimoto C, Furuichi T, Tashiro Y (2001) What factors affect the size of chimpanzee parties in the Kalinzu Forest, Uganda? Examination of fruit abundance and number of estrous females. Int J Primatol 22:947–959

Hashimoto C, Tashiro Y, Hibino E, Mulavwa MN, Yangozene K, Furuichi T, Idani G, Takenaka O (2008) Longitudinal structure of a unit-group of bonobos: male philopatry and possible fusion of unit-groups. In: Furuichi T, Thompson J (eds) The bonobos: behavior, ecology, and conservation. Springer, New York, pp 107–119

Hohmann G (2001) Association and social interactions between strangers and residents in bonobos (*Pan paniscus*). Primates 42:91–99

Hohmann G, Fruth B (2003) Lui Kotal – a new site for field research on bonobos in the Salonga National Park. Pan Afr News 10:25–27

Hohmann G, Gerloff U, Tautz D, Fruth B (1999) Social bonds and genetic ties: kinship, association and affiliation in a community of bonobos (*Pan paniscus*). Behaviour 136:1219–1235

Idani G (1990) Relations between unit-groups of bonobos at Wamba, Zaïre: encounters and temporary fusions. Afr Stud Monogr 11:153–186

Kano T (1982) The social group of pygmy chimpanzees (*Pan paniscus*) of Wamba. Primates 23:171–188

Kano T (1986) The social organization of pygmy chimpanzee and common chimpanzee: similarities and differences. Paper presented at International Biological Award Symposium, Kyoto

Kano T (1992) The last ape: pygmy chimpanzee behavior and ecology. Stanford University Press, Stanford, CA

Kano T (2001) Counter strategies against potential infanticide?: reconsideration of social characteristics of *Pan paniscus*. Primate Res 17:223–242 (in Japanese with English abstract)

Kano T, Bongoli L, Idani G, Hashimoto C (1996) The challenge of Wamba. Ecta Animali 8(96):68–74

Kuroda S (1979) Grouping of the pygmy chimpanzees. Primates 20:161–183

Malenky RK, Wrangham RW (1994) A quantitative comparison of terrestrial herbaceous food consumption by *Pan paniscus* in the Lomako Forest, Zaire, and *Pan troglodytes* in the Kibale Forest, Uganda. Am J Primatol 32:1–12

Mulavwa MN, Furuichi T, Yangozene K, Yamba-Yamba M, Motema-Salo B, Idani G, Ihobe H, Hashimoto C, Tashiro Y, Mwanza NN (2008) Seasonal changes in fruit production and party size of bonobos at Wamba. In: Furuichi T, Thompson J (eds) The bonobos: behavior, ecology, and conservation. Springer, New York, pp 121–134

Mulavwa MN, Yangozene K, Yamba-Yamba M, Motema-Salo B, Mwanza NN, Furuichi T (2010) Nest groups of wild bonobos at Wamba: selection of vegetation and tree species and relationships between nest group size and party size. Am J Primatol 72:575–586

Muller MN, Kahlenberg SM, Thompson ME, Wrangham RW (2007) Male coercion and the costs of promiscuous mating for female chimpanzees. Proc R Soc Lond B 274:1009–1014

Nishida T (1968) The social group of wild chimpanzees in the Mahali Mountains. Primates 9:167–224

Nishida T (1972) Preliminary information of the pygmy chimpanzees (*Pan paniscus*) of the Congo basin. Primates 13:415–425

Nishida T (1979) The social structure of chimpanzees of the Mahale Mountains. In: Hamburg DA, McCown ER (eds) The great apes. Benjamin/Cummings, Menlo Park, CA, pp 73–121

Nishida T, Corp N, Hamai M, Hasegawa T, Hiraiwa-Hasegawa M, Hosaka K, Hunt KD, Itoh N, Kawanaka K, Matsumoto-Oda A, Mitani JC, Nakamura M, Norikoshi K, Sakamaki T, Turner L, Uehara S, Zamma K (2003) Demography, female life history, and reproductive profiles among the chimpanzees of Mahale. Am J Primatol 59:99–121

Parish AR (1994) Sex and food control in the uncommon chimpanzee: how bonobo females overcome a phylogenetic legacy of male dominance. Ethol Sociobiol 15:157–179

Parish AR (1996) Female relationships in bonobos (*Pan paniscus*): evidence for bonding, cooperation, and female dominance in a male-philopatric species. Hum Nat 7:61–96

Reynolds V (2005) The chimpanzees of the Budongo Forest: ecology, behaviour, and conservation. Oxford University Press, New York

Savage-Rumbaugh ES, Wilkerson BJ (1978) Socio-sexual behavior in *Pan paniscus* and *Pan troglodytes*: a comparative study. J Hum Evol 7:327–344

Schwarz E (1929) Das Vorkommen des Schimpansen auf dem linken Kongo-Ufer. Rev Zool Bot Afr 16:425–426

Stevens JMG, Vervaecke H, de Vries H, van Elsacker L (2007) Sex differences in the steepness of dominance hierarchies in captive bonobo groups. Int J Primatol 28:1417–1430

Stumpf RM, Boesch C (2005) Does promiscuous mating preclude female choice? Female sexual strategies in chimpanzees (*Pan troglodytes verus*) of the Taï National Park, Côte d'Ivoire. Behav Ecol Sociobiol 57:511–524

Sugiyama Y (1999) Socioecological factors of male chimpanzee migration at Bossou, Guinea. Primates 40:61–68

Sugiyama Y (2004) Demographic parameters and life history of chimpanzees at Bossou, Guinea. Am J Phys Anthropol 124:154–165

Surbeck M, Mundry R, Hohmann G (2011) Mothers matter! Maternal support, dominance status and mating success in male bonobos (*Pan paniscus*). Proc R Soc Lond B 278:590–598

Susman RL (1984) The pygmy chimpanzee: evolutionary biology and behavior. Plenum Press, New York

Suzuki S, Hill DA, Sprague DS (1998) Intertroop transfer and dominance rank structure of nonnatal male Japanese macaques in Yakushima, Japan. Int J Primatol 19:703–722

Takahata Y, Suzuki S, Okayasu N, Hill D (1994) Troop extinction and fusion in wild Japanese macaques of Yakushima Island, Japan. Am J Primatol 33:317–322

Thompson-Handler N, Malenky RK, Badrian N (1984) Sexual behavior of *Pan paniscus* under natural conditions in the Lomako Forest, Equateur, Zaïre. In: Susman RL (ed) The pygmy chimpanzee: evolutionary biology and behavior. Plenum, New York, pp 347–368

Tutin CEG (1979) Mating patterns and reproductive strategies in a community of wild chimpanzees (*Pan troglodytes schweinfurthii*). Behav Ecol Sociobiol 6:29–38

van Elsacker L, Vervaecke H, Verheyen RF (1995) A review of terminology on aggregation patterns in bonobos (*Pan paniscus*). Int J Primatol 16:37–52

Vervaecke H, de Vries H, van Elsacker L (2000) Dominance and its behavioral measures in a captive group of bonobos (*Pan paniscus*). Int J Primatol 21:47–68

Watts DP (1998) Coalitionary mate guarding by male chimpanzees at Ngogo, Kibale National Park, Uganda. Behav Ecol Sociobiol 44:43–55

White FJ (1988) Party composition and dynamics in *Pan paniscus*. Int J Primatol 9:179–193

White FJ, Wrangham RW (1988) Feeding competition and patch size in the chimpanzee species *Pan paniscus* and *Pan troglodytes*. Behaviour 105:148–164

White FJ, Waller MT, Cobden AK, Malone NM (2008) Lomako bonobo population dynamics, habitat productivity, and the question of tool use. Am J Phys Anthropol 135(suppl 46):222

Wrangham RW (1979) Sex differences in chimpanzee dispersion. In: Hamburg DA, McCown ER (eds) The great apes. Benjamin, Cummings, Menlo Park, CA, pp 481–489

Wrangham RW (1986) Ecology and social relationships in two species of chimpanzee. In: Rubenstein DI, Wrangham RW (eds) Ecological aspects of social evolution: birds and mammals. Princeton University Press, Princeton, pp 352–378

Wrangham RW (1993) The evolution of sexuality in chimpanzees and bonobos. Hum Nat 4:47–79

Wrangham RW (2000) Why are male chimpanzees more gregarious than mothers? A scramble competition hypothesis. In: Kappeler PM (ed) Primate males: causes and consequences of variation in group composition. Cambridge University Press, Cambridge, pp 248–258

Wrangham RW, Chapman CA, Clark-Arcadi AP, Isabirye-Basuta G (1996) Social ecology of Kanyawara chimpanzees: implications for understanding the costs of great ape groups. In: McGrew WC, Marchant LF, Nishida T (eds) Great ape societies. Cambridge University Press, Cambridge, pp 45–57

Part VI
Summary

Chapter 19
Long-Term, Individual-Based Field Studies

Tim Clutton-Brock

19.1 Introduction

This book syntheses the result of 18 long-term individual-based studies of primates, of which the shortest has spanned over 10 years and the longest now exceeds 50 years. Its production is timely, for primate studies have now reached a stage where individual-based data extending over several decades are available for a wide range of species spanning all the continents where primates occur and are being used to explore an increasing range of fundamental questions in ecology, evolutionary biology and behaviour (Robbins 2010).

Although some of the longest running individual-based field studies of mammals have involved primates, primate studies are neither the longest running field studies of vertebrates nor the most extensive. The first long-term field studies of vertebrates that could recognise and monitor the life histories of large samples of individuals were of blue tits and great tits in Holland (Kluijver 1951) and Britain (Lack 1954, 1966) and a similar approach was subsequently extended to the studies of other passerine birds (Fig. 19.1). Most early studies of birds followed the lead of Kluijver and Lack and used records of individual life histories to extend research on population dynamics and demography. Individuals were normally recognised from leg rings and were rarely habituated to close observation. The same approach was subsequently extended to the studies of other passerine birds (Grant 1986; Grant et al. 2001; Smith et al. 2006), seabirds (Dunnet et al. 1979), waders (Harris 1970; Ens et al. 1995), waterfowl (Cooke and Rockwell 1988; Scott 1988) and raptors (Newton 1985, 1986).

The first systematic field studies of mammals began in the 1930s, but long-term studies that tracked the behaviour of individuals over part or all of their lifespan were not initiated until three decades later, starting in the late 1950s and early 1960s

---

T. Clutton-Brock (✉)
Department of Zoology, University of Cambridge, Cambridge, UK
e-mail: thcb@cam.ac.uk

P.M. Kappeler and D.P. Watts (eds.), *Long-Term Field Studies of Primates*, 437
DOI 10.1007/978-3-642-22514-7_19, © Springer-Verlag Berlin Heidelberg 2012

Fig. 19.1 Six of the longest running individual-based studies of birds, showing the date when systematic data began to be collected (great tits, Wytham Wood, UK; fulmar petrels, Orkney, Scotland; Bewick's swans, Slimbridge, UK; scrubjays, Florida, USA; acorn woodpeckers, California, US; song sparrows, Mandarte Island, Canada). All six studies continue today

Fig. 19.2 Six of the longest running studies of mammals: chimpanzees, Gombe, Tanzania; *yellow-bellied marmots*, Colorado, USA; African lions, Serengeti, Tanzania; savannah baboons, Amboseli, Kenya; bighorn sheep, Alberta, Canada; *red deer*, Rum, Scotland

(Fig. 19.2). Several of the first studies were of primates (Itani 1958; Kawamura 1958; Schaller 1963; DeVore 1965; van Lawick-Goodall 1968; Kummer 1968) but a similar approach was soon extended to the studies of ungulates and other large herbivores (Geist 1971; Douglas-Hamilton 1973; Clutton-Brock et al. 1982; Festa-Bianchet 1989), carnivores (Kruuk 1972; Schaller 1972) and rodents (Armitage 1975; Hoogland 1995) and later, cetaceans (Connor and Smolker 1985; Mann et al. 2000), bats (Racey 1982; Kerth 2008a, b) and marsupials (Jarman and Southwell 1986).

19 Long-Term, Individual-Based Field Studies

Compared to mammals, many birds have the great advantage that adults can be caught and marked with relative ease, nests and chicks are commonly accessible and many species are diurnal and relatively visible. In addition, many hole-nesting species, like the tits, can be induced to breed in artificial nest sites that can be designed to maximise the ease of observation at the nest. Relatively few wild mammals offer the same range of advantages. Many species are partly nocturnal and are relatively difficult to observe unless individuals have been habituated to the presence of humans. Permanent, visible marks are often difficult to produce or are undesirable for other reasons. In some groups of mammals, it is possible to catch and mark unhabituated individuals repeatedly (including some bats (Kerth 2008a, b), rodents (Schradin 2005), nocturnal lemurs (Kappeler 1997) and pinnipeds (Campagna et al. 1988; Le Boeuf and Reiter 1988)), but many others (including many diurnal primates) have to be habituated to close observation before it is possible to monitor behaviour and life histories of individuals on a regular basis. Habituation is almost always time consuming to establish and costly to maintain and limits the number of individuals and groups that can be sampled, constraining both the range of questions that can be asked and the generality of conclusions. As a result, it is unsurprising that detailed field studies of birds preceded those of mammals and that, even today, mammalian field studies are less advanced than those of birds.

In contrast to early studies of birds, which focussed primarily on population ecology, many of the early individual-based studies of mammals concentrated on describing and investigating the structure and organisation of societies. However, as field studies developed, these contrasts have disappeared. Long-term, individual-based studies of birds and mammals are now commonly used both to explore ecological processes and to investigate the costs and benefits of different phenotypic traits or behavioural strategies. In addition, an increasing number of studies of birds are exploring the structure of social groups and the development of social relationships between individuals (Koenig and Dickinson 2004). Like studies of mammals, some now rely on habituating individuals to close observation and on recognition using natural markings, while others have habituated individuals to close observation to a point where it is possible to weigh them repeatedly. Conversely, several field studies of mammals now monitor multiple groups and several 100 individuals, marking individuals with tags, dye-marks or transponders and individual-based studies are now commonly used to investigate the dynamics and demography of populations as well as the organisation and structure of social groups.

Few people who initiate long-term field studies set out with the objective of working at a particular site on a particular species for a protracted period and most long-term studies start with limited objectives and short time horizons and only develop into lengthy research programmes because of the opportunities they provide. One consequence is that long-term studies commonly develop on a hand-to-mouth basis and little time is spent considering alternative courses or strategic decisions. However, as anyone who has run a long-term field study knows only too well that they are demanding, time-consuming, expensive and commonly absorb a large proportion of the working lives of the people that run them, so it is important to be clear about the reasons for conducting and maintaining them.

In this chapter, I focus on some of the strategic issues affecting long-term studies. In the second section, I examine the benefits of long-term studies, extending several of the arguments outlined by Kappeler et al. in the first chapter. It is also important to recognise the strengths and limitations of particular systems and to adapt research objectives to match them, for if politics is the art of the possible, research is the art of the soluble (Medawar 1967) and there are few prizes for battling unsuccessfully against overwhelming odds. In the third section, I provide a brief (and subjective) evaluation of the particular strengths and some of the limitations of research on primates and argue that future studies of primates will need to recognise both the constraints that working on primates commonly involve and the opportunities they provide. Finally, there are a number of dilemmas that many long-term studies are either facing already or are likely to face in the future. Some of these were discussed at the VII. Freilandtage, ranging from the need to prevent disease transmission to the protection of field staff against bandits, but many of the most fundamental ones were never discussed and some of them have yet to be confronted. In the final section, I briefly review some of these issues.

19.2 The Benefits of Long-Term Studies

Longitudinal studies of animal populations fall into two main groups (Clutton-Brock and Sheldon 2010a). First there are studies at the population level that measure the size, structure and distribution of particular populations but do not depend on individually marked or recognisable animals. Second, there are studies based on data collected from recognisable (or marked) individuals, like all of those represented in this book. Since laboratory-based scientists (including a substantial proportion of referees) often fail to appreciate the important differences between them, it is important to be clear about their differences.

Long-term studies at the population level can measure the size and stability of populations, the relative numbers of animals of different age and sex, the size, composition and movements of groups, the timing of breeding seasons and the proportion of juveniles that survive (Lack 1966; Riney 1982). Where individuals can be aged reliably before or after death, they can investigate age-related changes in fecundity, rearing success and survival (Sinclair 1977). Protracted long-term research is necessary partly because many of the most important ecological and evolutionary processes affecting populations – including the demographic processes controlling animal numbers and the evolutionary processes generating adaptation – commonly occur over multiple years or decades rather than across hours, weeks or months, and partly because the ecological factors affecting animal numbers fluctuate over time and extreme circumstances or events can have long lasting consequences. Studies at the population level provided the base for the first investigations of the regulation of animal numbers (Lack 1966) and have continued to play an important role in research on the demography and dynamics of animal populations and into the environmental factors that affect recruitment, dispersal and survival (Hanski et al. 1991; Sinclair and Arcese 1995; Newton 1998; Lawler et al. 2006). More recently, they have played an important role in documenting the effects of global climate change and investigating their consequences (Crick and Sparks 1999; Cotton 2003; Edwards and Richardson 2004; Langvatn et al. 2004; Lawler et al. 2006; Thackeray et al. 2010).

Although population level studies have an important part to play in ecological research and will continue to be the normal level of investigation for studies of many taxonomic groups, they have important limitations, especially for research into the ecology and behaviour of long-lived iteroparous organisms with overlapping generations (Clutton-Brock and Sheldon 2010a). In particular, they are unable to investigate the extent or consequences of individual differences in reproductive success, which are pronounced in most natural populations and can have important consequences for population dynamics (Clutton-Brock 1988; Newton 1989). In addition, they often have difficulty in distinguishing between the effects of changes in fecundity, survival and dispersal (or immigration) and in identifying the reasons for changes in each of these parameters. Their estimates of age-related changes in reproductive success and survival are often compromised by the shorter life spans of inferior phenotypes, with the result that superior individuals are over-represented in older age groups and the effects of age are underestimated. Although they can assess the distribution of genotypes, they can seldom reconstruct the detailed kinship structure of social groups reliably. Finally, they are not in a position to explore the wide range of social mechanisms operating within groups that affect the distribution of breeding success, emigration and survival among group members.

Long-term studies that are able to monitor the development and life histories of individually recognisable animals are able to investigate many ecological and evolutionary process that are inaccessible to population level studies, as Kappeler et al. described in Chap. 1. In particular, they make it possible to measure individual differences in growth, behaviour and reproductive success throughout the lifespan, generating estimates of the magnitude of individual differences and providing insight into the causes of demographic change (Clutton-Brock 2001; Clutton-Brock and Sheldon 2010a, b). They also permit investigation of relationships between events or decisions at one stage of the lifespan and subsequent growth, reproduction and survival, providing insight into the costs and benefits of different strategies and the evolution of trade-offs between reproduction and survival. They allow comparison of these effects between categories of individuals, based either on phenotype or genotype, so that it is possible to measure the strength of interactions between environmental and genetic factors (Kruuk and Hill 2008). Where they extend over multiple generations, genetic pedigrees can be constructed (Pemberton 2008) and used to measure both the heritability of traits (Kruuk et al. 2000, 2002), the extent to which environmental factors have trans-generational effects (Albon et al. 1987; Clutton-Brock 2004) and the evolution of phenotypic plasticity (Brommer et al. 2005; Nussey et al. 2005, 2007; Clutton-Brock and Sheldon

2010a). Finally, they provide a basis for accurate descriptions of the social environment that individuals live in (including variation in social organisation, mating systems and social structure) and for investigating the causes and consequences of variation in social behaviour (Silk 2007; Clutton-Brock and Sheldon 2010a).

Because of the range of questions that they can investigate, many long-term, individual-based studies come to involve multiple scientists working on different but complementary aspects of biology, providing opportunities to investigate relationships between social parameters and encouraging interdisciplinary research. One practical consequence of the development of multi-faceted research teams working on the same animals is that individual projects can maintain unusually high rates of publication which commonly increase with the duration of the research. To confirm our impression of this effect, Kelly Moyes and I carried out a survey of the output of long-term individual-based field projects based in the UK. Our sample included records for 51 long-term (>10 years) individual-based studies of which the longest had been running for more than 50 years (Fig. 19.3a) and included most of the well-known long-term studies in Britain. Almost all of these studies were productive – but a quarter of them had unusually high rates of publication, producing multiple papers each year and large numbers over the total period they had been running (Fig. 19.3b). Analysis of publication trajectories of the studies in our sample shows that, in general, the longest running studies are the most productive. In the majority of cases, the first publications only appear after studies have been

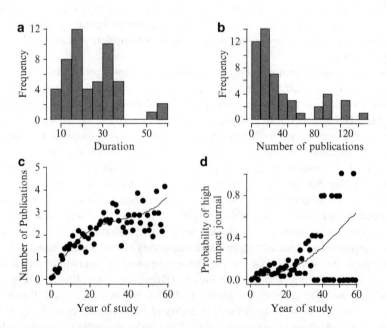

Fig. 19.3 Output of 51 long-term, individual-based studies based in the UK (**a**) frequency of studies of different duration; (**b**) total publications per study; (**c**) publications per year of study; (**d**) probability of producing a paper in *Science, Nature, PNAS* or *Current Biology*

19 Long-Term, Individual-Based Field Studies

running for several years but, subsequently, annual rates of publication increase rapidly – and continue to do so for at least two decades (Fig. 19.3c). In addition, the probability that studies will publish at least one paper per year in a high-impact journal (defined for the purposes of this study as a paper in *Science*, *Nature*, *PNAS* or *Current Biology*) increases with time (Fig. 19.3d), suggesting that the originality of the work increases rather than decline with the duration of studies. We believe that analyses of this kind could play an important role in convincing funding agencies of the role played by long-term field studies in the development of research in ecology, evolutionary biology and animal behaviour and of the importance of maintaining them and that there would be benefits in extending the analysis shown in Fig. 19.3 to include a wider range of studies.

One conclusion to be drawn from surveying the results of long-term field studies is the need for careful choice of species. While long-term data will (almost) always generate broader opportunities and more credible results than short-term ones, species differ in the extent to which they are likely to justify the effort and cost of long-term research. There are substantial benefits to focussing on species where relatively large numbers of individuals and groups can be recognised and monitored; where it is feasible to catch, mark and weigh individuals at regular intervals; where samples of skin, blood, faeces and urine can be collected to establish paternity and monitor hormonal variation; where visibility is good and it is possible to monitor the breeding cycles of individuals throughout the year; where emigrating individuals can be followed to ascertain their fate; where the quality and availability of food supplies can be monitored; where it is feasible and permissible to carry out experiments in natural populations; and where life spans are sufficiently short that it is feasible to track several cohorts throughout their lives (Clutton-Brock and Sheldon 2010a). Few, if any, mammals meet all of these criteria but some meet more than others – and there are some species (including many primates) that meet very few of them. This is not to suggest that only the most tractable species should always be the target of long-term research, for there are other criteria that affect research priorities, but it is important to recognise these limitations and to adjust the objectives of research programmes in relation to what is feasible.

19.3 Primates and Other Animals

The studies in this book provide an opportunity to compare long-term studies of primates with those of other animals and to identify both their strengths and their limitations. In his opening chapter, Kappeler et al. review the history of primate studies and the subsequent chapters provide extensive support for their claims. Perhaps the single greatest achievement of long-term primate studies has been their documentation and investigation of diversity in social behaviour – in particular, of social organisation in different species, of social structure within groups, of social relationships between individuals and of their effects on feeding behaviour development, survival and breeding success (Smuts et al. 1986; Campbell et al. 2007).

In this area, long-term studies of primates that have been able to habituate individuals to close observation have led the field, providing data of greater depth and resolution and field experiments of greater sophistication than are available for any other group of mammals. Comparisons of research on social behaviour in primates with studies of rodents (Wolff and Sherman 2007), bats (Kerth 2010), ungulates (Jarman 1974; Jarman and Kruuk 1996), carnivores (Gittleman 1989; Creel and Creel 2002), cetaceans (Mann et al. 2000) and macropods (Jarman 2000) emphasise how relatively far advanced studies of social behaviour in primates still are. It is no accident that many of the most novel discoveries have come from research on cercopithecine monkeys (and baboons and macaques in particular) where research projects have been running for several decades and individuals can be habituated to close observation (Silk 2007; Alberts and Altmann 2011). Other areas where primate field studies have generated pioneering studies and produced ground breaking results include research on behavioural development; on the structure and content of signals; on food choice and differences in feeding behaviour within and between species; and on ranging behaviour and selective habitat use.

In contrast, the difficulties of studying many primates in their natural habitats have constrained research on other topics. The need for habituation often limits the number of groups that can be sampled and has restricted research on the dynamics of populations and on the effects of population density and of density-independent parameters with the result that demographic research is less developed in primates than in rodents or ungulates (Sinclair and Arcese 1995; Hanski et al. 2001), where more extensive data sets have made it possible to sample larger numbers of individuals and groups, to test predictive models of population size (Coulson et al. 2001) and to explore the effects of changes in global climate (Thackeray et al. 2010; Moyes et al. 2011). The relatively long life spans of most primates have delayed accurate assessments of age-related changes in reproduction and survival, with the result that research on the causes of variation in life-history parameters (including pre-natal effects of environmental variation and of variation in geno-type) are more developed in other animals (Kruuk and Hill 2008). Until the recent development of non-invasive sampling methods, the difficulties of catching and anaesthetising individuals have also impeded research on reproductive physiology and development as well as on the genetic structure of populations. In particular, pedigree-based studies spanning multiple generations than can assess the heritabi-lity of traits, measure the intensity of selection and compare selection coefficients between the sexes or between animals of different ages, which are available for some other vertebrates (Pemberton 2008), have not yet been developed in primates. Finally, with the important exception of research on vocal communication (where research on primates has led research on other mammals), opportunities for experiments and, in particular, for experiments that investigate trade-offs between life-history parameters are usually limited with the result that studies that are able to investigate the causal basis of observed relationships is low.

As the duration of primate field studies increases and sample sizes rise, the range of questions that can usefully be investigated will increase and, in future, we can expect to see primate studies exploring many of the ecological, physiological and evolutionary questions that are currently being investigated in other animals where larger and more complete data are available. In particular, the unusual opportunities for close observation and the detailed knowledge of kinship and social relations that primate studies allow will provide important opportunities to investigate the way in which aspects of the physical and social environments interact to affect breeding success and survival, as well as the extent to which behavioural and developmental strategies are modified to respond to these effects.

19.4 Problems and Dilemmas

Long-term studies of primates face a wide range of problems and dilemmas, some practical, others more conceptual (Clutton-Brock and Sheldon 2010b). Most long-term studies necessarily involve the collection of data by many individuals requiring detailed protocols and regular checks on accuracy and consistency. In addition, the very large data sets they generate require careful organisation and regular updating. Modern methods of data storage and customised databases have an important role to play (see Alberts and Altmann 2012; Wilson 2012) but new dilemmas continue to emerge. For example, the recent tendency for many journals and some granting agencies to require authors to provide public access to all data used in published papers (Whitlock 2011) is likely to present difficulties to many long-term projects. In addition, all long-term projects will eventually face the problem of transferring expertise, data and control between successive principal investigators.

One particular dilemma facing anyone running a long-term field study is the extent of reliance on observational data versus field experiments. It is scarcely necessary to emphasise the need for field experiments to determine the causal basis of relationships and studies of the evolution of behaviour in other animals rely increasingly on experimental studies (Krebs and Davies 1993). However, while some experiments can be performed without jeopardising records of individual life histories, others have inevitable effects on the life histories of experimental animals that may affect other members of their social groups. One way of alleviating this difficulty is to conduct research on two or more separate populations and to restrict experiments to one of them – but this is only possible where habituation is rapid or unnecessary and, more commonly, it is necessary to decide between the benefits of field experiments and their costs to naturalistic records of individual life histories.

Long-term primate studies also face a range of conceptual dilemmas, many of them related to the aims and priorities of primatology itself. Is the principal aim of primate field research is to use the unusual opportunities that primates provide to investigate ecological and evolutionary questions of general importance both to primates and to other animals? Or is it, rather, to extend the breadth of knowledge of the ecology and evolution of primates despite the limitations involved? Or a bit of both? This issue is important because it affects answers to a range of more specific questions about research priorities that primatologists are likely to face both in determining their own research priorities and as referees. For example, what relative priority should be given to extending taxonomic coverage by descriptive studies of previously unstudied species versus to detailed research on more specific questions in species whose behaviour and ecology has already been extensively described? How important are repeated studies of the same species across a range of habitats? And what priority should be given to extending longitudinal studies of particular populations?

There may be no general answers to these questions but there are some obvious guidelines. The highest priority should be given to research that is likely to answer novel questions relevant to understanding the biology of both primates and other animals. The case for initiating the studies of previously unstudied species depends both on their feasibility and on the probability that they will confirm, refute or extend explanations of interspecific differences, so that the highest priority is typically for studies of taxonomically or ecologically divergent species. Repeated studies of the same species in contrasting habitats are important to establish the extent and causes of intraspecific variation but need to cover a sufficient number of populations to demonstrate credible relationships between ecological and behavioural parameters. Finally, a very high priority should be given to longitudinal studies (and especially to long-term studies of the most accessible species) because of the wide range of questions that only long-term, individual-based studies can answer.

References

Alberts SC, Altmann J (2012) The Amboseli Baboon Research Project: forty years of continuity and change. In: Kappeler PM (ed) Long-term field studies of primates. Springer, Heidelberg

Albon SD, Clutton-Brock TH, Guiness FE (1987) Early development and population dynamics in red deer. II. Density-independent effects and cohort variation. J Anim Ecol 56:69–81

Armitage KB (1975) Social behavior and population dynamics of marmots. Oikos 26:341–354

Brommer JE, Merila J, Sheldon BC, Gustafsson L (2005) Natural selection and genetic variation for reproductive reaction norms in a wild bird population. Evolution 59:1362–1371

Campagna C, Le Boeuf BJ, Cappozzo HL (1988) Group raids: a mating strategy of male southern sea lions. Behaviour 105:224–248

Campbell CJ, Fuentes A, MacKinnon KC, Panger M, Bearder SK (2007) Primates in perspective. Oxford University Press, Oxford

Clutton-Brock TH (1988) Reproductive success: studies of individual variation in contrasting breeding systems. University of Chicago Press, Chicago

Clutton-Brock TH (2001) Sociality and population dynamics. In: Press MC, Huntly NJ, Levin S (eds) Ecology: achievement and challenge. Blackwell, Oxford, pp 47–66

Clutton-Brock TH (2004) The causes and consequences of instability. In: Clutton-Brock TH, Pemberton JM (eds) Soay sheep: dynamics and selection in an island population. Cambridge University Press, Cambridge, pp 276–310

Clutton-Brock TH, Sheldon BC (2010a) Individuals and populations: the role of long-term, individual-based studies of animals in ecology and evolutionary biology. Trends Ecol Evol 25:562–573

19 Long-Term, Individual-Based Field Studies

Clutton-Brock TH, Sheldon BC (2010b) The seven ages of *Pan*. Science 327:1207–1208

Clutton-Brock TH, Guinness FE, Albon SD (1982) Red deer: the behavior and ecology of two sexes. Edinburgh University Press, Edinburgh

Connor RC, Smolker RA (1985) Habituated dolphins (*Tursiops* sp.) in western Australia. J Mammal 66:398–400

Cooke F, Rockwell RF (1988) Reproductive success in a lesser snow goose population. In: Clutton-Brock TH (ed) Reproductive success: studies of individual variation in contrasting breeding systems. University of Chicago Press, Chicago, pp 237–250

Cotton PA (2003) Avian migration phenology and global climate change. Proc Natl Acad Sci USA 100:1219–1222

Coulson T, Catchpole EA, Albon SD, Morgan BJT, Pemberton JM, Clutton-Brock TH, Crawley MJ, Grenfell BT (2001) Age, sex, density, winter weather, and population crashes in Soay sheep. Science 292:1528–1531

Creel S, Creel NM (2002) The African wild dog: behavior, ecology, and conservation. Princeton University Press, Princeton

Crick HQP, Sparks TH (1999) Climate change related to egg-laying trends. Nature 399:423–424

DeVore I (1965) Primate behavior: field studies of monkeys and apes. Holt Rinehart and Winston, New York

Douglas-Hamilton I (1973) On the ecology and behavior of the Lake Manyara elephants. East Afr Wildlife J 11:401–403

Dunnet GM, Ollason JC, Anderson A (1979) A 28-year study of breeding fulmars *Fulmarus glacialis* (L.) in Orkney. Ibis 121:293–300

Edwards M, Richardson AJ (2004) Impact of climate change on marine pelagic phenology and trophic mismatch. Nature 430:881–884

Ens BJ, Weissing FJ, Drent RH (1995) The despotic distribution and deferred maturity: two sides of the same coin. Am Nat 146:625–650

Festa-Bianchet M (1989) Individual differences, parasites, and the costs of reproduction for bighorn ewes (*Ovis canadensis*). J Anim Ecol 58:785–795

Geist V (1971) Mountain sheep: a study in behavior and evolution. University of Chicago Press, Chicago

Gittleman JL (1989) Carnivore group living: comparative trends. In: Gittleman JL (ed) Carnivore behavior, ecology, and evolution, vol 1. Cornell University Press, Ithaca, pp 183–207

Grant PR (1986) Ecology and evolution of Darwin's finches. Princeton University Press, Princeton

Grant PR, Grant BR, Petren K (2001) A population founded by a single pair of individuals: establishment, expansion and evolution. Genetica 112–113:359–382

Hanski I, Hansson L, Henttonen H (1991) Specialist predators, generalist predators, and the microtine rodent cycle. J Anim Ecol 60:353–367

Hanski I, Henttonen H, Korpimäki E, Oksanen L, Turchin P (2001) Small-rodent dynamics and predation. Ecology 82:1505–1520

Harris MP (1970) Territory limiting the size of the breeding population of the oystercatcher (*Haematopus ostralegus*) – a removal experiment. J Anim Ecol 39:707–713

Hoogland JL (1995) The black-tailed prairie dog: social life of a burrowing mammal. University of Chicago Press, Chicago

Itani J (1958) On the acquisition and propagation of a new food habit in the natural group of the Japanese monkey at Takasaki-Yama. Primates 1:84–98 [in Japanese with English summary]

Jarman PJ (1974) The social organisation of antelope in relation to their ecology. Behaviour 48:215–267

Jarman PJ (2000) Males in macropod society. In: Kappeler PM (ed) Primate males: causes and consequences of variation in group composition. Cambridge University Press, Cambridge, pp 21–33

Jarman PJ, Kruuk H (1996) Phylogeny and spatial organisation in mammals. In: Croft DB, Ganslosser U (eds) Comparison of marsupial and placental behaviour. Filander Verlag, Fürth, pp 80–101

Jarman PJ, Southwell C (1986) Grouping, associations and reproductive strategies in eastern grey kangaroos. In: Rubenstein DR, Wrangham RW (eds) Ecological aspects of social evolution. Princeton University Press, Princeton, pp 399–428

Kappeler PM (1997) Determinants of primate social organization: comparative evidence and new insights from Malagasy lemurs. Biol Rev 72:111–151

Kawamura S (1958) The Matriarchal social order in the minoo-B group – a study on the rank system of Japanese macaque. Primates 1:149–156 [in Japanese with English summary]

Kerth G (2008a) Animal sociality: bat colonies are founded by relatives. Curr Biol 18:R740–R742

Kerth G (2008b) Causes and consequences of sociality in bats. BioScience 58:737–746

Kerth G (2010) Group decision-making in animal societies. In: Kappeler PM (ed) Animal behaviour: evolution and mechanisms. Springer, Heidelberg, pp 241–265

Kluijver HN (1951) The population ecology of the great tit *Parus m. major* L. Ardea 39:1–135

Koenig WD, Dickinson JL (2004) Ecology and evolution of cooperative breeding in birds. Cambridge University Press, Cambridge

Krebs JR, Davies NB (1993) An introduction to behavioural ecology, 3rd edn. Blackwell, Oxford

Kruuk H (1972) The spotted hyaena: a study of predation and social behavior. University of Chicago Press, Chicago

Kruuk LEB, Hill WG (2008) Introduction. Evolutionary dynamics of wild populations: the use of long-term pedigree data. Proc R Soc Lond B 275:593–596

Kruuk LEB, Clutton-Brock TH, Slate J, Pemberton JM, Brotherstone S, Guinness FE (2000) Heritability of fitness in a wild mammal population. Proc Natl Acad Sci USA 97:698–703

Kruuk LEB, Slate J, Pemberton JM, Brotherstone S, Guinness FE, Clutton-Brock TH (2002) Antler size in red deer: heritability and selection but no evolution. Evolution 56:1683–1695

Kummer H (1968) Social organization of hamadryas baboons: a field study. University of Chicago Press, Chicago

Lack D (1954) The natural regulation of animal numbers. Clarendon, Oxford

Lack D (1966) Population studies of birds. Oxford University Press, Oxford

Langvatn R, Mysterud A, Stenseth NC, Yoccoz NG (2004) Timing and synchrony of ovulation in red deer constrained by short northern summers. Am Nat 163:763–772

Lawler JJ, White D, Neilson RP, Blaustein AR (2006) Predicting climate-induced range shifts: model differences and model reliability. Global Change Biol 12:1568–1584

Le Boeuf BJ, Reiter J (1988) Lifetime reproductive success in northern elephant seals. In: Clutton-Brock TH (ed) Reproductive success: studies of individual variation in contrasting breeding systems. University of Chicago Press, Chicago, pp 344–362

Mann J, Connor RC, Tyack PL, Whitehead H (2000) Cetacean societies: field studies of dolphins and whales. University of Chicago Press, Chicago

Medawar PB (1967) The art of the soluble. Methuen, London

Moyes K, Nussey DH, Clements MN, Guiness FE, Morris A, Morris S, Pemberton JM, Kruuk LEB, Clutton-Brock TH (2011) Advancing breeding phenology in response to environmental change in a wild red deer population. Global Change Biol 17:2455–2469

Newton I (1985) Lifetime reproductive output of female sparrowhawks. J Anim Ecol 54:241–253

Newton I (1986) The sparrowhawk. T. and A. D. Poyser, Calton, UK

Newton I (1989) Lifetime reproduction in birds. Academic, London

Newton I (1998) Population limitation in birds. Academic, London

Nussey DH, Clutton-Brock TH, Elston DA, Albon SD, Kruuk LEB (2005) Phenotypic plasticity in a maternal trait in red deer. J Anim Ecol 74:387–396

Nussey DH, Wilson AJ, Brommer JE (2007) The evolutionary ecology of individual phenotypic plasticity in wild populations. J Evol Biol 20:831–844

Pemberton JM (2008) Wild pedigrees: the way forward. Proc R Soc Lond B 275:613–621

Racey PA (1982) Ecology of bat reproduction. In: Kunz TH (ed) Ecology of bats. Plenum Press, New York, pp 57–104

Riney T (1982) Study and management of large mammals. Wiley, New York

Robbins MM (2010) Long-term field studies of primates: considering the past, present, and future. Evol Anthropol 19:87–88

Schaller GB (1963) The mountain gorilla: ecology and behavior. University of Chicago Press, Chicago

Schaller GB (1972) The Serengeti lion: a study of predator-prey relations. University of Chicago Press, Chicago

Schradin C (2005) When to live alone and when to live in groups: ecological determinants of sociality in the African striped mouse (*Rhabdomys pumilio*, Sparrman, 1784). Belg J Zool 135(suppl):77–82

Scott DK (1988) Reproductive success in Bewick's swans. In: Clutton-Brock TH (ed) Reproductive success: studies of individual variation in contrasting breeding systems. University of Chicago Press, Chicago, pp 220–236

Silk JB (2007) The adaptive value of sociality in mammalian groups. Philos Trans R Soc Lond B 362:539–559

Sinclair ARE (1977) The African buffalo: a study of resource limitation of populations. University of Chicago Press, Chicago

Sinclair ARE, Arcese P (1995) Serengeti II: dynamics, management and conservation of an ecosystem. University of Chicago Press, Chicago

Smith JNM, Marr AB, Hochachka WM (2006) Life history: patterns of reproduction and survival. In: Smith JNM, Keller LF, Marr AB, Arcese P (eds) Conservation and biology of small populations: the song sparrows of Mandarte Island. Oxford University Press, New York, pp 31–41

Smuts BB, Cheney DL, Seyfarth RM, Wrangham RW, Struhsaker TT (1986) Primate societies. University of Chicago Press, Chicago

Thackeray SJ, Sparks TH, Frederiksen M, Burthe S, Bacon PJ, Bell JR, Botham MS, Brereton TM, Bright PW, Carvalho L, Clutton-Brock TH, Dawson A, Edwards M, Elliott JM, Harrington R, Johns D, Jones IS, Jones JT, Leech DI, Roy DB, Scott WA, Smith M, Smithers RJ, Winfield IJ, Wanless S (2010) Trophic level asynchrony in rates of phenological change for marine, freshwater and terrestrial environments. Glob Change Biol 16:3304–3313

van Lawick-Goodall J (1968) The behaviour of free-living chimpanzes in the Gombe Stream Reserve. Anim Behav Monogr 1:161–311

Whitlock MC (2011) Data archiving in ecology and evolution: best practices. Trends Ecol Evol 26:61–65

Wilson ML (2012) Long-term studies of the chimpanzees of Gombe National Park, Tanzania. In: Kappeler PM (ed) Long-term field studies of primates. Springer, Heidelberg

Wolff JO, Sherman PW (2007) Rodent societies: an ecological and evolutionary perspective. University of Chicago Press, Chicago

Erratum to:

Chapter 4
Long-Term Lemur Research at Centre Valbio, Ranomafana National Park, Madagascar

**Patricia C. Wright, Elizabeth M. Erhart, Stacey Tecot,
Andrea L. Baden, Summer J. Arrigo-Nelson, James Herrera,
Toni Lyn Morelli, Marina B. Blanco, Anja Deppe, Sylvia Atsalis,
Steig Johnson, Felix Ratelolahy, Chia Tan, and Sarah Zohdy**

P.M. Kappeler and D.P. Watts (eds.), *Long-Term Field Studies of Primates*,
DOI 10.1007/978-3-642-22514-7_4, © Springer-Verlag Berlin Heidelberg 2012

In this chapter, the following sentences are incorrect:

Section 4.3.4 Reproductive Hormones, 2nd paragraph:

While DHT levels were higher in males than in females, there was no significant sex difference in testosterone levels. Similar testosterone results were found in *M. rufus*.

Section 4.4.4 Habitat Disturbance, 4th paragraph:

In a comparative study of stress hormones in adult *E. rubriventer* in selectively logged versus minimally logged sites, patterns of cortisol excretion were similar in both sites, but those in the undisturbed site showed little response to variation in food availability and rainfall. In contrast, at the disturbed site, fecal cortisol levels were significantly higher when fruit was scarce (parturition and early lactation) compared with when fruit was abundant (prebreeding season).

The corrected versions are:

Section 4.3.4 Reproductive Hormones, 2nd paragraph:

While DHT levels were higher in males than in females, the relationship between sex and testosterone level varied across seasons, with each sex excreting higher levels at different times (Tecot et al. in prep.). Similar testosterone results were found in *M. rufus* (Zohdy et al. 2010).

P.M. Kappeler and D.P. Watts (eds.), *Long-Term Field Studies of Primates*,
DOI 10.1007/978-3-642-22514-7_20, © Springer-Verlag Berlin Heidelberg 2012

Section 4.4.4 Habitat Disturbance, 4th paragraph:

In a comparative study of stress hormones in adult *E. rubriventer* in selectively logged versus minimally logged sites, patterns of cortisol excretion were similar in both sites, but those in the disturbed site showed little response to variation in food availability and rainfall. In contrast, at the undisturbed site, fecal cortisol levels were significantly higher when fruit was scarce (parturition and early lactation) compared with when fruit was abundant (prebreeding season).

The online version of the original chapter can be found under
DOI 10.1007/978-3-642-22514-7_4

Erratum to:

Chapter 13
The 30-Year Blues: What We Know and Don't Know About Life History, Group Size, and Group Fission of Blue Monkeys in the Kakamega Forest, Kenya

Marina Cords

P.M. Kappeler and D.P. Watts (eds.), *Long-Term Field Studies of Primates*,
DOI 10.1007/978-3-642-22514-7_13, © Springer-Verlag Berlin Heidelberg 2012

In this chapter, figure 13.5 is incorrect.

The corrected version is:

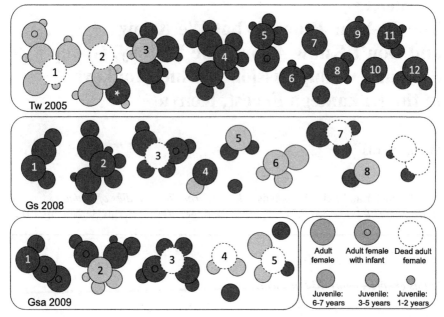

Fig. 13.5 Maternal kinship, rank, and group fission. For each of three fissions, individuals are represented as *circles*, with *shading* (*light vs. dark grey*) indicating group membership after fission, and size proportional to age (see legend). To indicate maternal relatedness, a large circle (mother) overlaps the circles representing her offspring. Matriline rank decreases from left to right, indicated by integer values (1 = highest). Matriline rank was derived from dyadic agonistic interactions among adult females, 9–12 months before fission; matrilines represented only by juveniles therefore have no rank and are randomly placed. Individual marked by *asterisk* is discussed in text

The online version of the original chapter can be found under
DOI 10.1007/978-3-642-22514-7_13

Index

A
Accipiter henstii (Henstii's goshawk), 84
Activity pattern
  cathemeral, 72
  diurnal, 72
  nocturnal, 72
Adult sifaka, 52
Affiliation, 28–30, 33
Africa, 49
Agarwood, 240, 241
Age at first birth, 111
Age at first reproduction, 252, 297, 298
  for each sex, 153
Aggression, 28–33, 37, 75–77, 81, 84–85,
    300, 302, 304, 305
  rates, 55
Aggressive, 58
Agonistic interactions, 266
Aloewood, 240
Alouatta palliate, 166
Alpha female, 28, 29
  predation risk, 194
  recruitment, 194
Alpha male tenures, 150
Alpha status
  challenges, 196, 198, 199, 205
  wounded, 205
Alternative strategy, 244
Altmann, S., 261–281
Alto's Group, 263, 264
Amboseli baboon population, 263, 278
Amboseli Baboon Research Project (ABRP),
    261–281
Amboseli National Park, 262
Amphibians, 51

Androgens, 78
Anecdotal reports, 241
ANGAP.*See* Association National pour le
    Géstion des Aires Protegées (ANGAP)
Anjaranantenaina, S., 26
Ankoba, 23, 25, 26, 33, 35
Anthropogenic change
  cattle, 56
  dogs, 56
  domestic animals, 56
Anthropogenic disturbance, 75, 80, 88
Antipredator behavior, 83
Apes, 245, 251, 253
Aquilaria, 241
  A. crassna, 240
A-record, 361
Assamese macaques, 7–8
Association National pour le Géstion des Aires
    Protegées (ANGAP), 69
Ateles geoffroyi, 166
Attraction of immigrant males, 209
Avahi laniger (*A. peyrierasi*), 72
Average lifespan, 252
Aye–aye *(Daubentonia madagascariensis)*, 71

B
BABASE, 264, 271
Baboon-proof, 364
Baboons, 206, 250
Bamboo lemur
  golden *(Hapalemur aureus)*, 69, 71
  greater *(Prolemur simus)*, 68, 69, 71
  grey gentle *(Hapalemur griseus)*, 71
Banana feeding, 359, 361

452   Index

Bard, K.A., 28
Bat, 26
Bealoka, 23, 27
Bealoka Reserve, 22, 23
Behavioral development, 7
Behavioral flexibility, 245
Behavioral plasticity, 125–136
Behavioral sampling methods, 267
Behavioral variation, 8
Berenty Private Reserve, 68
Best-male hypothesis, 250
Beza Mahafaly, 26, 30, 45–62
Beza Mahafaly Lemur Biology Project, 56
Beza Mahafaly Reserve, 46–52
Beza Mahafaly Special Reserve, 68–69, 103
Biodiversity, 68–70, 89, 90, 239
Biomass, 88
Birds, 27, 51
Birth(s), 107–112, 114, 116, 117
    rate, 297, 305, 306
    records, 252
    season, 46, 54, 58, 59, 62, 79
    seasonality, 153–154, 172, 173
Black and white ruffed lemurs *(Varecia
    variegata)*, 71
Blood parasites, 74
Blumenfeld-Jones, K.C., 23, 25, 27, 36
Boa constrictor *(Sanzinia madagascariensis)*, 83
Body mass, 74, 79, 81, 87, 253, 297
Bonobos, 5, 413–429
    cohesive grouping, 423–424
    life history, 420–429
    population structure, 417–420
    social status, 429
Brazil, 126, 128
B-record, 362–364, 375
Breeding season, 77, 79, 86
Brown mouse lemur *(Microcebus rufus)*, 71,
    77, 80
Budnitz, N., 23, 27

C
Camera traps, 75
Canines, 241
Captive populations, 7
Capture, 74, 75
Capuchin, 206
Cavigelli, S.A., 29
Cebus
    C. capucinus, 166, 173
    C. nigritus, 172
    female philopatry, 169

    females also disperse, 169
    Lomas Barbudal, 173
    male dispersal, 169
Census data, 53–55
Censuses, 52, 54
Centre ValBio (International Centre pourle
    Valorization de Biodiversity), 69
Cercopithecus, 402
    C. mitis, 297
Charles-Dominique, P., 26
Charrier, A., 26
Cheirogaleus major *(Cheirogaleus crossleyi)*,
    72, 83
Chimpanzees, 5, 9, 250, 313–332, 339–352,
    357–378
Chrysophyllum albidum, 318, 331
Civet, 26
Claire de Heaulme Foulon, 23
Climate, 24, 31, 32, 36, 37, 70, 74–75, 80
Climate change, 68, 90
Climatic fluctuations, 10
Cognition, 70, 241
Collaboration, 128, 136
Colobines, 215–218, 223, 231, 232
Co-migration, 147
Communication, 26, 28, 30, 70, 77, 84, 85
Community based-model, 254
Community ecology, 83–89
Community model, 241, 254
Community size, 322–324
Comparative field studies, 136
Competition, 68, 75, 84, 85, 89, 302–305, 307
Competitive regimes, 216
Conception, 245, 248, 253, 254
Confuse paternity, 245, 249, 250
Congeneric, 89
Conservation, 6, 10, 13, 14, 68–70, 77, 89–91,
    126–129
    clubs, 70
    community, 417
    project, 387
Contest, 85
Copulations, 248, 249, 251, 253
Copulatory plugs, 30
Co-resident, 111, 112
Cortisol, 88, 90
Cortisol-Fitness Hypothesis, 88
Costa, 141, 142, 145, 146, 155, 156
Costa Rican
    ACG, 167, 168, 178, 181
    Guanacaste, 166, 167, 182
    Santa Rosa, 165–182
    white-faced, 166, 169, 170, 172–175

Costa Rica, population size, 171, 175–177
Courtship, 348, 349
Critically endangered species (IUCN), 68, 72, 88
Crocodile, 51
Crowley, H.M., 22, 26, 27
Cryptoprocta ferox (fossa), 26, 84
CS7, 107
Cultural behavior, 347–351
Cyperus latifolius, 390

D
Dainis, K., 23, 35
Dapporto, L., 28
Database management, 75
Data collection, 51–54, 61
  methods, 144–145, 361–364
Data deficient species (IUCN), 68, 72
Data management, 12, 13
Daubentonia madagascariensis, 71, 72, 83, 88
Decline of the fever tree woodlands, 262–263
Deforestation, 403
de Heaulme, 22, 23, 36
Demographic, 127, 129, 130, 134–136
  changes, 342–345
  cycles, alpha position, 209
  data, 60
  structures, 116
Demography, 47, 55, 58–61, 70, 87, 88, 90, 91, 127, 129–135, 146–154, 238, 239
Dental ecology, 56–57
Dental tooth casts, 74
DFPis emigrated, 109
Diet, 265
Disease(s), 11, 27
  ecology, 375–376
Dispersal, 70, 77, 78, 91, 125, 127, 131–136, 244, 253, 271, 272, 278, 279
  age at death, 170
  females are more likely to, 174
  infant, 170, 174
  males, 169, 174, 179
  natal, 169, 172–174
  observational studies, 135
  patterns, 146–149
  Santa Rosa, 169–175
Dispersing, 174
Dolins, F., 29
Dominance, 7, 76, 78
  rank, 28, 304
Dong Phayayen–Khao Yai Forest Complex (DPKY), 239

Drought, 24, 27, 36, 46, 55, 59, 61
Dry deciduous forest, 48
Dry season, 90
Dubovick, 29
Duke University, 69
Dwarf lemurs, 71
Dynamic community model, 254

E
Eaglewood, 240
Ecological crises, 9
Ecology, 241
Ectoparasites, 80
Effective population size, 264
Egalitarian
  population dynamics, 125–136
  population's density, 129, 131, 133–135
E group, 415
Elanga, 415
El Niño years, 24, 36, 90
Elongated vulva, 252
Emigration, 294
Encroachment, 240
Endangered species (IUCN), 68, 72, 88
ENSO (southern oscillation), 90, 91
EPCs, 251, 254
Estradiol, 78
Estrous synchrony, 58, 78
Ethogram, 346–347, 352
Eulemur, 22, 26, 36
Eulemur fulvus rufus (Eulemur rufifrons) (red-fronted brown lemur), 71
Eulemur rubriventer (red bellied lemur), 71
Evolutionary forces, 238, 245
Exaggerated sexual swellings, 250
Exhibits, 116
Exotic species, 22, 26
Extract DNA from feces, 271, 274
Extract metabolites of steroid hormones, 271
Extra-pair copulations, 77, 87, 237

F
Faecal progestogen, 250
Fallback strategies, 402, 403
Fauna of Beza Mahafaly, 51
Fecal glucocorticoid, 58
Fecal samples, 143, 145
Fecundity, 296
Feeding, 70, 79, 83, 87, 348, 350–351
  ecology, 322–324
  patterns, 398

454   Index

Female, 74, 76–79, 81, 85–87,
    90, 91
  aggregation, 424–429
  defense polygyny, 217, 218, 231
  dispersal, 216, 218, 224, 231
  dominance, 27, 28, 31–32, 37, 56
  dominance rank, 268, 270
  life history, 413–429
  mate choice, 245
  reproductive ecology, 324–325
  reproductive strategies, 245–251
  take-overs, 254
  transfer, 108–110, 117, 118
Female-female, 78, 79
Fertile periods, 245, 250, 254
Fertile phase, 248
Fertility, 54, 59–61
Fetal losses, 272
Fichtel, C., 26
Field experimentation, 5
Field manual, 222
Firearms, 241
First conception, 253, 254
Fission, 143, 146–148, 155–157,
    289–308
  group, 196, 198–203, 206–209
Fission–fusion social systems, 423
Fission–fusion societies, 372
Fitness determinants, 7
Flexibility of behaviors, 76
Flexible mating behavior, 251
Floating, 244
Flora
  Euphorbiaceae, 50
  Mimosaceae, 50
Flora of Beza Mahafaly, 48–50
Flowers, 83
Folivore paradox, 216
Follicular, 252
Food availability, 59
Food defense hypothesis, 331
Foraging, 265
  strategies, 385–406
  techniques, 156–160
Forest, 101–118
  composition, 220
  history, 289–308
  plantations, 290, 291
Fossa, 26, 79, 84
Fossil primates, 57
Fruit, 74, 76, 82, 83, 85–89
Fund, 46
Fusion, 199–203, 207, 299, 300

G
Galidia elegans (ring-tailed mongoose), 84
Gallery forest, 46, 48, 50
Ganyamulume groups, 389, 390, 392
Garber, P.A., 28
Gene flow, 263–264
Genetic relatedness, 60
Genetic structure, 59, 60, 149–151
Génin, 26
Gestation, 59, 90, 170, 172, 223, 253
Global positioning system (GPS), 365
Golden bamboo lemur, 69, 71
Gombe
  ecosystem, 360
  National Park, Tanzania, 357–378
  research infrastructure, 364
  Stream Research Centre, 358–365
Gombe West, 362
Goodman, S.M., 26, 27
  paradise flycatcher, 27
Gorilla beringei graueri, 386
Gorilla Organization, 404
Gorilla tourism, 387
Gould, L., 25, 28, 30, 31
Graded signal hypothesis, 250, 251
Grappling, 249
Grauer's gorillas, 385–406
Great apes, 251
Greater bamboo lemur, 69, 71
Grey gentle bamboo lemur, 71
Grooming, 348–351
Grooming (self-grooming), 87
Group dynamics, 226–232, 245
Group
  extinctions, 108
  fission, 29, 289–308
  fusion, 228, 229
  life cycle, 185–210, 2300–232
  living, 102, 117, 299
    costs and benefits, 299
  sizes, 29, 108, 117, 251, 289–308
    and composition, 76
  travel, 302
Growth, 108, 114–115
Guenon, 289, 292, 297–299, 307

H
Habitat
  change, 292
  disturbance, 56, 79, 80, 87–88, 91
  saturation, 131–133
Habituation, 5, 11, 12, 221–222, 224, 226, 227,
    290, 293, 294

Index

Hamadryas baboons, 5
Hanuman langurs, 5
Hapalemur aureus.See Golden bamboo lemur
Hapalemur griseus.See Grey gentle bamboo
  lemur
Hapalemur simus (Prolemur simus).See
  Greater bamboo lemur
Harassment hypothesis, 329
Harrier, 102, 113, 116
Health, 70, 74, 79–80, 89
  and disease ecology, 56
Henst's goshawk *(Accipiter henstii)*, 84
Hepatocystis, 280
Hibernation, 70
History strategy, 48, 62
Hladik, C.M., 26
Hole-nesting species, 439
Home ranges, 53, 54, 60
Hood, L.C., 29, 31
Hook's Group, 263, 264
Hormonal responses, 58, 62
Hormonal studies, 78
Hormone levels, 58, 62
Hostile takeovers, 77
Human population pressure,
  403–406
Hunting, 240, 292, 315, 320, 321, 328–332
Hybrids, 278
Hygienic baheviors, 349–350
Hylobates lar, 219
Hylobatidae, 237

I
Ibid, 9
Ichino, S., 29–31, 34, 36
Identification collar, 54
Illegal logging, 218
Imbalance of power hypothesis, 330, 375
Immigrate, 244
Immigrating, 198, 205
Immigration, 76, 77, 81, 89
Immobilize baboons, 279
Inbreeding, 144, 150, 160
  avoidance, 110, 116
Incest, 209
Incest avoidance
  mortality, 204
  solitary living, 205
Incestuous mating, 242
Individual-based field studies, 437–446
Individual-centric, 238
Individual identification, 5, 12, 53–54
Individual recognition, 293

Infanticides, 31, 46, 58, 62, 80–81, 109, 113,
  116, 152, 153, 186, 190, 200–202, 204,
  207–209, 216, 217, 229, 245, 249, 251,
  393–396
Infants, 28–31, 34
  mortality, 80, 88, 253
  survival, 305, 306
Infrastructure, 141–146
In press, 266, 273, 274
Insects, 51
Institute for the Conservation of Tropical
  Environments (ICTE), 69–70, 75
Intellectual reservoirs, 238
Interahamwe, 389
Interbirth intervals (IBI), 111, 116, 152, 170,
  172, 176, 221, 223, 249, 253, 294–298,
  305, 367–368
  males and females regularly, 177
Intercommunity
  aggression, 320, 321, 366, 375
  violence, 372
Interference, 84, 85
Intergroup
  aggression, 330–331, 373–375
  dynamics, 372–375
  encounters, 244, 299, 307
  transfer, 420–423
International, 128, 136
  collaborations, 129
Inter-observer reliability, 145
Interspecific, 85
  aggression, 84–85
Intestinal parasites, 80
Introduced species, 22, 26, 33–37
Intruders, 248
Isotope ecology, 57

J
Japanese macaques, 5, 8
Jeanne, 262, 265, 267, 268, 272, 276
Jeanne's study of baboon mothers and
  infants, 268
Jolly, A., 21–37
Jones, K.C., 30
Jump-fights, 30
Juvenile mortality, 253
Juvenile period, 253

K
Kaboko groups, 389, 390
Kahama community, 331, 361, 369, 372, 373

456 Index

Kahuzi-Biega National Park (KBNP), 385–406
Kahuzi gorillas, life history, 391–396
Kanyawara–Ngogo comparisons, 315, 332
Kappeler, P.M., 3–14, 26, 31, 33
Karpanty, S.M., 27
Kasekela community, 361, 362, 368, 370, 372–374
Kenyan field team, 264
Khao Yai, 237–254
Khao Yai National Park, 238, 240, 252
Kibale National Park, 313–332
Kinship, 187, 194, 209, 300, 301, 304
  structure, 441
Kirindy, 26, 101–118
Kirindy Forest, 68
Klopfer, P.H., 30
Koechlin, J., 29
Koyama, N., 27–29, 31, 34, 37

L

Lactating females, 249, 250
Lactation, 31, 32, 59, 88, 90, 251
Lam Takhong river, 239
Lane, L., 28
Law enforcement, 240
Lemur catta, 22, 25, 46, 51–53
Lemurs, 5
Lepilemur leucopus, 26, 51
Lepilemur microdon (Lepilemur mustilinus), 71, 72, 83, 88
Lethal aggression, 76
Leucaena leucocephala, 22, 33–34
Lewis, R.J., 31
Life cycle of Phayre's leaf monkey groups, 231
Life cycles, 186, 189, 207, 209, 210
Life history, 4, 6, 7, 9–11, 46, 48, 51, 52, 54, 55, 57–62, 70, 87, 91, 125, 129–135, 237–254, 289–308, 324–325, 345–346, 351, 365–369
  data, 51, 54
  population expansion, 129, 130
  slow, 294, 308
  speed, 298
  strategies, 127
Lifespan, 78, 81–82, 91, 252, 295, 296, 298
Life table, 117
Lifetime reproductive success, 78, 79
Lizards, 51
Lodge Group, 263, 275
Longevity, 6, 9, 102, 103, 114
Loud songs, 254
Luo Scientific Reserve, 413–429
Luteal, 252

M

Macaca
  M. assamensis, 219
  M. fuscata, 422
  M. leonine, 219
  M. mulatta, 219
Macaques, 206, 250
Madagascar
  cattle, 48
  domestic animals, 56
  flora, 48–50
  goats, 48
  Tamarindus indica, 48, 57
Madagascar boa *(Sanzinia madagascariensis)*, 83, 84
Madagascar National Parks (MNP), 69, 70
Maeshe groups, 387, 394
Mahafaly Special Reserve, 103
Mahale, 339–352
Mai hom, 240
Malagasy fauna, 51, 59
Malaza, 23, 25, 26, 33, 35
Male
  baboon, 270, 275, 276, 278
  change, 248, 249
  dispersal, 58, 169, 174, 179, 227, 231, 232
  dominance rank, 269, 273–275
  fitness, 60, 62
  influxes, 307
  life expectancy, 205
  maturation, 270
  politics, 329
  reproductive, 60
  residence, 215–232
  takeovers, 216, 218, 227, 249, 306, 307
    Piliocolobus tephrosceles, 228
  transfers, 30, 58, 77
Male–male competition, 245, 250, 254, 314
Male–male social relationship, 326
Male-philopatric species, 413–429
Mandrare River, 22, 27, 36
Mangevo field site, 70, 81, 86
Mantled howler monkeys, 5
Many-male hypothesis, 250
Map of Ranomafana National Park, 73
Masataka, N., 28
Mate choice, 58, 60
Mate polyandrously, 250, 254
Maternal dominance rank, 270, 271, 280
Mating, 27, 28, 30, 32, 103, 111, 113, 118, 242, 245, 248, 249, 251
  activity, 248, 251
  behavior, 269, 272, 275

competition, 58, 60
season, 55, 59, 60, 62
strategies, 321, 327, 332
systems, 216, 217, 237, 245, 442
Matrilines, 28, 59, 186, 189, 192–194, 202, 206–209
group structure, 206
McGeorge, L.W., 27
Meat
scrap hypothesis, 328
sharing, 328–330
transfers, 315, 328, 329, 331
Medicinal plants, 70
Menarche, 252
Menstrual cycles, 248, 250, 252
Mertl, A.S., 28
Mertl-Millhollen, A.S., 23, 27, 28, 30
Microcebus
griseorufus, 26, 51
murinus, 26
rufus (brown mouse lemur), 68, 71, 72, 74, 83, 88
Microchips, 74
Microhabitats
Alluaudia procera, 50
Cedrelopsis grevei, 50
Commiphora spp, 50
Euphorbia spp, 50
flora, 50
Migration, 76, 146–149, 152, 153, 156
Miles, W., 29
Milne-Edwards' sifaka *(Propithecus edwardsi)*, 71
Mistletoe, 83
Mitumba community, 361–363, 370, 373
Moist evergreen forest, 240
Molecular paternity data, 254
Monandrous, 248, 250
Monitoring Guide for the Amboseli Baboon Research Project, 267
Monogamous, 245
Monogamy, 241, 245, 254
Morphology, 70
Morphometric data, 74
Mortality, 80–81, 83, 88, 103, 113, 116, 118, 151–153, 170–172, 295, 298
life expectancy, 171
Mo Singto, 239
Mountain gorillas, 253
Mouse lemurs, 6, 251
Canis lupus, 51
Carnivores, 51
Cryptoprocta ferox, 51

domestic dog, 51
Familiaris, 51
Felis, 51
Viverricula indica, 51
Mubalala groups, 387
Mulder, R.A.
coua *Coua cristata*, 27
Multi-female, 242, 243
Multi-female/single-male, 243
Multi-male, 242, 244, 245, 248, 250
Multi-male groups, 217, 218, 225, 226, 231, 232, 393–395, 397
Multi-male/single-female, 243–245
Multiple mating, 245
Mushamuka groups, 387, 392, 394–396

N
Nakamichi, M., 28, 31, 33
Namorona River, 69
Natal
dispersal, 244
group, 253
Natural selection, 186, 207, 209
Neotropics
Acacia rovumae, 50
Euphorbia tirucalli, 50
flora, 50
Salvadora augustifolia, 50
Tamarindus indica, 50
New one-or multi-male groups, 231
New World monkey, 125
Niche separation, 25, 26, 35
Nocturnal lemurs, 70, 72, 76
Noninvasive analyses, 128
Non-natal males
birth rates, 202
group splits, 202
immigration, 198, 207
matriline, 209
replacement, 198, 207
Norscia, I., 26, 28, 35
Northern muriqui *(Brachyteles hypoxanthus)*, 125–136
Novel behaviors, 8
Nursing, 242, 249
Nutritional ecology, 70, 82–83
Nycticebus bengalensis, 219

O
Obvious-ovulation hypothesis, 250
O'Connor, S., 23, 25, 35

Oda, R., 26, 28
Oesophagostomum, 365
Oestrus, 30
Offspring survival, 60
Okamoto, M., 30
Old World monkeys, 251
One-male groups, 225, 226, 232
Ontogeny, 7
Operational sex ratio, 134
Optimal diet, 266
Opuntia, 27, 33
Orangutans, 6, 11, 12, 251
Oust, 254
Ovarian cycle, 248
Ovarian function, 248
Overlapping areas, 244, 254
Ovulation, 249–251

P

Pair-living, 241, 242, 245, 250
  species, 68
Pairmate, 250
Pairs, 241, 243–245
Pairwise dyadic affinity, 326
Palagi, E., 26, 28, 33, 35
Pan troglodytes schweinfurthii, 386
Papio cynocephalus, 262, 278
Parasites, 46, 56, 62, 70, 74, 79–80
Parasitism, 27
Paternal sisters from non-kin, 276
Paternity, 107, 112, 118
  analysis, 272, 275
  certainty, 250
  confusion, 251
  probability, 245
Patrolling, 315
Pereira, M.E., 30, 31
Peri-ovulatory period (POP), 368
Periovulatory periods, 248
Per se, 6
Phenology, 105, 107
Philopatric, 59
Philopatry, group structure, 206
Pitts, A., 26
Plasticity, 125–136
Play, 28, 31, 32, 349, 350
Poachers, 240, 241
Pole Pole Foundation (POPOF), 404
Polyandrous groups, 245, 250
Polyandrous mating, 245, 250
Polyboroides radiatus (Madagascar harrier
  hawk), 84
Polyspecific associations, 85
Population, 23, 26, 27, 29, 30, 34, 35, 37,
  125–136

censuses, 53–54
density, 8, 10, 84–86, 88–89, 91, 322–324
ecology, 76–83
genetics, 47, 51, 58–61
growth, 176–178
size, 369–371
Post-reproductive, 253
Predation, 70, 75, 79, 80, 82–85, 89, 91,
  107–109, 113, 116–118
  avoidance, 216
Predation rate
  lability, 206
  socially tolerant, 206
Predator–prey dynamics, 315
Predators, 171, 218, 292
  black kite, 27
  Buteo madagascariensis, 27
  harrier hawk, 27
  on lemurs
    carnivores, 83
    raptors, 83
    snakes, 83
  Madagascar buzzard, 27
  Milvus migrans, 27
  polyboroides radiatus, 27
Pregnancy, 78, 248–251
Presbytis, 216, 217, 224, 225, 232
  P. thomasi, 215
Pride, R.E., 29, 30, 32
Primary males, 248–250
Primate field blogs, 238
Primates, 443–445
Pristine, 317
Probability of conception, 245
Progestogen, 250, 252
Progressions, 29
Prolemur simus (greater bamboo lemur),
  69, 71
Propithecus, 102, 103, 106, 116
Propithecus edwardsi.See Milne Edwards
  sifaka
Propithecus verreauxi, 46, 51, 52
Protected zone, 52
Proximity, 249
Pseudo-estrus, 428
Pteropus rufus, 26

R

Radio telemetry collars, 73
Rafidinarivo, E., 26, 27, 32, 33
Rainfall, 48, 59, 74, 75, 88, 90, 105, 111
Rainforest, 69, 72, 80–82, 84, 85, 87, 90
Rambeloarivony, H., 23, 29
Ranging, 27, 28

Index 459

Ranomafana National Park, 67–91
Raptors, 83, 84, 91
Rasamimanana, H.R., 25–27, 32, 33, 35
Rasoloarison, R.M., 26
Rattus, 26
Razafindramanana, J., 23, 26, 35
Receptivity, 249, 251
Red-bellied lemur *(Eulemur rubriventer)*, 71, 87
Referential model, 314
Reforestation, 70
Reliable-quality indicator hypothesis, 250
Replacement, 186, 189, 198–202, 207–209
Reproduction, 70, 75, 78, 85, 86, 90, 132,
    252, 253
Reproductive hormones, 70, 78
Reproductive lifespan, 60
Reproductive physiology, 248
Reproductive Strategies
Reproductive strategies, 237–254
Reproductive success, 78–79, 82, 87, 266, 268,
    274, 275, 279
Reproductive termination, 296
Reproductive units, 244
Resin, 240
Resource, 69, 76, 85
    defense polygyny, 231–232
Rhesus monkeys, 5
Rica, 141, 142, 155, 156
Richard, A.F., 25, 26, 32, 45–62
Ringtailed lemur
    dental ecology, 56–57
    dental wear, 57
    flora, 53
    Tamarindus indica, 57
    tooth loss, 57
Ring tailed mongoose *(Galidia elegans)*, 84
Ripe fruits, 390, 391, 394, 397–399, 402
Rival reduction hypothesis, 330
Ruffed lemur *(Varecia variegata)*, 71, 85
Russell, R.J., 26

S
Sahoby Marin Raharison, 23
Sanzinia madagascariensis (Madagascar tree
    boa), 84
Savanna baboons, 5
Scent marking, 28, 30, 77
Scramble competition, 216
Screaming, 249
Seasonality
    Lomas, 173
    Lomas Barbudal, 173
    natal, 174
    secondary, 174

Seasonal reproduction (breeding), 85, 86
Secondary
    dispersal, 174
    estimated population sizes, 176
    Lomas Barbudal, 173
    males, 244, 248–250
    population recovery, 177, 179
    status, 244
Seeds, 82, 83, 91
    dispersal, 70
Selective logging, 105
Semi-deciduous forest, 316
Semi-solitariness, 244
Semi-solitary, 243
Semnopithecus entellus, 215
Senescence, 78, 82, 189, 205
Sex ratios, 170, 174, 179, 181
    secondary/breeding, 174
Sexual activity, 249, 250
Sexual behavior, 242
Sexual interest, 251
Sexually polyandrous, 248
Sexual maturity, 133, 134, 242, 253
Sexual monogamy, 245
Sexual selection, 62, 249
Sexual selection theory, 254
Sexual signal, 251
Sexual swellings, 249–251, 254
Short-term, 251
Sifakas, 22, 24–26, 28, 31, 33, 46, 51–54,
    56–62, 101–118
    behavioral endocrinology, 57–58
    populations, 46, 54, 58, 59, 62
Simmen, B., 26, 32
Sinarundinaria alpina, 390
SIVcpz, 370, 375, 376
Slow life histories
    large brain size, 204
    reproductive strategy, 204
Snakes, 51
Social
    behaviour, 5–7, 241, 444
    conventions, 155–156
    dynamics, 238, 245, 254
    flexibility, 135, 237–254
    learning, 144, 154–160
    organization, 76, 91, 238, 241–245,
        248, 251
    organization and male residence pattern in
        Phayre's leaf monkeys, 215–232
    relationships, 324–325
    sphere, 254
    status, 241
Socio-ecological models, 8

460   Index

Socio-sexual monogamy, 254
Socio-sexual strategies, 245
Soma, T., 27, 34
Sportive lemur, 51
Sportive lemur *(Lepilemur microdon,*
    L. mustelinus), 71
Starvation, 404
Stinkfighting, 28
Stony Brook University (SBU), 69, 70, 75
Stress, 26
   hormones, 88, 90
Strong affiliative relationships have direct
    adaptive value for females, 268
Subgroups, 300. 304
Subordinate, 244, 248
Suckling infant, 249, 253
Sunset calls, 424
Survival and longevity differences, 276
Survivorship, 295
Sussman, R.W., 26, 28, 30, 35, 45–62
Swelling phase, 250
Sympatric chimpanzees, 385–406
Synchrony, 78, 79

T
Takahata, Y., 27, 29, 30
Takeovers, 46, 58, 62
   male, 174
   natal, 174
Talatakely field site, 72, 73, 79, 84, 86, 87
Tamarind, 24, 27, 36
Tamarindus indica, 27
Tandroy, 22
Targeted aggression, 77
Terpsiphone mutate, 27
Terrestrial vertebrate camera traps, 74–75
Territorial, 28–30, 32
Territoriality, 330–331
Territory, 23, 28–30, 32, 76, 77, 299–301, 308
Testes size, 74
Testosterone, 55, 62, 77, 78
The Four Year War, 373
~350 mm of rain per year, 265
Thymelaeaceae, 240
Time budgets, 302, 304
Torpor, 74
Tortoise, 51
Tourists, 22, 25
Trachypithecus, 216, 217, 223, 224, 226, 232
   phayrei crepusculus, 217
Traditions, 144, 154–160
Transects, 48, 50, 72, 73, 88
Transfer, 77, 79, 80, 254

Trans-generational effects, 441
Travel and group check-sheets, 362
Tree phenology, 70, 74, 75
Trio, 242
Tropical seasonal forest, 240
Turtle, 51

U
University of Antananarivo, 70
University of Fianarantsoa, 70
University of Helsinki, 70
Unprovisioned, 251
Uvaropsis congensis, 318, 331

V
Valohoaka field site, 70, 73, 79, 86, 87
Varecia variegata editorium (black and white
    ruffed lemur), 71
Vatoharanana field site, 70, 73, 79, 86, 87
Verreaux's sifakas, 6, 46, 51, 52, 57–61
Vital population statistics, 10
Vítoria, Brazil
   female-biased, 125
Vocal communication, 241
Vocalization, 70, 85
Vulnerable species (IUCN), 68, 72

W
Wamba, 413–429
Water, 24, 27, 32, 33, 35, 36
Weaning, 221, 223, 253
Wet season, 90
White-handed gibbons, 5, 237–254
Wildcat, 26
Woodchips, 240
Woolly lemur *(Avahi laniger, Avahi
   peyrieriasi)*, 71
World Heritage site (UNESCO), 89, 239
World Wildlife Fund (WWF), 46, 47
Wright, P.C., 25, 32
WWF.*See* World wildlife fund (WWF)

X
Xerocysios, 27
Xerophytic forest, 46, 48

Y
Yellow baboons, 278
Yellow-bellied marmots, 439

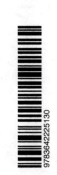